Paul T. McCormack
731-1027

OPTIMIZATION MODELS FOR PLANNING AND ALLOCATION:
TEXT AND CASES IN MATHEMATICAL PROGRAMMING

OPTIMIZATION MODELS FOR PLANNING AND ALLOCATION: TEXT AND CASES IN MATHEMATICAL PROGRAMMING

Roy D. Shapiro

Graduate School of Business Administration
Harvard University

JOHN WILEY & SONS
New York Chichester Brisbane Toronto Singapore

Library of Congress Cataloging in Publication Data:
Shapiro, Roy D.
Optimization models for planning allocation.
Includes index.
1. Management science—Case studies. 2. Linear programming—Case studies. I. Title.
T56.24.S53 1984 658.4'034 83-21878
ISBN 0-471-09468-4

Printed in the United States of America

10 9 8 7 6 5 4 3 2 1

DEDICATED TO DEBORAH, FOR WAITING

PREFACE

During the past decade, the teaching of management science has begun to shift away from a pedagogy based entirely on the theory toward one that emphasizes the application of models in actual practice. Rather than focus primarily on the theoretical development of algorithms, the latter approach stresses the practical considerations involved in the recognition and formulation of appropriate models and the subsequent analysis of the economic and managerial implications of the solution results.

Over the past three decades, the development of management science techniques has far outpaced our understanding of how and when and, in fact, whether to use them. The successful practice of management science requires the recognition that these concepts and techniques are applied to problems involving people—people who interact with each other and their environment within the confines of an oftentimes complex organizational structure. Thus, the study of management science is incomplete without exposure to actual decision problems faced by actual managers—problems fraught with the ambiguity and uncertainty that often accompany decisions made in the world of real organizations. The cases in this book provide this exposure.

For many students who are accustomed to textbook exercises, an approach that emphasizes problem identification as well as problem solution may seem disconcerting. For most students, the entire educational process has attuned them to finding answers (using well-defined methods) to precisely structured questions. In practice, decision problems can rarely be solved that easily. First, there is almost never a single right answer; second, determining what the problem really is often requires far more insight and ability than its analysis once it is precisely defined. Thus, providing the student with real situations that lacks rigid structure is an essential element in a management education.

Mathematical programming is one of the most widely used and certainly one of the most successful of management science methodologies. Over the nearly four decades since its inception, it has attracted the attention of managers, engineers, social scientists, and public policy makers as one of the most powerful and flexible tools available to support decision making. Hence an understanding of these tools is useful to all levels of management; yet due to their mathematical complexity, these are typically not well understood by the nonexpert. In view of this need the book is structured not to teach mathematics but to demystify the often elusive concepts underlying these optimization models.

Optimization Models for Planning and Allocation emphasizes the formulation of mathematical programming models, the use of computer output to gain insight as well as answers, and the discussion of the difficult implementation problems that often accompany the development and use of quantitative models in the corporate decision-making process. While the purpose of the text is not to train technicians, it is important that both line and staff management understand the technology that can be brought to bear in making decisions for planning and allocation. Thus, the material in the book has been structured to be suitable for a wide audience from top management who will use this technology to middle management who will initiate and/or manage the development of appropriate models to staff who will participate in the actual model design and development. Observations of the uses of quantitative models in corporate settings has reinforced the conviction that problems often result from the gap between line management and technical staff. Thus, the aim of this book is to educate managers who might effectively act as a liaison between these two groups, understanding *and* being able to communicate the needs of both. Both students with a great deal of technical training and experience and those with limited backgrounds can use the material and benefit from this orientation.

This focus, rather than the more traditional focus on theory and algorithms, is intensified by a sequence of *cases* that highlights both the formulation of optimization models and the subsequent analysis of the economic and managerial implications of the solution results. Typically, a case is a description of a real-life management situation that the student is expected to analyze, discuss in class, and suggest some recommendations for action. Rather than having students memorize facts and techniques, this pedagogical approach emphasizes learning by "doing," through the process of exposure to a series of cases covering a variety of different problems in different industries.

There are two reasons that the cases in the book were chosen. First, I want students to follow the progress of management science projects. Exposure to the choices real analysts face in designing models for actual situations is invaluable; it illustrates the benefits and pitfalls in the use of mathematical programming practice. Second, I think that students should have the opportunity to do their own model development in an environment as close to the real one as is possible in the classroom—an environment of people and thus personalities, of organizational hierarchy and thus political complexity. Developing management science approaches to a case situation forces the student to come to grips with these issues and to grapple with problem definition, implemention, and communication with different levels of the organization—as well as with methodology. Perhaps more importantly, these cases provide "hands-on" management science experience. Charles Gragg, one of the more outspoken and best remembered advocates of the use of cases in business education, wrote:

> *It can be said flatly that the mere act of listening to wise statements and sound advice does little for anyone . . . no amount of information, whether theory or fact, in itself improves insight or judgment or increases ability to act wisely under conditions of responsibility. . . . In the process of learning, the learner's dynamic cooperation is required. Such cooperation from students does not arise automatically, however. It has to be provided for and continually encouraged.*[1]

The cases herein provide this encouragement.

Optimization Models for Planning and Allocation has been designed for a graduate course in applied mathematical programming, especially for students who may soon be in a position to develop and/or implement such models. The book has *not* been designed for the "technical course in optimization theory" of the sort offered by many operations research or mathematics departments. However, enough underlying theoretical and conceptual development has been provided to allow future managers to understand the techniques. Nor is the text appropriate for those who have had no prior exposure to management science, although the text does, for the sake of completeness, review the initial conceptual underpinnings that would be presented in a first course such as "Introduction to Management Science" or "Quantitative Methods in Management." The cases, however, clearly the *raison d'etre* for this book, can be taught at several different levels. Many of these cases have been used successfully with audiences ranging from public policy makers and practicing business executives to students in MBA programs and in graduate operations research courses.

Apart from the desirability of a prior course in management science, which covers, among other topics, the basics of linear programming and the presumption of some familiarity with basic mathematical concepts (e.g., graphing functions, using equations), there are few prerequisites for using this book. I have tried to deal as little as possible with complex mathematics and have

[1]Charles Gragg, *Because Wisdom Can't Be Told,* distributed by the Intercollegiate Case Clearing House, ICCH No. 9-451-005.

structured the material so that neither experience with linear algebra, familiarity with vector and matrix notation, nor knowledge of calculus is required. I have assumed that students can handle equations containing summations, although I have used that notation as sparingly as possible. When uncertainty is explicitly brought into the picture, I presume at least a fleeting prior exposure to expected values and cumulative probability distributions, and use decision diagrams as occasional illustrations.

Chapter 1 briefly discusses a taxonomy of problem solving approaches and indicates how optimization models fit. Chapters 2 and 3 provide an in-depth discussion of linear programming, the topic that not only is most important in its own right, given the vast majority of actual applications, but that introduces the major features of all mathematical programming methodologies. Chapter 4 gives some insight as to how multiple criteria may be incorporated into the optimization process.

Chapters 5 through 9 offer a range of special mathematical programming structures which have found or are beginning to find "their place in the sun." Chapters 5 and 6 introduce combinatorial optimization, covering networks and integer programming. Chapters 7, 8, and 9 deal with more advanced topics: incorporating uncertainty into the optimization process, nonlinear optimization, and an approach for sequential decision problems—dynamic programming.

February 1983 *Roy Shapiro*

ACKNOWLEDGMENTS

I am greatly indebted to the authors of the cases included here; without their efforts, this book would not exist.

The case material was developed over the last decade at the Harvard Business School. The school administration and the Division of Research have my appreciation for their support. Several of my colleagues deserve special thanks, not only for their written contributions but also for their support and encouragement, particularly Stephen Bradley, John Bishop, Sherwood Frey, Michael Huguenin, Ramchandran Jaikumar, Richard Meyer, Joan Sapir, Arthur Schleifer, Sandra Skiba, Paul Vatter, and Tom Walker.

R. S.

CONTENTS

OPTIMIZATION MODELS FOR PLANNING AND ALLOCATION:
TEXT AND CASES IN MATHEMATICAL PROGRAMMING

Chapter 1

MODELS AND OPTIMIZATION

In today's world, it would be hard to find many successful managers, engineers, or social scientists who had not benefited from meaningful insights into the "systems" with which they deal which were made possible by the creation and use of *models*—those physical or mathematical abstractions that, although simplified, reflect the key interactions of the system variables. Indeed, one of the key truths of "modern management" involves the recognition that today's problem solving requires talents beyond those inherent in the combination of intuition and experience. Central to the material in this book is the concept of the model as an insight-producing aid to managerial decision making. In this vein, this prefatory chapter is meant to serve as a brief, but important, initial discussion of modeling in general and optimization modeling in particular, prior to the detailed study that begins in Chapter 2.

Since the dawn of civilization, we have constructed simplified models of our environment so as to understand that environment better. Modeling as a means of analysis and, ultimately, insight, has been with us since our earliest beginnings. The first model was, without a doubt, a physical model, constructed crudely. This early experimentation is a far cry from the analysis made possible today to the scientist, engineer, or manager armed with a mathematical model, an optimization algorithm, and a computer. The advances in management science and the phenomenal developments in computer hardware and software have created an environment where an understanding of the construction and use of models is an indispensible part of the education of a modern manager. There is scarcely an industry today that has not experienced the significant impact of model-based planning and analysis. Unfortunately, this impact has not always been positive. One key reason for this often referred to "failure of management science models" is the refusal of management to become involved in the modeling process. Models are too important to be left to analysts alone. The manager, if he or she wishes the model to be useful and *used*, must be part of the development. This requires some grounding in the "art" of modeling and the "science" of solution techniques. It is to this end that this casebook is devoted.

We begin by suggesting a taxonomy of decision-making or problem-solving approaches so as to see where the mathematical model fits in, and then how optimization fits in the continuum of mathematical models.

A TAXONOMY OF PROBLEM-SOLVING APPROACHES

A team of aircraft engineers needs to determine appropriate wing dimensions and shape for a new aircraft which they have designed. Perhaps the conceptually simplest approach to solving this problem is to build a variety of aircraft, utilizing a range of alternative wing configurations. These aircraft could then be subjected to extensive *in-flight* monitoring to determine the performance characteristics of each. This approach—*direct experimentation*—has the advantage of representational accuracy, that is, each wing being tested is the actual alternative under consideration. On the other hand, the time, expense, and difficulty that would be experienced in implementing this series of experiments would severely limit the number of alternatives that could be tested. As such, this approach would likely be deferred until all but a few alternative designs had been screened out.

A second approach, one utilized extensively in the engineering sciences, would be the construction of a variety of *physical models*. These models could then be subjected to a series of tests in simulated environments (e.g., a wind tunnel) to predict how the real wing might perform in flight. Note that this alternative offers a reduced level of representational accuracy. That is, in certain ways, the actual wing would no doubt act differently in flight than the model in a wind tunnel. On the other hand, this approach allows less time consuming and less costly evaluation of far more alternatives than would be possible with direct experimentation.

A third approach would be the formulation of a *mathematical model*. The mathematical model would replace with equations those characteristics of performance the physical model would reveal in the wind tunnel. For example, the mathematical model of wing performance would incorporate laws of motion, momentum, and air flow—in essence, equations that capture precisely how the *system variables* interact. Of the three approaches discussed, this would be the easiest, quickest, and least expensive to implement, especially with modern day computers. Unfortunately it would also require the greatest degree of simplification—it would provide the least representational accuracy.

Indeed, the major difficulties in using a mathematical model are (1) assuring that the model captures the key features of the system under study as well as the interrelationships among these features that determine causes and effects, and (2) interpreting the "answer" to a simplified picture in a way that will provide useful insight in dealing with the actual environment, which is far more complex than the model reflects.

The major benefit of the mathematical model is its ability to evaluate a vast array of alternatives in a fraction of the time and at a fraction of the cost necessary for the construction and testing of the same number of physical alternatives. Figure 1-1 illustrates these features.

While all three approaches might be feasible for the above example, the mathematical model may be the only feasible choice for the analysis of a variety of managerial or social problems. Consider, for example, the plight of a manager trying to determine an economically viable location for the firm's new factory. While it might be possible to actually construct a plant at a site under consideration and then observe the economic impacts of that choice, such a direct experimentation approach would be ludicrous. A physical model would be of no help, since the performance to be monitored is economic, not subject to physical laws. Instead, as illustrated by Figure 1-2, the manager might describe the interaction of the system variables via a set of equations describing—for example, transport of raw materials to the site, the production process, and the distribution of finished goods to the firm's warehouses, capturing the economics of each set of activities within those equations. Similarly, competitive and market forces might be modeled with a series of relationships that describe how customers or competitors might react to specified actions by the firm. This mathematical model might then be applied to a set of locations under consideration and, for each, would predict its economic performance over time.

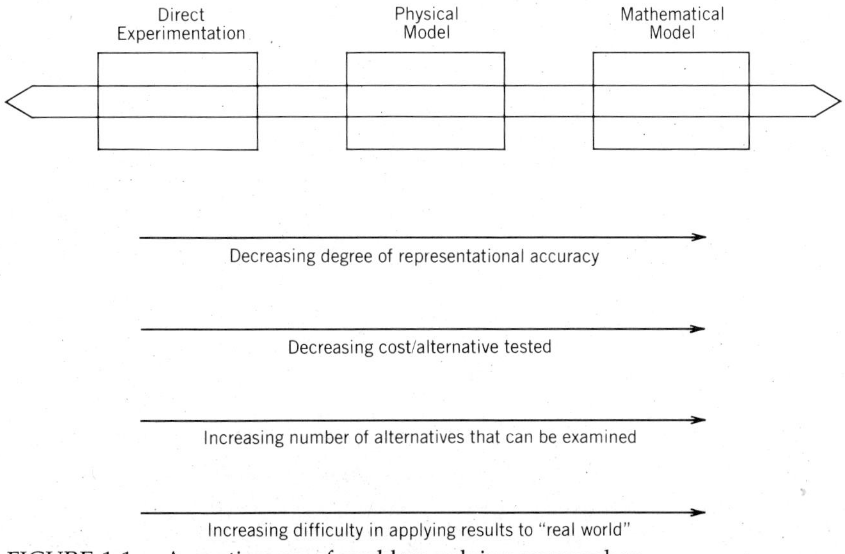

FIGURE 1-1. A continuum of problem-solving approaches.

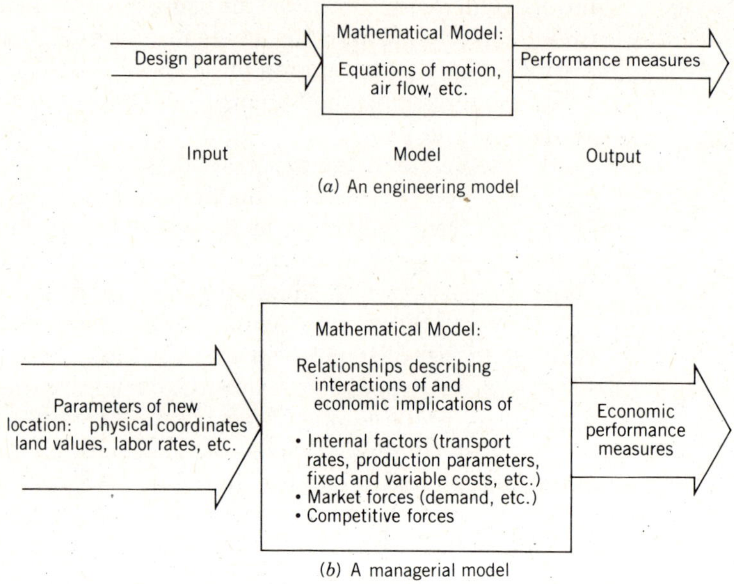

FIGURE 1-2. Input-output schematics for two models.

EVALUATION VERSUS OPTIMIZATION

A further categorization of problem solving approaches is helpful in understanding mathematical models. The preceding examples have been examples of *evaluative* models. That is, they operate by evaluating alternatives one at a time—they neither generate alternatives nor do they choose the "best." Those activities are undertaken by the human problem solver or decision maker. Simulations are perhaps the best known and most commonly used examples of evaluative models. They permit a great deal of flexibility, and a creative modeler with a good understanding of the underlying system can guarantee a relatively high level of representational accuracy. On the other hand, there is a practical limit to the number of alternatives that can be evaluated. Typically, evaluative models are used in a *satisficing* mode; that is, the analyst or manager will evaluate candidate plans until he or she is satisfied with the performance of one or more, rather than in an *optimizing* mode, where the best possible plan is determined. With an evaluative model, there can be no guarantee that the chosen alternative—the best alternative evaluated—is the best *possible* alternative. That guarantee requires the use of an optimization model.

Most simply, an optimization model is an evaluative model with added features. Those added capabilities are:

1. The ability to automatically generate new alternatives.

2. The ability to test whether a given alternative that has just been evaluated is the "best."

The first step in constructing an optimization model is the same as that necessary in constructing an evaluative model-the description of the system: the specification of the set of equations (or inequalities) that characterizes the interrelationship of key system variables. Most of the effort in any analysis will, in practice, be devoted to understanding the system and translating that understanding into quantitative form. This is as it should be; a model without understanding has little value.

The second step is the adoption of a measure of system effectiveness, a measure by which we may judge, "Plan A is *better* than Plan B." Sometimes, in a physical or simplified economic system, the criterion to be chosen is easy—efficiency, speed, cost, or profit. Often, however, in a social, political, or more realistic economic system, conflicting objectives make the choice of any one criterion difficult: Decreased inflation or decreased unemployment? A healthy economy or a healthy natural environment? Even for the individual firm, there are issues to be resolved—for example, short-term gain or long-term stability? Chapter 4 will suggest optimization methods to help ease this criterion-selection difficulty.

Step three is the choice of an algorithm that will optimize; that is, consistent with any restrictions which may be imposed by or on the system, as embodied by the mathematical description of that system (step one), the algorithm will choose that plan that, when evaluated along the dimensions suggested by the chosen measure of effectiveness (step two), achieves the best score of *all* possible feasible plans. We might think of optimization algorithms as operating as shown by the schematic flow chart of Figure 1-3. The process starts with an initial candidate plan, either user-generated or generated by the algorithm itself. The parameters (e.g., operating levels for the various activities) of the plan are fed to an evaluative model that, through the mathematics which describe the interrelationships among the system variables, "scores" the

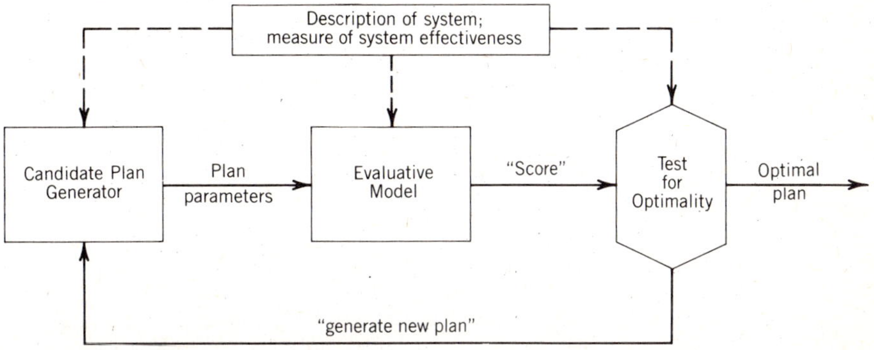

FIGURE 1-3. An optimization schematic.

plan with respect to the measure of system effectiveness specified. Next is a key to the optimization algorithm's success—a test for optimality, a test that determines whether any further improvement (with respect to the chosen criterion) is possible. If not, then the candidate plan under study is optimal. If improvement is possible, the algorithm cycles back and generates a new candidate plan, one that is created within the restrictions or guidelines embodied in the system description. Especially successful algorithms guarantee that the new plan will provide better performance (i.e., will score better) than the previous plan. Another key success factor is the (possibly empirical) guarantee that this cycling will end quickly.

THE MODELING PROCESS: A FRAMEWORK FOR ANALYSIS

This book focuses less on the theory behind the specific details of various optimization algorithms than on the practical considerations involved in the recognition and formulation of appropriate models and the subsequent analysis of the economic and managerial implications of the solution results. The schematic "bow tie" in Figure 1-4 illustrates our view of the modeling process: A necessary precursor to the formulation or use of a mathematical model is *understanding*: What drives the system under consideration? What are the key factors in predicting system behavior? What is the managerially relevant problem? What are the key measures of success? Only after these questions

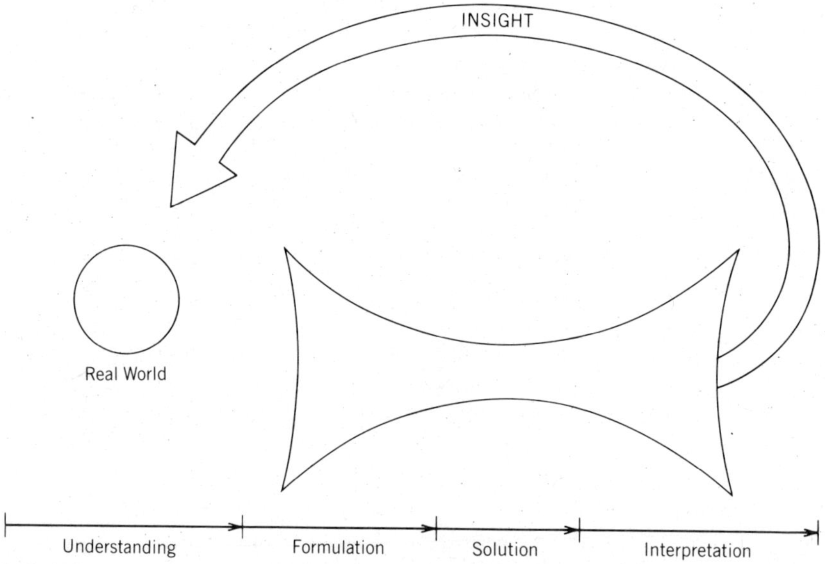

FIGURE 1-4. "Bow-tie" schematic of the modeling process.

have been answered can modeling begin. *Formulating* the model consists conceptually of extracting from complex, real-world situation those variables and relationships essential to the operation of the system of interest, and of narrowing down the multitude of variables and issues to those crucial to solving the problem at hand. This simplification process is key: not enough simplification and, at best, solution will be overly time-consuming and costly; at worst, solution techniques will prove to be unable to handle the model's complexity or size. Too much simplification and the ultimate users of the model results will be unwilling to use them due to the lack of realism and accuracy in the model specification.

Although, for ease of illustration, we often discuss formulation as separate from underlying relationships that drive the system, in practice the formulation exercise often greatly increases understanding. There are countless examples of model development in corporate settings where the true value of the model comes from the discipline imposed by formulating that model. Involved managers who must express their understanding in quantitative terms often find that their understanding of the system has increased. Questioning of underlying assumptions, quantifying perceptions, and specifying relationships increases insight, in and of itself.

Moving forward as shown in Figure 1-4, formulation is followed by solution. First, the selection of the appropriate solution methodology given the characteristics of the formulated system relationships; second, the collection of data—in practice, the most time-consuming effort; and, third, the actual running of the appropriate algorithm to produce a numerical "solution."

Altogether too many courses and textbooks stop at this point, having given short shrift to formulation and focussing on solution algorithms. In a managerial context this is disastrous. It is perhaps the source of the often heard complaint about the analyst who, after having disappeared for months to formulate a complex optimization model, drops a two-inch thick stack of computer output on the manager's desk, cheerfully exclaiming, "Here's the answer." The "bow-tie" in Figure 1-4 indicates, however, that at this point we are only halfway through the process—that *we have no answers, only solution results*. These solution results require interpretation. This interpretive task is, in some sense, a reversal of the formulation task. Whereas formulation requires that we narrow down the broad expanse of the real-world system, interpretation suggests broadening the narrow meaning of the solution results to apply them to the actual situation.[1] Interpretation involves recognizing the simplifications made and checking the solution results against them, verifying that nothing of import was lost in the translation. Interpretation is truly an art. Certainly the most difficult of the three modeling tasks, its goal is to take solution results—a set of numbers—and create insight into the management problem, insight that will be key in the ultimate management decision.

[1]This simplification or *narrowing* of the vast array of real world factors, followed by the interpretation or *expansion* of the numerical solution results, gives rise to the "bow-tie" shape of Figure 1-4.

MATHEMATICAL PROGRAMMING

The material in this book covers the class of optimization models that has had the widest application to management problems—mathematical programming models. These models, probably the best developed and most used of management science techniques—certainly the most successful—are concerned with allocating scarce resources to competitive activities in a way that is "best" for the enterprise as a whole. The resource allocation problem is one of the most important decision problems managers face—how to allocate productive capacity or scarce materials to a variety of possible products, how to allocate available finances among various capital investment opportunities, how to allocate manpower among possible R&D projects, and so forth. In view of the difficulties frequently encountered in making these decisions as well as their importance, managers are grateful for tools to bring to bear for aid in the decision process. This, perhaps, explains the success mathematical programming has had.

Note that the example applications in the paragraph above all have several characteristics in common:

1. All require a quantitative specification of *how much* of the available resources are to be allocated to each of the competing activities.
2. For each, there is implicitly some objective(s) to be optimized. One focus of this book is to help the manager make the objective or objectives *explicit*.
3. For each, there are constraints limiting the amount of resources which can be "consumed" by the competing activities as well as, possibly, additional constraints restricting the makeup of the portfolio of activities the manager can choose.

Formally, mathematical programming seeks to determine the values of certain *decision variables*[2] (alternatively, the levels of specified *activities*) subject to various restrictions *(constraints)* that are expressed as equalities or inequalities in terms of the decision variables. The constraints, in practice, reflect financial, availability, technological, organizational, or political considerations as well as a multitude of others. The values of the decision variables are determined so as to optimize (minimize or maximize) some chosen criterion that, when expressed in terms of the variables, is called the *objective function* (or simply the objective). The form of the objective, the constraints, and the values the decision variable may take determine the type of mathematical programming algorithm that need be used.

[2]The optimal values of the decision variables represent the optimal plan or *program*; thus the origin of the term "mathematical programming."

For example, if all mathematical relationships may be expressed by linear functions, the model is a linear programming model. Linear programming is the most strikingly successful of all management science applications. Its ease of use and the wealth of sensitivity information provided along with the optimal solution contribute to its success. Chapters 2 and 3 deal with these models by discussing, in some detail, the key features of linear programming. The format of these two chapters mirror the three tasks discussed in the preceding section—formulation (Chapter 2), solution, and interpretation (both covered in Chapter 3). Chapter 4 introduces multicriteria issues and ways of dealing with them.

In the remainder of the book we systematically relax the assumptions that underlie continuous linear programming. Chapters 5 and 6 focus on combinatorial optimization—situations in which we wish to optimize over discrete classes or sets of items, where the solution sought specifies the best combination or grouping of these discrete entities. In this case, "yes-no" variables are key. We make the transition between linear programming and combinatorial optimization by considering a class of optimization models that is a special case of both model categories—network models. These structures, in use since the late 1940s, are discussed in Chapter 5. Chapter 6 introduces integer programming, a class of optimization models of growing importance. Hitherto unsolvable except for relatively small models, the rapid increase in speed and decrease in cost which characterizes computer technology of the 1970s and 1980s has made the solution of larger and larger models possible.

Chapters 7, 8, and 9 deal, quite briefly, with a set of methodologies that, as yet, have had only a minor impact on business practice, but, as computer capabilities continue to increase and computer costs continue their descent, these now-advanced topics will find their way more and more into managerial application. All are tools of which the manager of tomorrow should be cognizant. Chapter 7 relaxes the assumption of deterministic coefficients and introduces uncertainty explicitly into the linear programming framework. Chapter 8 relaxes the assumption of linearity and briefly surveys the broad expanse of nonlinear programming. Chapter 9 considers sequential decision problems, both deterministic and stochastic, introducing dynamic programming.

As is said above, this book focuses on formulation and interpretation, not on algorithmic details. Nevertheless, the serious student of optimization models will want some background knowledge of the Simplex method, George Dantzig's 1947 discovery that forms the core of all mathematical programming; perhaps the most successful algorithm devised in the last half-century. Appendix A gives a brief treatment of this algorithm, at a level that requires little mathematical sophistication. Appendix B illustrates, for the interested reader, another important solution methodology—the transportation method. Like the Simplex method, this algorithm for solving the transportation problem—probably the most often used and successful special linear programming structure—has historical as well as practical significance. Appendix C discusses a structural aspect of mathematical programs which has tremendous significance—duality. The theory of duality is key in virtually all mathematical programming solution methodologies. Appendix D provides, for the practitioner, a necessary reference—a brief survey of commercially available mathematical programming packages and an overview of computational issues of which the future user should be aware.

As the Preface points out, in its sometimes oratorical style, the true value of this book lies not in its text—that material has appeared in other forms to date—but in its cases. These descriptions of actual business problems, people, and organizations give the student the real education—learning by doing.

Chapter 2

LINEAR PROGRAMMING: FORMULATION

There is little doubt, here in the 1980s, that linear programming, nearly 40 years after its discovery, has become a recognized tool in the continuing managerial quest for meaningful insights and better decisions.[1] Considerable experience over the past decades has demonstrated that the depth of managers' understanding and the quality of their decisions improve with the application of such models. Successful applications have been seen in manufacturing, transportation and warehousing, plant size and location decisions, inventory management, production planning and scheduling, new product development, advertising, cash and portfolio management, and a variety of short- and long-range planning models.

Chapters 2 and 3 serve as an overview of some of the key concepts and constructs of linear programming, mirroring the "bow-tie" framework illustrated by Figure 1-4. This chapter concentrates, in detail, on *formulation*, presenting several examples to give the spirit of linear programming and then discussing formulation concepts in general. Chapter 3 starts with an overview of linear programming *solution* methodology, utilizing a graphical approach to illustrate how the Simplex method operates. The algebraic discussion of the

[1] A 1970 IBM study of computer usage reported that 25% of all "scientific" computation is devoted to linear programming and closely related techniques. A 1977 survey of large private corporations (G. Thomas and J. DaCosta, "A Sample Survey of Corporate Operations Research," *Interfaces*, vol. 9, no. 4, August 1979) indicated that of the 150 firms responding, 79% used linear programming. Of management science tools, only simulation (at 84%) could claim more application.

Simplex algorithm is deferred to Appendix A for those readers who want more in-depth exposure to the algorithm. Chapter 3 then brings the computer into the picture, describing a mathematical programming package we will use throughout the book. Using the examples formulated in Chapter 2 as illustrations, Chapter 3 goes on to discuss how the computer solution can be used for extensive sensitivity analysis, thus giving valuable aid in the very important *interpretation* phase. Duality, perhaps the most important theoretical element in the solution of not only linear programs, but a wide span of other mathematical programming structures as well, is treated in Appendix C.

Recall from the previous chapter that a mathematical program is comprised of three major elements the modeler must consider in the formulation process: decision variables, constraints, and the objective function. Let us consider several examples, each somewhat more simplified than models used in practice, but each representative of an actual situation where a linear program would be used.

EXAMPLE 1: PRODUCTION PLANNING BY A POWER DRILL MANUFACTURER

A nationally known producer of home and industrial power tools, Breck and Dapper, Inc., has in recent years sold a line of power drills for home use. A power drill consists essentially of an electric motor encased in a plastic and metal housing. The four models in B & D's product line differ in size and composition of the housings and in the sizes of the electric motors. Models 1 and 2, the more traditional models, have casings with several all-metal parts, whereas models 3 and 4, the newer models with the smaller motors, are encased in mostly plastic housings. Breck and Dapper is vertically integrated and for years has been casting their own housings, doing all their own motor-winding, and, recently, drawing some of the wire needed in their motors.

A shortage of the copper alloy used both in the metal casing and in their wire production has caused some worry in the firm, and next quarter's plans will reflect this. Table 2-1 gives the resource requirements of the four models. Up to 16,000 pounds of plastic can be ordered and delivered in time for use, but no more than 5000 pounds of copper alloy will be available within the

TABLE 2-1

	Model 1	Model 2	Model 3	Model 4
Plastic Required (lb)	.82	.62	1.42	2.03
Copper Alloy Required in housing (lb)	.43	.69	.33	.20
Wire Required in Motor (yd)	15	16	9	9
Contribution (excluding wire cost)	$12.50	$11.30	$17.20	$19.90

quarter under consideration. Wire for motor winding is available from two sources. It can be produced internally at a cost of 14¢/yard from the copper alloy, on machines with the capacity to draw 80,000 yards of wire during the quarter. Each 100 yards of wire uses 3.6 pounds of copper alloy. Alternatively, it can be ordered from outside suppliers in virtually unlimited quantities at 29¢/yard. At the start of the quarter, 8000 yards of wire is expected to be in inventory.

The other materials required in the production of these drill models are in abundant supply and can be obtained easily. Management believes that all the drills they make can be sold; however, there is a commitment on the part of the marketing people that the newer models not "take over" the product line. They have indicated that, for customer satisfaction, at least as many units be produced of the model 1 and 2 type as are produced of the newer model 3 and 4 type.

This product-mix decision can be formulated as a linear program. To do so, we consider, in sequence, the three modeling elements:

Decision Variables

What does Breck and Dapper management need to know in order to plan their product mix for the coming quarter? By answering this question, we can generate the decision variables. In this case, a complete specification of next quarter's plan requires a *production plan*, including:

The number of model 1s to be produced (let us note this quantity by x_1).

The number of model 2s to be produced (denoted x_2).

The number of model 3s to be produced (x_3).

The number of model 4s to be produced (x_4).

and *purchase/manfacturing specifications* for the key raw material, copper wire:

The number of yards of copper wire to be purchased (denoted W_p).

The number of yards of copper wire to be internally manufactured (W_m).

Constraints

Next, we recognize that it is only the scarcity of resources (material, productive capacity) which keeps Breck and Dapper from producing an unlimited quantity of each of these models. These resource limitations, as well as the restriction imposed by marketing, are crucial in determining what production plans are possible. These limitations or *constraints* can be expressed algebraically in terms of the decision variables above. They are of three types.

(1) Resource Limitations. There are four resources whose availability is limited: two raw materials—plastic and copper alloy; a material that can be either purchased or internally produced—wire; and capacity for wire-drawing. Let us consider plastic availability first. We can write the constraint in words as

$$\text{amount of plastic used} \leq \text{amount of plastic available}$$

To express this algebraically, refer to Table 2-1 and note that, if x_1 represents the number of units of model 1 that are to be produced, then $.82x_1$ represents the plastic (in pounds) which will be consumed by that activity. Similarly, the plastic used in the production of models 2, 3, and 4 may be written as $.62x_2$, $1.42x_3$, and $2.03x_4$, respectively. Now, since the total plastic used will simply be the sum of the plastic used in each of the four manufacturing activities, we may write an expression for the total plastic used:

$$.82x_1 + .62x_2 + 1.42x_3 + 2.03x_4$$

The available plastic for the quarter has been given as 16,000 pounds, so the resource limitation constraint can be written

$$.82x_1 + .62x_2 + 1.42x_3 + 2.03x_4 \leq 16000$$

The other three constraints may be similarly generated:

$$\text{amount of copper alloy used} \leq \text{amount of copper alloy available}$$

$$.43x_1 + .69x_2 + .33x_3 + .20x_4 + .036W_m \leq 5000$$

(Note the inclusion of the W_m term since wire production is an activity that consumes copper.)

$$\text{amount of wire used} \leq \text{wire available}$$

The left-hand side, "wire available," however, is somewhat more complicated than before since

$$\text{wire available} = \text{wire produced} + \text{wire bought} + \text{wire on hand}$$

$$= W_m + W_p + 8000$$

The constraint, written in full, is hence

$$15x_1 + 16x_2 + 9x_3 + 9x_4 \leq W_m + W_p + 8000$$

Usual linear programming conventions have equations and inequality constraints written with all variables on the left-hand side and the constant on the right-hand side. Thus, this constraint would be rewritten

$$15x_1 + 16x_2 + 9x_3 + 9x_4 - W_m - W_p \leqslant 8000$$

Last, we need a constraint that recognizes the limited wire-drawing capacity. This is simply

$$W_m \leqslant 80000.$$

(2) Marketing's Policy Restriction.

production of models 1 and 2 ⩾ production of models 3 and 4:

$$x_1 + x_2 \geqslant x_3 + x_4$$

(3) Nonnegativity Constraints. Since we are dealing with a mathematical technique that has no inherent "sense of reasonableness," we must be careful to be precise. In particular, negative levels cannot be permitted—all decision variables must be restricted to be greater than or equal to zero—that is,

$$x_1 \geqslant 0$$

$$x_2 \geqslant 0$$

$$x_3 \geqslant 0$$

$$x_4 \geqslant 0$$

$$W_m \geqslant 0$$

$$W_p \geqslant 0$$

Objective Function

This situation is prototypical of that faced by most plants— the product mix decision will be based largely on economic factors. The objective will be to allocate resources so as to maximize profits. Since fixed costs will be unaltered by the allocation decision, the appropriate objective is the maximization of *contribution* to profits—that is, revenue minus variable costs. The contribution per unit for each of the models, excluding the cost of wire, was given in Table 2-1. Thus, since each model 1 earns \$12.50 for Breck and Dapper and x_1 denote the number of units of model 1 produced during the quarter, $12.50x_1$ represents the contribution (excluding wire cost) earned from the manufacture of model 1s. Similarly, $11.30x_2$ represents the contribution from model 2s and so forth. The decrease in contribution attributable to wire usage is \$.14 per yard of wire manufactured and \$.29 per yard of wire purchased. Hence, we may write the objective function (to be maximized) as

$$12.50x_1 + 11.30x_2 + 17.20x_3 + 19.90x_4 - .14W_m - .29W_p$$

Table 2-2 summarizes the three formulation steps.

TABLE 2-2

Decision Variables		Production Plan Wire Purchase/Mfg	x_1, x_2, x_3, x_4 W_p, W_m
Constraints	Plastic	$.82x_1 + .62x_2 + 1.42x_3 + 2.03x_4$	$\leq 16{,}000$
	Copper	$.43x_1 + .69x_2 + .33x_3 + .20x_4 + .036W_m$	$\leq 5{,}000$
	Capacity	W_m	$\leq 80{,}000$
	Wire	$15x_1 + 16x_2 + 9x_3 + 9x_4 - W_m - W_p$	$\leq 8{,}000$
	Marketing	$x_1 + x_2 - x_3 - x_4$	≥ 0
	Nonnegativity	$x_1, x_2, x_3, x_4, W_m, W_p \geq 0$	
Objective Function	Max	$12.50x_1 + 11.30x_2 + 17.20x_3 + 19.90x_4 - .14W_m - .29W_p$	

Often, linear programs are described succinctly (especially by computer) by "the tableau of detached coefficients," or what we will usually call simply the *tableau*. The tableau for Breck and Dapper is given in Table 2-3. Blank spaces are to be interpreted as zero coefficients. Nonnegativity is assumed. We will adhere to this convention in all tableaus throughout the book, unless explicitly stated otherwise.

The optimal solution to this problem, along with solutions for the succeeding examples, will be discussed in Chapter 3.

TABLE 2-3

x_1	x_2	x_3	x_4	W_m	W_p	Relation	Right-hand Side
.82	.62	1.42	2.03			$\leq$	16,000
.43	.69	.33	.20	.036		$\leq$	5,000
				1		$\leq$	80,000
15	16	9	9	−1	−1	$\leq$	8,000
1	1	−1	−1			$\geq$	0
12.50	11.30	17.20	19.90	−.14	−.29		Max

EXAMPLE 2: AN OPERATING PLAN FOR A CHEMICAL PLANT

Dowpont, Inc. has a plant that produces a proprietary chemical using the continuous process shown by the process flow diagram in Figure 2-1. Each "box" in the diagram represents a series of reactors and other equipment that convert input chemicals into outputs. Process 1 converts specified raw materials and energy generated internally into chemicals *A* and *B*. Process 2 converts chemical *A* into chemical *C*. Process 3 converts chemicals *B* and *C* into chemical *D*, which is the major product for this plant. Originally, *D* was the *raison d'etre* for this Dowpont plant—the intermediates *A*, *B*, and *C* were produced solely for use in producing *D*. However, there is a market for all three intermediates. In fact, the market prices for all three have risen sharply in recent years, while intense competition has kept the price of *D* relatively stable. The plant manager, having observed this trend, wonders whether he should plan to sell any or all of *A*, *B*, or *C* rather than continue to use all of it to produce *D*. In short, he needs help in devising an operating plan for next month.

Figure 2-2 depicts the input-output characteristics of each of the three processes. Note that processes 2 and 3 are weight-losing with respect to usable output. These yield coefficients are an essential part of the model to be designed.

The raw material for process 1 is easily obtainable at short notice and energy is in plentiful supply although expensive. The costs of energy, labor,

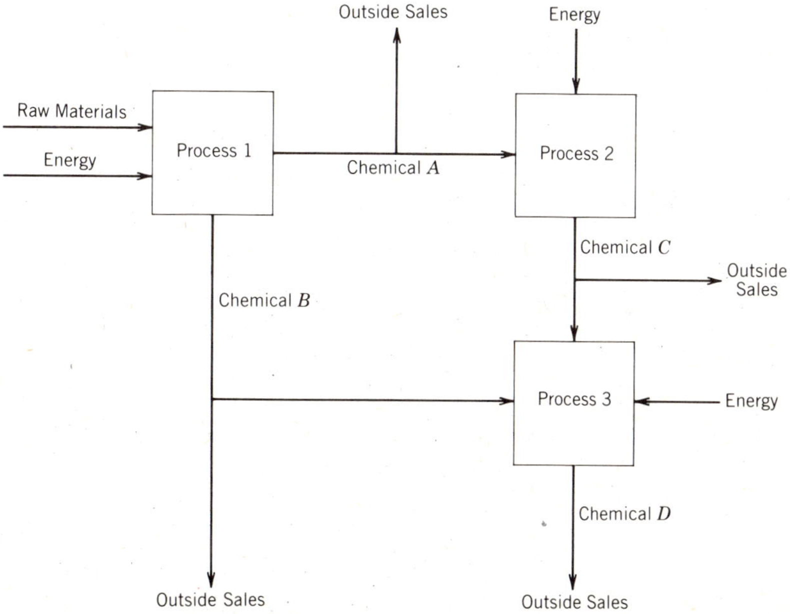

FIGURE 2-1. Dowpont's process flow.

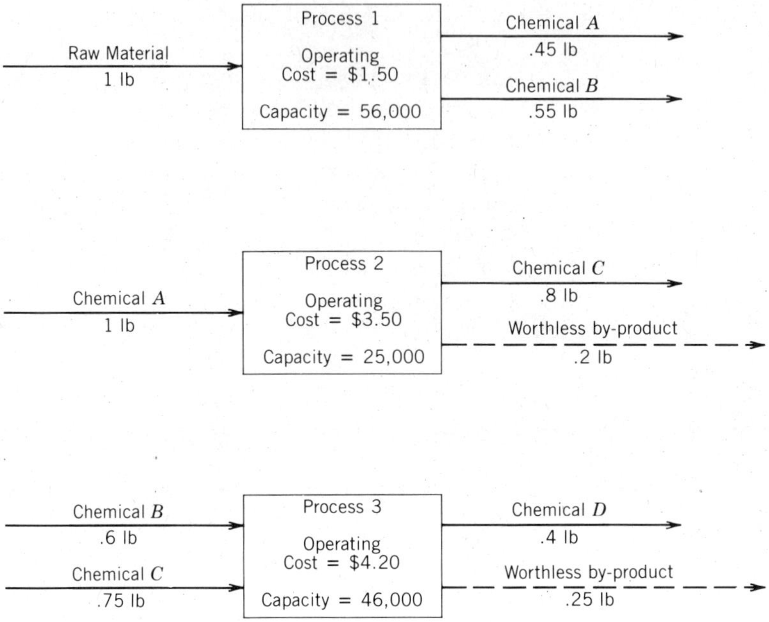

FIGURE 2-2. Dowpont's input-output characteristics.

maintenance, and so on are included in the operating cost figures given in Figure 2-2.

Decision Variables

The activities relevant in this problem are threefold:

1. Purchase of raw material—Let r denote the amount of raw material (in thousands of pounds) to be purchased for process 1 for the coming month.
2. Production for sale—Let S_A, S_B, S_C, S_D denote the number of thousand pounds *sold* of the four chemicals during the coming month.
3. Production for internal use—Let U_A, U_B, U_C denote the number of thousand pounds produced of A, B, and C for internal *usage* during the coming month.

Constraints

The constraints here are the physical flow restrictions or *material balance equations* implicit in Figures 2-1 and 2-2, along with the necessary capacity constraints. For each process, we can state material balance simply as

$$\text{What comes out} = \text{What goes in}$$

adjusted for the appropriate yield factor, since our model may as well ignore the worthless by-products. For example, consider process 1. Chemical *A*, the output of this process, may be used internally or sold outside, so, relative to *A*, we can write "what comes out" of process 1 as

$$U_A + S_A$$

The input to process 1 is simply r, and the weight of *A* output corresponds to 45% of the input weight, so, relative to *A*, "what goes in" is $.45r$. Material balance requires input and output to be equal, i.e.

$$U_A + S_A = .45r$$

Similarly,

$$U_B + S_C = .8U_A$$

For process 2, the input is the *A* used internally, U_A. The output *C* can be sold or retained and the weight of *C* out corresponds to 80% of the weight of *A* input. Algebraically,

$$U_C + S_C = .8U_A$$

For process 3, output *D* is only to be sold, and corresponds to 75% of the input, by weight. The input to process 3 is the combined weight of *B* and *C*. Thus,

$$S_D = .75(U_B + U_C)$$

However, the inputs, *B* and *C*, must be in the correct proportion—.6 : .4 or 1.5 : 1. That is, the weight of *B* input must be one-and-a-half times the weight of *C* input to process 3. Algebraically,

$$U_B = 1.5U_C$$

The capacity constraints can be written simply once one recognizes that the capacity figures are quoted in terms of input. Hence,

$$r \leqslant 60$$

$$U_A \leqslant 52$$

$$U_B + U_C \leqslant 46$$

Of course, all decision variables must be constrained to be nonnegative.

Objective Function

As in the preceding example, Dowpont's plant manager is concerned with maximizing the contribution to fixed costs. Fixed costs themselves will be unaltered by next month's operating plan and can, thus, be ignored in this decision problem. Consider the three classes of decision variables, and the monetary flows associated with each: (1) raw material is acquired at a purchase *cost*, (2) production for sale generates *revenue*, and (3) operation of the three processes causes the firm to incur *operating costs*. Taking the raw material cost and product revenues from Table 2-4 and the operating costs from Figure 2-2, and associating the operating costs, quoted per pound of input, with those inputs, the monetary flows associated with each variable can be calculated as shown in Table 2-5. Hence, the objective function is

$$-2.5r + 2.4S_A + 3.2S_B + 6.4S_C + 12.2S_D$$
$$- 3.5U_A - 4.2U_B - 4.2U_C$$

Summarizing the preceding formulation discussion, Dowpont's linear programming problem is

$$\begin{array}{ll}
\text{Max} & -2.5r + 2.4S_A + 3.2S_B + 6.4S_C + 12.2S_D - 3.5U_A - 4.2U_B - 4.2U_C \\
\text{Subject to} & -.45r + S_A + U_A = 0 \\
 & -.55r + S_B + U_B = 0 \\
 & S_C - .8U_A + U_C = 0 \\
 & S_D - .75U_B - .75U_C = 0 \\
 & U_B - 1.5U_C = 0 \\
 & r \leqslant 56 \\
 & U_A \leqslant 25 \\
 & U_B + U_C \leqslant 46 \\
 & r,\ S_A,\ S_B,\ S_C,\ S_D,\ U_A,\ U_B,\ U_C \geqslant 0
\end{array}$$

TABLE 2-4

	Market Price (per pound)
Raw material	\$1.00
Chemical *A*	2.40
Chemical *B*	3.20
Chemical *C*	6.40
Chemical *D*	12.20

TABLE 2-5

	(−) Purchase	(+) Revenue	Operating Costs (−) Process 1	Process 2	Process 3	Total
r	\$1.00		\$1.50			−\$2.50
S_A		\$2.40				2.40
S_B		3.20				3.20
S_C		6.40				6.40
S_D		12.20				12.20
U_A				\$3.50		−3.50
U_B					\$4.20	−4.20
U_C					4.20	−4.20

Table 2-6 gives the corresponding tableau.

TABLE 2-6

r	S_A	S_B	S_C	S_D	U_A	U_B	U_C	Relation	Right-Hand Side
−.45	1				1			=	0
−.55		1				1		=	0
			1		−.8		1	=	0
				1		−.75	−.75	=	0
						1	−1.5	=	0
1								≤	56
					1			≤	25
						1	1	≤	46
−2.5	2.4	3.2	6.4	12.2	−3.5	−4.2	−4.2		Max

EXAMPLE 3: A PORTFOLIO SELECTION PROBLEM

A portfolio manager in charge of bank portfolio has \$10 million to invest. The securities available for purchase, as well as their respective quality ratings, maturities, and yields, are shown below:

Bond Name	Bond Type	Quality Scales Moody's	Bank's	Years to Maturity	Yield	After-tax Yield
A	Municipal	Aa	2	9	8.6%	8.6%
B	Agency	Aa	2	15	10.8	5.4
C	Government	Aaa	1	4	10.0	5.0
D	Government	Aaa	1	3	8.8	4.4
E	Municipal	Ba	5	2	9.0	9.0

Note that municipal bonds are untaxed.

The bank places the following policy limitations on the portfolio manager's actions:

1. Municipal bonds must total no more than \$3 million.
2. The average quality of the portfolio cannot exceed 1.4 on the bank's quality scale. (Note that a low number on this scale means a high-quality bond.)
3. The average years to maturity of the portfolio must not exceed 5 years.

Assuming that the objective of the portfolio manager is to maximize after-tax earnings and that the tax rate is 50%, what bonds should he purchase? If it became possible to borrow up to \$1 million at 11% before taxes, how should the selection be changed?

Leaving the question of borrowed funds aside for the moment, the decision variables for this problem are simply the dollar amount of each security to be purchased:

x_A = Amount to be invested in bond A, in millions of dollars.

x_B = Amount to be invested in bond B, in millions of dollars.

x_C = Amount to be invested in bond C, in millions of dollars.

x_D = Amount to be invested in bond D, in millions of dollars.

x_E = Amount to be invested in bond E, in millions of dollars.

Now let us consider each of the restrictions of the problem. The portfolio manager has only a total of ten million dollars to invest; therefore,

$$x_A + x_B + x_C + x_D + x_E \leq 10$$

Further, of this amount no more than \$3 million must be invested in municipal bonds. Hence,

$$x_A + x_E \leq 3$$

The average quality of the portfolio, which is given by the ratio of the total quality to the total value of the portfolio, must not exceed 1.4:

$$\frac{2x_A + 2x_B + x_C + x_D + 5x_E}{x_A + x_B + x_C + x_D + x_E} \leq 1.4$$

Note that the inequality is less-than-or-equal-to, since a low number in the bank's quality scale means a high-quality bond. By clearing the denominator and rearranging terms, we find that this inequality is equivalent to the linear constraint:

$$0.6x_A + 0.6x_B - 0.4x_C - 0.4x_D + 3.6x_E \leq 0$$

The constraint on the average maturity of the portfolio is a similar ratio. The average maturity must not exceed 5 years:

$$\frac{9x_A + 15x_B + 4x_C + 3x_D + 2x_E}{x_A + x_B + x_C + x_D + x_E} \leq 5$$

equivalent to the linear constraint:

$$4x_A + 10x_B - x_C - 2x_D - 3x_E \leq 0$$

Note that the two ratio constraints are, in fact, nonlinear constraints, which would require sophisticated computational procedures if included in this form. However, multiplying both sides of each ratio constraint by its denominator (which must be nonnegative since it is the sum of nonnegative variables) transforms this nonlinear constraint into a simple linear constraint.

We must now determine the form of the objective function. Assuming that all securities are purchased at par (face value) and held to maturity and that the income on municipal bonds is tax-exempt, the after-tax earnings are given by:

$$.086x_A + .054x_B + .05x_C + .044x_D + .09x_E$$

We summarize this formulation in Table 2-7.

Now consider the additional possibility of being able to borrow up to \$1 million at 11% before taxes. Essentially, we can increase our cash supply above ten million by borrowing at an after-tax rate of 5.5%. We can define a new

TABLE 2-7

x_A	x_B	x_C	x_D	x_E	Relation	Right-Hand Side
1	1	1	1	1	$\leq$	10
1				1	$\leq$	3
.6	.6	− .4	− .4	3.6	$\leq$	0
4	10	−1	−2	−3	$\leq$	0
.086	.054	.05	.044	.09		Max

decision variable y as amount to be borrowed, in millions of dollars. There is an upper bound on the amount of funds that can be borrowed, and hence

$$y \leqslant 1$$

The cash constraint is then modified to reflect that the total amount purchased must be less than or equal to the cash that can be made available including borrowing:

$$x_A + x_B + x_C + x_D + x_E \leqslant 10 + y$$

Now, since the borrowed money costs 5.5% after taxes, the new after-tax earnings are:

$$.086x_A + .054x_B + .050x_C + .044x_D + .090x_E - .055y$$

We summarize the formulation when borrowing is allowed, as shown in Table 2-8.

TABLE 2-8

x_A	x_B	x_C	x_D	x_E	y	Relation	Limits
1	1	1	1	1	−1	⩽	10
					1	⩽	1
1				1		⩽	3
.6	.6	−.4	−.4	3.6		⩽	0
4	10	−1	−2	−3		⩽	0
.086	.054	.056	.044	.090	− .055		Max

EXAMPLE 4: A TRANSPORTATION PROBLEM

An agricultural cooperative, which buys wheat from surrounding farms, is faced with deciding to which of two mills to send its recently purchased 50,000 bushels. From the mills, the ground flour is to be sent to three local markets. The cooperative, as well as deciding how much wheat to send to each mill, must instruct each mill as to where to deliver the flour. The transport and milling cost are shown by Table 2-9 along with the demand for flour in each of the markets.

The decision to be made by the agricultural cooperative concern the appropriate levels of the *flows* in the system, flows of wheat *from* the cooperative *to* each of the mills, and flows of flour *from* each of the mills to each of the three markets. These flows are represented schematically by Figure 2-3. The cooperative's decision problem consists of determining the optimal flows along each

TABLE 2-9

Milling Cost ($/bushel of wheat)		Transport Cost ($/bushel)		
Mill 1	Mill 2	1. From Cooperative to:		
.13	.16	Mill 1	Mill 2	
		.31	.33	
Demand (bushels)		2. From to:	Mill 1	Mill 2
		Northburg	.18	.25
Northburg	10,000	Midburg	.22	.23
Midburg	15,000	Southburg	.27	.19
Southburg	25,000			

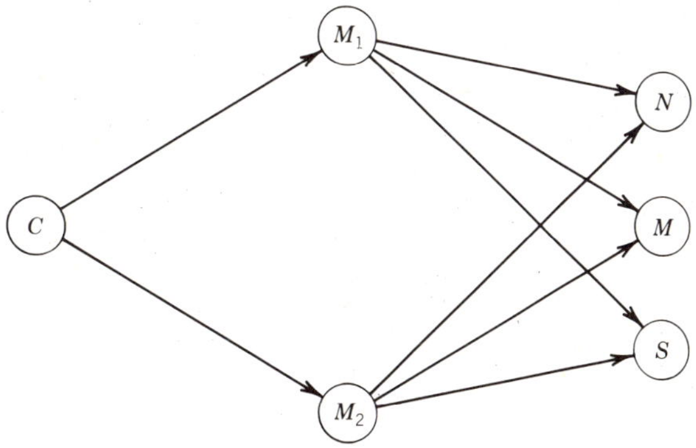

FIGURE 2-3. Possible flows.

of the paths illustrated in Figure 2-3. Algebraically, we can denote these decision variables by reference to the start and end of each path; for example, let

$C \rightarrow M_1$ Denote the flow of wheat (in bushels) between location C (the cooperative) and location M_1 (mill 1).

$C \rightarrow M_2$ Denote the flow of wheat (again, in bushels) between C and M_2.

$M_1 \rightarrow N$ Denote the flow of flour (in bushels) M_1 to market N (Northburg).

$M_1 \rightarrow M$ Denote the flow of flour from M_1 to market M (Midburg).

Similarly, we define flow variables

$$M_1 \rightarrow S$$

$$M_2 \rightarrow N$$

$$M_2 \rightarrow M$$

$$M_2 \rightarrow S$$

The constraints of this decision problem are simply those implicit in the physical flow, namely, at each location—cooperative, mills, and markets:

$$\text{Flow out} - \text{Flow in} = \text{Amount supplied at that location} \\ - \text{Amount drawn out at that location}$$

This type of constraint is often referred to as a *material balance constraint* or a *conservation of flow constraint*. For example, at the cooperative, "flow out" can be written as

$$C \rightarrow M_1 + C \rightarrow M_2$$

and the amount supplied at that location is 50,000 bushels. Since there is neither inflow nor wheat to be drawn *out* at this location, the material balance constraint can be written as

$$C \rightarrow M_1 + C \rightarrow M_2 = 50{,}000$$

At mill 1, outflow can be written as

$$M_1 \rightarrow N + M_1 \rightarrow M + M_1 \rightarrow S$$

inflow is simply $C \rightarrow M_1$. Hence, since there is neither additional supply, other than the incoming wheat, provided by the mill, nor is there any demand for flour to be drawn out of the system at the mill, the material balance constraint can be written

$$M_1 \rightarrow N + M_1 \rightarrow M + M_1 \rightarrow S - C \rightarrow M_1 = 0$$

we have assumed, for simplicity, that one bushel of wheat produces one bushel of flour.[2] Similarly, at mill 2, we can write

$$M_2 \rightarrow N + M_2 \rightarrow M + M_2 \rightarrow S - C \rightarrow M_2 = 0$$

At each of the three markets, there are inflows of flour and amounts to be drawn out of the system. For example, Northburg (N) receives as inflow

$$M_1 \rightarrow N + M_2 \rightarrow N$$

[2]This simplification poses no difficulty in practice since the units in which flour is measured can be defined so that one "unit" equals the amount of flour produced from one bushel of wheat.

and draws 10,000 bushels out. Thus,

$$-M_1 \to N - M_2 \to N = -10{,}000$$

expresses the constraint at that location.[3] Similarly,

$$-M_1 \to M - M_2 \to M = -15{,}000$$

and

$$-M_1 \to S - M_2 \to S = -25{,}000$$

are the constraints which correspond to the Midburg and Southburg markets.

Given that all demands are to be met, the agricultural cooperative wishes to set the levels of flow to minimize the sum total of milling and transport costs. These costs can be assigned to the various paths pictured in Figure 2-3, for example, as illustrated in Table 2-10. The choice to assign milling cost to the incoming paths was due to the fact that milling costs were quoted per bushel of incoming wheat.

Putting the data in Table 2-10 together with the material balance constraints from above, we arrive at the complete specification of the linear program that will solve the agricultural cooperative's flow allocation problem:

$$\begin{aligned}\text{Min}\quad & .44C \to M_1 + .49C \to M_2 + .18M_1 \to N + .22M_1 \to M \\ & + .27M_1 \to S + .25M_2 \to N + .23M_2 \to M + .19M_2 \to S\end{aligned}$$

Subject to

$$\begin{array}{rrrrrrrrl}
C \to M_1 & + C \to M_2 & & & & & & & = 50{,}000 \\
-C \to M_1 & & + M_1 \to N & + M_1 \to M & + M_1 \to S & & & & = 0 \\
& -C \to M_2 & & & & + M_2 \to N & + M_2 \to M & + M_2 \to S & = 0 \\
& & -M_1 \to N & & & -M_2 \to N & & & = -10{,}000 \\
& & & -M_1 \to M & & & -M_2 \to M & & = -15{,}000 \\
& & & & -M_1 \to S & & & -M_2 \to S & = -25{,}000
\end{array}$$

and all decision variables nonnegative. Table 2-11 shows the tableau.

The tableau reveals the special structure of this linear program:

1. All constraint coefficients are 0, + 1, or −1.
2. Every column has *exactly* one + 1 entry and one − 1 entry.

Linear programs with this special structure form a very important class of models called network optimization models. The name reflects the fact that a

[3]More commonly, this constraint might be written $M_1 \to N + M_2 \to N = 10{,}000$. To keep inflow and outflow sign conventions consistent, we retain the minus signs.

TABLE 2-10

Path	Milling Cost ($/bushel)	Transport Cost ($/bushel)	Total Cost ($/bushel)
$C \rightarrow M_1$	.13	.31	.44
$C \rightarrow M_2$	.16	.33	.49
$M_1 \rightarrow N$	—	.18	.18
$M_1 \rightarrow M$	—	.22	.22
$M_1 \rightarrow S$	—	.27	.27
$M_2 \rightarrow N$	—	.25	.25
$M_2 \rightarrow M$	—	.23	.23
$M_2 \rightarrow S$	—	.19	.19

TABLE 2-11

$C \rightarrow M_1$	$C \rightarrow M_2$	$M_1 \rightarrow N$	$M_1 \rightarrow M$	$M_1 \rightarrow S$	$M_2 \rightarrow N$	$M_2 \rightarrow M$	$M_2 \rightarrow S$	Relation	Right-Hand Side
1	1							=	50,000
−1		1	1	1				=	0
	−1				1	1	1	=	0
		−1			−1			=	−10,000
			−1			−1		=	−15,000
				−1			−1	=	−25,000
.44	.49	.18	.22	.27	.25	.23	.19		Min

network diagram like Figure 2-3 is sufficient to completely describe the structure of the flow allocation problem. These network models have been applied to a wide range of problems, most successfully in transportation and distribution. Much of their success emanates from the existence of a class of special network algorithms that exploit the structure illustrated by Table 2-11 and permit the solution of network models in a tiny fraction of the time that would be required to solve a general linear program of the same size. We will delve further, in detail, into the features of network models in Chapter 5.

EXAMPLE 5: A MULTIPERIOD AGGREGATE PRODUCTION SCHEDULING PROBLEM

The four examples just discussed all deal with a decision made for a single future time period. In a large number of situations with which managers must deal, there are crucial linkages between one time period and the next. In these cases, it often becomes necessary to consider the future in planning for the current period. In production planning problems such as the one to follow, the linkage is inventory.

Consider a plant that assembles videotape recorders, and wants to schedule the production facilities for the next year. There is a pronounced seasonal pattern in demand, as shown by Figure 2-4, so that a single time period model will not suffice. Instead, the plant manager wishes to model 12 one-month periods.

The assembly operation is quite labor intensive; the current workforce numbers 500. At this level, monthly production could be 4000 recorders; that is, each worker can assemble up to eight per month. The assembly line can produce no more than 7000 per month under any conditions, due to equipment limitations. Assembly workers can be hired from the surrounding locale, but the personnel department estimates that no more than 50 workers can be hired in any one month. Union contracts specify that no more than 10% of the workforce can be laid off in any one month, and there are severe penalties (e.g., unemployment compensation, lack of work benefits) that must be paid in the event of any layoff.

In short, the plant manager wishes to determine the "best" production schedule which will satisfy anticipated demand as given by Figure 2-4. One set of decision variables that will completely specify such a schedule is:

P_t : Number of recorders to be produced in month t. There are 12 of these variables; that is, t goes from 1 to 12.

I_t : Number of recorders in inventory at the beginning of month t. (Currently—at the beginning of month 1—there are 750 in stock; that is, $I_1 = 750$.)

W_t : Number of workers during month t.

H_t : Number of workers hired at the beginning of month t.

L_t : Number of workers laid off at the beginning of month t.

For all variables, $t = 1, 2, \ldots, 12$. I_{13} will denote the end-of-year inventory.

Using these decision variables we can generate the required constraints. For any month t:

1. Demand must be satisfied, so production plus the change in inventory, $I_t - I_{t+1}$, must equal demand:

$$P_t + I_t - I_{t+1} = D_t$$

where D_t denotes the demand in month t. (D_t can be read from Figure 2-4.)

2. Production cannot exceed the capabilities of the workforce.

$$P_t \leq 8W_t$$

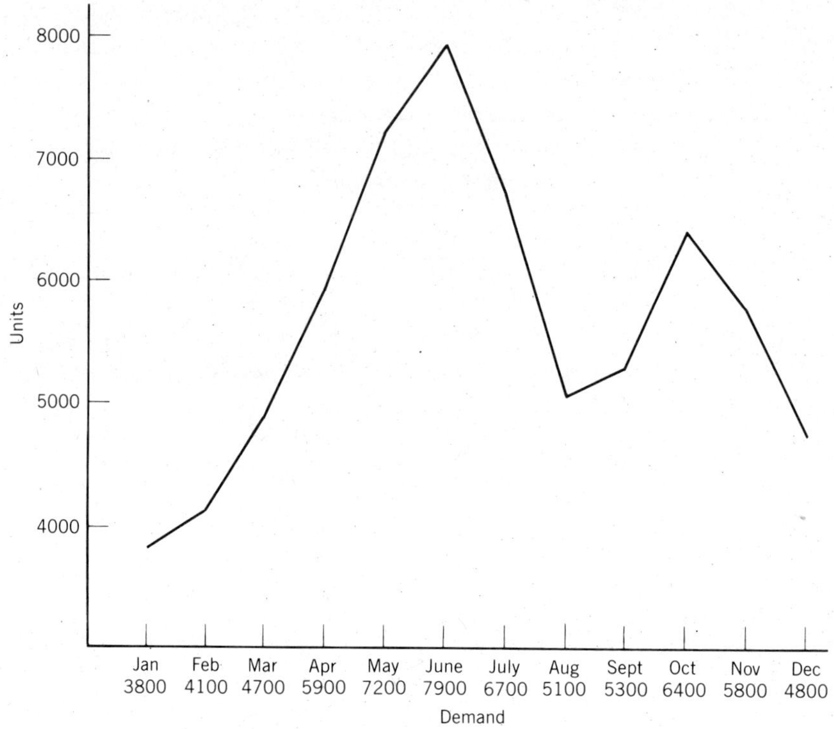

FIGURE 2-4. Seasonal demand.

(Note that an assumption implicit in this constraint is that there is no learning; in each month, each worker can assemble, on average, eight recorders. In particular, new workers are assumed to come on stream fully trained. In practice, we might wish to model the start-up of new workers by writing this constraint as

$$P_t \leqslant \alpha H_t + 8 (W_t - H_t)$$

where $\alpha < 8$ would denote the productive capability of new workers; $W_t - H_t$ is the workforce who have been working at least one month.)

3. Production cannot exceed assembly line capacity:

$$P_t \leqslant 7000$$

4. The available workforce is defined to be the previous month's workforce plus hires minus those laid off:

$$W_t = W_{t-1} + H_t - L_t$$

(Since the current workforce numbers 500, $W_1 = 500 + H_1 - L_1$.)

5. Personnel can hire no more than 50 new workers per month:

$$H_t \leqslant 50$$

6. No more than 10% of the existing month's workforce can be laid off:

$$L_t \leqslant .1 W_{t-1}$$

(For $t = 1$, $L_1 \leqslant .1 \times 500 = 50$.)

To the above we add the nonnegative constraints.

The plant manager wanted the "best" schedule subject to the constraints above. What does "best" mean? In this case, the objective is likely to be different from the first three examples. Since the formulation requires that demand be met, revenue is fixed and the plant manager can be concerned simply with keeping the costs as low as possible.[4] Let us list the costs associated with each variable:

P_t : Variable production costs (other than labor)—$100 per unit.

I_t : Inventory carrying costs—$7 per unit per month, to be applied to the average inventory level.

W_t : Labor costs—$1600 per worker, monthly.

H_t : Hiring costs, for example, recruiting, training—$300 per worker hired.

L_t : Layoff costs, for example, lack-of-work benefits—$500 per worker fired.

In practice, we might want to allow for the possibility of differential costs during the year, but we ignore such complexities. Given the notation above, we can write the objective function as

$$\text{Min} \sum_{t=1}^{12} \left[100P_t + 7\left(\frac{I_t + I_{t+1}}{2}\right) + 1600\,W_t + 300\,H_t + 500\,L_t\right]$$

[4]In practice, there will be multiple, possibly conflicting objectives in the plant manager's mind. For example, the plant manager, if he or she thinks long-term, may be concerned with workforce stability. Thus, he or she may wish to minimize layoffs over and above the effect that union penalties would have on keeping layoffs down. Chapter 4 will deal with ways of handling such multicriterion problems; for now, we ignore these issues.

We can summarize the model as below:

$$\text{Min} \sum_{t=1}^{12} (100P_t + 1600W_t + 300H_t + 500L_t + 7I_t) - 3.5I_1 + 3.5I_{13}$$

Subject to, for each $t = 1,2,\ldots,12$:

$$\begin{aligned}
P_t + I_t - I_{t+1} &= D_t \\
P_t - 8W_t &\leq 0 \\
P_t &\leq 7000 \\
W_t - W_{t-1} - H_t + L_t &= 0 \\
H_t &\leq 50 \\
-.1W_{t-1} + L_t &\leq 0
\end{aligned}$$

$$P_t,\ I_t,\ W_t,\ H_t,\ L_t \geq 0;\ D_t \text{ as in Figure 2-4}$$

To add one more layer of complexity to this example, we might recognize that in practice, the restrictive requirement that demand be met is often not appropriate. Instead, management might allow some demand to be unsatisfied and assess a penalty (loss of goodwill, lost competitive advantage, or simply lost contribution) to be incurred when stockouts occur.[5] In such a situation, we might add another set of decision variables.

$$S_t = \text{stockouts in period } t$$

and rewrite the first constraint as

$$P_t + I_t - I_{t+1} + S_t = D_t$$

The per unit penalty times S_t would then be appended to the objective function summation.

FORMULATION GUIDELINES

Formulation is an art rather than an exact science. Ability comes with practice and the recognition of similarities in different situations. The preceding examples were intended to give a sense of the kinds of problems amenable to a linear programming solution and to illustrate ways of thinking about decision variables, constraints, and objective function formulation. Let us now try to generalize.

[5]Of course, in practice, this penalty cost is typically very hard to assess. The ability to determine how sensitive the optimal solution is to the penalties attributed to stockouts becomes essential.

Decision Variables

One way of generating the appropriate set of decision variables for a given allocation problem is to use *activity analysis*. Simply put, the decision maker asks "In order for me to make the decision at hand, what are the activities whose operating levels I need to know?" For illustration, in Example 1, the plant manager is concerned with five production activities (one for each of the four models and one for the internal manufacture of wire) and a purchase activity (wire). The levels at which each of these six activities are to be operated in the coming quarter completely specify the information needed by the plant manager.

Often the decision problem can be characterized by the interrelationship of flows of material, energy, finished product, and/or information. Example 2, as illustrated by Figures 2-1 and 2-2, is such a situation. In this case, the decision variables represent the flows—that is, the activities are the *levels* of flow on each of the arcs of the process flow diagram (Figure 2-1). The main body of the constraints is the set of material balance (or conservation of flow) equations, taking into account the technological coefficients embodied by the input-output relationships (Figure 2-2).

Constraints

Let us try to characterize certain commonly occurring constraint types and their mathematical representations.

Resource Limitations. If activities whose levels are represented as x_1, $x_2, \ldots, x_n$ use a given amount of a scarce resource, a constraint is needed to represent the inequality

$$\text{Total amount of resource used} \leq \text{Total amount of resource available}$$

If a_i represents the amount of resource used by activity i, and b represents the total amount available, we may write

$$\sum_{i=1}^{n} a_i x_i \leq b$$

For example, a manufacturer produces products Z54, Z96, and Z99. Their manufacture requires 3, 2, and 6 pounds of iron, respectively. Currently, 2000 pounds of iron are in the factory's raw material inventory. Thus, if x_1 represents the number of Z54s produced monthly, x_2 the number of Z96s produced monthly, and x_3 the number of Z99s produced monthly, their values must be restricted as follows:

$$3x_1 + 2x_2 + 6x_3 \leq 2000$$

Often the supply of resource can be increased through purchase. Then,

Total amount of resource used $\leq$ Amount of resource currently available
+ Amount of resource purchased

If, in the above example, iron purchase is possible, we might define a new decision variable p, with

p = Number of pounds of iron purchased

and change the constraint given above to

$$3x_1 + 2x_2 + 6x_3 - p \leq 2000$$

In addition, it may be possible to produce "resource," as in the next constraint category.

Multistage Production Restrictions. Often the production of one item will require intermediate components manufactured by the same firm. Thus, resource limitations would apply to the components' resources, and the final product would be constrained by an inequality of the form

Amount of intermediate product used $\leq$ Amount of intermediate product made

or, since the intermediate component production consists, itself, of activities whose levels are to be determined by the linear program,

Amount of intermediate product used $-$ Amount of intermediate product
made ≤ 0

In the case where some amount of the intermediate product is available in work-in-process inventory, this becomes

Amount of intermediate product used $-$ Amount of intermediate product made
$\leq$ Amount of intermediate product available

For example, Table 2-12 gives the usage of components A and B in our products, and Table 2-13 describes the resource required by the components. For example, component A requires 2 pounds of iron and an hour of labor. The model Z54 requires 3 A components and 15 B components for assembly. Currently, there are 400 A components in inventory. If we denote by y_A and y_B the production of components A and B, respectively, then we need two resource limitation constraints,

$$2y_A + 11y_B \leq 2000$$

$$y_A + 14y_B \leq 4090$$

and two multistage production constraints:

$$3x_1 + 3x_2 \qquad - y_A \leq 400$$

$$15x_1 + 10x_2 + 23x_3 - y_B \leq 0$$

Demand Restrictions. Typically, decisions as to what and how much to produce depend on demand projections. If firm commitments exist for the sale of certain products, management may require that

$$\text{Amount of product made} = \text{Amount of product demanded}$$

Or, management may feel that it is possible to sell product over and above the contracted amount. This policy might give rise to

$$\text{Amount of product made} \geq \text{Amount of product promised}$$

In other situations, marketing personnel may establish a demand estimate, and management may want to insure that extra product not be manufactured. In this case, the policy might state that

$$\text{Amount of product made} \leq \text{Amount of product demanded}$$

For example, our manufacturer might project next month's demand to be 1000, including a firm commitment to deliver 100 to a steady customer. Two constraints might be introduced to reflect this situation:

$$x_1 + x_2 + x_3 \leq 1000$$

$$x_1 + x_2 + x_3 \geq 100$$

TABLE 2-12
Number of Components in Each Model

	Model		
Component	Z54	Z96	Z99
A	3	3	0
B	15	10	23

TABLE 2-13
Resources Consumed by Each Component

	Component		
Resource	*A*	*B*	Available
Iron (lb)	2	11	2000
Labor (hr)	1	14	4090

Quality/Mix Constraints. When intermediate products are blended to form a final product, it may be necessary to restrict the blend so that the final product has some desired characteristic. For example, gasoline blending stocks are mixed so that the final product has at least a certain octane; alloys are blended so that the final product has no less than a specified percentage of copper; as we saw in Example 3, a portfolio of stock is selected so that it has an average quality rating of at least a given amount. To illustrate, consider six metal alloys, characterized by their copper content below:

Alloy	A	B	C	D	E	F
Copper content	5%	6%	4%	11%	13%	7%

A firm has just received a special order for new alloy to contain at least 10% copper. Let x_A be the amount of alloy A (lb) used, x_B the amount of alloy B used, and so on. Then $.05x_A$ represents the copper contributed by alloy A, $.06x_B$ represents the copper contributed by alloy B, and so forth, and the amount of copper in the blend is

$$.05x_A + .06x_B + .04x_C + .11x_D + .13x_E + .07x_F$$

The *percentage* of copper in the blend is then

$$\frac{.05x_A + .06x_B + .04x_C + .11x_D + .13x_E + .07x_F}{\text{total weight of blend}}$$

$$= \frac{.05x_A + .06x_B + .04x_C + .11x_D + .13x_E + .07x_F}{x_A + x_B + x_C + x_D + x_E + x_F}$$

assuming no weight loss in the blending process. Since the blend is supposed to have at least 10% copper, the necessary constraint would be

$$\frac{.05x_A + .06x_B + .04x_C + .11x_D + .13x_E + .07x_F}{x_A + x_B + x_C + x_D + x_E + x_F} \geqslant .10$$

or, rearranging, after multiplying by the denominator,

$$-5x_A - 4x_B - 6x_C + x_D + 3x_E - 3x_F \geqslant 0$$

Management Policy Restrictions. Often, for a variety of reasons, management will constrain the values decision variables can take. Examples include:

- "Make at least twice as many model Z54s as the other types combined," that is,

$$x_1 \geq 2(x_2 + x_3)$$

or

$$x_1 - 2x_2 - 2x_3 \geq 0$$

- "Revenue received from the three models this month must be at least as great as last month's revenue." That is, if the per unit revenue for the three models is \$15, \$19, and \$29, and last month's revenue was \$9000, this month's production plan must satisfy

$$15x_1 + 19x_2 + 29x_3 \geq 9000$$

It is difficult to generalize these constraints because they can take any form. In fact, demand restrictions may be considered a subset of this class.

Flow/Yield Constraints. Suppose a chemical plant can represent part of its production process by the diagram in Figure 2-5. We need a constraint to tie together the various components—that is, a Material Balance or Conservation of Flow constraint:

$$r_1 + r_2 + x_A + x_B = r_3 + x_C$$

or

$$r_1 + r_2 - r_3 + x_A + x_B - x_C = 0$$

This flow constraint conserves mass; that is, if 10 pounds go into the reactor (made up of reagents 1 and 2 and chemicals *A* and *B*), then 10 pounds must come out (into chemical C and/or reagent 3). As we have seen, process yields are often not 100%. If the process depicted in Figure 2-5 was characterized by an 80% yield, the constraint would then be

$$.8(r_1 + r_2 + x_A + x_B) = r_3 + x_C$$

or

$$.8r_1 + .8r_2 - r_3 + .8x_A + .8x_B - x_C = 0$$

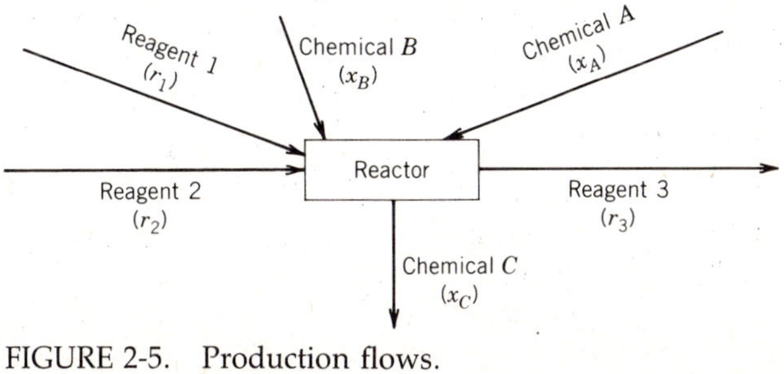

FIGURE 2-5. Production flows.

Notational Constraints. Often, for convenience, we denote by a single symbol a more complex algebraic expression. For example, if demand is sufficiently high,

$$\text{Stock-outs} = \text{Demand} - (\text{production and inventory})$$

or, in symbols,

$$S = d - p - i$$

If "S" appears in other constraints, the program must know how it is tied to other variables, so that the above constraint is added as a definition. In multiperiod linear programs such as that of Example 5, such definitions tie the variables from one period to the variables of the next; for example, for inventories I, production P and demand d:

$$I_{t+1} = I_t + P_t - d_t$$

or, for workers W, beginning of the period layoffs L and hires H,

$$W_{t+1} = W_t + H_{t+1} - L_{t+1}$$

Algebraic Manipulations

Often it is necessary to transform a given linear program in some way. In particular, some commercially available linear programming computer packages have conventions that must be adhered to; for example, only maximization problems can be solved, or only "less-than-or-equal-to" inequalities can be handled, or all decision variables are assumed to be nonnegative. Let us illustrate with a few occasionally needed algebraic manipulations:

- The sense of the required optimization may be changed by reversing the signs of all objective function coefficients; that is,

$$\text{Min} \quad 2x_1 - x_2 + 4x_3 \text{ is equivalent to}$$

$$\text{Max} \quad -2x_1 + x_2 - 4x_3$$

- The sense of any inequality may be changed by reversing the signs of all coefficients in the constraint and the sign of the right-hand-side value; that is,

$$4x_1 - 2x_2 + x_3 \geqslant 4 \quad \text{is equivalent to} \quad -4x_1 + 2x_2 - x_3 \leqslant -4$$

- Typically linear programming packages assume that variables are nonnegative. If a decision variable x is desired to be unrestricted in sign, it must then be transformed before the model can be run.[6] Define two new variables x^+ and x^-, both nonnegative, such that

$$x = x^+ - x^-$$

(Any number can be represented as the difference between two *nonnegative* numbers: $3 = 3 - 0$; $5 = 5 - 0$; $-2 = 0 - 2$; $-10 = 0 - 10$.) Whenever x appears in a constraint, it must be replaced by the above expression.

- Any inequality can be transformed into an equality by the addition of a new variable, called a *slack* variable in the case of a "$\leqslant$" constraint, a *surplus* variable in the case of a "$\geqslant$" constraint. If we desire to transform

$$x_1 + x_2 + 3x_3 \leqslant 4$$

to an equality, we write

$$x_1 + x_2 + 3x_3 + S = 4$$

For any values of x_1, x_2, and x_3 that satisfy the original inequality, there will be some nonnegative value of S such that the equality is satisfied. If the inequality is

$$x_1 + x_2 + 3x_3 \geqslant 4$$

we must write

$$x_1 + x_2 + 3x_3 - S = 4$$

in order to guarantee that S will be nonnegative.

- If a variable x has an associated lower bound, $x \geqslant l$, rather than include this explicitly as an additional constraint, some computation efficiency may be gained by substituting $(x' + l)$ for x wherever x appears in the formulation. Note that the new variable x' need not be constrained except by the automatic nonnegativity condition, since $x \geqslant l$ is equivalent to $x' \geqslant 0$.

[6] To refer back to an oft-used example, if I denotes current inventory, P current production, and D demand, we might write

$$x = I + P - D$$

If x is positive, it represents excess (beginning of next period's) inventory; if negative, x represents the amount of stockout.

Objective Functions

The difficulty in formulating single criterion objective functions is to determine just what it is that determines the "best" plan—that is, what the measure of effectiveness is. Once that is done, writing that objective algebraically in term of the problem's decision variables is usually straightforward. There is one oft-seen special case that warrants consideration.

Separable Linear Programming. This case, somewhat complex but of special interest, allows the piecewise linearization of a nonlinear objective function. Suppose we are concerned with a typical production planning problem where scarce resources are to be allocated to several products. Rather than estimating demand as is often done, we might tie price and demand together, explicitly recognizing that demand is a function of the price; that is, if we produce a large quantity of a product, we might expect to have to charge less per unit in order to move our inventory than if we had produced less. Typically, then, total revenue and consequently total contribution, increases (at a diminishing rate) as production increases—a nonlinear relationship shown schematically in Figure 2-6.

This can be handled directly via nonlinear programming, but as we will see in Chapter 8, the techniques are computationally complex and not easily obtainable in many commercially available mathematical programming packages. An alternative solution approach is to approximate the nonlinear function and use standard linear programming procedures. Let us consider an example.

Suppose we wish to maximize contribution and x, one of our decision variables, is the amount of Product X to be produced. Analogous to Figure 2-6*c*, Figure 2-7*a* depicts the total contribution as a function of the production level. We can approximate this relationship by the dashed lines in Figure 2-7*b*, redrawn as Figure 2-7*c*. Figure 2-7*c* may be interpreted as saying that the slope of each line segment equals the contribution per unit *in that range*. The approximation can be made as close as we wish by increasing the number of line segments, but, as we shall see, this increases the size of the subsequent LP, so we must judgmentally trade off the size of the LP with the accuracy of the approximation.

We may now incorporate this linearized relationship into the LP as follows:

1. Whenever the variable x occurs in the constraints of the LP, replace it by $x_1 + x_2 + x_3$, where

   x_1 = number of product X's to be produced *up to 100 units*,

   x_2 = *additional* number of Product X's to be produced between *100 and 300 units*,

   x_3 = additional number of Product X's to be produced *beyond 300 units*.

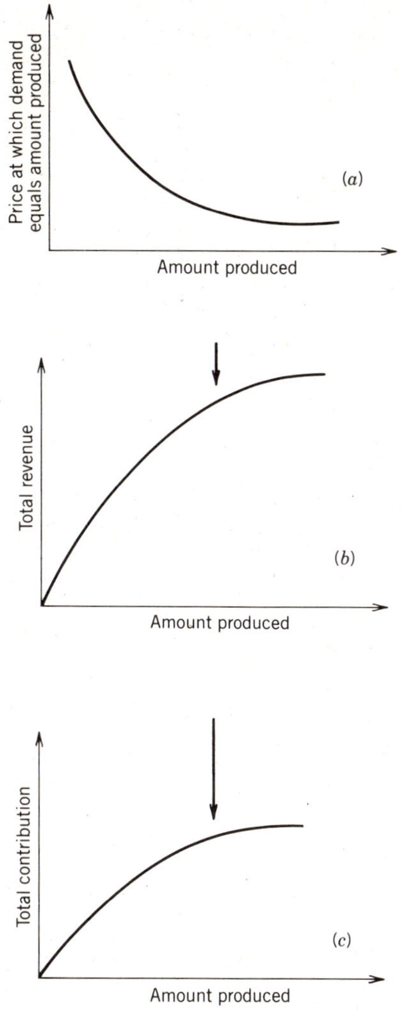

FIGURE 2-6. Nonlinear production relationships.

For example, if 225 units of Product X were to be produced, $x_1 = 100$; $x_2 = 125$; $x_3 = 0$. If 450 were to be produced, $x_1 = 100$; $x_2 = 200$; $x_3 = 150$. Note that in both cases,

$$\text{Total number of Product } X\text{'s} = x = x_1 + x_2 + x_3$$

2. Upper bound the new decision variables as appropriate; that is,

$$x_1 \leqslant 100$$

$$x_2 \leqslant 200$$

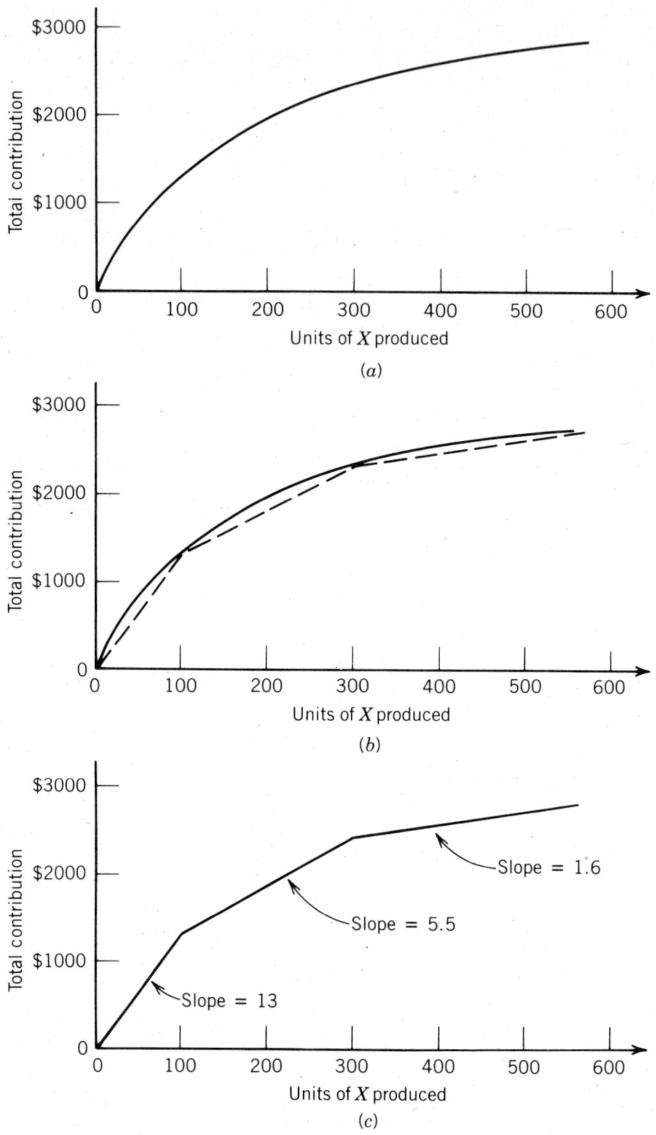

FIGURE 2-7. Linearizing the contribution function.

Note that: (1) x_3 is unconstrained; (2) x_2 is constrained to be less than or equal to 200, not 300, since the variable represents the number of *additional* units over the initial 100.

3. Add to the objective function a term that incorporates the contribution of Product X; that is,

$$13x_1 + 5.5x_2 + 1.6x_3$$

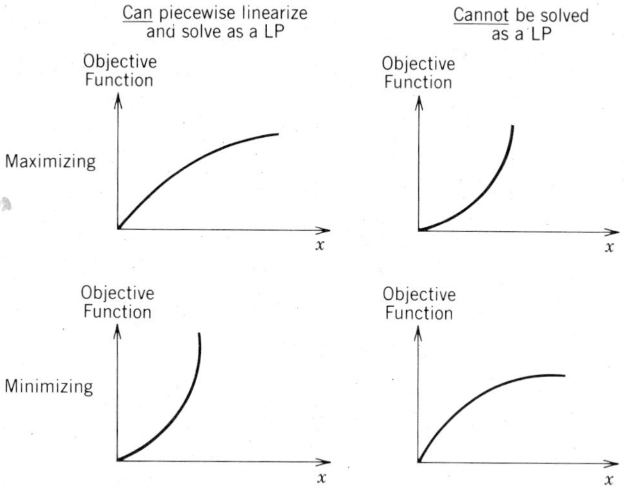

FIGURE 2-8. Will piecewise linearization permit an LP solution?

This procedure, called *separable linear programming*, may be followed for each of the products under consideration.[7] Note that we must replace each old variable by three new ones, corresponding to the number of linear segments used in the approximation.

One thing seems to be missing: we have nowhere required the linear program to "fill up" x_1 before starting on x_2 or to "fill up" x_2 before starting on x_3. That is, what prevents a solution of the form

$$x_1 = 0$$

$$x_2 = 0$$

$$x_3 = 200?$$

This would certainly be feasible, but would make no sense. The essence of the answer to this question is that the objective function coefficients *decrease* from x_1 to x_2 to x_3. Thus, since contribution is being maximized, the LP will want to get "as much of" x_1 as possible before using any x_2 and as much x_2 as possible before using any x_3. If instead of maximizing contribution, we were *minimizing* cost, the LP would try to get as much of x_3 as possible (since x_3 is unbounded,

[7] The term "separable" comes from the fact that the nonlinear objective function can be approximated as a sum of *separate* linear functions in each of the original variables. That is, we can write

$$\text{Max} \quad f(X, Y, Z)$$

as

$$\text{Max} \quad f_1(X) + f_2(Y) + f_3(Z)$$

where each f_i is linear.

x_1 and x_2 would always be 0), and this problem could not be solved as an LP.[8] Figure 2-8 summarizes this concept.[9]

Key Assumptions Underlying Linear Programming

Much of the preceding discussion of formulation guidelines applies equally well to all forms of mathematical programming. What is it that allows us to classify a mathematical program as a *linear* program? Clearly, from the name, we require that the objective function and all constraints be written as linear functions of the decision variables. That is, we assume that two characteristics will hold for all modeled relationships:

1. *Proportionality.* For each activity, the consumption of input and the associated value of the criterion are strictly proportional to the value of the decision variable. For example, if the manufacture of one product requires 3 pounds of iron, two products must require 6 pounds, ten products, 30 pounds, and so on. Similarly, if costs associated with the manufacture of this product are $1, then the costs associated with the manufacture of two products must be $2, with ten products, $10.
2. *Additivity.* The *total* consumption of resources must be the sum of the resources consumed by each of the individual activities. Similarly, the total value of the objective must be the sum of the contributions to the objective associated with each activity. If a firm would incur a cost of $1000 in the manufacture of model X13 alone and would incur a cost of $1000 in the manufacture of model X23 alone, it must incur a cost of $2000 in the manufacture of both models simultaneously.

Note that these two assumptions bar economies of scale and shared costs.

Linear programs, as conventionally formulated, carry two additional underlying assumptions:

3. *Divisibility.* An activity can be divided into fractional units; any decision variable may take on noninteger values.
4. *Deterministic.* The data associated with a linear program, the objective function coefficients; the available resources; and the constraint coefficients, are known constants.

[8]There is no linear way of expressing the constraint "$x_2 = 0$ unless $x_1 = 100$" without resorting to integer programming. We will look at this later in the text.

[9]Formally, in order to solve as a LP, the objective function need be *concave* if maximizing; *convex*, if minimizing. Chapter 8 will expand on these notions in its discussion of nonlinear programming.

Needless to say, in any real-world situation, these assumptions are violated to some degree. Manufacturing processes rarely exhibit true proportionality or additivity. By using a linear programming model the manager is expressing the sentiment that the model represents a reasonable approximation to the actual situation. Similarly, real data are rarely known with certainty. Instead, we use linear programs in cases where the amount of uncertainty is small compared to the quantities involved. Even then, sensitivity analysis is a critical element in exploring the effects of these uncertainties.

The assumption whose violation may be most serious is that of divisibility. Can 3.7 workers be hired, or 1.4 airplanes be bought?

We will investigate these matters later in the text, sequentially relaxing each of these assumptions and seeing what happens to the resulting mathematical programming structure. Relaxing divisibility produces an *integer program*, investigated in Chapter 6. Relaxing the deterministic assumption gives rise to *stochastic programming*, which we discuss in Chapter 7. Relaxing linearity (proportionality and additivity) gives *nonlinear programs*; Chapter 8 treats these in detail. Throughout, the essential job of the modeler is to abstract out of the complexities of real problems enough information to represent the situation realistically, yet achieve sufficient simplification to allow the model to be analyzed.

EXERCISES

1. A large-scale distributor of dressed meats which specializes in servicing the hotel market has recently received an order from a large hotel chain for 10,000 pounds of ground meatloaf containing ground beef, pork, and veal, and soybean paste. Based on flavor considerations as well as nutritional needs, the order specifies that:

a. The ground beef comprise between 40% and 60%, by weight, of the mix.

b. The ground pork comprise between 20% and 30%.

c. The ground veal comprise between 10% and 40%.

d. The ground pork, by weight, comprise no more than one and one-half times the weight of veal.

e. The mix must be at least 30% protein, and no more than 12% fat.

The table below gives the protein and fat composition, by weight, of the four ingredients:

	Fat	*Protein*	*Cost*
Beef	12	28	\$ .70
Pork	15	29	\$ .63
Veal	10	27	\$ .81
Soybean paste	8	36	\$1.22

An analysis based on prevailing stockyard prices, percent usable meat (recognizing waste in the dressing process) and labor costs indicated that the cost per pound of the four ingredients is as shown in the above table.

Formulate a linear program to determine the optimal composition of the meatloaf.

2. A manufacturer of consumer goods is concerned with the allocation of its advertising budget for the coming month among the available media—radio, television, magazines, newspapers, and outdoor advertising. The advertising budget for the month is $1 million and the firm wants to maximize the number of *exposures*, defined as people exposed to an advertisement. A trade magazine report indicates that the *rated* exposure (people per dollar of advertising expenditure) is:

TV	22
radio	12
magazines	15
newspapers	10
outdoor advertising	5

In addition to maximizing total exposures, there are conditions on exposures within certain target populations which must be met:

Market Segment	*Minimum*	*Maximum*
Youth	2,500,000	3,500,000
Women	4,000,000	—
College educated	3,500,000	—

The aforementioned trade magazine indicates, within these target populations, that rated exposure is:

Rated Exposure (people per dollar)

	Youth	Women	College Educated
TV	10	6	3
Radio	5	4	1
Magazines	1	7	4
Newspapers	0	2	5
Outdoor advertising	1	2	1

In addition, to assure coverage, the firm wants no more than 30% of the budget spent on any one medium and no more than half the budget spent on the airwaves (TV and radio). Formulate a linear program to determine an optimal advertising plan for this firm.

What are the key underlying assumptions for this formulation? Are any of them questionable?

3. A young firm has been surprisingly successful over the last year selling two models of home computer and now find themselves in the enviable position of being able to sell all they can produce. Unfortunately, however, this growth has severely strapped not only the firm's manufacturing capacity, but also the firm's ability to finance itself. Capacity, price, and cost data are given below:

	Production Cost	Sales Price	Required Hours in Production		
			Fabrication	Subassembly	Assembly
Model 1	$1750	$2800	20	24	16
Model 2	$1320	$2200	8	32	8

Production time available in the coming quarter is 5000 hours in fabrication, 8000 hours in subassembly, and 4000 hours in assembly.

The company can finance its production through two sources: (1) internal funds ($150,000 available for the coming quarter) and (2) short-term loan. The bank will loan the firm up to $100,000 per quarter at 10% as long as the company's *acid test ratio* is at least three to one.

The acid test ratio is defined as

$$\frac{\text{Cash on hand plus accounts receivable}}{\text{Accounts payable}}$$

Formulate this problem as a linear program.

4. An Israeli kibbutz operates three farms of comparable productivity. The output of each farm is limited both by the usable acreage and by the amount of water available for irrigation. The data for the upcoming season are the following.

Farm	Usable Acreage	Water Available (in acre feet)
1	400	1500
2	600	2000
3	300	900

The kibbutz is considering three crops for planting which differ primarily in their expected profit per acre and in their consumption of water. Furthermore, the total acreage that can be devoted to each of the crops is limited by the amount of appropriate harvesting equipment available.

Crop	Maximum acreage	Water Consumption (in acre feet per acre)	Expected Profit Per Acre
A	700	5	$400
B	800	4	$300
C	300	3	$100

In order to maintain a uniform workload among the farms, it is the kibbutz's policy that the percentage of the usable acreage planted must be the same at each farm. However, any combination of the crops may be grown at any of the farms. The kibbutz wishes to know how much of each crop should be planted at the respective farms in order to maximize expected profit. Formulate this as a linear programming problem.

5. A farmer has 100 acres that can be used for growing wheat or corn. The yield is 60 bushels per acre per year of wheat or 95 bushels of corn. Any fraction of the 100 acres can be devoted to growing wheat or corn. Labor requirements are 4 hours per acre per year, plus 0.15 hour per bushel of wheat and 0.70 hour per bushel of corn. Cost of seed, fertilizer, and so on, is 20 cents per bushel of wheat and 12 cents per bushel of corn. Wheat can be sold for $1.75 per bushel, and corn for $0.95 per bushel. Wheat can be bought for $2.50 per bushel, and corn for $1.50 per bushel.

In addition, the farmer may raise pigs and/or poultry. The farmer sells the pigs or poultry when they reach the age of one year. A pig sells for $40. He measures the poultry in terms of one "pig equivalent" (the number of chickens needed to bring in $40 at the time of sale). One pig requires 25 bushels of wheat or 20 bushels of corn, plus 25 hours of labor and 25 square feet of floor space. An equivalent amount of poultry requires 25 bushels of corn or 10 bushels of wheat, 40 hours of labor, and 15 square feet of floor space.

The farmer has 10,000 square feet of floor space. He has available per year 2000 hours of his own time and another 2000 hours from his family. He can hire labor at $1.50 per hour. However, for each hour of hired labor, .15 hour of the farmer's time is required for supervision. Formulate a linear program to determine how much land should be devoted to corn and how much to wheat, and in addition, how many pigs and/or poultry should be raised to maximize the farmer's profits.

6. You have been asked by the Secretary of Defense to help decide how many infantry platoons of soldiers, piloted aircraft, and manned battleships to deploy next month. You have on hand, already paid for, the following resources:

1000	Men who are trained to be in infantry platoons
10	Airplanes
5	Trained pilots for these airplanes
3	Battleships
1000	Sailors trained to serve on battleships

Each infantry platoon consists of 20 soldiers and requires 1 officer. These officers must be hired for \$1000. One officer must be hired for each platoon you deploy. Extra infantrymen may be hired for \$500 each.

Each airplane you send must be accompanied by three pilots. You may hire extra pilots for \$3000. You may buy extra airplanes for \$100,000.

Each battleship you send must have 100 sailors and 5 naval officers. You may hire naval officers for \$2500 each, extra sailors for \$500 each, and extra battleships may be bought for \$250,000.

For every platoon deployed you must also send one airplane with its three pilots and for every ten platoons you must send one battleship and its contingent of sailors and officers.

Each infantry platoon you send yields 1 unit of effectiveness in potential combat: each airplane you send yields 5 units of effectiveness; each battleship yields 10 units of effectiveness.

You must send enough forces, planes, and ships to yield 300 units of effectiveness for the next month.

Formulate a linear program to find the minimum cost set of forces to send in order to achieve the desired effectiveness.

7. A firm that specializes in custom blended alloys has received an order for 100 pounds of a special alloy containing 25% copper, 70% nickel, and 5% chromium. The firm has in stock five alloys whose values and compositions are shown below.

Alloy	Amount in Stock (lb)	Purchase Price (\$/lb)	Composition (% of) Copper	Nickel	Chromium	Scrap Value After Copper, Nickel, Chromium is Removed (\$/lb)
A	1000	.50	5	25	—	.15
B	100	1.00	—	25	10	.55
C	500	.80	75	—	—	.35
D	200	2.00	30	—	25	.90
E	1000	.40	—	10	2	.20

These alloys are regularly stocked and any amounts used will be repurchased for inventory.

a. Formulate a linear program to determine the least cost way of filling the order.

b. How does the formulation in (a) change if pure chromium can be purchased at \$5.00/lb?

8. Recall the production planning problem formulated as Example 5 of the text. Instead of changing workforce levels to match changing production, consider a firm whose personnel policy guarantees no layoffs. Instead, workers not needed can be temporarily idled (e.g., sent home one day per week) with pay. Any worker who works less than 80% of full-time is paid 80% salary. If a worker puts in more hours than 80% of full-time but less than 100% full-time, he or she is paid the percentage of salary equal to the percentage of full-time worked (e.g., 95% of full-time earns 95% salary). If production requirements call for more than 100% full-time, rather than hire new workers as was done in Example 5, the firm can pay time and a half for overtime, up to 20% additional hours. The graph below summarizes the relationship of labor cost to time worked.

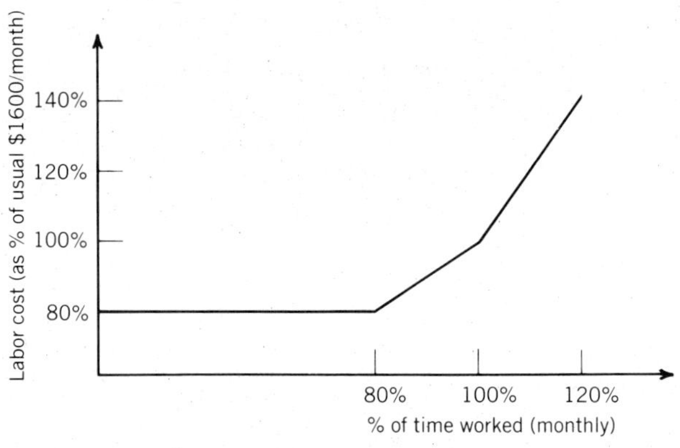

Assuming the same seasonal demand pattern as shown in Figure 2-4 of the text, formulate this firm's production planning problem.

9. A firm manufactures electronic test equipment requiring a key component purchased from another company. The demand for test equipment changes from quarter to quarter so production planning can be difficult. Unfortunately, component availability also changes from quarter to quarter, not always in concert with the seasonality of demand, as can be seen from the table below:

Quarter	Available Component Supply (units)	Demand for Test Equipment (units)
1	5000	1000
2	3000	4000
3	1000	3000
4	2000	1500

Inventory costs for storing finished test equipment are high and storage is limited. In fact, the firm wants to schedule its next year's production to minimize the carryover of finished product from one quarter to the next, while meeting demand.

Storage is available for either 10,000 components or 2000 units of finished product or any other combination in this 1 : 5 ratio (e.g., 1000 finished testers *and* 5000 components). Prior to the start of the first quarter, 3000 components are in stock. Management would like to have 3000 components in stock at the end of the year.

Formulate a model to help the firm determine how many testers to manufacture and how many components to purchase in each of the four quarters.

10. An important problem with which all airlines are faced is the determination of how much fuel an airplane should load at each stop. Since the efficiency of an airplane is related to its weight, an airplane fully loaded with fuel burns more fuel per mile in flight. This fact argues for planes filling up at each stop with just enough fuel to take them to their next stop. Unfortunately, since fuel prices vary quite substantially from one location to another, this policy of minimal fillups may be more costly than filling the tank at a very inexpensive location. Airlines need some way of achieving the appropriate balance, especially since at current fuel prices, fuel costs are, by far, the largest single item in airlines' costs.

Let us look at a small example: Consider an airplane that flies the following "rotation":

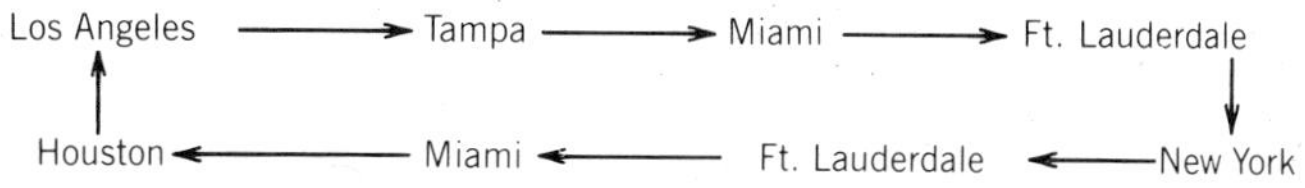

The table below indicates the fuel requirements for each leg and cost at each location:

City	Fuel cost/gal	Minimum Fuel (gal) Needed to Get to Next Stop	Additional Fuel (.01 gal) Burned *per gallon* of Fuel Loaded *over the Minimum*
Los Angeles	$1.236	330	4.604
Tampa	1.276	42	.585
Miami	1.299	5	.070
Ft. Lauderdale	1.204	151	2.105
New York	1.269	151	2.105
Ft. Lauderdale	1.204	5	.070
Miami	1.299	186	2.593
Houston	1.217	177	2.468

For example, in Los Angeles, the airplane could be loaded with 330 gallons, which would just take it to Tampa. Alternatively, in order to fill up with enough fuel in Los Angeles to get to Miami without refilling at Tampa, 330 + 42 + 42 × .04604 = 373.934 gallons would be required. The last term (42 × .04604) represents the extra fuel burned on the trip from Los Angeles to Tampa due to the presence of the additional 42 gallons loaded in L.A. Comparing costs for these two simple alternatives,

Alt. 1	Min. fill-up in L.A.:	330 × $1.236 =	$407.88
	Min. fill-up in Tampa:	42 × $1.276 =	53.59
			$461.47
Alt. 2	373.934 gallon fill-up in L.A.:	373.934 × $1.236 =	$462.18

shows the former to be less costly. The cost of the additional fuel (almost 2 gallons) needed for the larger fill-up in L.A. is greater than the cost savings due to less expensive fuel in L.A.

Formulate a linear program to determine the least-cost locations and fill-up amounts for the rotation above.

Chapter 3

LINEAR PROGRAMMING: SOLUTION AND INTERPRETATION

Now that we have looked at the formulation of linear programs in some depth, we are ready for the next step—solution. The emphasis of this text is on formulation and interpretation rather than on solution methodology. However, a brief discussion of the geometry of a linear programming model will highlight several important features of the Simplex method, the efficient solution procedure developed by George Dantzig in 1947. A more detailed discussion of the algebra that underlies the Simplex method is included as Appendix A.

A GEOMETRIC OVERVIEW

Let us consider an exceedingly simple example for illustrative purposes. A commercial printer has decided to offer two types of calendars for the upcoming season. Both can be produced on the same press. Each standard calendar, an item the printer has produced for several years, requires 8 square feet of paper and can be produced in only 10 minutes on the press. The second calendar, a deluxe version, was offered by the printer for the first time the previous year, and it was a surprising success. This deluxe version is smaller than the standard calendar, requiring only 4 square feet of paper, but the intricate detail and multicolor artwork requires 70 minutes of press time. In stock, the printer has 4800 square feet of paper. The production schedule indicates that 38,500 minutes of press time are available for the production of these calendars.

The trend in recent years' demand leads the printer to expect to sell no more than 500 of the standard calendars, on which he earns \$3.00 in contribution. He feels sure, however, that he could sell as many of the deluxe calendar, which produces \$5.00 in contribution, as he can produce. How many of each calendar type should the printer schedule into production?

This simple production planning problem can be formulated as a linear program with decision variables S, the number of standard calendars to be produced, and D, the number of deluxe calendars to schedule into production:

$$\begin{array}{lrr} \text{Max} & & 5S + 3D \\ \text{Subject to} & & \\ & \text{Paper:} & 8S + 4D \leq 4{,}800 \\ & \text{Press:} & 10S + 70D \leq 38{,}500 \\ & \text{Standard demand:} & S \leq 500 \\ & & S,D \geq 0 \end{array}$$

Since there are only two decision variables in this example, we can attack the problem graphically, since *any point (S,D) represents a production plan*. For example, each of the three points in Figure 3-1*a* represents an alternative plan:

A = produce 250 standard calendars; produce 250 deluxe calendars.

B = produce 350 standard calendars; produce 500 deluxe calendars.

C = produce 500 standard calendars; produce 450 deluxe calendars.

Associated with each point (plan) there is a contribution level that will result. That is, plan A will generate $250 \times \$3.00 + 250 \times \$5.00 = \$2000$ in contribution; plan B, \$3550; and plan C, \$3750. Clearly, plan C is best (highest in contribution); but, unfortunately, it violates the first constraint since it requires 5800 square feet of paper. Our goal is to identify, from the infinity of possible ordering plans, the feasible plan that generates the highest contribution.

The first step in our graphical solution procedure is to identify those solutions that are potentially optimal. Since the optimal solution must satisfy all the constraints, our search can be restricted to the set of constraint-satisfying solutions. We call a pair (S, D) that satisfies all the constraints a *feasible solution*. The feasible solution with the highest (in a minimization problem, the lowest) objective function value is the *optimal solution*.

To determine the set of feasible solutions, we can graph each constraint as a straight line. For example, in Figure 3-1*b*,we have graphed the line "$8S + 4D = 4800$"; only points on or below that line satisfy the constraint "$8S +$

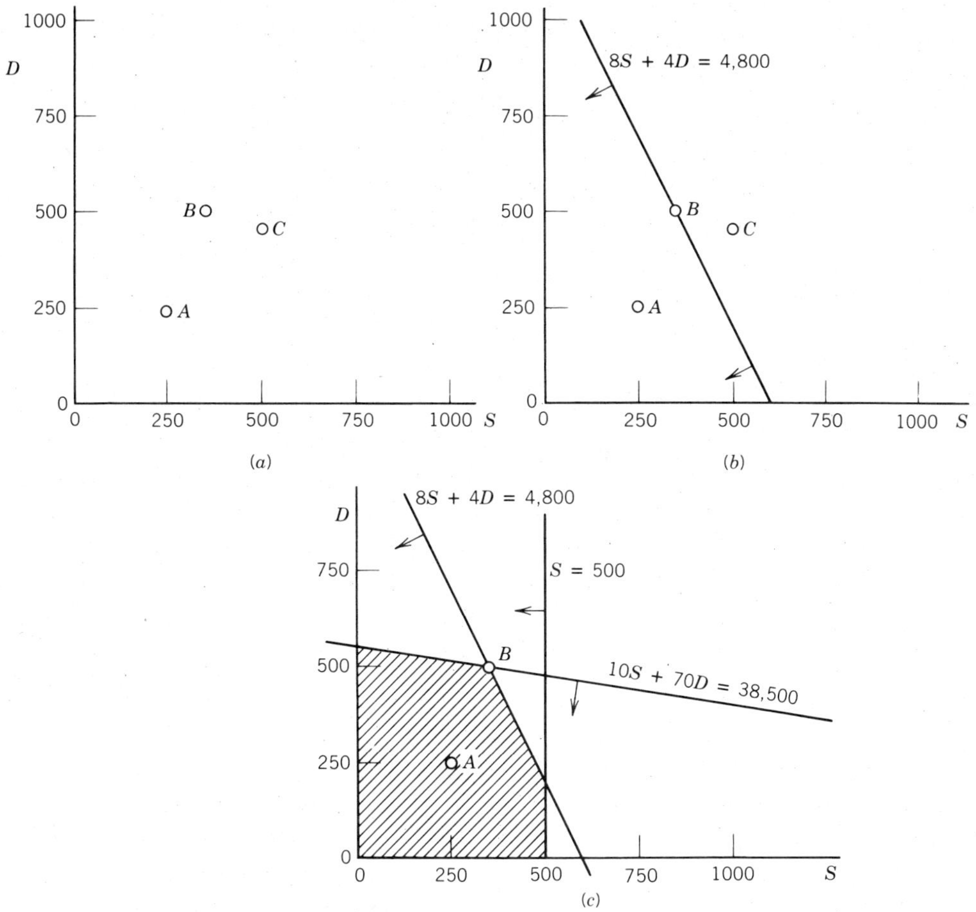

FIGURE 3-1. A graphical solution approach.

$4D \leqslant 4800$." For example $A = (250, 250)$ is below that line; it satisfies the constraint. On the other hand, $C = (500, 450)$ is above that line; indeed, $8 \times 500 + 4 \times 450 > 4800$. Hence, C is *infeasible*; A and B are feasible with respect to that constraint. Figure 3-1*c* shows all the constraints of the problem. The cross-hatched area, called the *feasible region*, represents all points (S, D) that are feasible for the printer's problem.

How can we find the optimal solution? First, observe that no *interior point* such as A can be optimal since we can increase the value of the objective function simply by moving upward or to the right until we encounter a constraint line. Similarly, a point *on* the constraint line (which we call a *boundary point* or an edge point) can be improved by moving to a *corner point* (or *vertex*).

Thus, *the optimal solution must be at a vertex.*[1] For a problem this small, we could conceivably list all the corner points, calculate the objective value for each, and choose the maximal one. Let us follow a slightly different procedure that mimics the operation of the Simplex method.

We choose any feasible point, say $S = 0, D = 0$, and look for a direction in which to move that will increase our objective.[2] We note that increasing S will do this, so we move rightward along the S-axis as far as possible, to the vertex (500,0). At this point, the value of the objective is $500 \times \$3 + 0 \times \$5 = \$1500$, indeed greater than the objective value (\$0) at the origin where we started. At this point, we again ask, "In what direction can we move so as to increase the objective function value?" Note that moving upward along the line $S = 500$ will increase D without decreasing S, thus increasing the objective function value. Hence, we move upward as far as we can until we reach the vertex at which $S = 500$ and $D = 200$. Here the objective equals $500 \times \$3 + 200 \times \$5 = \$2500$, indeed greater than our previous value. Again, we search for a feasible direction in which to move. From where we are, the only possibility is to move northwest along the line "$8S + 4D = 4800$." (Moving back in the direction from which we just came is clearly futile.) Unfortunately, moving in this direction increases D only at the cost of *decreasing* S. In fact, for every unit that D increases as we move in a northwesterly direction along this constraint, S decreases by ½ unit (since the slope of the line is −2). However, taking into account the objective function coefficients, we observe that a unit increase in D increases contribution by \$5 while a ½-unit decrease in S only causes a \$1.50 drop in contribution. Hence moving along this line will benefit us. We do so as far as possible (again, there is no point in stopping before we have to, since continued movement is beneficial) to the vertex at which $S = 350$ and $D = 500$. At this point, we could continue, moving leftward along the line $10S + 70D = 38{,}500$. Unfortunately, for each unit increase in D as we move along this constraint, S decreases by 7! This would increase contribution by \$5 but would cost us $7 \times \$3 = \$21$, clearly not worthwhile. Thus, we have finally come to rest. Further movement in any direction from this point will only decrease the objective, whose optimal value (at B) is $350 \times \$3 + 500 \times \$5 = \$3550$.

The Simplex method works exactly in this way, traveling from one corner point to the next until no direction exists for which movement will better the objective. In more than two dimensions, choosing the direction in which to move can be quite difficult, and making this decision is part of the "guts" of the Simplex algorithm. The reader may verify, however, using the "vertex-hopping" algorithm illustrated above, that if we had started by moving vertically along the D-axis, we would have some to rest at the same optimal point.

[1]There may be multiple optimal solutions (e.g., an entire edge could be optimal). At least one must be at a vertex.

[2]In practice, finding an initial feasible point is often quite difficult. The Simplex method contains a built-in procedure for accomplishing this.

Special Cases

There are three special situations that the graphical approach will help elucidate:

1. No Feasible Solution. Let us suppose that the printer expects an unusually large order from a major chain of bookstores and wants to insure that at least 1000 calendars are produced, in total. He removes the restriction that no more than 500 of the standard calendars be produced and wants to solve the new problem:

$$\begin{aligned} \text{Max} \quad & 5S + 3D \\ \text{Subject to} \quad & 8S + 4D \leq 4{,}800 \\ & 10S + 70D \leq 38{,}500 \\ & S + D \geq 1{,}000 \\ & S, D \geq 0 \end{aligned}$$

Note the replacement of "$S \leq 500$" by "$S + D \geq 1{,}000$" to reflect the production-minimum restriction. What happens when we try to solve this problem? The answer can be easily seen graphically (Figure 3-2). The new constraint requires that feasible points be located *above* the line "$S + D = 1{,}000$," but no such point simultaneously satisfies the other two constraints. There is simply no way to produce that many calendars, in total, as long as paper supplies or machine availability is limited as specified. In the language of mathematical programming, "There is no feasible solution."

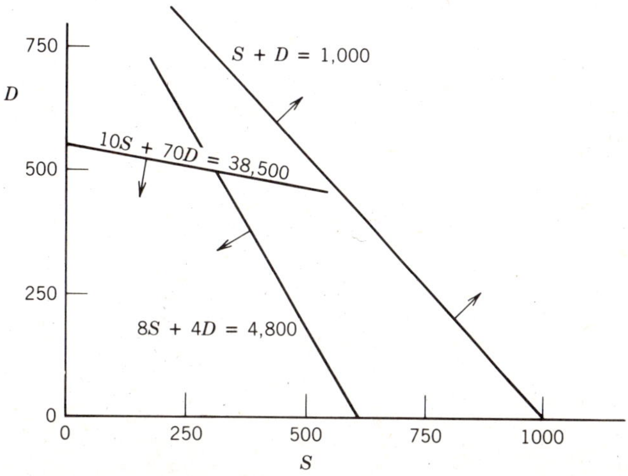

FIGURE 3-2. Infeasibility.

2. Alternative Optima. Suppose that, as part of a marketing campaign to sell the deluxe calendars, the printer reduces the price by \$3.50, thus cutting the contribution per unit to \$1.50. The new objective function to be maximized is thus

$$3S + 1.5D$$

Referring to Figure 3-3, how would our "vertex-hopping" algorithm proceed? As before, we would move from the initial feasible point 0 to point X, at which $S = 500$ and $D = 0$, and then to point $Y = (500, 200)$. Here, we are again confronted with the choice to move northwesterly along the line "$8S + 4D = 4800$." As before, for every unit that D increases, we lose ½ unit of S. Previously, the objective function coefficients indicated that this movement was worthwhile. Now that the objective function has changed, we need to reevaluate that decision:

$$1 \text{ unit increase in } D \longrightarrow \text{contribution} \nearrow \$1.50$$

$$\tfrac{1}{2} \text{ unit decrease in } S \longrightarrow \text{contribution} \searrow \$1.50$$

Hence, the net effect in moving along the line "$8S + 4D = 4800$" is to leave the value of the objective unchanged! In fact, any point along the edge between Y and B is as good as any other:

$$\text{At } Y, \text{ objective value} = 500 \times \$3 + 200 \times \$1.50 = \$1800.$$

$$\text{At } Y', \text{ objective value} = 450 \times \$3 + 300 \times \$1.50 = \$1800.$$

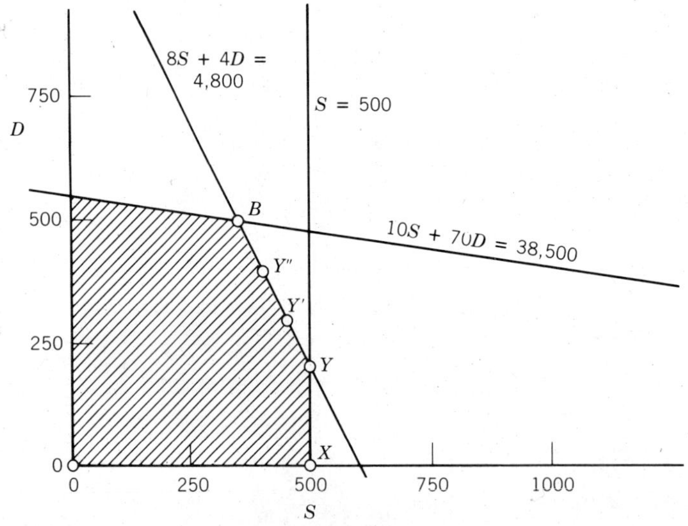

FIGURE 3-3. Alternative optima.

At Y'', objective value = 400 × \$3 + 400 × \$1.50 = \$1800.

At B, objective value = 350 × \$3 + 500 × \$1.50 = \$1800.

In the language of mathematical programming, we say that there exist *alternative optima* for this problem. In fact, for a *linear* program, since the edges of the feasible region are lines in two-space,[3] there are an infinite number of alternative optima—any point on the edge is an optimum. However, because of the vertex-choosing nature of the Simplex algorithm, either Y or B will chosen, depending on the initial direction selected. It is useful, however, for the analyst to recognize that an infinity of equally optimal choices actually exist.

3. Unboundedness. Consider another simple linear program:

$$\begin{array}{ll} \text{Max} & x_1 + 2x_2 \\ \text{Subject to} & -x_1 + x_2 \leqslant 400 \\ & x_1 + 8x_2 \geqslant 1600 \\ & x_1 \geqslant 500 \\ & x_1, x_2 \geqslant 0 \end{array}$$

As above, this problem can be solved graphically. Note, however, that the feasible region, shown in Figure 3-4, is unbounded. If we were to apply the

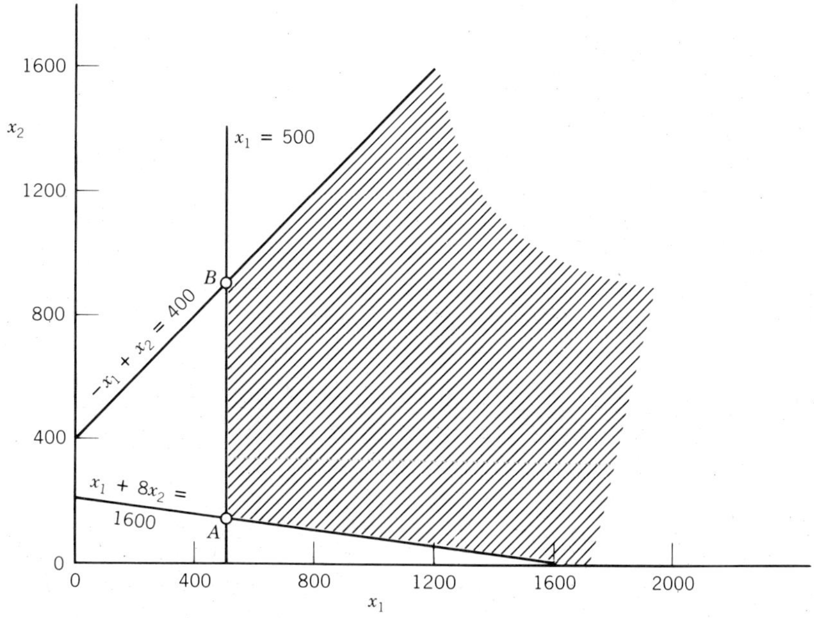

FIGURE 3-4. An unbounded feasible region.

[3]Planes in three-space, hyperplanes for a multi-dimensional (more than three dimensions) problem.

"vertex-hopping" algorithm as before, we might start at point A and move along the "$x_1 = 500$" line to point B. Now what? Moving northeast along the line "$-x_1 + x_2 = 400$" is beneficial since both x_1 and x_2 increase, but there is no vertex at which to stop. Instead, the objective function value can be increased without bound. As can be seen from the discussion in Appendix A, the Simplex algorithm has a simple built-in test to determine when unboundedness occurs.

INTERPRETING THE SOLUTION RESULTS GEOMETRICALLY

As the introductory comments in Chapter 1 have stressed, determining the optimal solution is only the beginning for the manager or analyst. Managers who find quantitative models most useful in the decision-making process are those who go far beyond simply looking for the "answer." The crucial task is interpretation of the solution and the use of that interpretation in gaining insight into the real-world problem at hand. Since the structure of the situation, as modeled by the LP, is reflected in constraint, right-hand side and objective function parameters, one can gain insight into underlying factors by looking at what happens as these change. The Simplex method, luckily, makes this task easy. In fact, one of the reasons for the success of linear programming as a managerially useful methodology is the easy availability, in addition to the optimal solution, of information essential for sensitivity analysis and useful in testing the results of alternative assumptions.

In this section, we discuss the information typically available from most commercially available LP packages, using the geometry of our printer's problem as illustration. We then introduce the computer package of which we will make extensive use in this text and show how this package can be used to solve and interpret each of the five examples formulated in Chapter 2.

Slack/Surplus in Constraints

An inequality that holds with equality at the optimum is called *binding*. A change in it will usually affect the value of the objective. In the case of a resource limitation, a binding constraint represents a resource that is totally used up. A *nonbinding* resource limitation has *slack*; that is, there is unused resource. Formally, for a "$\leq$" constraint of the form

$$\sum_{i=1}^{n} a_i x_i \leq b$$

$$\text{Slack} = b - \sum_{i=1}^{n} a_i x_i^*$$

where x_i^* are the optimal values of the decision variables. For example, returning to the printer's problem, consider the demand constraint "$S \leq 500$." As Figure 3-5 shows, the optimal solution has slack of 150 standard calendars. In other words, there is excess demand for 150 of these calendars over what will be produced in following the optimal plan. Note, however, that the optimum B lies *on* both the press time constraint line and the paper constraint line. This indicates no slack in these resources. That is, both resource constraints are binding. The demand constraint is nonbinding.

For a "$\geq$" constraint of the form

$$\sum_{i=1}^{n} a_i x_i \geq b$$

the *surplus* is defined as

$$\sum_{i=1}^{n} a_i x_i^* - b$$

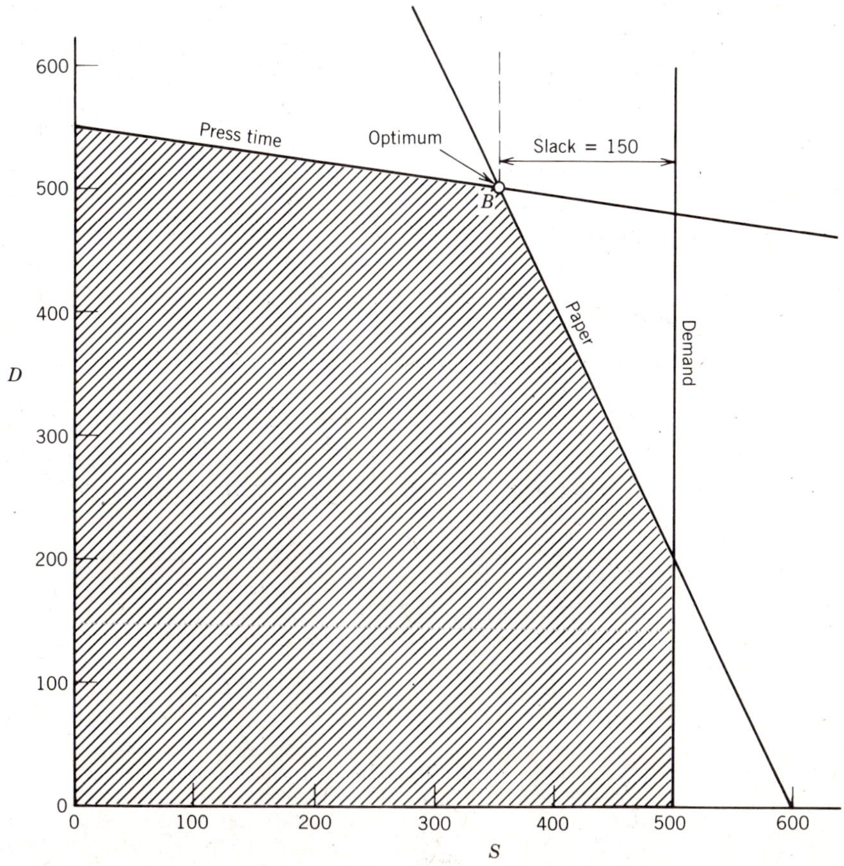

FIGURE 3-5. A nonbinding constraint.

For example, for a demand constraint that requires that certain promised deliveries will be made, the surplus represents the amount over the minimum that is produced.

Shadow Prices

One of the most difficult tasks in allocating resources to competing activities is the evaluation of the "true worth" of an incremental unit of a scarce resource (the *shadow price* of that resource). The problem is complicated by the fact that the value of that unit of resource can only be determined by first knowing *how* that extra resource will be used. The Simplex method makes shadow prices easily available.[4] Formally, we define the shadow price for a given constraint to be the change in the objective that would accompany a unit increase in that constraint's right-hand side value.[5] Operationally, for a resource limitation, the shadow price is the *value* of one additional unit of the specified resource. These "prices" may be used as such: If additional resource is available for purchase, the shadow price gives the maximum per unit amount that should be paid.[6] If the resource can be purchased for any less, then the objective function value will increase. Alternatively, if a new activity is introduced that would consume resources, the total price of those resources can be readily calculated. This total may then be compared to the contribution of the new activity to the objective to determine if the activity is worth introducing. This comparison is called *pricing out* an activity.

A geometric interpretation of shadow prices is given in Figure 3-6 for the printer's production problem above. For example, consider the paper supply restriction. What is the *value* of additional paper? Graphically, increasing the supply of paper consists of shifting the constraint line "$8S + 4D = 4800$" rightward. For example, if rather than 4800 square feet of paper, 5216 square feet were available,[7] the constraint would shift from its current position to the position shown in Figure 3-6*a*, corresponding to the line "$8S + 4D = 5216$." As this shift in the constraint line takes place, the optimal solution moves with it, along the line "$10S + 70D = 38{,}500$," coming to rest at the point B' where $S = 406$ and $D = 492$. The contribution from this new plan is $\$3 \times 406 + \$5 \times 492 = \$3678$, an increase of \$128 over our current optimum. On a per unit basis, this represents an increase of \$128/416 = \$.3077 per square foot of additional paper. This quantity is the *shadow price* for the paper supply constraint and represents the maximum amount that the printer should pay to increase his paper supply.

[4]See Appendix A for an algebraic illustration of why this is true.
[5]We speak somewhat loosely here of "a unit increase." Formally, the shadow price is the marginal rate of change of the objective with respect to the right-hand-side value. In calculus terms, if we denote the objective function by z and the right-hand-side value by b, the shadow price is $\partial z/\partial b$.
[6]Over and above the price for the resource *already* included in the objective function specification.
[7]This number has been chosen merely to simplify the arithmetic.

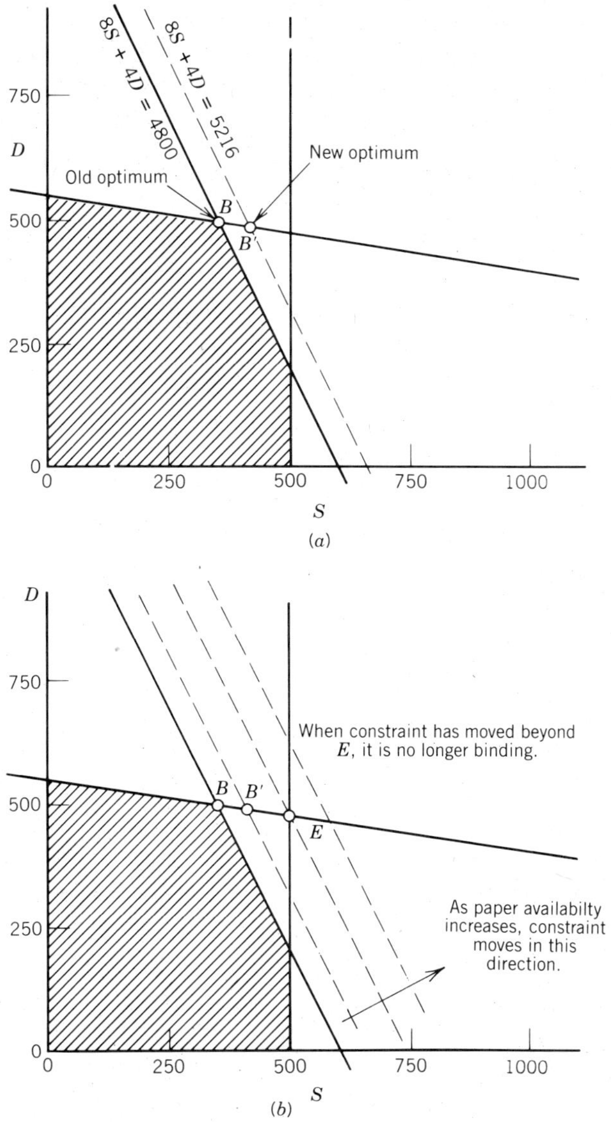

FIGURE 3-6. Increasing paper availability.

If we were to continue to shift the paper supply constraint upward, the objective function value would continue to increase at $.3077 per additional square foot acquired, due to the linearity of the constraints. As is clear from Figure 3-6*b*, this continues only until we reach point *E*. Continued movement of the constraint line beyond this point has no effect on the objective function value since the optimum, now defined by the two other constraints, remains at E. Thus, a shadow price may have a constant value only within a certain range of changes to that resource's supply.

To see, in more depth, the changes that can occur as resource availability changes, let us add another constraint to this problem. Suppose that not only are paper supply and machine availability resources to contend with, but also availability of a special ink used in the publishing process. In particular, each standard calendar requires 2.1 ounces of the ink; each deluxe calendar requires 4.7 ounces. The printer has, in stock, 3165 ounces that can be used in the production of the calendars. Adding a new constraint to reflect this new resource limitation generates the linear program formulated below and graphed in Figure 3-7:

$$
\begin{array}{lrl}
\text{Max} & 3S + 5D & \\
\text{Subject to} & 8S + 4D & \leq 4800 \\
 & 10S + 70D & \leq 38{,}500 \\
 & S & \leq 500 \\
 & 2.1S + 4.7D & \leq 3165 \\
 & S, D & \geq 0
\end{array}
$$

Note, from Figure 3-7 that this new resource limitation is not binding—the optimal production plan, as before, is 350 of the standard calendars and 500 of the deluxe. In fact, the *slack* for the ink supply constraint is 80—using the optimal production plan will result in 80 ounces of ink left in stock.

Now, let us return to the paper supply constraint "$8S + 4D \leq 4800$." As before, as the paper supply increases above 4800 square feet, the optimal solution (B) "slides" rightward along the machine availability constraint line. Again, as before, for each additional square foot of paper supplied, the objective function increases by \$.3077, the shadow price for this constraint. This continues until the paper supply constraint line reaches point B'. Further shifting of the paper supply constraint line will cause the optimal point to slide rightward along the ink supply constraint line. Since the slope of that constraint is different than the slope of the machine availability constraint line, the gain per additional square foot of paper will change from the \$.3077 experienced as we moved from B to B'. In summary, the shadow price of \$.3077 holds as long as the paper supply is below 5216 square feet; if between 5216 and 5800 (the amount required at B'') square feet, a shadow price of \$.1233 holds; and beyond 5800 square feet, additional paper has no value—its shadow price is zero. This process is summarized by Figure 3-7 and Table 3-1. In general, when we discuss the shadow price of a constraint, we must recognize that there are limits on the right-hand-side value beyond which the shadow price changes.[8] These limits are also available as output from most commercially available LP packages.

[8]If the right-hand-side value of a "$\leq$" constraint is increased beyond its upper limit, the constraint's shadow price will decrease. If it is decreased beyond the lower limit, the shadow price will increase. (The opposite holds for a "$\geq$" constraint.) In general, the amount of the increase or decrease can only be determined by rerunning the model.

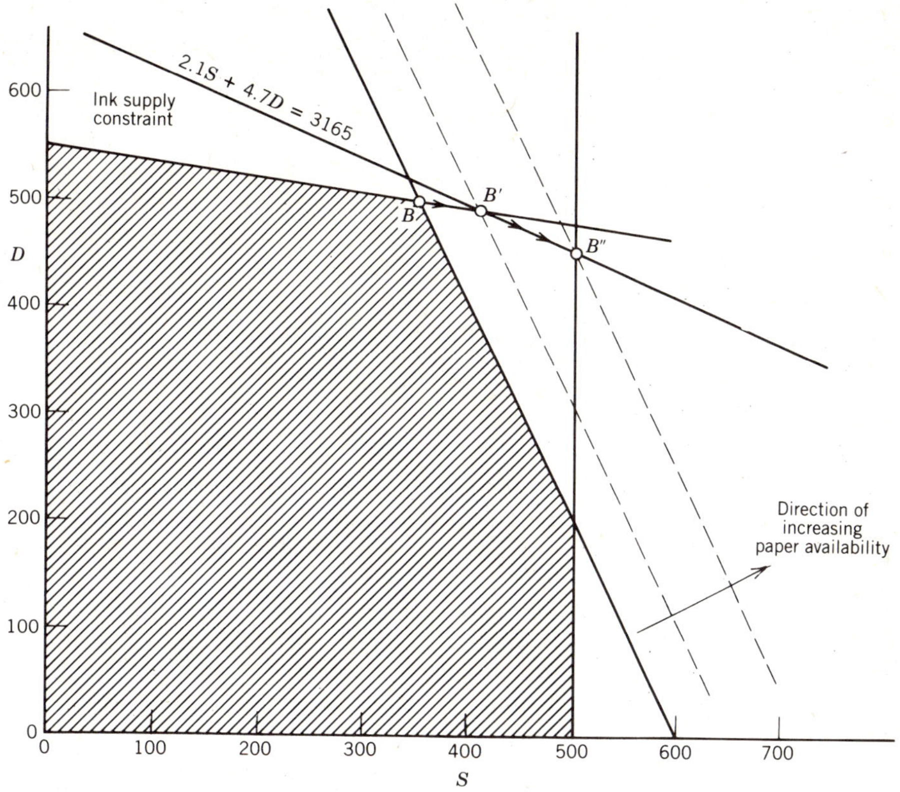

FIGURE 3-7. Increasing paper availability with an ink supply constraint added.

TABLE 3-1

Optimal Point (*S*,*D*)	Paper Supply (sq ft)	Objective Function Value ($)	Change in Objective Function	Change in Objective Function Value per square foot of Additional Paper
B = (350,500)	4800	3550	—	—
B′ = (406,492)	5216	3678	$128	$.3077
B″ = (500,450)	5800	3750	$ 72	$.1233

Note that nonbinding constraints have a zero shadow price. This makes sense; if there is unused resource, additional resource has no value.

Certain shadow prices have special importance—the shadow prices on the nonnegativity constraints. If an activity is at a zero level in the optimal solution, its *reduced cost* represents the cost of deviating from the optimal solution by running the activity at a nonzero level (to produce a full product line, to keep machinery or workers busy, etc.). Formally, the reduced cost for each such variable is the change in the objective that would occur if the value of that variable were to be changed from 0 to 1, corresponding to changing the nonnegativity constraint from $x \geqslant 0$ to $x \geqslant 1$. Note that, for our stationer's problem, both reduced costs are zero, since neither S nor D is at a zero level in the optimal solution. Changing the nonnegativity conditions

$$S \geqslant 0, D \geqslant 0$$

to

$$S \geqslant 1, D \geqslant 1$$

would thus have no effect on the optimal objective function value.

Solution's Sensitivity to the Objective

Typically the costs, revenues, and other data that go into the objective function coefficients are estimates rather than known constants. Thus, it is essential for the manager to be able to investigate the sensitivity of the decision to these coefficients. If we return to the printer's dilemma, we could have represented different values of the objective as the sequence of paralleled dashed lines shown in Figure 3-8*a*. Each of these lines (called *contours*) represent the set of points (S, D) for which the objective function has a single value. Note that as we move away from the origin, the objective function increases in value. Hence, to maximize contribution we would like to locate the optimal point on the "farthest-away-from-the-origin" objective function contour. Of course, we are limited by the bounded feasible region. For example, no point on the "contribution = \$4000" contour is feasible. As can be seen from Figure 3-8*a*, the best we can do is $S = 350$, $D = 500$ at B, on the objective function contour with value \$3550.

In general, the feasible point at which these lines have moved out the farthest will represent the optimal solution. Note that we could "tilt" this contour with the optimal point as fulcrum quite substantially before it would intersect any other point in the feasible region (Figure 3-8*b*). This "tilting" allows the calculation of limits on the objective function coefficients within which the optimum value of the decision variables remain unchanged. Algebraically, "tilting" corresponds to a change in one of the objective function coefficients relative to the other. For example, changing the contribution for standard calendars to \$4 corresponds to the new objective function contour shown in Figure 3-8*c*. Note that for this change, the optimal point remains the same. If, however, the contribution for standard calendars were to increase beyond \$10, say to \$11, then the corner point Y, at which $S = 500$ and $D = 200$, would become optimal (Figure 3-8*d*). The exact upper and lower limits for each objective function coefficient are part of the information available from most commercially available LP packages.

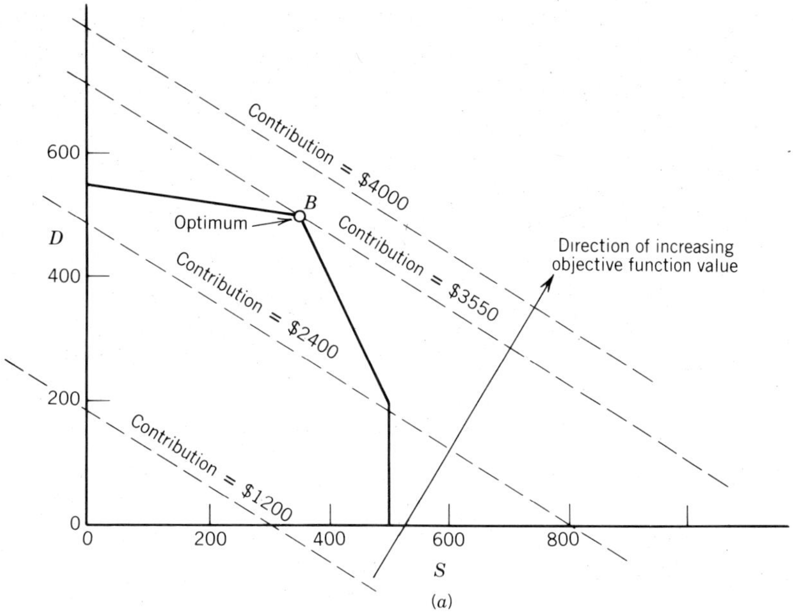

FIGURE 3-8(a). Contribution contours.

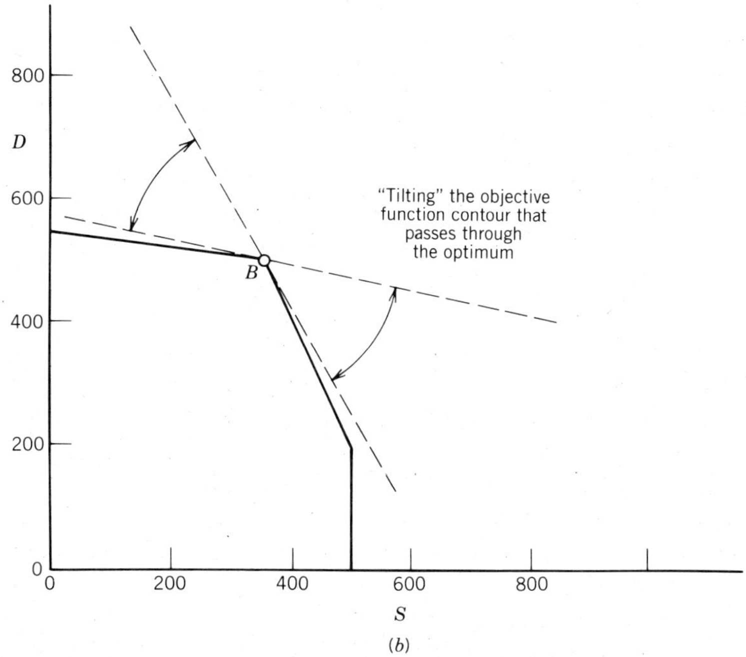

FIGURE 3-8(b). Tilting the optimal contribution contours.

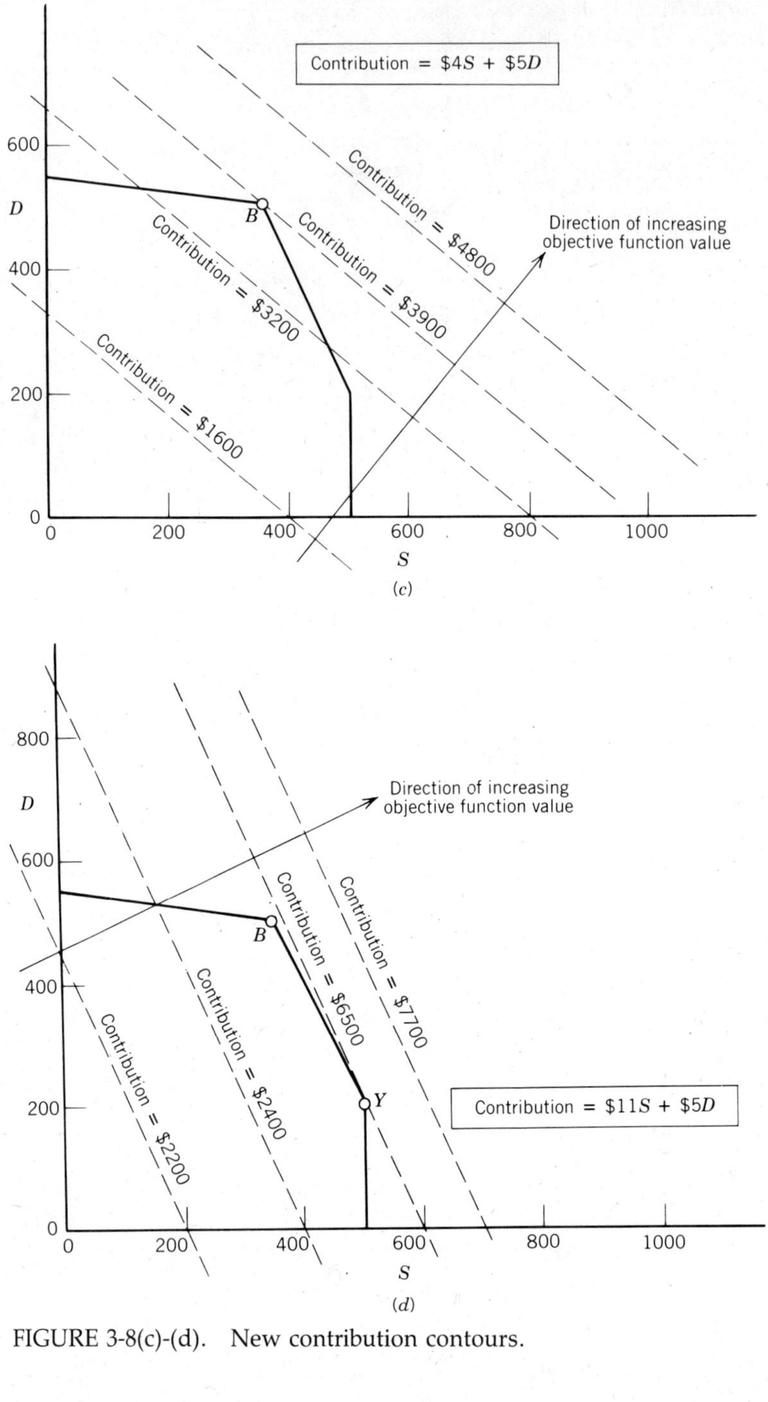

FIGURE 3-8(c)-(d). New contribution contours.

The 100% Rule

The above description indicates how changes in a *single coefficient* may be investigated. If several changes occur simultaneously, sensitivity analysis becomes more difficult. For example, if *both* of the objective function coefficients change, it is difficult to predict what will happen. If the objective function were to change to

$$1.5S + 2.5D$$

corresponding to both contributions being cut by 50%, then the slope of the objective function would remain unchanged, the objective function contours would remain unchanged and the optimal point would remain unchanged, as shown in Figure 3-9*a*.[9] If, on the other hand, both contributions were to be cut by $2.70, creating the new objective function

$$.3S + 2.3D$$

then, the optimal point would shift to $S = 0, D = 550$, as shown by Figure 3-9*b*.

Although, in general, it is difficult to determine whether simultaneous changes in objective function coefficients will cause the optimal solution to shift to a new vertex, there is one simple calculation which helps. This can be stated as *the 100% rule*:

> *If the sum of the coefficient changes (expressed as percentages of the way to the respective limits) is less than 100%, the optimal decision variables will remain the same. Mathematically, let the ith variable have an objective function coefficient c_i and assume that this coefficient can increase up to u_i without affecting the decision variable value at the optimum. If the ith objective function coefficient increases to c_i', we can say that whenever*
>
> $$\sum_i \left(\frac{c_i' - c_i}{u_i - c_i} \right) \leq 1$$
>
> *the optimal decision variable values will remain unchanged.*[10]

If this expression does not hold, then those values might change. Rerunning the revised program usually becomes necessary in order to determine whether the optimal solution will be different and, if so, how.

[9] As is clear from this example, it is true in general that if *all* objective function coefficients change by the same *percentage*, then the optimal solution remains unchanged.
[10] A similar expression holds for lower bounds or for combinations of changes toward upper and lower bounds.

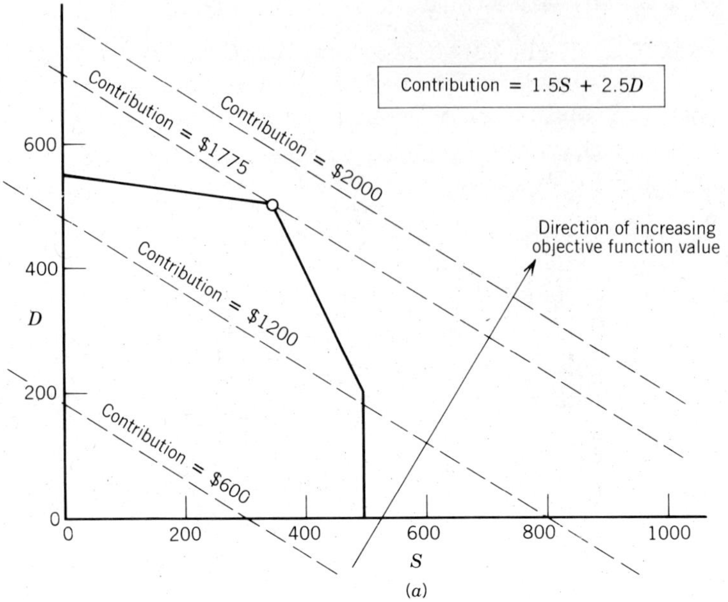

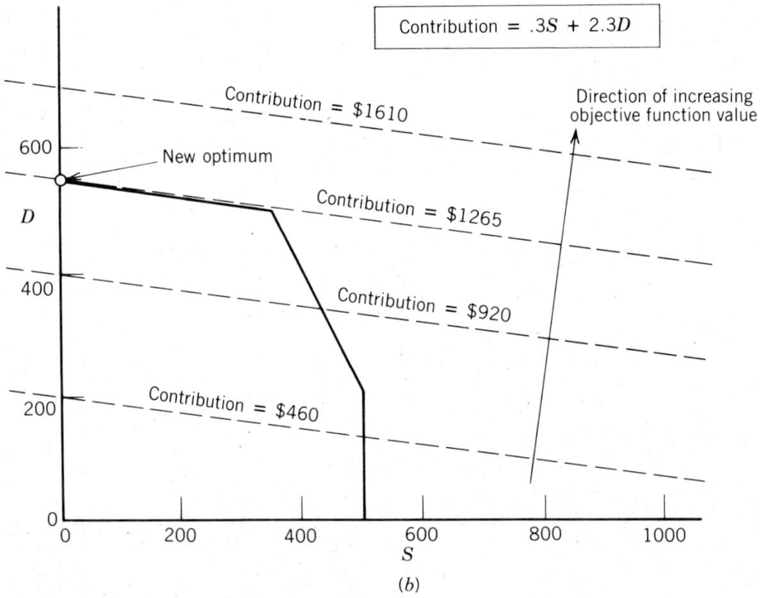

FIGURE 3-9. Contribution contours.

THE CMP PACKAGE

Throughout the remainder of this text we will make reference to the output from an interactive mathematical programming package called CMP (Conversational Mathematical Programming), developed at the Harvard Business School by Stephen P. Bradley. This output is typical of that available from most commercially available mathematical programming packages. In solving linear programs, the package gives the value of the objective function at the optimum and then will display, at the user's request, any or all of six output sections, as summarized by Table 3-2.

TABLE 3-2

CMP Section	Information Displayed	Comments
1	The *optimal values* of the decision variables	Decision variables not listed are at a zero level in the optimal solution.
2	*Slack* and *surplus* in constraints	The name of each constraint having slack is preceded by "+"; the name of each constraint having surplus is preceded by "−"; any constraint not included is binding (i.e., has zero slack or surplus).
3	*Shadow prices* on constraints	For each constraint listed, the value given is the change in the objective function that would accompany a unit increase in the right-hand-side value of that constraint. A unit *decrease* would cause a change in the objective function equal to the negative of the shadow price displayed. Any constraint not included is nonbinding and thus has zero shadow price. Note that every constraint must be included either in Section *2* (nonbinding) or in Section *3* (binding).
4	*Reduced costs* for decision variables	For each decision variable *not* listed in Section *1* and thus, at a zero level in the optimal solution, this section lists the penalty incurred by deviating from the optimum by forcing that variable to a unit level.

TABLE 3-2 (Continued)

CMP Section	Information Displayed	Comments
5	Ranges on objective function coefficients within which the optimal solution remains unchanged	The objective function coefficient for each decision variable is listed along with an upper and lower bound. If the current value of the coefficient were to change in such a way that the new coefficient value remained inside the interval represented by the upper and lower bounds, then the levels of the decision variables given in Section *1* would remain optimal. The objective function *value* would, of course, change. The 100% rule discussed earlier holds for multiple simultaneous changes in objective function coefficients.
6	Ranges on right-hand-side constants within which the shadow prices remain unchanged	Similar to Section *5*, but aimed at constraints and shadow price values. The right-hand-side value for each constraint is listed along with an upper and lower bound. If the right-hand-side value were to change, yet remain within the interval defined by the listed upper and lower bounds, then the set of optimal shadow prices as given in Section *3* would not change. The optimal values of the decision variables as well as the objective function value would, in general, change. The 100% rule, discussed above, holds for multiple simultaneous changes in right-hand-side values and for simultaneous changes of *both* right-hand-side values *and* objective function coefficients.

For example, running the printer's production problem on CMP produces the output shown in Figure 3-10. Note that CMP displays the formulation of the problem and allows the user to name the decision variables and constraints with mnemonics that are useful to the manager or analyst.

```
LINEAR: PRINTER'S CALENDAR PRODUCTION PROBLEM

OBJECTIVES        S          D

  CONTRIB         3          5

CONSTRAINTS       S          D         RELATION    AVAIL

  PAPER           8          4           LE         4800
  PRESS          10         70           LE        38500
  DEMAND          1          0           LE          500


MAXIMIZE OR MINIMIZE? MAX

OPTIMAL SOLUTION FOUND.
     CONTRIB         3550.00


*1* DECISION VARIABLES
  1. S             350.000
  2. D             500.000

*2* SLACK(+) AND SURPLUS(-) IN CONSTRAINTS
  3. +DEMAND      150.000

*3* SHADOW PRICES FOR CONSTRAINTS
  1. PAPER         .307692
  2. PRESS         .538462E-01

*4* REDUCED COSTS FOR DECISION VARIABLES

*5* RANGES ON COEFFICIENTS OF OBJECTIVE CONTRIB
     VARIABLE   LOWER BOUND   CURRENT VALUE   UPPER BOUND
  1. S             .71429        3.0000          10.000
  2. D             1.5000        5.0000          21.000

*6* RANGES ON VALUES OF RIGHT-HAND-SIDE AVAIL
     CONSTRNT   LOWER BOUND   CURRENT VALUE   UPPER BOUND
  1.  PAPER        2200.0        4800.0          5914.3
  2.  PRESS        19000.        38500.          84000.
  3.  DEMAND       350.00        500.00          UNBOUNDED

OUTPUT OPTION?
```

FIGURE 3-10. Printer's calendar production problem.

The solution given (Section *1*) corresponds to the one we found earlier, and Section *2* shows the slack on the "DEMAND" constraint to be 150 units. Note, as is indicated in Table 3-2 above, only nonzero entries are given. Thus, the implication is that both the "PAPER" and "PRESS" constraints have no slack; that is, they are binding.

Section *3* informs us, as we had already calculated, that the shadow price of paper is \$.3077. In addition, we discover that the shadow price on press time is \$.0538.[11] Thus the printer should be willing to pay up to approximately 5⅓¢

[11]The "E-01" following the press time shadow price is computer notation meaning "exponent = −1"; that is, the value is $.538462 \times 10^{-1}$. Similarly,

$$538462E - 03 = .538462 \times 10^{-3} = .000538462$$

$$.538462E + 04 = .538462 \times 10^{+4} = 5{,}384.62$$

per minute for additional press time over and above the press time cost that has been figured into the contribution calculations represented by the objective function coefficients.

Skipping to Section *6*, we see that the $.0377 shadow price on paper is valid until the availability of paper increases beyond 5914.3 square feet. Beyond that amount, the value of paper would be less. (Actually, it drops to zero as shown graphically by Figure 3-6, but in general, we would not know this without rerunning the problem with a new paper availability figure larger than 5914.3.) Similarly, the incremental value of paper remains $.3077 even if availability were to drop to 2200 square feet. Below this amount, the shadow price would increase. In the same way, the shadow price on press time is $.0538 as long as the available time remains between 19,000 and 84,000 minutes, quite a large range.[12]

Section *5* indicates the range of objective function coefficients within which the production plan given by Section *1* remains optimal. The contribution for standard calendars could drop as low as 71.43¢ or rise to $10.00 before a different solution would become optimal. Similarly, as long as the contribution for deluxe calendars remains between $1.50 and $21.00, no change in the optimum would occur. These large ranges tell us (as we might have suspected earlier from Figure 3-8) that the optimal production plan is quite insensitive to changes in prices and costs.[13]

What if both contributions were to change simultaneously? In general, this question is difficult to answer without rerunning the model. We can, however, use the 100% rule to garner some insight. For example, if the printer's costs were to increase so that both contributions would drop by $1.00, to $2.00 for standard calendars and $4.00 for deluxe, would the optimal solution given in Section *1* change? Applying the 100% rule, we calculate

$$\frac{1}{3 - .7143} + \frac{1}{5 - 1.5} = .723$$

In words, the sum of changes, expressed as percentages of the amount allowed for each coefficient separately, is 72.3%, less than 100%. Hence, the optimal solution remains unchanged.

Let us continue to illustrate the use of CMP for solution and interpretation by solving the five examples presented in Chapter 2.

Example 1: Production Planning by a Power Drill Manufacturer

CMP displays the formulation derived in Chapter 2 along with accompanying solution output (Figure 3-11). The reader should refamiliarize himself or

[12]Although the shadow prices remain unchanged as long as the right-hand-side value remains within the given ranges, the optimal production plan will change.

[13]Although the optimal solution remains the same as long as the objective function coefficients remain within the specified ranges, the shadow prices will change.

OBJECTIVES	MODEL-1	MODEL-2	MODEL-3	MODEL-4	WIRE-MAN	WIRE-PUR		
CONTRIB	12.5	11.3	17.2	19.9	-.14	-.29		
CONSTRAINTS	**MODEL-1**	**MODEL-2**	**MODEL-3**	**MODEL-4**	**WIRE-MAN**	**WIRE-PUR**	**RELATION**	**AVAIL**
PLASTIC	.82	.62	1.42	2.03	0	0	LE	16000
COPPER	.43	.69	.33	.2	3.6E-2	0	LE	5000
WIRE	15	16	9	9	-1	-1	LE	8000
WIRE-CAP	0	0	0	0	1	0	LE	80000
MKTG-MIX	1	1	-1	-1	0	0	GE	0

```
MAXIMIZE

OPTIMAL SOLUTION FOUND.
     CONTRIB                 160306.

ALL ITEMS NOT LISTED IN SECTIONS 1 - 4 HAVE THE VALUE ZERO.

*1* DECISION VARIABLES
  1. MODEL-1     6796.50
  3. MODEL-3     5524.64
  4. MODEL-4     1271.86
  6. WIRE-PUR    155116.

*2* SLACK(+) AND SURPLUS(-) IN CONSTRAINTS
  4. +WIRE-CAP 80000.0

*3* SHADOW PRICES FOR CONSTRAINTS
  1. PLASTIC     6.63514
  2. COPPER      10.3649
  3. WIRE        .290000
  5. MKTG-MIX -1.74770

*4* REDUCED COSTS FOR DECISION VARIABLES
  2. MODEL-2   -2.85784
  5. WIRE-MAN -.223135

*5* RANGES ON COEFFICIENTS OF OBJECTIVE CONTRIB
      VARIABLE    LOWER BOUND   CURRENT VALUE   UPPER BOUND
  1. MODEL-1       10.069          12.500          15.918
  2. MODEL-2     UNBOUNDED         11.300          14.158
  3. MODEL-3       15.558          17.200          25.150
  4. MODEL-4       13.310          19.900          21.989
  5. WIRE-MAN    UNBOUNDED        -.14000          .83135E-01
  6. WIRE-PUR     -.42139         -.29000          .00000

*6* RANGES ON VALUES OF RIGHT-HAND-SIDE AVAIL
      CONSTRNT    LOWER BOUND   CURRENT VALUE   UPPER BOUND
  1.   PLASTIC      14737.          16000.          22619.
  2.   COPPER       3536.8          5000.0          5428.6
  3.   WIRE        UNBOUNDED        8000.0          .16312E+06
  4.   WIRE-CAP     .00000          80000.         UNBOUNDED
  5.   MKTG-MIX    -2823.5          .00000          5882.4
```

FIGURE 3-11. Breck and Dapper's production problem.

herself with the problem as presented there. Let us consider each of the output sections in turn.

1 The maximal contribution ($160,306) occurs when the production plan specifies the manufacture of 6796 Model 1s, 5524 Model 3s, and 1271 Model 4s.[14] Model 2 is not to be produced this quarter (not surprising since it is so heavy on the use of copper alloy). Also, Breck and Dapper should plan to go out and purchase over 155 thousand yards of wire—none is to be made internally even though it costs 15¢ less per yard. Further inspection of the output will indicate why.

2 Since no wire is to be produced internally, there is enough slack wire-drawing capacity to produce 80,000 yards of wire. Breck and Dapper's plant manager should perhaps ask whether this capacity could be used elsewhere.

3 The value of plastic to Breck and Dapper is $6.64/lb (over and above the usual raw material cost that, supposedly, has already been taken into account in the contribution figures). This represents a large premium some part of which the firm might offer to suppliers for additional rush delivery. The copper alloy is worth $10.36/lb over the going rate. This explains why no wire is produced internally—each yard would consume .036 pounds of copper alloy, 37.3¢ (.036 × $10.36) in value. Rather than consume 37.3¢ in value, the firm is glad to pay 15¢ extra outside. At this inexpensive rate, maybe the firm should buy extra wire and melt it down! The third shadow price indicates that wire is worth 29¢/yard to the firm—not surprising.

The fourth shadow price is somewhat hard to interpret. Literally, it says that if the right-hand side of the last constraint was changed from 0 to 1, the total contribution would decrease by $1.74. What does it mean to restrict the production by

$$x_1 + x_2 - x_3 - x_4 \geqslant 1$$

(in words—the production of Models 1 and 2 must exceed the production of Models 3 and 4 by one unit)? In this case, we will refrain from interpreting the shadow price as a price. Instead, we can simply say that the marketing policy restriction is cutting the firm's contribution to some extent. Possibly, other constraints less binding might be considered.

The firm can use these shadow prices to "price out" new products. Suppose, for example, that a lightweight drill was proposed that required only .1 pounds of copper alloy, one pound of plastic, used the same motor as Models 3 and 4 (with 9 yards of wire for motor

[14]We will not worry about the effects of rounding here. See Chapter 6 for a full discussion of integer restrictions.

winding), and would contribute $9.90 to overhead. The value consumed by this new product could be calculated as below:

Plastic:	1 lb × $6.36/lb =	$ 6.36
Copper alloy:	.1 lb × $10.36/lb =	1.04
Wire:	9 yd × $ .29/yd =	2.61
		$10.01

Thus, the value consumed is greater than the product's contribution, and the product should not be considered this quarter.

4 If marketing personnel required a full product line, this could be accomplished only at a cost of $2.86/unit of Model 2 produced. Alternatively, if production personnel wanted the idle wire-drawing machinery to be utilized, the cost to the firm of this policy would be 22.3¢/yard. This amount could have been calculated earlier when we remarked that the firm would rather pay 15¢ extra per yard to an outside supplier rather than consume 37.3¢ of value in copper alloy (37.3 − 15 = 22.3).

5 The optimal decisions are relatively insensitive to the respective model contributions, so the firm need not worry if revenues or costs change slightly. Similarly, the cost of wire would have to increase by 45% (from 29¢ to 42.1¢) before any change need be made.

6 The economic values of the resources may depend on the amounts available. Plastic's value is unlikely to change, even if the 16,000-pound estimate is somewhat in error since the listed lower and upper bounds are 14,737 and 22,619, respectively. The copper alloy's value, however, can be expected to decrease somewhat (how much is uncertain until we rerun the LP) if 430 or more additional pounds can be obtained, bringing copper's right-hand-side value above the listed upper bound of 5429. Thus, if Breck and Dapper's plant manager were offered additional copper at a $9.00 premium, he could be absolutely sure that the purchase of 429 pounds would be beneficial. Any purchase beyond that could only be evaluated by rerunning the model with a new copper right-hand side.

Let us pursue this further: The plant manager recognizes that copper availability drives the operating plan. He wants to plan a few quarters ahead to guarantee that raw material supplies, particularly copper, are adequate. The current shadow price on copper indicates that he should be willing to pay a premium of up to $10.36 per pound (Section *3*) for larger supplies, but this "resource value" holds only if the copper supply is between 3537 pounds and 5429 pounds. The plant manager, however, has the opportunity to renegotiate copper purchases with new vendors and in the upcoming negotiation, would like to know the value of this resource over a wider range. That is, we would like to write the copper availability constraint as

$$.43x_1 + .69x_2 + .33x_3 + .2x_4 + .036W_m \leq 5000 + \Theta$$

where Θ represents the increase or decrease in the copper supply, over or under the current level of 5000 pounds.[15] The plant manager would like to know the following, all as a function of Θ, for a wide range of values of that parameter:

- The objective function value (contribution).
- The optimal operating plan.
- Resource values (shadow prices).

The Simplex method makes it very efficient to go from one solution (graphically, the optimal corner point) to another (an adjacent corner point) corresponding to a (big enough) increase or decrease in Θ.

Table 3-3 gives the response to changes in Θ, and Figure 3-12 illustrates how the plant's contribution for the quarter depends on copper supply. Note the series of ranges given. These are ranges on the right-hand-side values as given in Section *6*. Within each range, the shadow price is constant so that the objective function grows *linearly* with increasing Θ. Once past the end of the range, the shadow price decreases to a new value, generating a different (less steep) linear segment for Figure 3-12.

This example, as well as those to follow, illustrates the wealth of information available along with the optimal solution. Not only does the plant manager obtain a plan that can be used as the basis for the proposal for both the upcoming and future quarters, there is additional information, either immediately available or easily acquired, for production, engineering, and marketing personnel.

[15]The use of the *parameter* Θ has caused this type of analysis to be known as parametric programming.

TABLE 3-3

Copper Availability	Optimal Operating Plan	Contribution	Value of Additional Pound of Copper
0–210 lb	Produce Models 1 and 4; neither buy nor manufacture wire.	@210 lb copper available, $10,800	$51.43
210–3537 lb	Produce Models 1 and 4; buy wire as necessary.	@3537 lb $145,144	$40.28
3537–5429 lb	Produce Models 1, 3 and 4; buy wire as necessary.	@5429 lb $165,746	$10.36
5429–8309 lb	Produce Models 1 and 4; purchase *and* manufacture wire; slack capacity on wire-drawing.	@8309 lb $176,755	$ 4.17
8309–10,880 lb	Produce Models 1, 2, and 3; run wire-drawing full out, manufacture as much wire as possible; purchase remaining requirements.	@10880 lb $180,997	$ 1.65
10,880–20,686 lb	Produce Models 2 and 3; manufacture wire up to capacity; purchase as needed.	@20686 lb $186,195	$ .53
Over 20,686 lb	Produce 25,806 Model 2s using all plastic and 20,686 lb of copper—any additional supply unused; wire-drawing at capacity; purchase 324,900 yd of wire.	$186,195	$0

Example 2: An Operating Plan for a Chemical Plant

The optimal solution offers a contribution of $22,720 in the coming month if the plant is run as indicated in Section*1* of the solution output (Figure 3-13) and indicated on the flow chart in Figure 3-14. Note from Section *2* that maximal contribution is generated if process 2 is operated below its capacity. Since process 3 is operated at full capacity, if process 2 were to be pushed

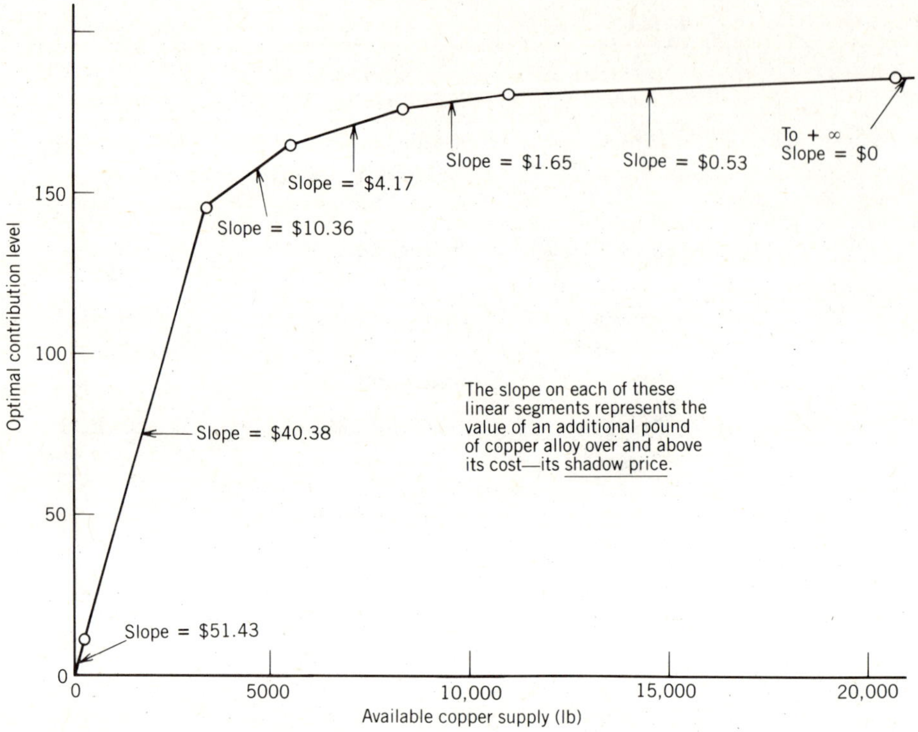

FIGURE 3-12. Optimal contribution level as copper supply increases.

further, any additional output (Chemical C) could only be sold, implying, per pound out,

Gain: Sell 1 lb of C		$= +6.40$
Loss: Input $1/.8 = 1.25$ lb of A;	$1.25 \times \$2.40$	$= -3.00$
Incur operating cost:	$1.25 \times \$3.50$	$= -4.375$
		$- .975$

Thus, as we expect, the optimal solution reflects the fact that using additional process 2 capacity to produce C for sale (requiring A as input) is not profitable. The linear programming solution saves us from having to do this (and similar) calculation. Even to ascertain the exact loss, we need not have done the above calculation explicitly. Section *4* reveals that the reduced cost of S_C—that is, the decrease in contribution which would occur if we ignore the suggested plan and sell one pound of C is, as we have computed, 97.5 cents.

The shadow prices, in Section *3*, on the material balance constraints are not of interest, but the shadow prices on process capacity are of value. If, for example, an additional pound of raw material could be pushed through process 1, it would be worth a premium of up to 34¢ over and above the current operating cost, which is already included in the objective function coefficients.

```
LINEAR: AN OPERATING PLAN FOR A CHEMICAL PLANT

OBJECTIVES     R       S-A     S-B     S-C     S-D     U-A     U-B     U-C

  CONTRIB    -2.5      2.4     3.2     6.4    12.2    -3.5    -4.2    -4.2

CONSTRAINTS    R       S-A     S-B     S-C     S-D     U-A     U-B     U-C    RELATION   RHS

  I/O-1-A    -.45       1       0       0       0       1       0       0       EQ       0
  I/O-1-B    -.55       0       1       0       0       0       1       0       EQ       0
  I/O-2        0        0       0       1       0     -.8       0       1       EQ       0
  I/O-3        0        0       0       0       1       0     -.75    -.75      EQ       0
  CAP-1        1        0       0       0       0       0       0       0       LE      56
  CAP-2        0        0       0       0       0       1       0       0       LE      25
  CAP3         0        0       0       0       0       0       1       1       LE      46
  RATIO-3      0        0       0       0       0       0       1     -1.5      EQ       0
     MAXIMIZE

     OPTIMAL SOLUTION FOUND.
           CONTRIB                22.7200
     *1* DECISION VARIABLES
       1. R           56.0000
       2. S-A         2.20000
       3. S-B         3.20000
       5. S-D         34.5000
       6. U-A         23.0000
       7. U-B         27.6000
       8. U-C         18.4000

     *2* SLACK(+) AND SURPLUS(-) IN CONSTRAINTS
       6. +CAP-2      2.00000

     *3* SHADOW PRICES FOR CONSTRAINTS
       1. I/O-1-A     2.40000
       2. I/O-1-B     3.20000
       3. I/O-2       7.37500
       4. I/O-3       12.2000
       5. CAP-1       .340000
       7. CAP3        .800003E-01
       8. RATIO-3     1.67000

     *4* REDUCED COSTS FOR DECISION VARIABLES
       4. S-C        -.975000

     *5* RANGES ON COEFFICIENTS OF OBJECTIVE CONTRIB
           VARIABLE   LOWER BOUND     CURRENT VALUE    UPPER BOUND
       1. R           -2.8400          -2.5000         UNBOUNDED
       2. S-A          1.6444           2.4000          2.5600
       3. S-B          2.5818           3.2000          3.3333
       4. S-C         UNBOUNDED         6.4000          7.3750
       5. S-D          12.093           12.200         UNBOUNDED
       6. U-A         -3.6600          -3.5000         -2.7200
       7. U-B         -4.3333          -4.2000         UNBOUNDED
       8. U-C         -4.4000          -4.2000          .15845E+07

     *6* RANGES ON VALUES OF RIGHT-HAND-SIDE RHS
           CONSTRNT   LOWER BOUND     CURRENT VALUE    UPPER BOUND
       1.  I/O-1-A    -2.2000           .00000         UNBOUNDED
       2.  I/O-1-B    -3.2000           .00000         UNBOUNDED
       3.  I/O-2      -1.6000           .00000          18.400
       4.  I/O-3      -34.500           .00000         UNBOUNDED
       5.  CAP-1       51.111           56.000          .85750E+08
       6.  CAP-2       23.000           25.000         UNBOUNDED
       7.  CAP3        .29564E-04       46.000          50.000
       8.  RATIO-3    -4.0000           .00000          8.0000
```

FIGURE 3-13. Dowpont's optimal operating plan.

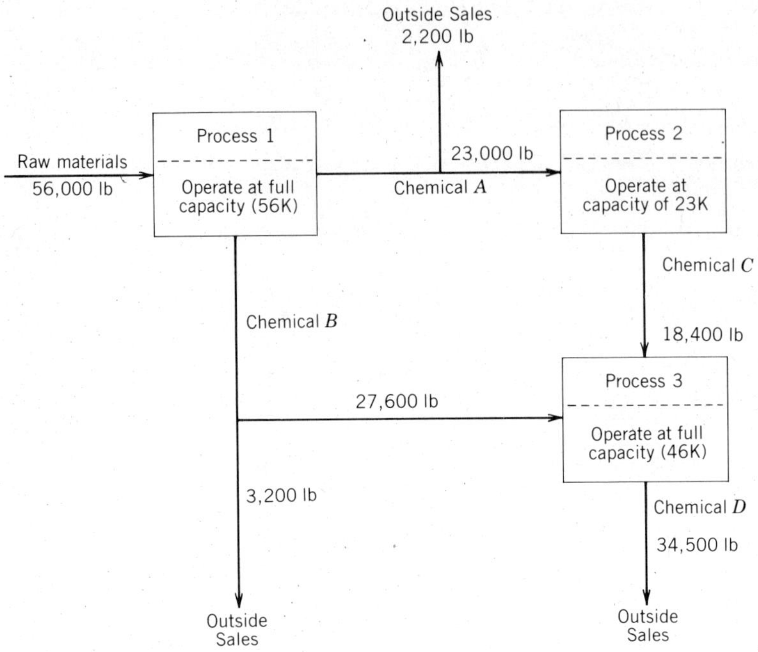

FIGURE 3-14. Dowpont's optimal operating plan.

That is, if Dowpont could crank up process 1 beyond its rated capacity (e.g., by overtime maintenance), even if that additional capacity would cost \$1.84 per input pound, a 23% cost increase, it would be profitable to do so.[16] Similarly, Section *3* tells us that additional process 3 capacity would be worth an 8¢ premium. Apparently, the market prices for *A* and *B* are sufficiently high that process-1 capacity is a more valuable resource.

For curiosity, let us pursue process-3 capacity. Why is extra capacity worth an 8¢ premium? The answer depends on how Dowpont would operate if process 3 could take another pound of input:

1. Sell .75 lb *D*: .75 × \$12.20	=	\$9.15
2. Operate process 3: operating cost	=	−4.20
(a) requires .6 lb *B*—divert from outside sales .6 × \$3.20	=	−1.92
(b) requires .4 lb *C*—operate process 2		
(i) requires .5 lb *A*—divert from outside sales .5 × \$2.40	=	−1.20
(ii) operating cost: .5 × \$3.50	=	−1.75
		+\$0.08

[16]\$1.84 = \$1.50 (the original operating cost) + \$0.34 (the premium indicated by the shadow price).

This calculation, which requires following flow back through the process, might be too tedious to justify for a large, real-world, chemical process. The fact that this information can be provided routinely by the LP solution methodology is key in that methodology's success.

The plant manager, in formulating this model, had initial concerns about the fact that the market prices for the plant's intermediate chemicals *A*, *B*, and *C* had been rising, and that final output *D*'s relative profitability was in question. The output indicates that the most profitable course of action is still to produce as much *D* as possible, pushing the intermediates through the process as far as possible. Section *5* indicates, however, that the plant manager's fears are justified—the optimal operating plan is quite sensitive to those market prices. Small increases (7% for *A*, 4% for *B*) in the prices for intermediates *A* or *B* or a small decrease (1%) in *D*'s price would change the optimal solution, lowering the output of *D* to be sold. To see how much would require rerunning the model. For example, if the price of each intermediate were to rise 8¢/lb, Figure 3-15 shows the new optimal solution—the new optimal operating mix has processes 2 and 3 idle, with process 1 running full steam to produce *A* and *B* for sale! This example strikingly points out one of the problems in using linear programming in a dynamically changing environment. Since the solution procedure operates by jumping from one corner point to another, when the optimal solution changes, it may jump to a very different plan, as in this case, rather than change slowly and continuously. Does this mean Dowpont's plant manager has to monitor market prices, ready to totally change the plant's

```
OPTIMAL SOLUTION FOUND.
     CONTRIB                    23.5200
*1* DECISION VARIABLES
  1. R              56.0000
  2. S-A            25.2000
  3. S-B            30.8000
  5. S-D            .224113E-04
  6. U-A            .150204E-04
  7. U-B            .178814E-04

*2* SLACK(+) AND SURPLUS(-) IN CONSTRAINTS
  6. +CAP-2         25.0000
  7. +CAP3          46.0000

*3* SHADOW PRICES FOR CONSTRAINTS
  1. I/O-1-A        2.48000
  2. I/O-1-B        3.28000
  3. I/O-2          7.47500
  4. I/O-3          12.2000
  5. CAP-1          .420000
  8. RATIO-3        1.67000

*4* REDUCED COSTS FOR DECISION VARIABLES
  4. S-C           -.995000
  8. U-C           -.199996E-01
```

FIGURE 3-15. Dowpont's optimal operating plan with higher market prices for intermediates.

```
LINEAR: A PORTFOLIO SELECTION PROBLEM

OBJECTIVES     X-A        X-B        X-C        X-D        X-E

  RETURN        9E-2      5.4E-2      5E-2     4.4E-2      9E-2

CONSTRAINTS    X-A        X-B        X-C        X-D        X-E       RELATION     RHS

  BUDGET        1          1          1          1          1          LE         10
  GOV'T         1          0          0          0          1          LE          3
  QUALITY      .6         .6        -.4        -.4        3.6          LE          0
  MATURITY      4         10         -1         -2         -3          LE          0

MAXIMIZE

OPTIMAL SOLUTION FOUND.
     RETURN              0.605455

ALL ITEMS NOT LISTED IN SECTIONS 1 - 4 HAVE THE VALUE ZERO.

*1* DECISION VARIABLES
  1. X-A        2.18182
  3. X-C        7.36364
  5. X-E        .454545

*2* SLACK(+) AND SURPLUS(-) IN CONSTRAINTS
  2. +GOV'T     .363636

*3* SHADOW PRICES FOR CONSTRAINTS
  1. BUDGET     .605455E-01
  3. QUALITY    .127273E-01
  4. MATURITY   .545454E-02

*4* REDUCED COSTS FOR DECISION VARIABLES
  2. X-B        -.687273E-01
  4. X-D        -.545455E-03

*5* RANGES ON COEFFICIENTS OF OBJECTIVE RETURN
     VARIABLE   LOWER BOUND   CURRENT VALUE   UPPER BOUND
  1. X-A         .60000E-01    .90000E-01      .93000E-01
  2. X-B        UNBOUNDED      .54000E-01      .12273
  3. X-C         .49520E-01    .50000E-01      .90000E-01
  4. X-D        UNBOUNDED      .44000E-01      .44545E-01
  5. X-E         .78000E-01    .90000E-01      .21000

*6* RANGES ON VALUES OF RIGHT-HAND-SIDE RHS
     CONSTRNT   LOWER BOUND   CURRENT VALUE   UPPER BOUND
  1.  BUDGET     .11921E-06    10.000          11.379
  2.  GOV'T      2.6364        3.0000         UNBOUNDED
  3.  QUALITY   -2.0000        .00000          1.1429
  4.  MATURITY  -12.000        .00000          2.6667
```

FIGURE 3-16. A portfolio selection problem.

operation as soon as relative prices change enough? Not necessarily! As Section *4* from Figure 3-15 indicates, running process 3 to produce *D* would worsen the objective by 2¢/lb of *C* produced and used. Hence, the *old* operating plan under this *new* environment would generate a reduced contribution—reduced by 18,400 × .02 = $368. If the disruption caused by the plant reorientation is significant enough, the plant manager may choose to forego the additional $368 this month, and wait until next month to institute a new operating plan.

Again, the solutions generated are not "answers." They merely provide a guide for action. The use of linear programming allows the manager to assign an economic gain or loss to alternative courses of action.

Example 3: A Portfolio Selection Problem

The output for the first problem, neglecting the availability of borrowed funds, is given in Figure 3-16. The optimal portfolio, which consists of $2.18 million invested in bond *A*, $7.36 million in *C*, and $455 thousand in *E*, will generate over $605,000 in after tax interest per year (a return of 6.05%). Section *2* shows that all constraints, except the restriction on government bonds, are binding. Section *4* indicates that poor portfolio selection—namely, investing in *B* or *D*—would reduce the anticipated return by 6.87¢ per dollar invested in *B* and .055¢ per dollar invested in *D*. In other words, *B* would be a very poor choice indeed; *D* is marginally unattractive. (We would expect a very small increase in *D*'s return would bring it into the optimal portfolio.)

Section *5* gives information on the sensitivity of the optimal portfolio to the anticipated returns:

A: Insensitive on the lower side—*A*'s return would have to drop from 9% to 6% before the portfolio would change; sensitive on the high side—an increase to 9.3% would change the portfolio selection (presumably increasing the amount of *A* purchased).

B: Very insensitive.

C: Very sensitive on the low side—a tiny drop would change the optimal portfolio (presumably reducing the amount of *C* purchased, perhaps in favor of *D*); insensitive on the high side.

D: Very sensitive on the high side; note that the increase necessary to bring *D* into the optimal portfolio, .055¢ per dollar invested ($.04455 − .044) is equal to *D*'s reduced cost.

E: Very insensitive on the high side—the after-tax return would have to more than double before *E*'s participation in the portfolio would increase; on the low side, *E*'s return would have to drop from 9% to 7.8%.

```
LINEAR: BOND PORTFOLIO - WITH BORROWING

OBJECTIVES     X-A      X-B      X-C      X-D      X-E      BORROW

 RETURN        9E-2     5.4E-2   5E-2     4.4E-2   9E-2     -5.5E-2

CONSTRAINTS    X-A      X-B      X-C      X-D      X-E      BORROW   RELATION   RHS

 BUDGET         1        1        1        1        1        -1       LE        10
 GOV'T          1        0        0        0        1         0       LE         3
 QUALITY        .6       .6       -.4      -.4      3.6       0       LE         0
 MATURITY       4        10       -1       -2       -3        0       LE         0
 LIMIT          0        0        0        0        0         1       LE         1

             LINEAR     6 VAR,     5 CON,     1 OBJ,     1 RHS
          BOND PORTFOLIO - WITH BORROWING

MAXIMIZE

OPTIMAL SOLUTION FOUND.
     RETURN                0.611000
*1* DECISION VARIABLES
  1. X-A          2.40000
  3. X-C          8.10000
  5. X-E          .500000
  6. BORROW       1.00000

*2* SLACK(+) AND SURPLUS(-) IN CONSTRAINTS
  2. +GOV'T        .100000

*3* SHADOW PRICES FOR CONSTRAINTS
  1. BUDGET       .605455E-01
  3. QUALITY      .127273E-01
  4. MATURITY     .545454E-02
  5. LIMIT        .554545E-02

*4* REDUCED COSTS FOR DECISION VARIABLES
  2. X-B          -.687273E-01
  4. X-D          -.545455E-03

*5* RANGES ON COEFFICIENTS OF OBJECTIVE RETURN
     VARIABLE   LOWER BOUND   CURRENT VALUE   UPPER BOUND
  1. X-A         .64583E-01    .90000E-01      .93000E-01
  2. X-B        UNBOUNDED      .54000E-01      .12273
  3. X-C         .49520E-01    .50000E-01      .90000E-01
  4. X-D        UNBOUNDED      .44000E-01      .44545E-01
  5. X-E         .78000E-01    .90000E-01      .21000
  6. BORROW     -.60545E-01   -.55000E-01     UNBOUNDED

*6* RANGES ON VALUES OF RIGHT-HAND-SIDE RHS
     CONSTRNT   LOWER BOUND   CURRENT VALUE   UPPER BOUND
  1.  BUDGET     -1.0000        10.000         10.379
  2.  GOV'T       2.9000        3.0000        UNBOUNDED
  3.  QUALITY    -2.2000        .00000         .31429
  4.  MATURITY  -13.200         .00000         .73333
  5.  LIMIT       .00000        1.0000         1.3793
```

FIGURE 3-17. The bond portfolio problem with borrowing.

The first lines of Section *3* and *6* give us directly useful information. The shadow price on the budget constraint says, not unexpectedly, that each additional dollar available for investment would generate 6.05¢ in after-tax return. The fact that this equals the optimal portfolio return indicates that this new dollar is to be invested in *A*, *C*, and *E* in the same proportions as in the original optimal portfolio. Section *6* tells us that this 6.05% return on additional funds is possible up to a total investment of $11,379,000—that is, *up to an additional $1.379 million*. Hence, our second problem is solved without reformulation and resolving! If one million additional dollars is available at 5.5% interest after taxes, then we will borrow the full amount to invest in our optimal portfolio, earning 0.55% net (6.05 − 5.5%) for a total gain of $5,500. Just to check, Figure 3-17 gives the output for the reformulated problem of Example 3. As expected, the additional million has been invested in the same proportions:

2.4 = 21.82% of $11M

8.1 = 73.64% of $11M

.5 = 4.55% of $11M

and the return is $610,955, which is $5,500 more than before. Also, as we would expect, the shadow price on borrowed funds availability is .55%, true up to an additional $379,000.

Example 4: A Transportation Problem

The formulation and solution of the agricultural cooperative's problem is shown by the CMP output of Figure 3-19 and the network representation of Figure 3-18. Note that only the paths which have positive flow have been indicated.

Since this problem, unlike those above, has been formulated with only equality constraints, there are no slacks or surpluses. This set of equalities produces another problem—individual shadow prices are impossible to interpret. Recall that each constraint is a material balance equality at a location or node of the network. Adding one unit to any right-hand side will put the

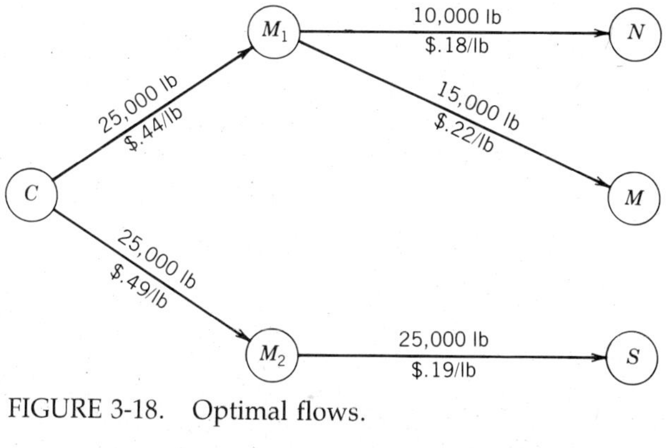

FIGURE 3-18. Optimal flows.

LINEAR: AGRICULTURAL COOP'S TRANSPORT PROBLEM

OBJECTIVES	C-TO-M1	C-TO-M2	M1-TO-N	M1-TO-M	M1-TO-S	M2-TO-N	M2-TO-M	M2-TO-S
COST	.44	.49	.18	.22	.27	.25	.23	.19

CONSTRAINTS	C-TO-M1	C-TO-M2	M1-TO-N	M1-TO-M	M1-TO-S	M2-TO-N	M2-TO-M	M2-TO-S	RELATION	RHS
NODE-C	1	1	0	0	0	0	0	0	EQ	50000
NODE-M1	-1	0	1	1	1	0	0	0	EQ	0
NODE-M2	0	-1	0	0	0	1	1	1	EQ	0
NODE-N	0	0	-1	0	0	-1	0	0	EQ	-10000
NODE-M	0	0	0	-1	0	0	-1	0	EQ	-15000
NODE-S	0	0	0	0	-1	0	0	-1	EQ	-25000

```
MINIMIZE

OPTIMAL SOLUTION FOUND.
    COST                     33100.0
*1* DECISION VARIABLES
  1. C-TO-M1    25000.0
  2. C-TO-M2    25000.0
  3. M1-TO-N    10000.0
  4. M1-TO-M    15000.0
  8. M2-TO-S    25000.0

*2* SLACK(+) AND SURPLUS(-) IN CONSTRAINTS

*3* SHADOW PRICES FOR CONSTRAINTS
  1. NODE-C     .490000
  2. NODE-M1    .500000E-01
  3. NODE-M2    .000000
  4. NODE-N    -.130000
  5. NODE-M    -.170000
  6. NODE-S    -.190000

*4* REDUCED COSTS FOR DECISION VARIABLES
  5. M1-TO-S    .300000E-01
  6. M2-TO-N    .120000
  7. M2-TO-M    .600000E-01

*5* RANGES ON COEFFICIENTS OF OBJECTIVE COST
      VARIABLE   LOWER BOUND   CURRENT VALUE   UPPER BOUND
  1. C-TO-M1      .41000        .44000          .50000
  2. C-TO-M2      .43000        .49000          .52000
  3. M1-TO-N     UNBOUNDED      .18000          .30000
  4. M1-TO-M     UNBOUNDED      .22000          .28000
  5. M1-TO-S      .24000        .27000         UNBOUNDED
  6. M2-TO-N      .13000        .25000         UNBOUNDED
  7. M2-TO-M      .17000        .23000         UNBOUNDED
  8. M2-TO-S     UNBOUNDED      .19000          .22000

*6* RANGES ON VALUES OF RIGHT-HAND-SIDE RHS
      CONSTRNT   LOWER BOUND   CURRENT VALUE   UPPER BOUND
  1.  NODE-C      50000.        50000.         UNBOUNDED
  2.  NODE-M1     .00000        .00000          25000.
  3.  NODE-M2     .00000        .00000         UNBOUNDED
  4.  NODE-N     -10000.       -10000.          .00000
  5.  NODE-M     -15000.       -15000.          .00000
  6.  NODE-S     -25000.       -25000.          .00000
```

FIGURE 3-19. Agricultural coop's transport problem.

network out of balance, causing the Simplex algorithm to report no feasible solution. Note, in fact, that Section *6* indicates that, in one direction for each constraint, the shadow prices become invalid as soon as each right-hand side changes. However, we can interpret shadow prices by considering *simultaneous* right-hand-side changes. For example, if we were to "create" an additional bushel of flour at mill M_1 and simultaneously increase the demand at N by a bushel, what would happen? Referring to Figure 3-18, it is clear that this new unit would be shipped directly from M_1 to N, at an additional cost of 18¢. In terms of right-hand-side changes this new "shipment" is caused by an *increase* in the right-hand side of "NODE-M_1" (corresponding to the injection of a unit at M_1) and a *decrease* in the right-hand side of "NODE-N" (corresponding to the withdrawal of an additional unit at N). According to Section *3*, this results in a cost increase of

At M_1	5¢
At N	−(−13¢)
Total	18¢

as we calculated above.

Let us repeat this for node M_1 and an increased demand at S:

At M_1	5¢
At S	−(−19¢)
Total	24¢

This case is somewhat different than the above. In this situation, the optimal way to ship this unit is *not* directly from M_1 to S. From the set of optimal paths in Figure 3-18, it can be seen that such direct shipment is not desirable (otherwise that path would have been given positive flow in the optimal solution). Instead, the "shipment" of one unit from M_1 to S is done by:

1. Ship one less unit from

   C to $M_1 \longrightarrow$ save 44¢

2. Ship one more unit from

   C to $M_2 \longrightarrow$ cost 49¢

3. Ship one more unit from

   M_2 to $S \longrightarrow$ cost 19¢

   Net additional cost = 24¢

If we *were* to ship the unit directly from M_1 to S, it would cost 27¢, the transport cost on that path. Direct shipment costs us 3¢ more than the method indicated above. Note, from Section *4*, that this 3¢ is exactly the reduced cost for using the $M_1 \rightarrow S$ path.

We can generalize this process to generate the additional cost that would be incurred if we wanted to move one unit from any node to any other—simply subtract the destination node's shadow price from the origin node's shadow price. For the network above:

$C \rightarrow M_1$	49¢ − 5¢ = 44¢
$C \rightarrow M_2$	49¢ − 0 = 49¢
$C \rightarrow N$	49¢ − (−13¢) = 62¢
$C \rightarrow M$	49¢ − (−17¢) = 66¢
$C \rightarrow S$	49¢ − (−19¢) = 68¢
$M_1 \rightarrow N$	5¢ − (−13¢) = 18¢
$M_1 \rightarrow M$	5¢ − (−17¢) = 22¢
$M_1 \rightarrow S$	5¢ − (−19¢) = 24¢
$M_2 \rightarrow N$	0 − (−13¢) = 13¢
$M_2 \rightarrow M$	0 − (−17¢) = 17¢
$M_2 \rightarrow S$	0 − (−19¢) = 19¢

Note that, for every pair for which the shipment path(s) is in the optimal network of Figure 3-18, the above cost is simply the actual per unit transport cost.When the direct path is not optimal, the above cost corresponds to the best possible retracing. In those cases, the difference between the above costs and the direct transport cost is the reduced cost for that path—that is, the cost of forcing one unit along that path:

Path	"Best Cost"	Direct Cost	Reduced Cost
$M_1 \rightarrow S$	24¢	27¢	3¢
$M_2 \rightarrow N$	13¢	25¢	12¢
$M_2 \rightarrow M$	17¢	23¢	6¢

Example 5: A Multiperiod Aggregate Production Scheduling Problem

The model for this problem is far larger than those formulated for the first four examples, requiring 73 constraints on 61 decision variables. The tableau is comprised of 4,453 entries, too many to try to reproduce here. In spite of the far greater size of this model, however, it is not necessarily more structurally complex. The time structure inherent in the formulation allows us to think of the tableau in monthly blocks, as shown schematically in Figure 3-20. Each month has its own set of (five) constraints with linkage to future months through inventory and to past months through workforce levels. That is, the block for month i is[17]

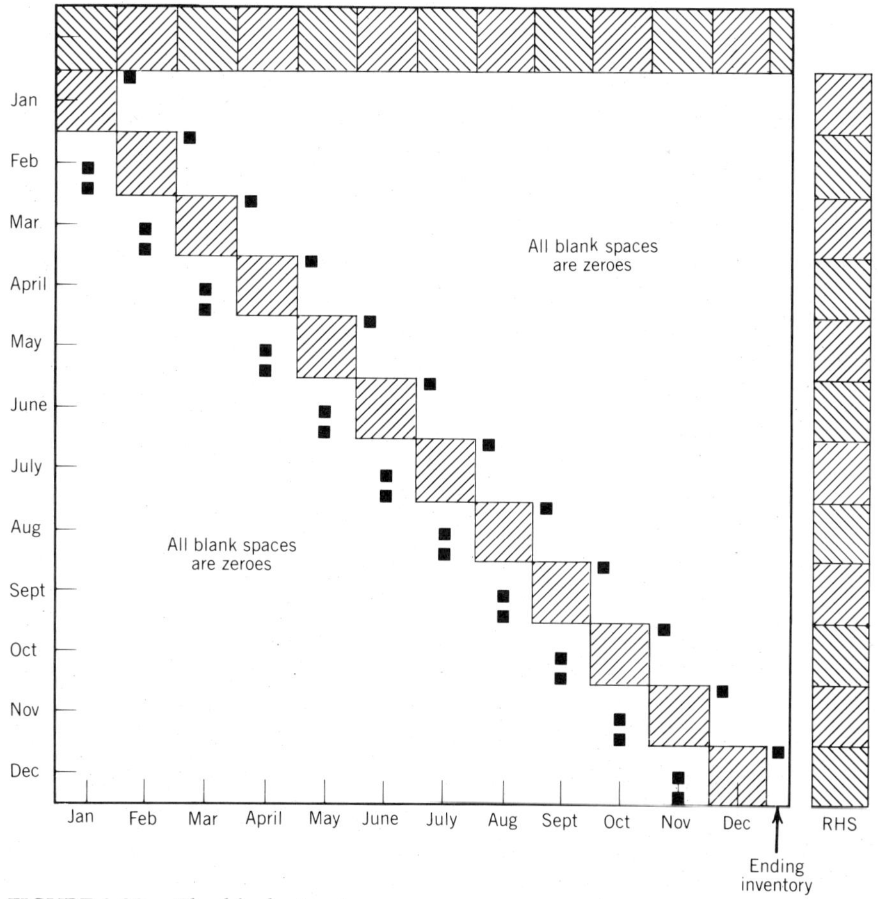

FIGURE 3-20. The block structure.

[17]The blocks for January and December are slightly different to accommodate starting and ending positions.

	W_{i-1}	P_i	I_i	W_i	H_i	L_i	I_{i+1}		Right-Hand Side
Min		100	7	1600	300	500			
P/DEM− i		1	1	0	0	0	−1	=	D_i
W/CAP− i		1	0	−8	0	0		≤	0
P/CAP− i		1	0	0	0	0		≤	7000
WORKER− i	−1	0	0	1	−1	1		=	0
HIRING− i		0	0	0	1	0		≤	50
UNION− i	−.1	0	0	0	0	1		≤	0

The solution output is given in Figure 3-21. The optimal production plan (costing \$20.3 million) in Section *1* of the CMP output, represented schematically in Figure 3-22, is quite flat in contrast to the marked seasonality of demand. Rather than trying to follow demand, the optimal program uses inventory build-ups (see Section *1* and Figure 3-22) to satisfy peak demand. Note that production never goes above 83% of equipment capacity (7000 units) although demand in May and June are above capacity limits. Thus, the manager thinking of expanding capacity to meet this peak receives a clear message here.

Figure 3-22 shows a gradual build-up in January, February, and March toward the virtually steady production that runs from March through October and a gradual fall-off late in the year. Figure 3-23 shows the optimal workforce levels (from output Section *1*) to achieve this production rate. Section *2* indicates that this workforce plan bumps up against the hiring limits only in February and March, and against union regulations only in the last month. There are important messages for personnel policy in this information:

1. Resources for recruiting are particularly important in the first quarter—that is where the effort should be spent. Section *3* indicates that production costs would drop \$54 for every worker (over the limit of 50) who could be recruited for February and \$351 for every additional worker who could be recruited for March.

2. Union restrictions on layoffs do not seem to be at all burdensome for the coming year. Only in December would the program like to lay off more than 10% of the workforce, and with new forecasts for the next year's demand, this might well disappear. Hence, in further negotiations with the union, this is not an issue to be concerned with.

```
                LINEAR   61 VAR,    73 CON.
              PRODUCTION SCHEDULING PROBLEM

MINIMIZE

OPTIMAL SOLUTION FOUND.
     COST                0.203229E+08

ALL ITEMS NOT LISTED IN SECTIONS 1 - 4 HAVE THE VALUE ZERO.

*1* DECISION VARIABLES
  1. P1         5021.43
  2. I1         750.000
  3. W1         627.679
  4. H1         2.67858
  6. P2         5421.43
  7. I2         1971.43
  8. W2         677.679
  9. H2         50.0000
 11. P3         5821.43
 12. I3         3292.86
 13. W3         727.679
 14. H3         50.0000
 16. P4         5821.43
 17. I4         4414.29
 18. W4         727.679
 21. P5         5821.43
 22. I5         4335.71
 23. W5         727.679
 26. P6         5821.43
 27. I6         2957.14
 28. W6         727.679
 31. P7         5821.43
 32. I7         878.571
 33. W7         727.679
 36. P8         5755.56
 38. W8         719.444
 40. L8         8.23412
 41. P9         5755.56
 42. I9         655.556
 43. W9         719.444
 46. P10        5755.56
 47. I10        1111.11
 48. W10        719.444
 51. P11        5333.33
 52. I11        466.667
 53. W11        666.667
 55. L11        52.7778
 56. P12        4800.00
 58. W12        600.000
 60. L12        66.6667

*2* SLACK(+) AND SURPLUS(-) IN CONSTRAINTS
  3. +P/CAP-1  1978.57
  5. +HIRING-1 47.3214
  6. +UNION-1  62.5000
  9. +P/CAP-2  1578.57
 12. +UNION-2  62.7679
 15. +P/CAP-3  1178.57
 18. +UNION-3  67.7679
```

```
21. +P/CAP-4   1178.57
23. +HIRING-4  50.0000
24. +UNION-4   72.7679
27. +P/CAP-5   1178.57
29. +HIRING-5  50.0000
30. +UNION-5   72.7679
33. +P/CAP-6   1178.57
35. +HIRING-6  50.0000
36. +UNION-6   72.7679
39. +P/CAP-7   1178.57
41. +HIRING-7  50.0000
42. +UNION-7   72.7679
45. +P/CAP-8   1244.44
47. +HIRING-8  50.0000
48. +UNION-8   64.5337
51. +P/CAP-9   1244.44
53. +HIRING-9  50.0000
54. +UNION-9   71.9444
57. +P/CAP-10  1244.44
59. +HIRES-10  50.0000
60. +UNION-10  71.9444
63. +P/CAP-11  1666.67
65. +HIRES-11  50.0000
66. +UNION-11  19.1667
69. +P/CAP-12  2200.00
71. +HIRES-12  50.0000
```

```
*3* SHADOW PRICES FOR CONSTRAINTS
 1. P/DEM-1     293.286
 2. W/CAP-1    -193.286
 4. WORKER-1   -300.000
 7. P/DEM-2     300.286
 8. W/CAP-2    -200.286
10. WORKER-2   -353.714
11. HIRING-2   -53.7143
13. P/DEM-3     307.286
14. W/CAP-3    -207.286
16. WORKER-3   -351.429
17. HIRING-3   -51.4286
19. P/DEM-4     314.286
20. W/CAP-4    -214.286
22. WORKER-4   -293.143
25. P/DEM-5     321.286
26. W/CAP-5    -221.286
28. WORKER-5   -178.857
31. P/DEM-6     328.286
32. W/CAP-6    -228.286
34. WORKER-6   -8.57146
37. P/DEM-7     335.286
38. W/CAP-7    -235.286
40. WORKER-7    217.714
43. P/DEM-8     293.000
44. W/CAP-8    -193.000
46. WORKER-8    500.000
49. P/DEM-9     300.000
50. W/CAP-9    -200.000
52. WORKER-9    444.000
55. P/DEM-10    307.000
56. W/CAP-10   -207.000
58. WKER-10     444.000
61. P/DEM-11    314.000
62. W/CAP-11   -214.000
64. WKER-11     500.000
67. P/DEM-12    221.944
68. W/CAP-12   -121.944
70. WKER-12     624.444
72. UNION-12   -124.444
73. BEG-INV    -289.786
```

```
*4* REDUCED COSTS FOR DECISION VARIABLES
 5. L1          800.000
10. L2          853.714
15. L3          851.429
19. H4          6.85713
20. L4          793.143
24. H5          121.143
25. L5          678.857
29. H6          291.429
30. L6          508.571
34. H7          517.714
35. L7          282.286
37. I8          49.2857
39. H8          800.000
44. H9          744.000
45. L9          56.0000
49. H10         744.000
50. L10         56.0000
54. H11         800.000
57. I12         99.0556
59. H12         924.444
61. I13         228.944
```

5 RANGES ON COEFFICIENTS OF OBJECTIVE COST

	VARIABLE	LOWER BOUND	CURRENT VALUE	UPPER BOUND
1.	P1	92.167	100.00	101.50
2.	I1	UNBOUNDED	3.5000	.10000E+16
2.	I1	-.10000E+16	3.5000	UNBOUNDED
3.	W1	1537.3	1600.0	1612.0
4.	H1	237.33	300.00	312.00
5.	L1	-300.00	500.00	UNBOUNDED
6.	P2	91.000	100.00	101.50
7.	I2	-.83333	7.0000	8.5000
8.	W2	1528.0	1600.0	1612.0
9.	H2	UNBOUNDED	300.00	353.71
10.	L2	-353.71	500.00	UNBOUNDED
11.	P3	-73.500	100.00	101.50
12.	I3	2.5000	7.0000	7.7500
13.	W3	212.00	1600.0	1612.0
14.	H3	-.13652E+12	300.00	351.43
15.	L3	-351.43	500.00	UNBOUNDED
16.	P4	98.000	100.00	122.50
17.	I4	1.3750	7.0000	7.5000
18.	W4	1584.0	1600.0	1780.0
19.	H4	293.14	300.00	UNBOUNDED
20.	L4	-293.14	500.00	UNBOUNDED
21.	P5	98.000	100.00	122.50
22.	I5	-.50000	7.0000	7.6667
23.	W5	1584.0	1600.0	1780.0
24.	H5	178.86	300.00	UNBOUNDED
25.	L5	-178.86	500.00	UNBOUNDED

26.	P6	98.000	100.00	122.50
27.	I6	-4.2500	7.0000	8.0000
28.	W6	1584.0	1600.0	1780.0
29.	H6	8.5715	300.00	UNBOUNDED
30.	L6	-8.5715	500.00	UNBOUNDED
31.	P7	98.000	100.00	122.50
32.	I7	-15.500	7.0000	9.0000
33.	W7	1584.0	1600.0	1780.0
34.	H7	-217.71	300.00	UNBOUNDED
35.	L7	217.71	500.00	UNBOUNDED
36.	P8	89.500	100.00	239.50
37.	I8	-42.286	7.0000	UNBOUNDED
38.	W8	1516.0	1600.0	2716.0
39.	H8	-500.00	300.00	UNBOUNDED
40.	L8	484.00	500.00	584.00
41.	P9	79.000	100.00	121.00
42.	I9	-3.5000	7.0000	146.50
43.	W9	1432.0	1600.0	1768.0
44.	H9	-444.00	300.00	UNBOUNDED
45.	L9	444.00	500.00	UNBOUNDED
46.	P10	58.000	100.00	110.50
47.	I10	-3.5000	7.0000	146.50
48.	W10	1264.0	1600.0	1684.0
49.	H10	-444.00	300.00	UNBOUNDED
50.	L10	444.00	500.00	UNBOUNDED
51.	P11	-9.7500	100.00	114.00
52.	I11	-7.0000	7.0000	116.75
53.	W11	722.00	1600.0	1712.0
54.	H11	-500.00	300.00	UNBOUNDED
55.	L11	416.00	500.00	584.00
56.	P12	-128.94	100.00	199.06
57.	I12	-92.056	7.0000	UNBOUNDED
58.	W12	624.44	1600.0	2392.4
59.	H12	-624.44	300.00	UNBOUNDED
60.	L12	-8280.0	500.00	612.00
61.	I13	-221.94	7.0000	UNBOUNDED

6 RANGES ON VALUES OF RIGHT-HAND-SIDE RHS

	CONSTRNT	LOWER BOUND	CURRENT VALUE	UPPER BOUND
1.	P/DEM-1	3650.0	3800.0	6100.0
2.	W/CAP-1	-2300.0	.00000	150.00
3.	P/CAP-1	5021.4	7000.0	UNBOUNDED
4.	WORKER-1	577.68	625.00	627.68
5.	HIRING-1	2.6786	50.000	UNBOUNDED
6.	UNION-1	.00000	62.500	UNBOUNDED
7.	P/DEM-2	3950.0	4100.0	6750.0
8.	W/CAP-2	-2650.0	.00000	150.00
9.	P/CAP-2	5421.4	7000.0	UNBOUNDED
10.	WORKER-2	-55.208	.00000	3.1250
11.	HIRING-2	.00000	50.000	53.125
12.	UNION-2	-62.768	.00000	UNBOUNDED
13.	P/DEM-3	4550.0	4700.0	7350.0
14.	W/CAP-3	-2650.0	.00000	150.00
15.	P/CAP-3	5821.4	7000.0	UNBOUNDED
16.	WORKER-3	-28.819	.00000	3.7500
17.	HIRING-3	21.181	50.000	53.750
18.	UNION-3	-67.768	.00000	UNBOUNDED
19.	P/DEM-4	5750.0	5900.0	8550.0
20.	W/CAP-4	-2650.0	.00000	150.00
21.	P/CAP-4	5821.4	7000.0	UNBOUNDED

22.	WORKER-4	-19.213	.00000	4.6875
23.	HIRING-4	.00000	50.000	UNBOUNDED
24.	UNION-4	-72.768	.00000	UNBOUNDED
25.	P/DEM-5	7050.0	7200.0	9350.0
26.	W/CAP-5	-2650.0	.00000	150.00
27.	P/CAP-5	5821.4	7000.0	UNBOUNDED
28.	WORKER-5	-14.410	.00000	6.2500
29.	HIRING-5	-.47684E-06	50.000	UNBOUNDED
30.	UNION-5	-72.768	.00000	UNBOUNDED
31.	P/DEM-6	7750.0	7900.0	10550.
32.	W/CAP-6	-2650.0	.00000	150.00
33.	P/CAP-6	5821.4	7000.0	UNBOUNDED
34.	WORKER-6	-11.528	.00000	9.3750
35.	HIRING-6	.00000	50.000	UNBOUNDED
36.	UNION-6	-72.768	.00000	UNBOUNDED
37.	P/DEM-7	6550.0	6700.0	9350.0
38.	W/CAP-7	-2650.0	.00000	150.00
39.	P/CAP-7	5821.4	7000.0	UNBOUNDED
40.	WORKER-7	-9.6065	.00000	18.750
41.	HIRING-7	.00000	50.000	UNBOUNDED
42.	UNION-7	-72.768	.00000	UNBOUNDED
43.	P/DEM-8	3833.3	5100.0	5297.6
44.	W/CAP-8	-197.62	.00000	1266.7
45.	P/CAP-8	5755.6	7000.0	UNBOUNDED
46.	WORKER-8	-8.2341	.00000	64.534
47.	HIRING-8	.00000	50.000	UNBOUNDED
48.	UNION-8	-64.534	.00000	UNBOUNDED
49.	P/DEM-9	4033.3	5300.0	5497.6
50.	W/CAP-9	-197.62	.00000	1266.7
51.	P/CAP-9	5755.6	7000.0	UNBOUNDED
52.	WORKER-9	-12.351	.00000	63.889
53.	HIRING-9	-.95367E-06	50.000	UNBOUNDED
54.	UNION-9	-71.944	.00000	UNBOUNDED
55.	P/DEM-10	5133.3	6400.0	6597.6
56.	W/CAP-10	-197.62	.00000	1266.7
57.	P/CAP-10	5755.6	7000.0	UNBOUNDED
58.	WKER-10	-24.702	.00000	31.944
59.	HIRES-10	.00000	50.000	UNBOUNDED
60.	UNION-10	-71.944	.00000	UNBOUNDED
61.	P/DEM-11	5333.3	5800.0	5997.6
62.	W/CAP-11	-197.62	.00000	466.67
63.	P/CAP-11	5333.3	7000.0	UNBOUNDED
64.	WKER-11	-52.778	.00000	19.167
65.	HIRES-11	-.47684E-06	50.000	UNBOUNDED
66.	UNION-11	-19.167	.00000	UNBOUNDED
67.	P/DEM-12	4693.8	4600.0	5085.0
68.	W/CAP-12	-285.00	.00000	106.15
69.	P/CAP-12	4800.0	7000.0	UNBOUNDED
70.	WKER-12	-35.625	.00000	13.269
71.	HIRES-12	.00000	50.000	UNBOUNDED
72.	UNION-12	-13.269	.00000	35.625
73.	BEG-INV	.00000	750.00	900.00

FIGURE 3-21. An aggregate production planning problem.

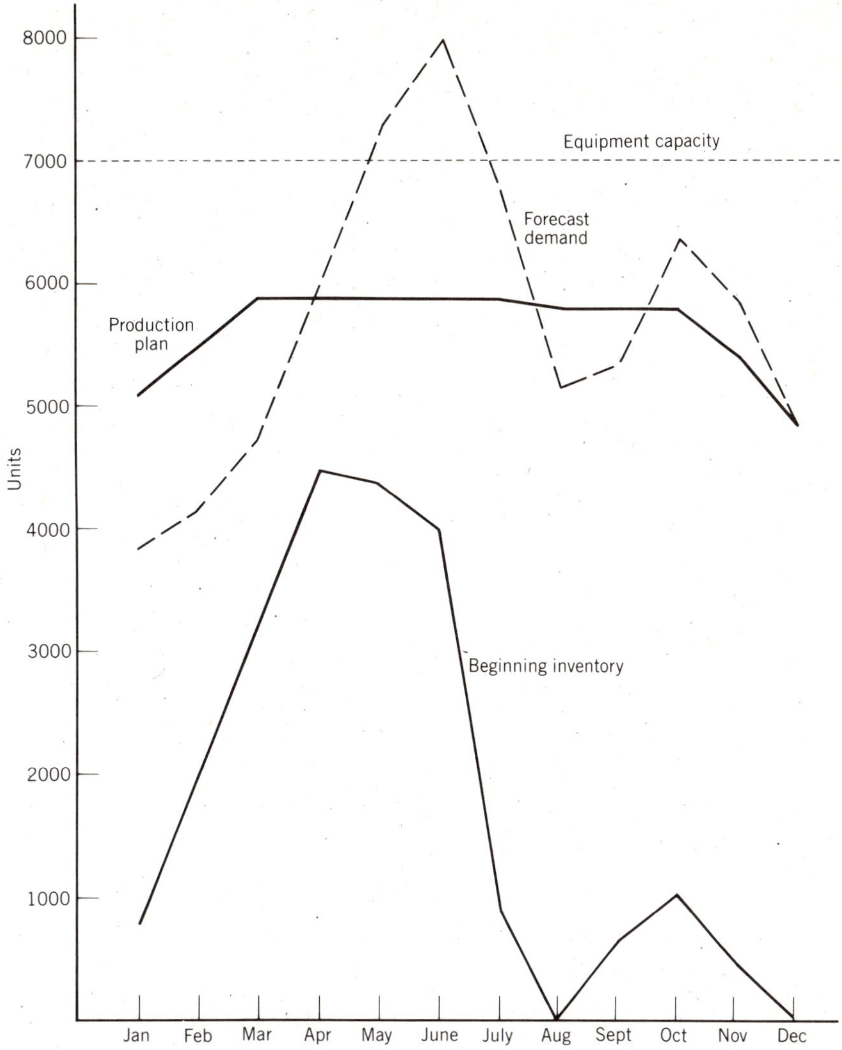

FIGURE 3-22. The optimal production plan.

Section *3* has other useful information. Many firms use *average* cost figures to make incremental decisions. Shadow prices are *marginal* costs that take into account the value of the resources consumed in production—not only the value of raw materials and labor, which remain constant from month to month, but also the impact on past and future constraints, an impact whose cost/value changes from month to month. The shadow price on the production/demand (P/DEM $-$ *i*) constraints represent, by definition, the change in objective function causes by a unit increase in the right-hand side—in other words,

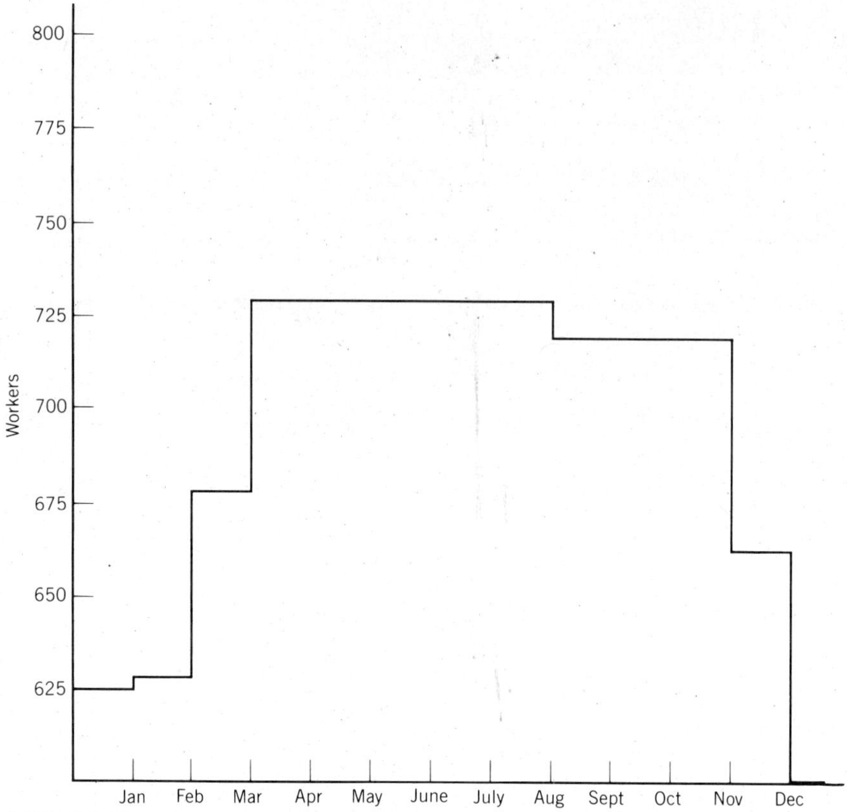

FIGURE 3-23. Optimal workforce levels.

the cost of producing an additional unit. Figure 3-24 shows how this marginal cost changes during the year, reaching a high of $335 per unit in July, and dropping to only $222 per unit in December. Note how the marginal cost curve follows the demand curve, indicating that the marginal cost of production is highest when the productive resources are most strained—that is, during periods of peak demand. Thus, if the sales manager were approached by a customer who would agree to take an additional 100 units (well within Section *6*'s ranges) at $300 apiece, the sale should be made (on economic grounds) only if he or she agrees to take the merchandise in January, August, or December.

The shadow prices on worker capacity (W/CAP − *i*) gives the "value of productivity" formally, the decrease in production cost possible if the *existing* workforce could produce one extra unit during the month. These shadow prices are simply $100 less than the shadow prices graphed in Figure 3-24. Again, as expected, productivity is more valuable when the productive resources are being strained. The CMP output allows us to put an exact dollar figure on those values.

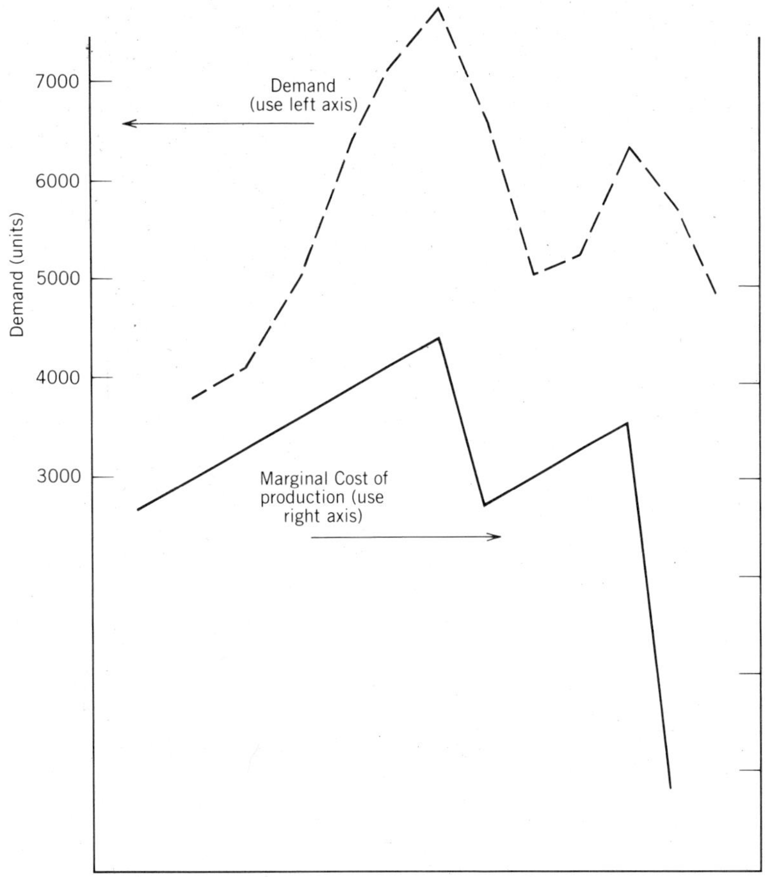

FIGURE 3-24. Marginal production cost.

EXERCISES

*1. A company produces and sells two kinds of liquor: blended whiskey and bourbon. The company purchases intermediate products in bulk, purifies them by repeated distillation, mixes them, and bottles the final product under their own brand names. In the past, the firm has always been able to sell all that it produced.

The firm has been limited by its machine capacity and available cash. The bourbon requires 3 machine hours per bottle while, due to additional blending requirements, the blended whiskey requires 4 hours of machine time per

*This problem was adapted by Stephen Bradley from the case *Rectified Liquors* written by V. L. Andrews.

bottle. There are 20,000 machine hours available in the current production period. The direct operating costs, which are principally for labor and materials, are \$3.00 per bottle of bourbon and \$2.00 per bottle of blended whiskey. The working capital available to finance labor and material is \$4000; however, 45% of the bourbon sales revenues and 30% of the blended-whiskey sales revenues from production in the current period will be collected during the current period and be available to finance operations. The selling price to the distributor is \$6 per bottle of bourbon and \$5.40 per bottle of blended whiskey.

- **a.** Formulate a linear program that maximizes contribution subject to limitations on machine capacity and working capital.
- **b.** What is the optimal production mix to schedule?
- **c.** Can the selling prices change without changing the optimal production mix?
- **d.** Suppose that the company could spend \$400 to repair some machinery and increase its available machine hours by 2000 hours. Should the investment be made?
- **e.** What interest rate could the company afford to pay to borrow funds to finance its operations during the current period?

2. Solve each of the following LPs graphically. In each, there is something unusual about the solution. What is it?

(a) Max $5x_1 - x_2$

Subject to

$$\begin{aligned} -x_1 + x_2 &\leqslant 4 \\ 4x_1 + x_2 &\leqslant 8 \\ x_2 &\geqslant 6 \\ x_1, x_2 &\geqslant 0 \end{aligned}$$

(b) Max $x_1 + 3x_2$

Subject to

$$\begin{aligned} -3x_1 + x_2 &\leqslant 2 \\ x_1 - x_2 &\geqslant 2 \\ x_1, x_2 &\geqslant 0 \end{aligned}$$

(c) Max $2x_1 + 1.5x_2$

Subject to

$$\begin{aligned} x_1 + 2x_2 &\leqslant 4 \\ 4x_1 + 3x_2 &\leqslant 12 \\ 2x_2 &\leqslant 3 \\ x_1, x_2 &\geqslant 0 \end{aligned}$$

Situation (c) is quite common for certain classes of real-world problems. For problems too large to solve graphically, how could you tell from the computer output (e.g., CMP output) that this situation was occurring? Why would it be important to be able to do so?

3. Consider a machine shop with lathes, drill presses, and milling machines. There are four products that can be made, each of which has a different contribution:

Product	*Contribution per unit*
1	$10.48
2	14.60
3	16.68
4	8.36

The availability of equipment depends on the available number of workers. The resultant number of hours available for the next month of each is as follows:

Machine	*Hours*
Lathes	1000
Drill presses	4000
Milling machines	2500

One unit of product 1 requires:

1.5 hours of lathe time
1 hour of drill press time
1.5 hours of mill time

For the second item, the hours are respectively 1, 5, and 3; for the third 2.4, 1, and 3.5; and for the fourth 1, 3.5, and 1.

The production foreman has formulated an LP to help him plan next month's production. The model and resultant output is shown as follows:

```
LINEAR: PRODUCTION MODEL


OBJECTIVES     PROD1        PROD2        PROD3        PROD4

  CONTRIB    10.48        14.6        16.68        8.36

CONSTRAINTS    PROD1        PROD2        PROD3        PROD4      RELATION     HOURS

  LATHES        1.5           1           2.4          1          LE        1000
  DRILLS          1           5             1        3.5          LE        4000
  MILLING       1.5           3           3.5          1          LE        2500



MAXIMIZE OR MINIMIZE? MAX

OPTIMAL SOLUTION FOUND.
      CONTRIB        12737.1


ALL ITEMS NOT LISTED IN SECTIONS 1 - 4 HAVE THE VALUE ZERO.

*1* DECISION VARIABLES
  1. PROD1       147.059
```

```
    2. PROD2          750.000
    4. PROD4          29.4118

*2* SLACK(+) AND SURPLUS(-) IN CONSTRAINTS

*3* SHADOW PRICES FOR CONSTRAINTS
  1. LATHES       3.90706
  2. DRILLS       .484706
  3. MILLING      2.75647

*4* REDUCED COSTS FOR DECISION VARIABLES
  3. PROD3        -2.82929

*5* RANGES ON COEFFICIENTS OF OBJECTIVE CONTRIB
     VARIABLE   LOWER BOUND  CURRENT VALUE   UPPER BOUND
  1. PROD1        9.0181         10.480         12.540
  2. PROD2        9.4558         14.600         22.414
  3. PROD3      UNBOUNDED        16.680         19.509
  4. PROD4        6.9867          8.3600        11.047

*6* RANGES ON VALUES OF RIGHT-HAND-SIDE HOURS
     CONSTRNT   LOWER BOUND  CURRENT VALUE   UPPER BOUND
  1.  LATHES       944.44        1000.0         2500.0
  2.  DRILLS       3916.7        4000.0         4625.0
  3.  MILLING      1666.7        2500.0         2538.5

OUTPUT OPTION? N
```

a. What is the optimal production mix? What contribution can the foreman anticipate if he produces that?

b. The foreman had naively supposed that the LP, mechanistic as it is, would support making a lot of product 3, the machine shop's highest contribution item. To his surprise, the optimal solution has no product 3 produced; indeed, some product 4, the shop's *lowest* contribution item, is to be produced instead. Why?

c. The foreman can have Personnel hire temporary machine-shop workers at $2.50 per hour more than they are usually paid. Should he? How many? Where should they be assigned?

d. Engineering has suggested a new product 5, which requires one half hour of lathe time, four hours of drill press time, and 2 hours of milling machine time. Marketing feels sure that this product can be sold at a price high enough to net the firm $10 in contribution. Should they produce any product 5?

e. The foreman has just been informed of a price increase in a raw material used for products 1 and 2. This increase will result in a decrease in contribution for both products, for 1 to $9.50; for 2, to $13.00. Will the optimal product mix change? Will the optimal level of contribution?

4. A producer of electronic test equipment has a factory that produces its most successful product line, which consists of three models: the Z-100, the Z-200 and the ZX-1000. These models are produced in two departments, fabrication in which equipment chasses are fabricated, largely from sheet steel, and assembly. The labor (in hours) required by each of the three models in each of the two departments is shown below:

	Z-100	Z-200	ZX-1000	Present Capacity
Fabrication	4	0	11.5	6000
Assembly	2.4	3.4	2.4	2000

The Z-200 does not go through fabrication, but is assembled into a purchased cabinet.

The firm receives $60 in contribution for each Z-100 it produces and sells, $60 for each Z-200 and $100 for each ZX-1000.

Demand is sufficiently strong that the firm can produce whatever it sells. However, due to a strike that has hit one of its suppliers, the factory is short of X37s, a key component in each of the three models. The Z-100 requires 4 X37s, and Z-200 requires 6, and the ZX-1000 requires 9. Currently, 4000 components are in stock.

To determine the optimal production mix, the plant manager has formulated the LP shown. The resultant CMP output is also given.

```
LINEAR: ELECTRONIC TEST EQUIPMENT

OBJECTIVES    Z-100       Z-200     ZX-1000

  CONTRIB        60          60        100

CONSTRAINTS   Z-100       Z-200     ZX-1000      RELATION    AVAIL

  FAB             4           0       11.5          LE        6000
  ASS'Y         2.4         3.4        2.4          LE        2000
  X37             4           6          9          LE        4000
MAXIMIZE

OPTIMAL SOLUTION FOUND.
     CONTRIB                 55333.3

ALL ITEMS NOT LISTED IN SECTIONS 1 - 4 HAVE THE VALUE ZERO.

*1* DECISION VARIABLES
  1. Z-100        700.000
  3. ZX-1000      133.333

*2* SLACK(+) AND SURPLUS(-) IN CONSTRAINTS
  1. +FAB         1666.67

*3* SHADOW PRICES FOR CONSTRAINTS
  2. ASS'Y        11.6667
  3. X37          8.00000
```

```
*4* REDUCED COSTS FOR DECISION VARIABLES
  2. Z-200      -27.6667

*5* RANGES ON COEFFICIENTS OF OBJECTIVE CONTRIB
      VARIABLE   LOWER BOUND   CURRENT VALUE   UPPER BOUND
  1. Z-100         44.444         60.000          100.00
  2. Z-200       UNBOUNDED        60.000          87.667
  3. ZX-1000       60.000         100.00          135.00

*6* RANGES ON VALUES OF RIGHT-HAND-SIDE AVAIL
      CONSTRNT   LOWER BOUND   CURRENT VALUE   UPPER BOUND
  1.   FAB         4333.3         6000.0        UNBOUNDED
  2.   ASS'Y       1066.7         2000.0          2400.0
  3.   X37         3333.3         4000.0          5111.1
```

a. What is the optimal product mix? What contribution should the firm expect by producing this mix?

b. A small local electronics contractor has offered to sell the factory X37s at $12 apiece—twice the normal price. Should the firm agree to the purchase? If so, how much additional contribution can the firm expect, if 1000 are purchased?

c. The sales manager insists that the loss of customer goodwill, which would accompany dropping Z-200s from the line, might be significant. She suggests producing at least 100 of this model. What effect would this have on the firm's contribution?

d. The manager of engineering disagrees with the sales manager's assessment. He has suggested redesigning the Z-200 to make it profitable. His staff has worked out a new design requiring 4 hours of fabrication, 1.2 hours of assembly, and 4 X37s. How much contribution would the firm need to earn per unit to make this redesigned Z-200 profitable?

e. The sales manager has seen a competitor's design for the new generation Z-300. It would require only one hour of fabrication, 3.6 hours of assembly and 3 X37s. She feels sure that it could be sold at a price which would earn the firm $70 in contribution. Should they produce it?

f. A neighboring manufacturer of medical test equipment, also strained by capacity, has offered to pay our firm $10 per hour if we will assemble some of their models. They will provide the necessary wiring components and chasses. Should the plant manager agree to this request?

5. A foundry that specializes in producing custom blended alloys has received an order for 1,000 pounds of an alloy containing at least 5% chromium and not more than 5% iron. Four types of scrap can be easily acquired that can be blended to produce the order. The cost and metal characteristics of the four scrap types are given below:

	Scrap Type 1	2	3	4
Chromium	7%	3.2%	—	8.2%
Iron	40%	80%	60%	32%
Cost/100 lb	$33	$22	$18	$49

The purchasing manager has formulated and run the LP shown below.

```
LINEAR: FOUNDRY'S BLENDING PROBLEM

OBJECTIVES   SCRAP-1    SCRAP-2    SCRAP-3    SCRAP-4

  COST          33         22         18         49

CONSTRAINTS  SCRAP-1    SCRAP-2    SCRAP-3    SCRAP-4    RELATION    REQ

  TOT-LBS        1          1          1          1         EQ        10
  CHROM          7        3.2          0        8.2         GE        50
  IRON          40         80         60         32         LE       500



OPTIMAL SOLUTION FOUND.
     COST                 282.941



ALL ITEMS NOT LISTED IN SECTIONS 1 - 4 HAVE THE VALUE ZERO.

*1* DECISION VARIABLES
  1. SCRAP-1    6.47059
  2. SCRAP-2    1.47059
  3. SCRAP-3    2.05882

*2* SLACK(+) AND SURPLUS(-) IN CONSTRAINTS

*3* SHADOW PRICES FOR CONSTRAINTS
  1. TOT-LBS    23.8824
  2. CHROM      1.86275
  3. IRON      -.980392E-01

*4* REDUCED COSTS FOR DECISION VARIABLES
  4. SCRAP-4    12.9804
```

```
*5* RANGES ON COEFFICIENTS OF OBJECTIVE COST
     VARIABLE    LOWER BOUND   CURRENT VALUE   UPPER BOUND
  1. SCRAP-1      26.750         33.000          43.442
  2. SCRAP-2      3.0000         22.000          24.857
  3. SCRAP-3      12.737         18.000          27.500
  4. SCRAP-4      36.020         49.000         UNBOUNDED

*6* RANGES ON VALUES OF RIGHT-HAND-SIDE REQ
     CONSTRNT    LOWER BOUND   CURRENT VALUE   UPPER BOUND
  1.  TOT-LBS     9.0278         10.000          10.714
  2.  CHROM       35.000         50.000          60.500
  3.  IRON        457.14         500.00          610.53
```

a. What is the optimal blend? How should they price the order?
b. Explain the right-hand sides of constraints 2 and 3.
c. How good a deal would the purchasing manager need to get on scrap type 4 before he would be willing to buy it for this order?
d. Upon further investigation, the purchasing manager finds that scrap type 1 is currently being sold at $40/lb, substantially more than he had originally thought. How should his purchase change? How should his pricing for this order change?
e. How much would the purchasing manager be willing to pay for pure chromium?
f. The customer has indicated that she might be willing to negotiate raising the ceiling on iron for a more favorable price. How should the sales manager plan this upcoming negotiation?

***6.** Jean-Pierre Levesque has recently been named the Minister of International Trade for the new nation of New France. In connection with this position, he has decided that the welfare of the country (and his performance) could best be served by maximizing the net dollar value of the country's exports for the coming year. (The net dollar value of exports is defined as exports *less* the cost of all materials imported by the country.)

The area that now constitutes New France has traditionally made three products for export: steel, heavy machinery, and trucks. For the coming year, Jean-Pierre feels that they could sell all that they could produce of these three items at existing world market prices of $900/unit for steel, $2500/unit for machinery, and $3000/unit for trucks.

In order to produce one unit of steel with the country's existing technology, it takes 0.05 units of machinery, 0.08 units of trucks, two units of ore purchased on the world market for $100/unit, and other imported materials costing $100. In addition, it takes .5 man-years of labor to produce each unit of steel. The steel mills of New France have a maximum usable capacity of 300,000 units/year.

*This problem was adapted by Stephen Bradley from the case *Land of Milk and Honey* written by Steven C. Wheelwright.

To produce one unit of machinery requires .75 units of steel, 0.12 units of trucks, and 5 man-years of labor. In addition, $150 of materials must be imported for each unit of machinery produced. The practical capacity of the country's machinery plants is 50,000 units/year.

In order to produce one unit of trucks, it takes one unit of steel, 0.10 units of machinery, three man-years of labor, and $500 worth of imported materials. Existing truck capacity is 550,000 units/year.

The total manpower available for production of steel, machinery, and trucks is 1,200,000 men/year.

To help Jean-Pierre in his planning, he had one of his staff formulate a linear program. The model and output are given below.

Variables:

X_1 = Steel production for export (EXSTEE),
X_2 = Machinery production for export (EXMACH),
X_3 = Truck production for export (EXTRUC),
X_4 = Total steel production (TOSTEE),
X_5 = Total machinery production (TOMACH),
X_6 = Total truck production (TOTRUC).

Constraints:

Steel output	(STEEL)
Machinery output	(MACHIN)
Truck output	(TRUCK)
Steel capacity	(CAPSTE)
Machinery capacity	(CAPMAC)
Truck capacity	(CAPTRU)
Manpower available	(AVAMAN)

```
TITLE: NEW FRANCE INTERNATIONAL TRADE
PROCEED, DISPLAY, OR REJECT? DISPLAY
```

OBJECTIVES:

	EXSTEE	EXMACH	EXTRUC	TOSTEE	TOMACH	TOTRUC
EXPORTS	900.0	2500.	3000.	-300.0	-150.0	-500.0

CONSTRAINTS:

	EXSTEE RELATION	EXMACH CAPACITY	EXTRUC	TOSTEE	TOMACH	TOTRUC
STEEL	-1.000	.0000	.0000	1.000	-.7500	-1.000
	EQ	.0000				
MACHIN	.0000	-1.000	.0000	-.5000E-01	1.000	-.1000
	EQ	.0000				
TRUCK	.0000	.0000	-1.000	-.8000E-01	-.1200	1.000
	EQ	.0000				
CAPSTE	.0000	.0000	.0000	1.000	.0000	.0000
	LE	.3000E+06				
CAPMAC	.0000	.0000	.0000	.0000	1.000	.0000
	LE	.5000E+05				
CAPTRU	.0000	.0000	.0000	.0000	.0000	1.000
	LE	.5500E+06				
AVAMAN	.0000	.0000	.0000	.5000	5.000	3.000
	LE	.1200E+07				

```
MAXIMIZE OR MINIMIZE? MAX

OPTIMAL SOLUTION FOUND.
     EXPORTS      0.490625E+09

OUTPUT OPTION? EXTENDED

ALL ITEMS NOT LISTED IN SECTIONS 1 - 4 HAVE THE VALUE ZERO.

*1* DECISION VARIABLES
  2. EXMACH       8750.00
  3. EXTRUC       232500.
  4. TOSTEE       300000.
  5. TOMACH       50000.0
  6. TOTRUC       262500.

*2* SLACK(+) AND SURPLUS(-) IN CONSTRAINTS
  6. +CAPTRU     287500.
  7. +AVAMAN     12500.0

*3* SHADOW PRICES FOR CONSTRAINTS
  1. STEEL     -2250.00
  2. MACHIN    -2500.00
  3. TRUCK     -3000.00
  4. CAPSTE     1585.00
  5. CAPMAC     302.500

*4* REDUCED COSTS FOR DECISION VARIABLES
  1. EXSTEE    -1350.00

*5* RANGES ON COEFFICIENTS OF OBJECTIVE EXPORTS
     VARIABLE  LOWER BOUND  CURRENT VALUE  UPPER BOUND
  1. EXSTEE    UNBOUNDED       900.00         2250.0
  2. EXMACH      2218.6        2500.0         13067.
  3. EXTRUC      1650.0        3000.0         3347.7
  4. TOSTEE     -1885.0       -300.00        UNBOUNDED
  5. TOMACH     -452.50       -150.00        UNBOUNDED
  6. TOTRUC     -1850.0       -500.00        -96.667

*6* RANGES ON VALUES OF RIGHT-HAND-SIDE CAPACITY
     CONSTRNT  LOWER BOUND  CURRENT VALUE  UPPER BOUND
  1.   STEEL      -4166.7       .00000        .23250E+06
  2.   MACHIN     UNBOUNDED     .00000        8750.0
  3.   TRUCK      UNBOUNDED     .00000        .23250E+06
  4.   CAPSTE      47283.       .30000E+06    .30357E+06
  5.   CAPMAC      41860.       50000.        54545.
  6.   CAPTRU     .26250E+06    .55000E+06   UNBOUNDED
  7.   AVAMAN     .11875E+07    .12000E+07   UNBOUNDED
```

Levesque has asked you to help him with the following questions:

a. What is the optimal production and export mix for New France? What would be the net dollar value of exports under this plan?
b. What do the first three constraint equations (STEEL, MACHIN, and TRUCK) represent? Why are they equality constraints?
c. The optimal solution suggests that New France produce 50,000 units of machinery. How are those units to be utilized during the year?
d. What would happen to the value of net exports if the world market price of steel increased to $1225/unit and the country chose to export one unit of steel?
e. If New France wants to identify other products it can profitably produce and export, what characteristics should those products have?
f. There is a chance that Jean-Pierre may have $500,000 to spend on expanding capacity. If this investment will buy 500 units of truck capacity, 1000 units of machinery capacity, or 300 units of steel capacity, what would be the best investment?

g. If the world market price of the imported materials needed to produce one unit of trucks were to increase by $400, what would be the optimal export mix for New France, and what would be the dollar value of their net exports?

h. The Minister of Defense has recently come to Jean-Pierre and said that he would like to stockpile (inventory) an additional 10,000 units of steel during the coming year. How will this change the constraint equation STEEL, and what impact will it have on net dollar exports?

i. A government R&D group has recently come to Jean-Pierre with a new product. Product *X*, that can be produced for export with 1.5 man-years of labor and 0.3 units of machinery for each unit produced. What must Product *X* sell for on the world market to make it attractive for production?

j. How does this particular formulation deal with existing inventories at the start of the year and any desired inventories at the end of the year?

k. With existing uncertainties in international currencies, there is some chance that the New France dollar will be revalued upward relative to world markets. If this happened, the cost of imported materials would go down by the same percent as the devaluation, and the market price of the country's exports would go up by the same percent as the revaluation. Assuming that the country can always sell all it wishes to export, how much of a revaluation could occur before the optimal solution would change?

*7. Shortly after the beginning of the oil embargo, the Eastern District Director of the Government Office of Fuel Allocation was concerned that he would soon have to start issuing monthly allocations specifying the amounts of heating oil that each refinery in the district would send to each city to ensure that every city would receive its quota.

Since different refineries used a variety of alternative sources of crude oil, both foreign and domestic, the cost of products at each refinery varied considerably. Consequently, under current government regulations, the prices that could be charged for heating oil would vary from one refinery to another. To avoid political criticism, it was felt that, in making allocations of supplies from the refineries to the cities, it was essential to maintain a reasonably uniform average refinery price for each city. In fact, the Director felt that the average refinery price paid by any city should not be more than 3% above the overall average.

Another complication had arisen in recent months, since some of the emergency allocations of supplies to certain cities could not be delivered due to a shortage in tanker capacity. If deliveries of the allocated supplies were to be feasible, the limited transportation facilities on some of the shipping routes would have to be recognized. Finally, it would be most desirable to maintain

*This problem was adapted by Stephen Bradley from the case *Holden Consulting Co.* written by Basil A. Kalymon.

distribution costs as low as possible under any allocation plan, if charges of government inefficiency were to be avoided.

Data for a simplified problem with three refineries and four cities are given below.

Refinery	December Availability (barrels)	Price at Refinery ($ per barrel)
A. Norton	95,000	10.53
B. Chatham	63,000	9.39
C. Eastport	116,000	12.43
Average Refinery Price $11.07		

City	December Quotas (barrels)
1. Westville	55,000
2. Bridgeton	73,000
3. Brookfield	105,000
4. Dorval	38,000

	Shipping Costs ($ per barrel) City			
Refinery	1	2	3	4
A	0.10	0.16	0.32	0.28
B	0.20	0.34	0.30	0.12
C	0.34	0.38	0.22	0.18

	Capacities (thousands of barrels per month) City			
Refinery	1	2	3	4
A	*	25	*	*
B	*	20	20	*
C	25	*	*	*

*No effective limit.

The allocation problem can be formulated as a linear program with decision variables:

$A - n$ = Barrels of heating oil (in thousands) to be supplied from Refinery A to City n, where $n = 1, 2, 3$, or 4.

$B - n$ = Barrels of heating oil (in thousands) to be supplied from Refinery B to City n, where $n = 1, 2, 3$, or 4.

$C - n$ = Barrels of heating oil (in thousands) to be supplied by Refinery C to City n, where $n = 1, 2, 3$, or 4.

PMAX = A value higher than the average refinery price paid by any of the four cities.

There are four types of constraints that must be satisfied:

1. Supply constraints at each refinery.
2. Quota restrictions for each city.
3. Average price restrictions which impose an upper limit to the average refinery price that will be paid by any city.
4. Shipping capacity limits on some of the routes.

All the constraints are straightforward except possibly the average price reductions. To restrict the average refinery price paid by City 1 to be less than the variable PMAX, the following expression is used:

$$\frac{(10.53)(A-1) + (9.39)(B-1) + (12.43)(C-1)}{55} \leqq \text{PMAX}$$

or

$$0.1915(A-1) + 0.1701(B-1) + 0.2260(C-1) - \text{PMAX} \leqq 0$$

Such a restriction is included for each city, to provide a uniform upper bound on average prices. The value of PMAX is limited by the prescribed maximum of $11.40 = 11.07 \times 1.03$ by the constraint

$$\text{PMAX} \leqq 11.40$$

The tableau and the resultant CMP output is given below:

	$A-1$	$A-2$	$A-3$	$A-4$	$B-1$	$B-2$	$B-3$	$B-4$	$C-1$	$C-2$	$C-3$	$C-4$	PMAX	Relation	Right-hand side	
REF-A	1.0000	1.0000	1.0000	1.0000										≦	95.0000	Supply Constraints
REF-B					1.0000	1.0000	1.0000	1.0000						≦	63.0000	
REF-C									1.0000	1.0000	1.0000	1.0000		≦	116.0000	
CIT-1	1.0000				1.0000				1.0000					=	55.0000	Quota Restrictions
CIT-2		1.0000				1.0000				1.0000				=	73.0000	
CIT-3			1.0000				1.0000				1.0000			=	105.0000	
CIT-4				1.0000				1.0000				1.0000		=	38.0000	
AP-CIT-1	0.1915				1.701				0.2260				−1.0000	≦	0.0000	Average Price Restrictions
AP-CIT-2		0.1442				0.1286				0.1703			−1.0000	≦	0.0000	
AP-CIT-3			0.1003				0.0894				0.1184		−1.0000	≦	0.0000	
AP-CIT-4				0.2771				0.2471				0.3271	−1.0000	≦	0.0000	
PRICELIM													1.0000	≦	11.4000	
ROUTE-A2		1.0000												≦	25.0000	Shipping Capacity Limits
ROUTE-B2						1.0000								≦	20.0000	
ROUTE-B3							1.0000							≦	20.0000	
ROUTE-C1									1.0000					≦	25.0000	
Shipcost	0.1000	0.1600	0.3200	0.2800	0.2000	0.3400	0.3000	0.1200	0.3400	0.3800	0.2200	0.1800	0.0000	=	*(min) Objective*	

```
MAXIMIZE OR MINIMIZE? MIN

OPTIMAL SOLUTION FOUND.
        SHIPCOST       61.6563

OUTPUT OPTION? USUAL

ALL ITEMS NOT LISTED IN SECTIONS 1 - 4 HAVE THE VALUE ZERO.

*1* DECISION VARIABLES
  1. A-1          54.4253
  2. A-2          15.6023
  3. A-3          24.9724
  5. B-1          .574694
  6. B-2          14.9803
  7. B-3          20.0000
  8. B-4          27.4450
 10. C-2          42.4174
 11. C-3          60.0276
```

```
 12. C-4           10.5550
 13. PMAX          11.4000

*2* SLACK(+) AND SURPLUS(-) IN CONSTRAINTS
  3. +REF-C       3.00000
 11. +AP-CIT-4 1.16580
 13. +ROUTE-A2 9.39768
 14. +ROUTE-B2 5.01967
 16. +ROUTE-C1 25.0000

*3* SHADOW PRICES FOR CONSTRAINTS
  1. REF-A      -.232518
  2. REF-B      -.600000E-01
  4. CIT-1       .341718
  5. CIT-2       .461679
  6. CIT-3      2.39515
  7. CIT-4       .180000
  8. AP-CIT-1 -.480411E-01
  9. AP-CIT-2 -.479616
 10. AP-CIT-3 -18.3712
 12. PRICELIM -18.8988
 15. ROUTE-B3 -.392764

*4* REDUCED COSTS FOR DECISION VARIABLES
  4. A-4         .332518
  9. C-1         .913941E-02

*5* RANGES ON COEFFICIENTS OF OBJECTIVE SHIPCOST
     VARIABLE  LOWER BOUND  CURRENT VALUE  UPPER BOUND
  1. A-1        .27482E-01     .10000       .10935
  2. A-2        .15065         .16000       .23252
  3. A-3        .74861E-01     .32000       .13630E+06
  4. A-4       -.52518E-01     .28000      UNBOUNDED
  5. B-1       UNBOUNDED       .20000       .27252
  6. B-2        .32000         .34000       .35494
  7. B-3       UNBOUNDED       .30000       .69276
  8. B-4        .10559         .12000       .14000
  9. C-1        .33086         .34000      UNBOUNDED
 10. C-2        .18615         .38000       .40000
 11. C-3       -.13630E+06     .22000       .55252
 12. C-4        .16000         .18000       .19441
 13. PMAX      UNBOUNDED       .00000      18.899

*6* RANGES ON VALUES OF RIGHT-HAND-SIDE RHS
     CONSTRNT  LOWER BOUND  CURRENT VALUE  UPPER BOUND
  1.  REF-A       92.000        95.000        104.40
  2.  REF-B       60.000        63.000        73.555
  3.  REF-C       113.00        116.00       UNBOUNDED
  4.  CIT-1       46.660        55.000        58.000
  5.  CIT-2       70.415        73.000        74.229
  6.  CIT-3       103.56        105.00        106.23
  7.  CIT-4       27.445        38.000        41.000
  8.  AP-CIT-1  -.86750         .00000        14.196
  9.  AP-CIT-2  -.20932         .00000        .44014
 10.  AP-CIT-3  -.14516         .00000        .17010
 11.  AP-CIT-4  -1.1658         .00000       UNBOUNDED
 12.  PRICELIM    11.315        11.400        11.568
 13.  ROUTE-A2    15.602        25.000       UNBOUNDED
 14.  ROUTE-B2    14.980        20.000       UNBOUNDED
 15.  ROUTE-B3    14.994        20.000        25.865
 16.  ROUTE-C1  -.23842E-06     25.000       UNBOUNDED

OUTPUT OPTION? NO
```

a. Determine the detailed allocations that might be used for December.

b. Evaluate the average refinery prices paid by each city under these allocations.

c. Determine the best way to utilize an additional tanker-truck capacity of 10,000 barrels per month.

d. Discuss the inefficiency in distribution costs resulting from the average refinery price restrictions.

e. Evaluate the applicability of the model in the context of a full allocation system.

f. Discuss any general insights into the fuel allocation problem provided by this model.

Red Brand Canners (A)

On Monday, September 13, 1983, Mr. Mitchell Gordon, Vice-President of Operations, asked the Controller, the Sales Manager, and the Production Manager to meet with him to discuss the amount of tomato products to pack that season. The tomato crop, which had been purchased at planting, was beginning to arrive at the cannery, and packing operations would have to be started by the following Monday. Red Brand Canners was a medium-sized company that canned and distributed a variety of fruit and vegetable products under private brands in the western states.

Mr. William Cooper, the Controller, and Mr. Charles Myers, the Sales Manager, were the first to arrive in Mr. Gordon's office. Dan Tucker, the Production Manager, came in a few minutes later and said that he had picked up Produce Inspection's latest estimate of the quality of the incoming tomatoes. According to their report, about 20% of the crop was Grade "A" quality and the remaining portion of the 3,000,000-pound crop was Grade "B."

Gordon asked Myers about the demand for tomato products for the coming year. Myers replied that they could sell all of the whole canned tomatoes they could produce. The expected demand for tomato juice and tomato paste, on the other hand, was limited. The Sales Manager then passed around the latest demand forecast, which is shown in Exhibit 1. He reminded the group that the selling prices had been set in light of the long-term marketing strategy of the company, and potential sales had been forecasted at these prices.

EXHIBIT 1
Demand Forecasts

Product	Selling Price per Case	Demand Forecast (Cases)
24 2½ whole tomatoes	$4.00	800,000
24 2½ choice peach halves	5.40	10,000
24 2½ peach nectar	4.60	5,000
24 2½ tomato juice	4.50	50,000
24 2½ cooking apples	4.90	15,000
24 2½ tomato paste	3.80	80,000

Bill Cooper, after looking at Myer's estimates of demand, said that it looked like the company "should do quite well (on the tomato crop) this year." With the new accounting system that had been set up, he had been able to compute the contribution for each product, and according to his analysis the incremental profit on the whole tomatoes was greater than for any other tomato product. In May, after Red Brand had signed contracts agreeing to purchase the grower's production at an average delivered price of 6 cents per pound, Cooper had computed the tomato products' contributions (see Exhibit 2.)

Dan Tucker brought to Cooper's attention that, although there was ample production capacity, it was impossible to produce all whole tomatoes as too small a portion of the tomato crop was "A" quality. Red Brand used a numerical scale to record the quality of both raw produce and prepared products. This scale ran from zero to ten, the higher number representing better quality. Rating tomatoes according to this scale, "A" tomatoes averaged nine points per pound and "B" tomatoes averaged five points per pound. Tucker noted that the minimum average input quality for canned whole tomatoes was eight and for juice it was six points per pound. Paste could be made entirely from "B" grade tomatoes. This meant that whole tomato production was limited to 800,000 pounds.

EXHIBIT 2
Product Item Profitability

Product	24 2½ Whole Tomatoes	24 2½ Choice Peach Halves	24 2½ Peach Nectar	24 2½ Tomato Juice	24 2½ Cooking Apples	24 2½ Tomato Paste
Selling Price	$4.00	$5.40	$4.60	$4.50	$4.90	$3.80
Variable Costs:						
Direct Labor	1.18	1.40	1.27	1.32	.70	.54
Variable OHD	.24	.32	.23	.36	.22	.26
Variable Selling	.40	.30	.40	.85	.28	.38
Packaging Mat'l	.70	.56	.60	.65	.70	.77
Fruit[a]	1.08	1.80	1.70	1.20	.90	1.50
Total Variable Costs	3.60	4.38	4.20	4.38	2.80	3.45
Contribution	.40	1.02	.40	.12	1.10	.35
Less Allocated OHD	.28	.70	.52	.21	.75	.23
Net Profit	.12	.32	(.12)	(.09)	.35	.12

[a]Product usage is given as follows.

Product	*Pounds per case*
Whole Tomatoes	18
Peach Halves	18
Peach Nectar	17
Tomato Juice	20
Cooking Apples	27
Tomato Paste	25

Gordon stated that this was not a real limitation. He has been recently solicited to purchase 80,000 pounds of Grade "A" tomatoes at 8½ cents per pound and at that time had turned down the offer. He felt, however, that the tomatoes were still available.

Myers, who had been doing some calculations, said that although he agreed that the company "should do quite well this year," it would not be by canning whole tomatoes. It seemed to him that the tomato cost should be allocated on the basis of quality and quantity rather than by quantity only as Cooper had done. Therefore, he had recomputed the marginal profit on this basis (see Exhibit 3), and from his results, Red Brand should use 2,000,000 pounds of the "B" tomatoes for paste, and the remaining 400,000 pounds of "B" tomatoes and all of the "A" tomatoes for juice. If the demand expectations were realized, a contribution of $48,000 would be made on this year's tomato crop.

EXHIBIT 3

Marginal Analysis of Tomato Products

Z = Cost per pound of A tomatoes in cents

Y = Cost per pound of B tomatoes in cents

1. $(600{,}000 \text{ lb} \times Z) + (2{,}400{,}000 \text{ lb} \times Y) = (3{,}000{,}000 \text{ lb} \times 6)$

2. $\dfrac{Z}{9} = \dfrac{Y}{5}$

$Z = 9.32$ cents per pound

$Y = 5.18$ cents per pound

Product	Canned Whole Tomatoes	Tomato Juice	Tomato Paste
Selling Price	$4.00	$4.50	$3.80
Variable Cost (excl. tomato costs)	2.52	3.18	1.95
	$1.48	$1.32	$1.85
Tomato Cost	1.49	1.24	1.30
Marginal Profit	($ .01)	$ .08	$ .55

Delta Oil Refinery

On January 15, 1979, Mr. Sam Watson, head of the Operations Analysis Division of Delta Oil Company, was preparing a production plan for the company's refinery in Louisiana.

BACKGROUND

Petroleum refining is one of the largest manufacturing industries in the United States. In 1979, there were over 300 refineries in the U. S. processing a total of more than 18 million barrels of crude oil a day. Crude oil is a complex mixture of chemical compounds. The refining process separates crude oil into components that eventually yield gasoline, fuel oil, asphalt, jet fuel, lubricating oil, and many other petroleum products. Refineries strive to operate at peak economic efficiency, taking due account of the demand for these products.

At the end of 1978, Mr. Chauncey Andrews, Vice President of Operations at Delta Oil Company, had examined the gasoline production process at the Louisiana refinery. He expressed some concern about the manner in which gasoline was separated from crude oil and about the allocation of end products between regular and premium gasoline. Andrews asked Sam Watson to study the situation further and formulate a plan for gasoline production.

THE REFINING PROCESS

A highly simplified representation of the gasoline refining process at Delta's refinery appears in Figure 1. Gasoline is produced from crude oil by either a distillation process alone or by a distillation process followed by a catalytic cracking process. The outputs of these processes are blended to obtain different grades of gasoline.

The distillation process at Delta's refinery separates gasoline from other components by heating crude oil under pressure until the gasoline vaporizes. The vapors are then collected and cooled in a condenser to produce distillate. Delta obtains crude oil at a cost of $12.00 a barrel. The distillation tower uses 5.2 barrels of crude oil to produce one barrel of distillate and 4.2 barrels of other petroleum by-products. Delta currently sells these other products for $11.80 a barrel. Delta's distillation tower can produce up to 60,000 barrels of distillate per day at an operating cost of $3.90 per barrel of distillate output. Some distillate will be blended into gasoline products; some will become "feedstock" for the catalytic cracker.[1]

The catalytic cracking process utilizes high temperatures to break (or "crack") heavy hydrocarbon compounds into lighter compounds. This process produces high-quality gasoline stock from the feedstock. Delta's catalytic cracker requires 2.2 barrels of distillate to produce one barrel of gasoline stock and 1.2 barrels of petroleum by-products. These by-products currently sell for $19.48 a barrel. The catalytic cracker can produce up to 20,000 barrels of

[1]Input to the catalytic cracker will be referred to as either "distillate" or "feedstock"; output from the catalytic cracker is called "gasoline stock" or, alternatively, "cracker stock."

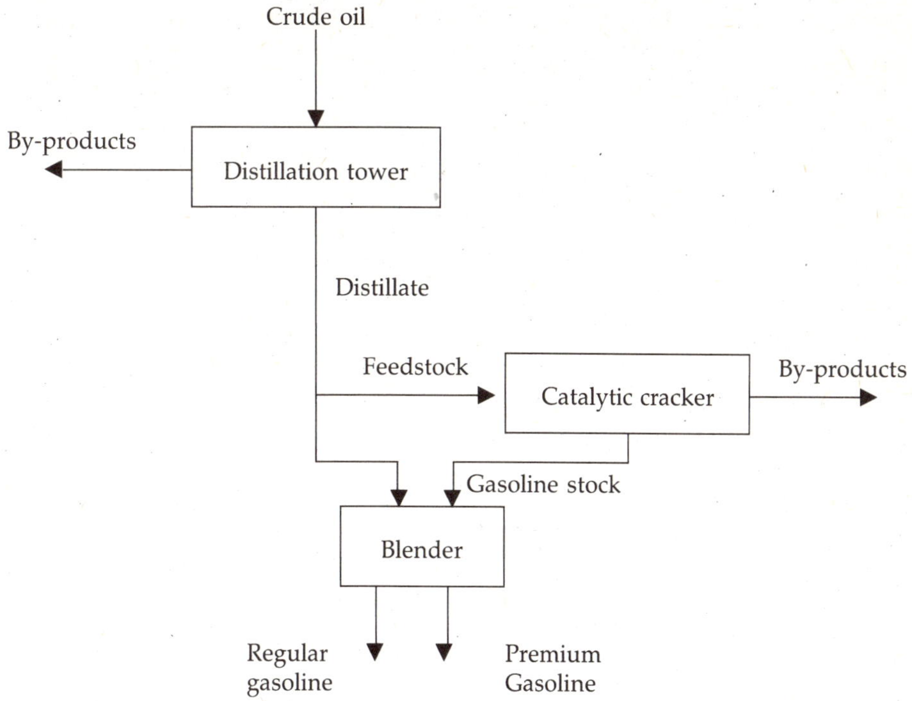

FIGURE 1. Gasoline Refining Process (simplified)

gasoline stock per day at an operating cost of $4.60 per barrel output. The total variable costs of distillate and cracker stock are calculated in Exhibit 1.

The octane rating of a gasoline measures its ability to burn smoothly without exploding. The distillate produced by Delta's distillation tower has an octane rating of 84, while the gasoline stock from the catalytic cracker has an octane rating of 94. Delta's gasoline refining process ends by blending distillate and cracker stock to form regular and premium gasoline, which have required octane ratings of at least 86 and 92, respectively. There is no volume or octane loss in blending, the variable cost of blending is negligible. For example, a barrel of gasoline made by blending a half barrel of distillate with a half barrel of cracker stock has an octane rating of 89 (that is, $\frac{1}{2} \times 84 + \frac{1}{2} \times 94$). Delta sells its regular gasoline for $20.10 per barrel and its premium gasoline for $22.90 per barrel.

The final gasoline products are pumped from the Delta Oil refinery in Louisiana via two small pipelines to several storage and distribution facilities in the eastern United States. One pipeline carries only regular gasoline, the other carries only premium gasoline. Each pipeline can handle up to 25,000 barrels a day. The market has recently been such that Delta could sell as much of either gasoline as its pipelines allowed, and Watson expected this situation to continue in the near future.

EXHIBIT 1
Variable Costs of Inputs to Blending Process

One Barrel Distillate (octane rating 84)

5.2 barrels of crude oil @ \$12.00	\$62.40
Operation of distillation tower	3.90
	\$66.30
Less 4.2 barrels of by-products @ \$11.80	49.56
	\$16.74

One Barrel Gasoline Stock from Catalytic Cracker (octane rating 94)

2.2 barrels of distillate @ \$16.74	\$36.83
Operation of catalytic cracker	4.60
	\$41.43
Less 1.2 barrels of by-products @ \$19.48	23.37
	\$18.06

As a first attempt to determine a production plan that would allocate distillate and cracker stock to the two grades of gasoline in a way which would maximize contribution, Watson formulated the linear program shown in Exhibit 2. The solution output is shown in Exhibit 3.

EXHIBIT 2
Watson's Linear Programming Model

Decision Variables

DIS-REG: Number of thousand barrels of *dis*tillate used to make *reg*ular gasoline.
DIS-PRE: Number of thousand barrels of *dis*tillate used to make *pre*mium gasoline.
CAT-REG: Number of thousand barrels of gasoline stock from the *cat*alytic cracker used to make *reg*ular gasoline.
CAT-PRE: Number of thousand barrels of gasoline stock from the *cat*alytic cracker used to make *pre*mium gasoline.

	DIS-REG	DIS-PRE	CAT-REG	CAT-PRE
CONTRIB	3.360	6.160	2.040	4.840

CONSTRAINTS:

	DIS-REG	DIS-PRE	CAT-REG	CAT-PRE	RELATION	RHS
DIST-CAP	1.000	1.000	2.200	2.200	LE	60.00
CATC-CAP	.0000	.0000	1.000	1.000	LE	20.00
REG-PIPE	1.000	.0000	1.000	.0000	LE	25.00
PRE-PIPE	.0000	1.000	.0000	1.000	LE	25.00
REG-OCT	-2.000	.0000	8.000	.0000	GE	.0000
PREM-OCT	.0000	-8.000	.0000	2.000	GE	.0000

EXHIBIT 3

```
MAXIMIZE OR MINIMIZE? MAX

OPTIMAL SOLUTION FOUND.
     CONTRIB       154.293

OUTPUT OPTION? E

ALL ITEMS NOT LISTED IN SECTIONS 1 - 4 HAVE THE VALUE ZERO.

*1* DECISION VARIABLES
  1.  DIS-REG    11.7333
  2.  DIS-PRE    4.26667
  3.  CAT-REG    2.93333
  4.  CAT-PRE    17.0667

*2* SLACK(+) AND SURPLUS(-) IN CONSTRAINTS
  3.  +REG-PIPE 10.3333
  4.  +PRE-PIPE 3.66667

*3* SHADOW PRICES FOR CONSTRAINTS
  1.  DIST-CAP   2.42667
  2.  CATC-CAP   .434667
  5.  REG-OCT   -.466667
  6.  PREM-OCT  -.466667

*4* REDUCED COSTS FOR DECISION VARIABLES

*5* RANGES ON COEFFICIENTS OF OBJECTIVE CONTRIB
     VARIABLE    LOWER BOUND  CURRENT VALUE  UPPER BOUND
  1. DIS-REG      1.0850        3.3600         3.5263
  2. DIS-PRE      5.1084        6.1600         42.560
  3. CAT-REG     -7.0600        2.0400         2.7053
  4. CAT-PRE      4.5771        4.8400         13.940

*6* RANGES ON VALUES OF RIGHT-HAND-SIDE RHS
     CONSTRNT    LOWER BOUND  CURRENT VALUE  UPPER BOUND
  1.   DIST-CAP    49.000       60.000         67.750
  2.   CATC-CAP    16.837       20.000         21.774
  3.   REG-PIPE    14.667       25.000         UNBOUNDED
  4.   PRE-PIPE    21.333       25.000         UNBOUNDED
  5.   REG-OCT    -22.000       .00000         62.000
  6.   PREM-OCT   -22.000       .00000         32.000

OUTPUT OPTION? N
```

Krebs Wire Company

The Krebs Wire Company was founded in 1931 primarily to capitalize on the telephone company's expanding need for high quality color-coded wire, which was growing because of increasing demand for telephone services.

Source: This case was prepared by Ronald S. Frank. Harvard Business School Case 9-174-030.

Since that time the Krebs Wire Company has produced a variety of wire coatings, other wire products, and unrelated molded plastic components. In 1982 a sizable portion of their business remains in specially coated wire. Krebs Wire has only one production facility for coated wire located in eastern Vermont, and has sales over much of the northeastern United States.

Krebs Wire is an intermediate processor in that it purchases uncoated wire in standard gauges and then applies the various coatings according to its customers' specifications. Krebs Wire has essentially two basic products—standard inexpensive plastic and the higher quality teflon. The two coatings come in a variety of colors but these are changed easily by introducing different dyes into the basic coating liquid.

The production process at Krebs Wire is a very modern and highly automated one. The uncoated wire arrives in large reels that are put on spindles at one end of the plant. The wire is continuously drawn off each successive reel over traverse guides and through a coating bath containing either the standard plastic or the quality teflon. The wire is then pulled through an extruder where it is smoothed to uniform thickness. Finally, it is hardened by a sequence of four electric drying ovens. The wire is then reeled up on spools similar to those it arrived on. Different dyes are added to the coating liquid during the process to produce the various colored wires ordered.

The production facilities at Krebs Wire consists of basically two independent wire trains, one engineered by the Kolbert Engineering Corporation and the other purchased secondhand from the Loomis Wire Company. Both the standard plastic and the quality teflon types of wire can be produced on either process train. Production of the quality teflon is a slower process because it takes a longer time to dry. The different production rates in tons per day are given below:

Process Train	*Plastic*	*Quality-Teflon*
Kolbert	40 tons/day	35 tons/day
Loomis	50 tons/day	42 tons/day

It has been the traditional view of the management at Krebs Wire to view production rates in terms of daily tonnage as opposed to reels per day or other production measures. The respective contributions in dollars per day are:

	July and August	
Process	*Plastic*	*Quality-Teflon*
Kolbert	$525/day	$546/day
Loomis	$580/day	$590/day

The purchasing agent from one of the customers of Krebs Wire has recently requested a special run of 20 tons of corrosion-resistant wire to be delivered at the end of August. Krebs has made this product before and has found that it can be produced only on the Loomis machine due to technical restrictions. The Loomis machine can produce this special wire at a rate of 40 tons per day. It seems that the purchasing agent will pay about $12 per ton although this is still open to negotiation. After carefully checking the availability of raw materials for the special coating, Krebs discovered to their dismay that they were stocked out and could not replenish their stock before the 1st of August.

Planning at Krebs Wire is usually done on a monthly basis. However, since most employee vacations are scheduled over the two summer months, the management feels that production planning for the two summer months should be combined so as to facilitate vacation scheduling. Each month the process trains must be shut down for scheduled maintenance, so that the total days available for production per month are as follows:

Process Train	*July*	*August*
Kolbert	26 days	26 days
Loomis	28 days	27 days

The scheduling process is further complicated by the fact that over the two summer months the total number of full days available for production is limited to 102 machine days due to vacation schedules. Fortunately the labor force at Krebs Wire is trained to operate either of the two trains and there is some possibility that some employees might be persuaded to take their vacations in June or September.

The amounts of wire that the management feels it can sell in the coming two months are:

Product	*July*	*August*
Plastic coated	1200 tons	1400 tons
Quality-teflon	800 tons	900 tons

Both types of wire may be stored for future delivery. Space is available in Krebs Wire's own warehouse which has a capacity of 20 tons. The inventory and carrying costs in dollars per ton for wire produced in July and delivered in August are:

Product	*Inventory and Carrying Costs*
Plastic coated	$1.00/ton
Quality-teflon	$1.20/ton

Due to a commitment with respect to the warehouse capacity for other products in September, it is not possible to stock any color coded wire in inventory at the end of August. There exists a possibility that additional warehouse space could be leased at a cost of $2.00 per ton for either wire above their respective inventory and carrying costs.

To help production planning for the two summer months, Krebs Wire management has called in a consultant from a well known West Coast business school. He has formulated and solved a linear program (see Exhibits 1 and 2) in order to determine the production schedule that will maximize contribution from these products.

After studying the production process in general, the consultant suggested that there was a 75% probability that he could increase the number of days available on each production train by one-quarter of a day each month. He felt that this could be achieved by increasing the efficiency of the maintenance procedures currently employed. He noted that this would require studying the maintenance procedure for which he would charge a fee of $150. The first open day in his schedule for performing this study would be July 29.

Management reaction to the results of the linear program was generally favorable. However, the managers of the two functional areas involved, production and sales, each had questions to which they hoped the consultant would respond. The production manager was not quite sure how the output from the linear program in Exhibit 2 related to the specific details of scheduling his machines and switching around employee vacations. The sales manager was visibly concerned by the fact that demand for plastic in the month of August was not being fully satisfied. Because of this, he felt that future sales might be affected.

EXHIBIT 1

A Formulation of the Problem as a Linear Program. Objective: Maximize Contribution

Coefficients for Contribution												
July						August						
525 K-P	580 L-P	546 K-T	590 L-T	−1.0 INV-P	−1.2 INV-T	525 K-P	580 L-P	546 K-T	590 L-T	Relation	Limit	Constraint Name
1	0	1	0	0	0	0	0	0	0	LE	26	K-Days-July
0	1	0	1	0	0	0	0	0	0	LE	28	L-Days-July
40	50	0	0	−1	0	0	0	0	0	LE	1200	P-DEM-July
0	0	35	42	0	−1	0	0	0	0	LE	800	T-DEM-July
0	0	0	0	1	1	0	0	0	0	LE	20	Ware-Capacity
0	0	0	0	0	0	1	0	1	0	LE	26	K-Days-Aug
0	0	0	0	0	0	0	1	0	1	LE	27	L-Days-Aug

(continued)

EXHIBIT 1 (Continued)
A Formulation of the Problem as a Linear Program. Objective: Maximize Contribution

July						August						
K-P	L-P	K-T	L-T	INV-P	INV-T	K-P	L-P	K-T	L-T	Relation	Limit	Constraint Name
0	0	0	0	1	0	40	50	0	0	LE	1400	P-DEM-Aug
0	0	0	0	0	1	0	0	35	42	LE	900	T-DEM-Aug
1	1	1	1	0	0	1	1	1	1	LE	102	Max-Days-Total

Index
K: Kolbert
L: Loomis
P: Plastic
T: Teflon
INV: Inventory
DEM: Demand maximum
Max-Days-Total: The maximum number of days available for production due to vacation constraints.

EXHIBIT 2
Solution of the Linear Program in Exhibit 1

```
MAXIMIZE OR MINIMIZE? MAX

OPTIMAL SOLUTION FOUND.
VALUE OF OBJECTIVE CONTRIBU IS     56839.0

*1* DECISION VARIABLES
 1. K-P-J      26.0000
 2. L-P-J      3.20000
 4. L-T-J      19.5238
 6. INV-T      20.0000
 7. K-P-A       857143
 8. L-P-A      27.0000
 9. K-T-A      25.1429

*2* SLACK(+) AND SURPLUS(-) IN CONSTRAINTS
 2. +L-DAY-J   5.27619
 8. +P-DEM-A   15.7143
10. +M-DAY-T   .276191

*3* SHADOW PRICES FOR CONSTRAINTS
 1. K-DAY-J    61.0000
 3. P-DEM-J    11.6000
 4. T-DEM-J    14.0476
 5. WAR-CAP    12.2476
 6. K-DAY-A    525.000
 7. L-DAY_A    580.000
 9. T-DEM-A    .600000

*4* REDUCED COSTS FOR DECISION VARIABLES
 3. K-T-J     -6.66667
 5. INV-P     -1.64762
10. L-T-A     -15.2000
```

(continued)

EXHIBIT 2 (Continued)

```
ALL ITEMS NOT LISTED HAVE THE VALUE ZERO.

*5* RANGES ON COEFFICIENTS OF OBJECTIVE CONTRIBU
     VARIABLE  LOWER BOUND   CURRENT VALUE  UPPER BOUND
 1.  K-P-J       518.33        525.00        UNBOUNDED
 2.  L-P-J       00000         580.00        588.33
 3.  K-T-J     UNBOUNDED       546.00        552.67
 4.  L-T-J       582.00        590.00        UNBOUNDED
 5.  INV-P     UNBOUNDED      -1.0000        64762
 6.  INV-T      -2.8476       -1.2000        UNBOUNDED
 7.  K-P-A       467.33        525.00        537.67
 8.  L-P-A       564.80        580.00        UNBOUNDED
 9.  K-T-A       533.33        546.00        603.67
10.  L-T-A     UNBOUNDED       590.00        605.20

*6* RANGES ON VALUES OF RIGHT-HAND-SIDE LIMIT
     CONSTRNT  LOWER BOUND   CURRENT VALUE  UPPER BOUND
 1.  K-DAY-J     19.405        26.000        27.381
 2.  L-DAY-J     22.724        28.000        UNBOUNDED
 3.  P-DEM-J     1040.0        1200.0        1213.8
 4.  T-DEM-J    -20.000        800.00        811.60
 5.  WAR-CAP     .00000        20.000        31.600
 6.  K-DAY-A     25.143        26.000        26.276
 7.  L-DAY_A     .00000        27.000        27.276
 8.  P-DEM-A     1384.3        1400.0        UNBOUNDED
 9.  T-DEM-A     886.25        900.00        930.00
10.  M-DAY-T     101.72        102.00        UNBOUNDED
```

Mitchell Enterprises (A)

Mr. Gordon Mitchell, President of Mitchell Enterprises, had called a special meeting of the company's Investment Review Committee in early December 1975. Members of that committee were Mr. Charles Gilbert, the Treasurer; Ms. Roberta Phillips, the Controller; and Mr. Paul Chesler, Special Assistant to Mr. Mitchell. The committee had spent its last meeting reviewing different methods for evaluating investment projects.

At that time, Mr. Gilbert had felt that it would be appropriate to use some method that took into account the value of funds over the entire project life, and Ms. Phillips had suggested the use of discounting to accomplish this. Unfortunately, the committee could not agree on a hurdle rate that reflected the company's current financial position. Indeed, they had found it hard to accept any rate that would stay constant over the life of projects that might be considered. They had considered using a 10% hurdle rate for after-tax cash flows, typical of practice in their industry, but had not felt comfortable about that figure. Mr. Chesler had suggested that by use of linear programming they could determine a portfolio of projects and decide on the amount to be invested in each. In addition, he knew that a linear programming solution would help determine what hurdle rate or rates would be appropriate for the company. Overall, there had been much argument and little progress because they had been discussing the methods in abstract terms, and finally Mr. Mitchell suggested that Mr. Chesler prepare a list of projects that were typical of the projects they had evaluated that year. This list was to be circulated and used to focus the discussion at the next meeting. Subsequently, Mr. Chesler had circulated the attached memoranda.

MEMORANDUM

To: Investment Review Committee — December 5, 1975
From: Mr. Paul Chesler
Subject: Hypothetical Investment Projects

As you know, Mr. Mitchell has asked me to prepare a list of typical investment projects for us to consider in our next meeting. The table below describes five projects that might compete for our investment dollar. The table below shows the cash flow that will result from investing one dollar. Project A is a two-year investment available at the beginning of 1976, which pays 30 cents per dollar invested at the end of the first year and returns an additional dollar per dollar invested at the end of the second. At most $500,000 can be invested in A. Project B is identical to A except that it is available a year later. C is a one-year investment available only at the beginning of 1976, which pays $1.10 per dollar invested at the end of that year. Project D is a three-year investment available at the beginning of 1976, which pays $1.75 per dollar invested at the beginning of 1979. E will become available at the beginning of 1978 and will, after a year, pay $1.40 per dollar invested. Project E is limited to a maximum investment of $750,000. Of course, the cash we receive from any of these projects may be reinvested in others that are available at the time. In addition, we could obtain 6% via short-term bank accounts for any money not invested in a given year.

Cash Flow per Dollar Invested

	Project				
	A	B	C	D	E
1976	−1.00	0	−1.00	−1.00	0
1977	+ .30	−1.00	+1.10	0	0
1978	+1.00	+ .30	0	0	−1.00
1979	0	+1.00	0	+1.75	+1.40
Limit	$500,000	$500,000	None	None	$750,000

For the purpose of discussion I am assuming we want to put $1 million of our money into some mix of these projects at the beginning of 1976 but no more thereafter, although we will reinvest throwoffs. All cash received on January 1, 1979 will be withdrawn.

The table belows gives the results of discounting these projects as we had considered in our last meeting.

Project	Net Present Value @ 10%[a]	Internal Rate of Return(%)
A or B	$.099	16.1
C	$.000	10
D	$.315	20.5
E	$.273	40

[a]A hypothetical hurdle rate similar to ones we have seen used in the past. NPV's are calculated to the beginning of the year in which the outflow occurs.

In addition, I have formulated this investment portfolio problem as a linear program. The formulation and the results of the computer run will be forthcoming.

Mitchell Enterprises (B)

MEMORANDUM

To: Investment Review Committee — December 8, 1975
From: Mr. Paul Chesler
Subject: Linear Programming Formulation and Results

As I remarked in my memorandum of last Friday, it is possible to formulate the portfolio problem proposed by Mr. Mitchell as a linear program. This formulation is shown in Exhibit 1. Notice that since there are no non-cash expenses, post-tax cash flow is strictly proportional to pre-tax flow. For this reason we can justify using pre-tax accumulated cash as the quantity to be maximized; this is equivalent to maximizing after-tax cash. The results of this weekend's computer run are shown in Exhibit 2.

EXHIBIT 1
Linear Programming Formulation

Constraints	Variables A	B	C	D	E	Bank 76	Bank 77	Bank 78		Cash Outlay ($1000)
1/1/76 Investment	1.00	0	1.00	1.00	0	1.00	0	0	=	1000
1/1/77 Investment	− .30	1.00	−1.10	0	0	−1.06	1.00	0	=	0
1/1/78 Investment	−1.00	− .30	0	0	1.00	0	−1.06	1.00	=	0
Limit on Cash invested in A	1	0	0	0	0	0	0	0	⩽	500
Limit on Cash invested in B	0	1	0	0	0	0	0	0	⩽	500
Limit of Cash invested in E	0	0	0	0	1	0	0	0	⩽	750
Objective Function Payout at horizon (1/1/79)	0	1.00	0	1.75	1.40	0	0	1.06		

EXHIBIT 2

```
OPTIMAL SOLUTION FOUND.
     PAYOUT                  1797.60
*1* DECISION VARIABLES
  1. A            500.000
  4. D            500.000
  5. E            659.000
  7. BANK-77      150.000

*2* SLACK(+) AND SURPLUS(-) IN CONSTRAINTS
  5. +B-LIMIT     500.000
  6. +E-LIMIT     91.0000

*3* SHADOW PRICES FOR CONSTRAINTS
  1. INV-76       1.75000
  2. INV-77       1.48400
  3. INV-78       1.40000
  4. A-LIMIT      .952000E-01
```

(continued)

EXHIBIT 2 (Continued)

```
*4* REDUCED COSTS FOR DECISION VARIABLES
  2. B          -.640000E-01
  3. C          -.117600
  6. BANK-76    -.176960
  8. BANK-78    -.340000

*5* RANGES ON COEFFICIENTS OF OBJECTIVE PAYOUT
     VARIABLE  LOWER BOUND   CURRENT VALUE  UPPER BOUND
  1. A         -.95200E-01    .00000        UNBOUNDED
  2. B         UNBOUNDED     1.0000          1.0640
  3. C         UNBOUNDED      .00000          .11760
  4. D          1.6324       1.7500          1.8452
  5. E          1.3278       1.4000          1.5009
  6. BANK-76   UNBOUNDED      .00000          .17696
  7. BANK-77   -.64000E-01    .00000          .10691
  8. BANK-78   UNBOUNDED     1.0600          1.4000

*6* RANGES ON VALUES OF RIGHT-HAND-SIDE CASH
     CONSTRNT  LOWER BOUND   CURRENT VALUE  UPPER BOUND
  1.  INV-76     500.00        1000.0        UNBOUNDED
  2.  INV-77    -150.00        .00000         85.849
  3.  INV-78    -659.00        .00000         91.000
  4.  A-LIMIT    .00000        500.00         569.04
  5.  B-LIMIT    .00000        500.00        UNBOUNDED
  6.  E-LIMIT    659.00        750.00        UNBOUNDED
```

Lowell Steel Corporation

The Lowell Steel Corporation, a major producer of steel in the United States, was located in a small town on the Shenango River about 16 miles from Youngstown, Ohio. The company and its subsidiaries manufactured and sold hot and cold rolled carbon steel strip, stainless and alloy steel strip, steel sheets, carbon and alloy seamless tubing, galvanized and other coated products, and certain miscellaneous items.

In 1955 the company began an extensive program of expansion and integration. The program was designed: (1) to enlarge the company's steel ingot producing capacity, and (2) to secure manufacturing subsidiaries which could further process steel the company had been selling as hot rolled strip and thereby enable the company to "take its products to market at higher levels of price." In April 1956, management was considering possible directions of further expansion. In addition, they were trying to decide whether to increase the operating rate by 10%.

Source: This case was prepared by Richard S. Rosenbloom.

EXHIBIT 1
Photograph of Lowell Works (Blast Furnace and Open Hearth Shop)

MANUFACTURING FACILITIES

The Lowell Works had blast furnaces, steelmaking, and rolling mill facilities on a single site (see photograph in Exhibit 1). The two 800-ton blast furnaces had rated annual capacities over 250,000 tons of hot metal each, for a total annual capacity at the Lowell works in excess of 500,000 tons of hot metal. Fourteen open hearth furnaces could produce a million tons of steel ingots in a year. A 36-inch blooming mill, 24-inch bar mill, and 18-inch bar mill reduced the ingots to billets, blooms, and slabs for further processing at Lowell, or shipment to a subsidiary mill or a customer. All of these facilities had been constructed before 1940.

At the end of March 1956, the two blast furnaces at the Lowell Works were being operated as close to capacity as was possible and together were producing from 45,000 to 46,000 tons of pig iron per month. This metal was charged hot to the open hearth furnaces. The company was operating ten open hearths, each of which had an output of about 7,000 tons of steel ingots per month. The hot metal and scrap charge to the open heart furnace was made up of about 58% pig iron and 42% scrap metal. Each furnace was tapped every 10 to 11 hours and produced slightly over 100 tons per heat.

The company had found that the optimum ratio of molten iron in the metal charge for the open hearth furnace was somewhere between 62% and 65%, a higher proportion of molten iron resulted in certain process difficulties; a lower proportion increased the time per heat because of the larger amount of cold scrap which had to be melted. Under normal circumstances the furnaces were never operated with less than 50% or more than 70% molten iron in the charge. Above 70%, the furnace reaction became so violent that it was somewhat uncontrollable. Below 50%, the furnace reaction removed too much of the carbon from the metal, and it became necessary to go to the cost and trouble of adding carbon to the melt. Moreover with more than 50% cold scrap in the charge, the melting times became excessively long.[1]

The Lowell Company's sales of ingots and semifinished steel products (plates, slabs and bars) averaged approximately 17,000 tons/month in the spring of 1956. Of this amount, just over 7,100 tons were sold as ingots and almost 9,900 tons were sold as semi-finished steel. In addition, semi-finished steel was sold to a subsidiary, Perkins Tube and Steel Company. (See below). Most of the company's semifinished steel output, however, was converted to hot rolled strip at the Lowell Works. In 1956, the company was operating four hot strip mills. The continuous mill was used for strip ranging between 5″ and 22″ in width, the 9-inch mill for strip between 1″ and 2¼″ in width, and the 10-inch mill for strip between 2″ and 5¼″ in width.

More than half of the hot strip was further processed on the company's cold finishing equipment at Lowell or in subsidiary plants. About 11,500 tons per month were being cold rolled at the Lowell Works and another 9,000 tons transferred to subsidiaries. The remaining strip (18,700 tons per month) was sold to customers in many different industries for fabrication into a very wide range of end products.

SUBSIDIARIES

The activities carried on by the several subsidiary companies may be summarized as follows:

The Perkins Tube and Steel Company operated a tube plant in Detroit and a cold rolling mill in Dearborn. The tube plant secured about 20% of its requirements of semifinished steel rounds[2] from the Lowell Works and purchased the remainder on the open market. Requirements from all sources in early 1956 averaged 7,500 tons per month. The Lowell Works did not have sufficient equipment to produce all of the different sizes and types of rounds required in the tube plant. Additional equipment was to be installed for rolling rounds, but it was anticipated that the tube plant would always buy some of its steel from other companies. The output of the tube plant was sold primarily to the automobile and oil industries.

The cold rolling plant in Dearborn secured all of its requirements of hot rolled strip from the Lowell Works. Monthly shipments to Dearborn were averaging 5,000 tons of hot rolled strip in early 1956. About 80% to 90% of the

[1] 1.12 tons of hot metal and scrap input were needed for 1 ton of steel output regardless of percentage scrap.

[2] These rounds did not go through the hot strip mill.

output of this plant was sold to customers in the automobile industry and the remainder to miscellaneous steel stamping companies.

The Maynard Steel Corporation operated cold reducing and cold finishing mills in Cleveland. The requirements of hot rolled strip 4,000 tons per month, were secured from the Lowell Works. The finished products of the company were sold to a large number of relatively small customers in all sections of the country. The most important item in the company's line was box strapping.

MANUFACTURING COSTS

The Lowell Steel Corporation was not a "low-cost producer of tonnage steel" in the same sense as were the larger concerns in the industry.The company's production operations in 1956 were integrated to a greater degree than were those of the other specialized, small independent steel producers. It did not, however, own sources of raw materials, ore mines, ore boats, loading docks, or railroads; ownership of these properties was quite common among the United States Steel Corporation, the Bethlehem Steel Corporation, the Republic Steel Corporation, and other large producers. In some instances, the Lowell Steel Corporation purchased its raw materials from the larger concerns (or their subsidiaries) and used their transportation facilities to assemble the materials at its plants. Moreover, the equipment in its plants was smaller in size and operated at slower speeds than the equipment maintained by some of the big steel companies. Illustrative manufacturing costs for the winter of 1956 are presented in Exhibit 2.

EXHIBIT 2

Manufacturing Costs—March 1956

Blast furnace operations

Materials charged (to furnace no. 2 during March)

	Tons	Cost per Ton	Total Cost
Iron ore	39,683	$ 9.04	$ 358,734
Scrap, dust, cinder, miscellaneous	5,547		35,167
Limestone	9,445	2.80	26,446
Coke	23,096	19.53	451,065
Subtotal			$ 871,412
Labor and other direct operating costs (not including depreciation)			142,592
Total Cost			$1,014,004
Tons of hot metal produced			22,971
Net cost per ton of hot metal			$ 44.14

(continued)

EXHIBIT 2 (Continued)

Open hearths

Cost per ton of ingots
(based on 10 furnaces producing 70,000 tons per month)

Metals Charged	Tons Charged per Ingot Ton Produced	Cost per Ton of Charge	Cost per Ton Ingots Produced
Hot metal from blast furnace	.650	$44.14	$28.69
Steel scrap	.470	48.00	22.56
Ore and miscellaneous metals	.060		1.74
Subtotal			$52.99
Cost above metal (labor, operating expenses, molds, fluxes, etc., excluding depreciation)			19.78
Net cost per ton of ingots			$72.77

Blooming billet and bar mills

Cost per ton of slabs produced

Gross Metal Input (84.7% yield)	
1.18 ingot tons × $72.77 per ton	$85.87
Scrap credit (13.7% salvaged, at $48 per ton)[a]	(7.76)
Net metal cost	$78.11
Cost above metal	7.24
Net cost of slabs	$85.35

Continuous hot rolling mill

Cost per ton of hot rolled strip

Gross Metal Input (93.6% yield from slabs)	
1./936 tons × $85.35	$ 91.20
Scrap credit (6% salvaged, at $48 per ton)[b]	(3.08)
Net metal cost	$ 88.12
Cost above metal	14.43
Net cost of hot rolled strip	$102.55

[a]13.7% of input is salvaged 1.18 × 1.37 × $48.

[b]6% of input is salvaged 1.07 × .06 × $48.

As a result of these circumstances, the Lowell Steel Corporation operated at somewhat of a cost disadvantage with respect to the big steel producers on large tonnage orders for standard grades of steel. The company was in a favorable position to compete, however, for small volume orders on certain specialty items. It was thus necessary for the company to "discriminate" in its selling activities; that is, to seek out customers whose particular requirements were such that the company could capitalize on the flexibility of its small plant, the capabilities of the particular equipment it operated, and the "know-how" it had acquired over a great many years for handling certain special grades of steel.

For example, a customer might wish to buy hot rolled strip 18¼ inches in width and .090 inches in thickness. The Lowell Steel Corporation could roll such steel in its continuous hot mill at a rate of perhaps 500 tons per 8-hour shift. A big producer with large equipment, however, would probably be able to roll the same steel in 37-inch or 54-inch widths at a rate of perhaps 1,000 to 2,000 tons per 8-hour shift. The large widths could then be quickly slit to the desired dimension of 18¼ inches. In a situation of this type the conversion costs of the big producer would unquestionably be lower than those of the Lowell Steel Corporation.

The Lowell Steel Corporation was in a favorable position to compete, however, if the customer wished to buy in a quantity which was too small to justify the large producer in setting up his mills. The company was also in a favorable position if the customer insisted on having "mill" rather than "sheared" edges on the steel,[3] because then the big producer could handle the order only by running it in single thickness with his equipment operating at less than full capacity. Similarly, if the customer wished a very high carbon steel or unusually close tolerances on the gauge, the Lowell Steel Corporation might be in a favorable position to handle the order, because high carbon steels were difficult to roll in wide widths and the problems in controlling tolerances tended to increase as the width of the strip increased.

Circumstances analogous to those noted for hot strip existed in the case of the many other different grades and finishes of steel which the company sold. As a result, it was the policy of the company to seek out customers and markets where the manufacturing requirements were such that it would enjoy a competitive advantage or at least compete on equal terms with the larger producers. Conversely, the company made no attempt to solicit business in such markets as automobile sheet steel where the contracts were let for very large tonnages of standard grades and finishes.

During early 1956 steel prices were such that the Lowell Steel Corporation could have met competitive prices and secured a fair rate of return in almost any market. In the long run, however, the executives believed that the company's interests would best be served by developing those markets which the large producers were not too well equipped to serve. As a result of its sales policies, the company had established a position as one of the most important producers in the field of special and coated steels. It was a major producer of high carbon steel and ranked as perhaps the largest producer of cutlery and cooperage steels in the United States.

[3] A "mill" edge was the edge resulting from normal hot rolling operations; a "sheared" edge was the edge left after shearing operations.

SITUATION IN APRIL 1956

The steel market was strong in early 1956 and the Lowell Company benefited from the high operating rate and firm product prices. In early April, management was considering whether to authorize the start-up of Open Hearth Furnace #11, which would add 7,000 ingot tons to monthly output. Steel scrap prices had recently risen to $50 per ton in the Shenango Valley and many observers were expecting a further rise of 5% to 10% in the next 30 days. The management was certain that any additional production could readily be sold in ingot form for further processing by other mills at or near the prevailing market price of $80 per ton. The market for semifinished steel seemed strong and it seemed likely that existing customers would be interested in additional tonnage during the summer. The Lowell Company was presently realizing an average revenue of $95 per ton on semifinished steel sales. The most attractive market was for hot rolled sheet and strip. Sales revenues averaged $140 per ton of strip, but the sales manager was uncertain whether additional volume for these products could be generated within the next few months.

A furnace that is ready for start-up must be heated for about 24 hours before the first steelmaking heat can begin. Three heats are usually required to reach the normal cycle. Cost of the light-up is $2,200, of which ½ is for fuel. Once a furnace has been started up, it is kept in operation for a campaign of 450 heats. This campaign is divided into two parts. After approximately 250 heats, the furnace roof must be rebuilt. This entails shutting the furnace down and relining the roof with refractory materials. The furnace is fired up again and run for another 200 heats. It is then shut down and completely rebuilt—end walls, sides, and roof. The floor is not rebuilt; it is patched after every heat, and will last about 20 years. The checker chamber (where air going to the furnace is heated) is cleaned of soot and its top layer of bricks is replaced. To rebuild the roof of a 100-ton furnace after 250 heats takes 5 days around-the-clock, from gas off to light-up, and costs $60,000. The cost of completely rebuilding a furnace is $150,000 and takes 9 to 10 days.

Furnace #11 at the Lowell Works had been completely rebuilt and was ready for start-up in early April. The Lowell Company's practice was to amortize the costs of rebuilding a furnace by a per ton charge against all ingots subsequently produced in that furnace. This charge was included in the "cost above metal" for the open hearths.

Scott Paper Company

Scott Paper Company, founded in 1879, has extensive manufacturing and distribution installations in the U.S. and abroad; in addition, the company

Source: This case was prepared by Judith Selvidge, based on research by Judith Selvidge and Tom Steiner, under the supervision of Professor Richard F. Meyer. Harvard Business School Case 1-171-061.

owns nearly 3 million acres of timberland. Sales in 1969 were $731 million (up from $514 million 5 years earlier) with net income of $60 million. The greatest percentage of the sales volume comes from the Packaged Products Division (PPD); the balance of the company's business comes from the other Divisions, chiefly from the S. D. Warren Division; the Plastic Coating Corporation, a subsidiary; the Foam and Container Division; and the International Division.

The Packaged Products Division (PPD) of Scott Paper Company manufactures and markets a variety of consumer and industrial paper products, including paper towels, napkins, facial and toilet tissue, and waxed paper. Wood pulp, the principal raw material used in making the paper products, comes from another division of Scott or from open market purchases and is processed at PPD paper mills. This processing, illustrated in Exhibits 1 and 2, begins at the huge paper making machines. Output from these machines (rolls of about seven feet long and five feet in diameter called "parent rolls") goes to rewinders and decorators where, depending on the particular product being made, the paper is printed, embossed, and so on. The paper is cut to size, wrapped, and packed in cases in the finishing operation and then moved to the

EXHIBIT 1
How to Make Paper

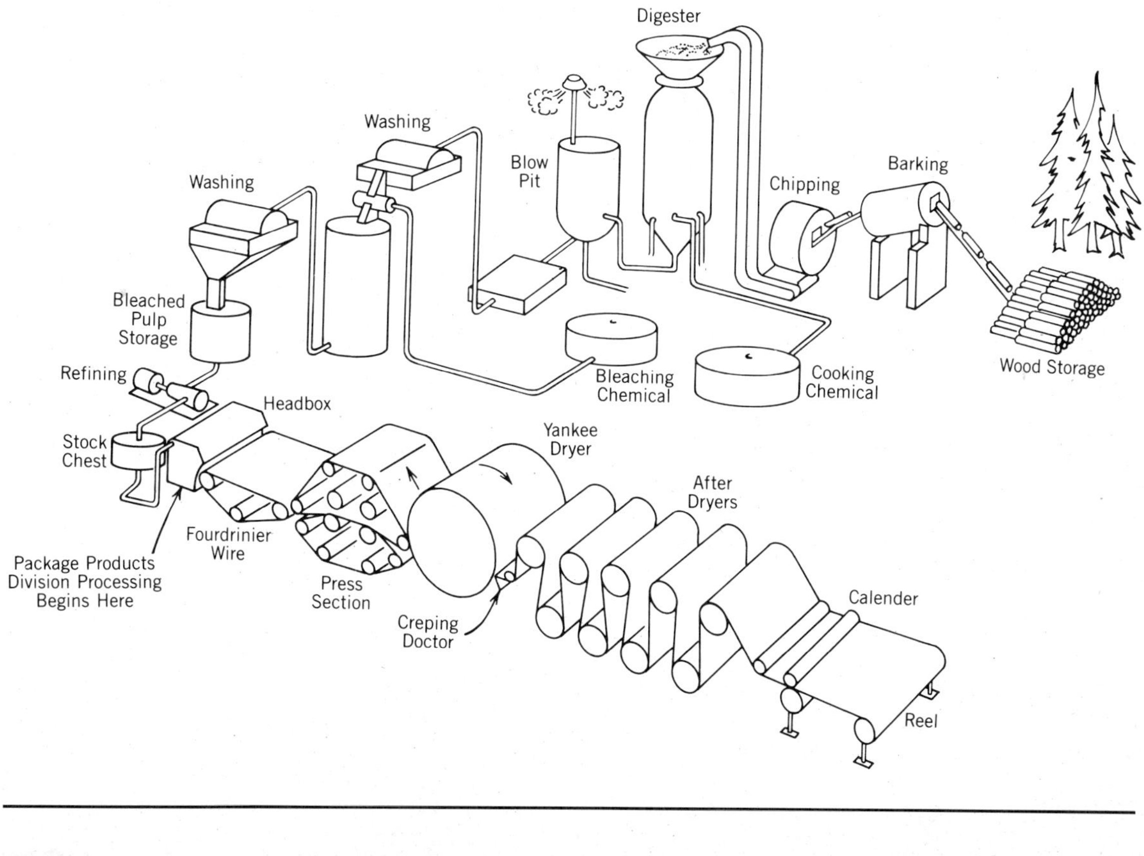

EXHIBIT 2
Finishing

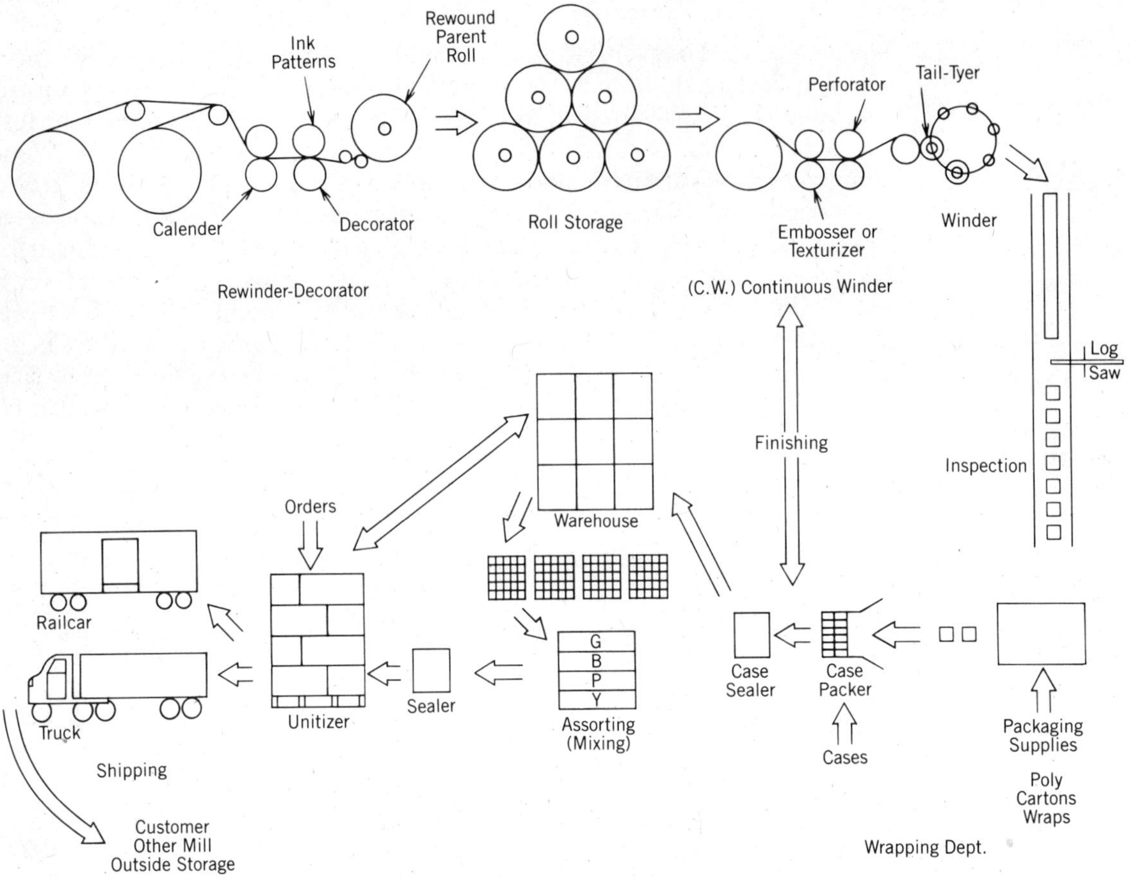

mill's distribution center and placed in finished products inventory. From the inventory, some product is shipped to other distribution centers (not all Scott products are made at each mill); these shipments are called "stock transfers." The rest of the product is shipped in van or carload lots directly to customers. If the lot contains a single product type, it is a "solid" shipment. If not, a "mixed" shipment.

Paper-making is a capital-intensive industry due to the high cost ($12–$15 million) of paper machines. One of the major pre-occupations of the PPD management, therefore, is the efficient scheduling of work on the different machines at different mills to fulfill the forecast production requirements.

These machines are not uniform. On the contrary, each machine (Scott has 40 paper machines, 30 rewinders and rewinder-decorators, and 75 finishing machines) has different production rates for different kinds of paper and products. The production required is broken down by sales area, each served by a particular distribution center. The Scheduling Department of PPD then divides the work among the various mills trading off production economies (due to efficient machine/brand combinations) against transportation costs of shipping the output to distant distribution centers and taking into account the variable production rates on different machines. There are seven mills with associated distribution centers, and three distribution centers that are warehouses only (see Exhibit 3).

In the middle 1950s Scott was a pioneer in the use of computers to aid in solving this scheduling-distribution problem. The corporate Operations Research group designed, built, and patented a special analogue computer in which the different paper machine-rewinder-decorator-finisher-distribution center-sales area combinations were represented by electrical circuits. By sending a current (corresponding to total annual production) through the system of circuits and then measuring the output current at the end of every branch, a

EXHIBIT 3
Mill and Distribution Centers

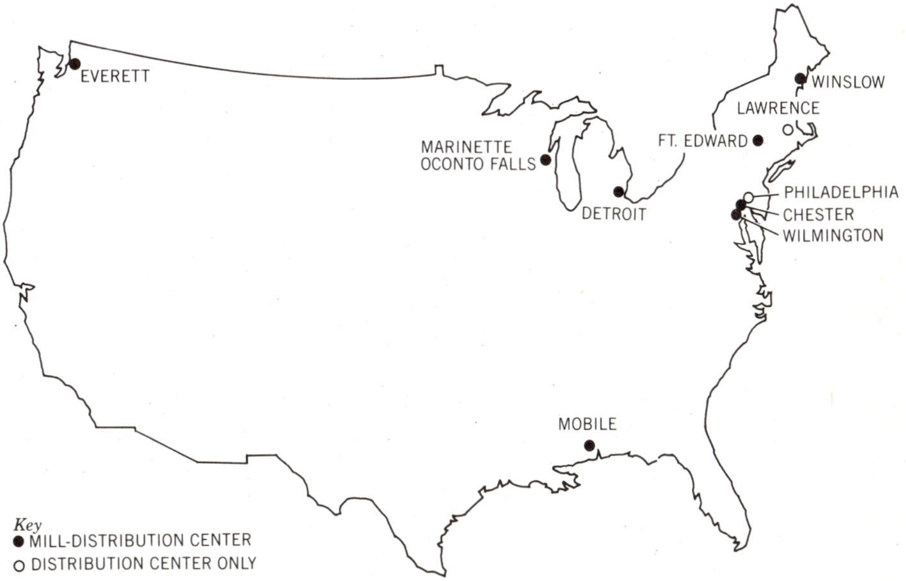

series of numbers was obtained each of which could be translated into *annual* tons of paper products produced on that particular machine combination and delivered to a customer via the specified distribution center. By applying voltages (representing production rates) at the nodes of the circuits (representing the paper machines) the ranges of permissible production rates could be explored by the model. The OR group had the responsibility for running the computer and for recording and interpreting the computer solution. The resulting recommendations were sent to the Scheduling Department in PPD where monthly schedules for each mill were drawn up by hand. In addition to determining the monthly targets or quotas for each mill—these quotas are subject to negotiation between the Scheduling Department and the mills—the Scheduling Department controls or reschedules the assignments during the month as necessary to account for machine breakdowns, changes in level of demand for a product, and so on. This rescheduling accounts for about 60% of the work of the department.

During the 1960s Scott's business increased significantly. This increase in sales volume, number of mills, and particularly the proliferation of products placed a strain on the capacity of the analogue computer and on the ability of the Scheduling Department to convert annual production amounts into monthly output by brands. Management suspected that greater automation of scheduling would increase the PPD contribution to profits with the existing machines, postpone the need for investment in new paper machines, and eliminate the vulnerability of the company under the current system where essentially only one man was experienced enough to perform hand scheduling.

In 1967 the idea of using linear programming (LP) techniques and a large digital computer to deal with the scheduling problem was considered. The proponents of this approach thought that optimization methods applied to the allocation of machine time could automatically provide the best production schedule for each month. The corporate OR group prepared a detailed technical proposal explaining how they would develop such an LP solution; higher management, however, eventually decided to set up a separate team of three men, one from the OR group and two from PPD, to carry out this LP task. The team was temporarily assigned to the Management Information Services (MIS) group in PPD.

Chuck MacFarlan, the OR representative on the team, was an electrical engineer who had played a major role in the development of the company's analogue computer. His experience, typical of the OR group which had been a part of the Research Division, came mainly from one-time studies of problems referred to the group from all over the company. These studies ranged from designing a "woodlands model" of tree growth to participating in new mill location decisions. As work on this new project began, Mr. MacFarlan took on the job of learning about LP and about the general-purpose computer program developed by IBM for solving LP problems.

The second team member came from the PPD accounting department and was given primary responsibility for the many decisions about how to calculate and assign variable manufacturing costs for the new computer model. He also began to define the division's goals in terms of different objective functions for the LP. After a few months of association with the project, however, he was promoted and left the project. An analyst/programmer with EDP experience was the third member of the team. He was to bring to the project a familiarity with the "systems" side of setting up operational problems for solution on a digital computer.

An initial contact was made with the prospective users of the computer model, the Scheduling Department. They were asked to list all the information required or desirable in the monthly schedules which the computer would prepare. However, the precise scope of the model—in terms of the restrictions, assumptions, and objectives—was not determined in advance of the programming.

During the next 12 months Mr. MacFarlan, with the help of a student working part time for Scott Paper, developed and programmed the model, which came to be called the Monthly Production Scheduling (MPS) system. Mr. MacFarlan's detailed knowledge of the production process enabled him to work independently, while the systems analyst coordinated the preparation of data input to the program.

MPS generates monthly production and distribution plans for each mill to satisfy estimated sales demand and inventory requirements. Potential manufacturing savings (due to selection of efficient machine/brand combination) are traded off against stock transfer costs. The LP matrix consists of 463 rows and 2871 columns broken down as follows:

Constraints	*No. of Rows*
Demand (sales forecasts) by distribution center and brand (in 0000 cases)	230
Machine capacities (paper machine, decorator, finisher, rewinder) (in days)	140
Mill totals by product (average monthly production for each product at each mill—not all products are produced at every mill) (in 0000 cases)	93

Activities	*No. of Columns*
Production rate (of brand A on machines B, C, & D at Mill E) (in days/0000 cases)	155
Stock transfer (23 brands among 10 distribution centers) (0000 cases)	2070
Beginning and ending inventories (at each distribution center) (0000 cases)	460
Slack variables (production for promotional purposes) (0000 cases)	186

The MPS considers the coming six-month period. Initially a computer solution is calculated for the six-month interval, viewed as a single period. (This requires from 15–27 hours of machine time on an IBM 360/40.) Then using this optimal solution as a starting point the matrix solution processor is rerun for each month separately (given the monthly sales forecast) to generate the monthly production schedules. These subsequent runs take about one hour of computing time each.

The model covers only a limited part of the PPD activities. It starts at the pulp end of the business with an incomplete description of pulp availability: although the *cost* of pulp is included in the formulation, there is no limit on the available supply. At the other end of the business, the distribution of products is taken only as far as stock transfers to distribution centers rather than on through to the customer.

By the middle of 1968 the Scheduling Department had been given the first examples of MPS output; Mr. MacFarlan had begun work on other projects including potential extensions of MPS; and the systems analyst was preparing to leave the MIS group, having been promoted to a job in his field of specialization, transportation.

At this point another programmer, Jack Lazlow, was assigned to the MPS project to handle the final details of getting the program into everyday use. He had been peripherally connected with earlier work on MPS, having written programs to generate sales forecasts which formed part of the input to the MPS system. As he began his new assignment, Mr. Lazlow was surprised to learn that a great deal of work remained to be done in explaining the new system and its output to the Scheduling Department. The head of the Scheduling Department, referring to the samples of MPS output, pointed out that many of the machine assignments in the computer solution did not seem to make any sense. He wanted an explanation of how these "funny" numbers were obtained. Mr. Lazlow realized that he himself did not understand the program well enough to describe how the machine assignments were computed, and so he began to learn about LP in general and in particular about the way this scheduling problem had been formulated. His efforts were hampered by the nearly total lack of documentation of the work done so far. Most helpful to him during this period was Manny Spenser, the part-time programmer who had assisted Mr. MacFarlan during the writing of the MPS system. Mr. MacFarlan was caught up in other things at this time and, although he was the person who most thoroughly understood the LP theory and its implementation in the MPS program, he quickly lost patience when trying to explain it to others, especially the users.

Concurrently with Mr. Lazlow's effort, the head of the Scheduling Department and his new assistant were making a lot of hand calculations in an attempt to try to figure out or disprove the assignments of the MPS output. As the two parties to the project at this stage, the users and the MPS programmers, increased their understanding of the system they met frequently and engaged in fairly heated disscusions—mainly about *details* of the system, but also touching upon basic questions—of what the system did and what it was supposed to do. The Scheduling Department, far from considering MPS an improvement on the old system, was generally critical of the results. They noticed several errors in the model formulation ranging from steps left out in the production process (the capacity of the rewinders had not been included as a constraint) to nonlinearity in the constraint equations. Apart from the purely technical mistakes the Scheduling Department also found the output format to be difficult to use: the solution was given in LP terminology—"activities," "shadow prices," and so on—buried in a lot of irrelevant material. Furthermore, they were uncomfortable with the LP *method* in which the details of the calculation took place inside the system with only the final solution being available for scrutiny. The schedulers were used to seeing how a particular schedule developed, allowing them to check the "reasonableness" of the answer. One source of considerable misunderstanding between the users and the programmers was the precise form of objective function specified in the LP. Originally a minimum-cost formulation had been tried, but before long this was changed to a maximum-contribution-to-profit formulation, allowing excess capacity to be applied toward the production of high-profit items for promotional purposes. The users grasped in a general way the "objective" of the LP system but a great deal of discussion and analysis of results took place before they were made aware of the importance of the exact formulation and, particularly, the provision for filling the excess capacity.

As the months passed Mr. Lazlow was disturbed to realize that as their understanding of the MPS system grew, the users seemed to put more emphasis on showing what was wrong with the program than on finding ways of making it work. Although the computer solution was produced for each month, the Scheduling Department apparently did not base its decisions on the MPS figures.

Energy Research and Development Administration

INTRODUCTION

The problems of energy supply have been the subject of much thought, discussion and research over the past few years. Although the problems are not at present acute, they are chronic because the global supply of fossil fuels is finite. The question of the quantity of fossil fuel resources remaining and the associated future demand on them has raised substantial controversy. There is, nonetheless, a general consensus that our fossil sources need to be effectively and efficiently managed and that the continued development of alternative energy forms is vital. The Energy Research and Development Administration (ERDA) was established to assist in this function.

ERDA brings together many diverse federal energy research and development (R&D) activities that will be required to develop alternative energy options which have commercial potential. About 80% of all federal energy R&D effort is handled by ERDA. With respect to the research and development of the nation's energy alternatives, ERDA's efforts are intended only to be supplementary to those of private industry. Perhaps the most important aspect of ERDA's responsibility is to carry out the catalytic development of economically viable technologies within, or their transfer to, the energy industry. The role of ERDA, therefore, is that of a short-, medium-, and long-term option generator for the national energy system.

In 1975, ERDA published "A National Plan for Energy Research, Development and Demonstration: Creating Energy Choices for the Future." The purpose of this plan is to expedite and help the transition from an energy system dependent upon a narrow base of diminishing resources, to a system which can rely upon a broader spectrum of less limited or unlimited energy sources. The Plan identified a need for a more integrated analysis of the range of energy programs and problems, which would allow for more extensive cross-comparisons among energy technologies. To assist in this integrated analysis, ERDA has used a mathematical programming model which determines the optimal allocation of competing energy sources among specific categories and helps to evaluate the impact of the introduction of selected new technologies into the U.S. energy system.

This model is a linear programming analysis of the Reference Energy System (RES). The RES is an aggregate representation of the nation's energy system in a framework within which alternative policies and technical strategies may be evaluated. In the RES the U.S. energy system is represented as a network diagram that includes all of the possible paths through which energy, supplied by a number of resources, may be utilized to satisfy the needs of society (see Exhibit 1). The LP model determines the minimum cost allocation of the energy resources through the RES network to satisfy energy demands. The economic, environmental and other impacts of new energy technologies or sources may be evaluated by including them in the LP model, thereby recalculating the energy flows, costs, and emissions.

THE MODEL

In the RES approach, the energy system is divided into two parts—the supply and the demand chains. The steps in the supply chain include extraction, refining, conversion, storage, transportation and distribution. The utilizing devices (space heaters, air conditioners, internal combustion engines, etc.)

EXHIBIT 1

Reference Energy System—Year 1985

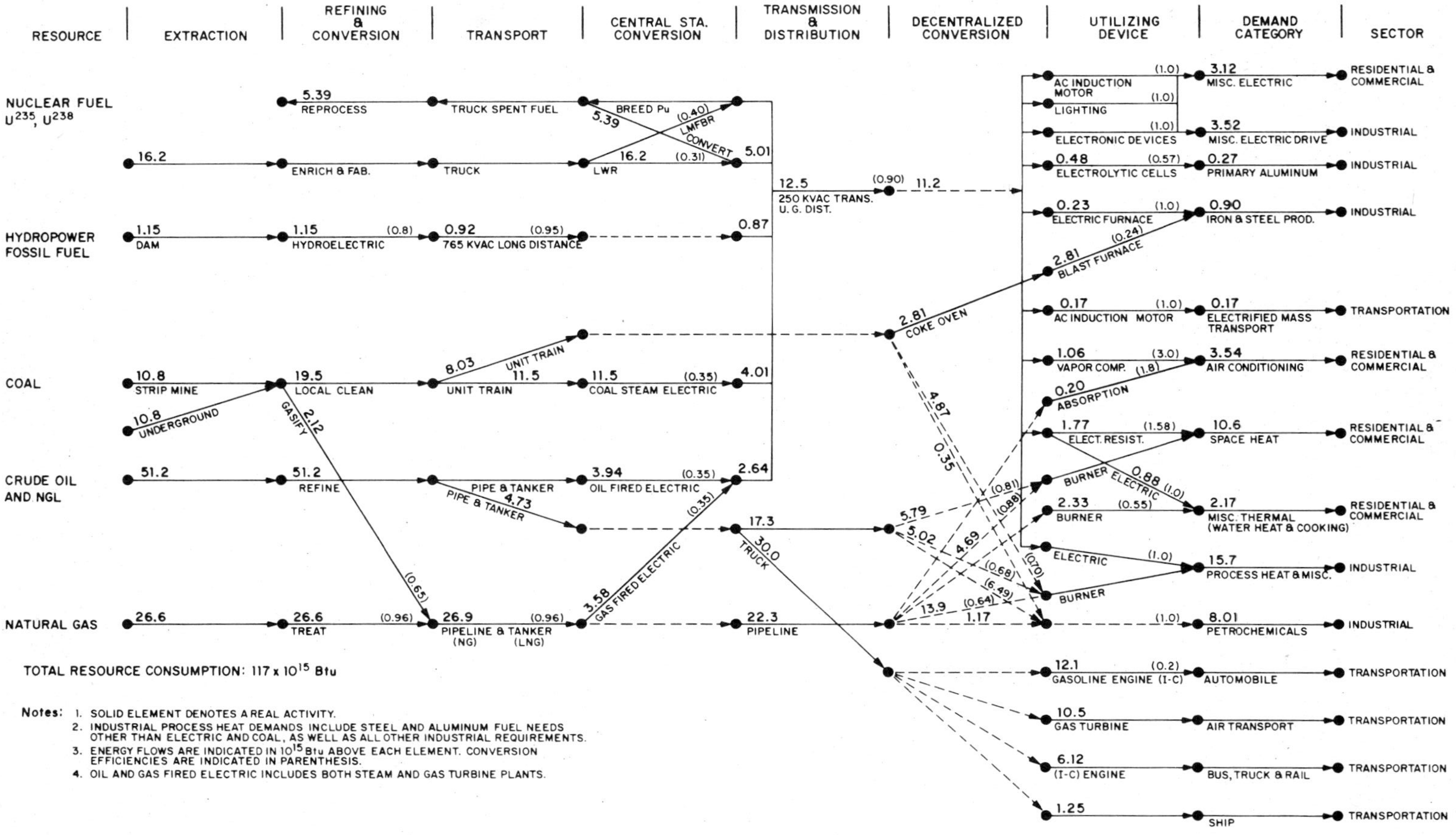

TOTAL RESOURCE CONSUMPTION: 117×10^{15} Btu

Notes:
1. SOLID ELEMENT DENOTES A REAL ACTIVITY.
2. INDUSTRIAL PROCESS HEAT DEMANDS INCLUDE STEEL AND ALUMINUM FUEL NEEDS OTHER THAN ELECTRIC AND COAL, AS WELL AS ALL OTHER INDUSTRIAL REQUIREMENTS.
3. ENERGY FLOWS ARE INDICATED IN 10^{15} Btu ABOVE EACH ELEMENT. CONVERSION EFFICIENCIES ARE INDICATED IN PARENTHESIS.
4. OIL AND GAS FIRED ELECTRIC INCLUDES BOTH STEAM AND GAS TURBINE PLANTS.

are included in the demand chains. Each link in the chain corresponds to a physical process, characterized by a conversion efficiency, capital and operating costs and emissions of pollutants, wherever applicable.

ENERGY DEMANDS AND SUPPLIES

The energy requirements of the system are divided into categories such as industrial, commercial, residential and transportation. These sectors are then divided into subcategories, since it is within the end-use subcategories where alternative energy technologies and fuel substitutions occur. For example, in the residential sector, energy demand is initially calculated by measuring the average number of degree days for heating and cooling per year, and calculating the energy necessary to heat or to cool an average dwelling unit per heating or cooling degree day. An aggregate energy requirement for heating and cooling is then derived from the above analysis and by considering the number of dwelling units in the region and the conversion efficiencies of each end-use technology. A national demand is then determined from the regional figures.

The national demand in the commercial, transportation and industrial sectors is determined in a similar fashion. These sectors are also disaggregated when calculating total energy demand. Exhibit 2 shows how the national energy demand is broken down by sector and end-use. The mathematical model uses similar consumption figures in the formulation of constraints requiring sufficient energy to be provided for all of these demands.

The demand for energy can be satisfied by a variety of energy resources or supplies. At present, coal, oil and natural gas are the most important, satisfying approximately 31%, 43% and 21% respectively of all the nation's energy demands. Hydro power and nuclear power account for the difference. From

EXHIBIT 2
Energy Consumption in the United States—1968

Sector and End Use	Consumption (10^{12}BTU)	% of National Total
Residential		
Space Heating	6,675	11.0
Water Heating	1,736	2.9
Cooking	637	1.1
Clothes Drying	208	0.3
Refrigeration	692	1.1
Air Conditioning	427	0.7
Other	1,241	2.1
Total	11,616	19.2

(continued)

EXHIBIT 2 (Continued)

Sector and End Use	Consumption (10^{12}BTU)	% of National Total
Commercial		
Space Heating	4,182	6.9
Water Heating	653	1.1
Cooking	139	0.2
Refrigeration	670	1.1
Air Conditioning	1,113	1.8
Feedstock	984	1.6
Other	1,025	1.7
Total	8,765	14.4
Industrial		
Process Steam	10,132	16.7
Electric Drive	4,794	7.9
Electrolytic Processes	705	1.2
Direct Heat	6,929	11.5
Feedstock	2,202	3.5
Other	198	0.3
Total	24,960	41.2
Transportation		
Fuel	15,058	24.9
Raw Materials	146	0.3
Total	15,184	25.2
National Total	60,526	100.0

the current estimates of the availability and accessibility of energy resources, estimates can be made as to the quantity of energy resources that may be recovered annually, at or near existing costs with current technology. Projections of these estimates can be made so that we may derive curves which depict future levels of the amounts of each of the fossil fuels annually recoverable. Such projections are speculative, and estimates give an indication of the relative abundance of the non-renewable resources and represent the levels of resource supply in the linear programming model.

PEAK LOAD DEMAND FOR ELECTRICITY

Electricity cannot be stored on a large scale for use at a later period and should be generated according to the demand cycle, which varies on a daily, weekly and seasonal basis. The demand or load represents the rate of consumption of energy. A typical annual load duration curve is shown in Exhibit 3, in which the base load is considered as 65% of peak power. The requirement

EXHIBIT 3
Typical Load—Duration Curve

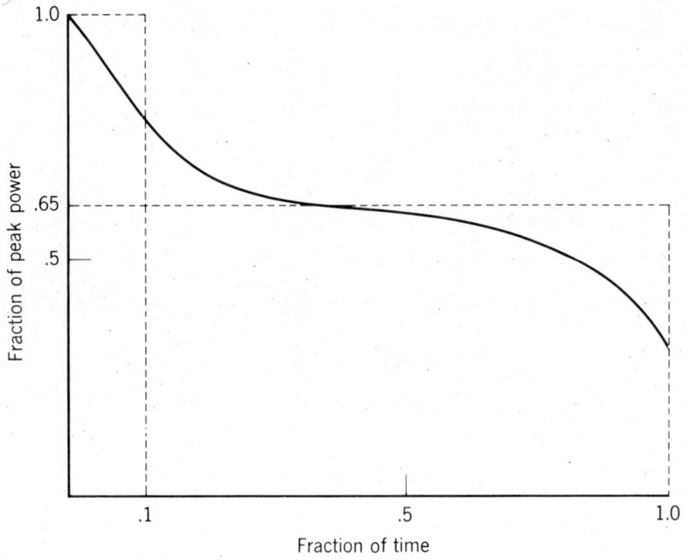

Note:

$$\frac{\text{Peak}}{\text{Load}} = \frac{\text{Area of top rectangle}}{\text{Total area}} = \frac{.1 \times .35}{.1 \times .35 + .65} = \frac{.035}{.685} = 5\%$$

Therefore peak load is 5% of total energy required.

above this rate of consumption is designated as "peak-load." The total energy delivered during the year represents the area under the curve. The area under the curve above the horizontal dotted line represents the energy supplied during the peak load period.

The types of electrical generating equipment used for meeting the base load and peak load are normally different. For example, base load service is provided by high efficiency, fossil-fueled steam-electric plants, nuclear plants and hydraulic power plants. These types of plants have high capital costs and low operating costs and usually operate for the entire year. Peak load service, however, is usually provided by oil or gas turbines and diesel generators, which have relatively low capital costs and high operating (fuel) costs. These may be operated for less than 10% to 15% of the annual period. Turbines can also be operated to meet some of the base load requirement. Sometimes pumped storage facilities are used in conjunction with hydraulic power plants for meeting the peak load.

While the peak load devices have to be capable of meeting the peak demand, on an average over the year, the energy supplied by them is only a small fraction of the total electrical energy used. In Exhibit 3, it is only 5%. The peak load requirement not only depends on the time cycle, but also on the characteristics of the end use equipment.

SUPPLY AND DEMAND COEFFICIENTS

The overall efficiency of generation and utilization of energy is represented in the model by two factors called the supply and demand coefficients. The supply coefficient represents the reciprocal of conversion efficiency associated with the conversion of the energy resource (for example, oil or coal) into the intermediate energy form (for example, electricity). This includes the efficiency of all steps in the supply chain. As the efficiency is generally lower than 1, the supply coefficients are greater than 1.

The demand coefficients represent the relative efficiency of conversion of the intermediate energy to useful work by the end use equipment. As they are only measures of relative efficiency, they can be higher than 1.

The energy conversion and utilization process may be represented schematically as follows.

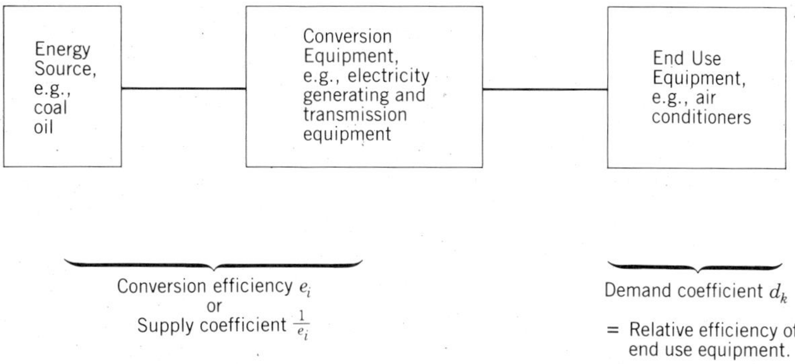

Let us consider an example in which oil is used for space heating in two different ways. For simplicity take the extraction, refining and transport efficiency for oil to be 1.0. Oil can be used directly in space heating or for generation of electrical energy, which is used in turn for space heating. Oil electric generators have a conversion efficiency of 0.3125 (i.e., that amount of oil, which under direct combustion will produce one BTU of thermal energy, can be converted to an amount of electricity which will produce only .3125 BTU of thermal energy.) Hence the supply coefficient for oil electric generators

$$= \frac{1}{.3125} = 3.2$$

The demand coefficients depend on the type of house being heated. If electricity is used in a "standard house," the demand coefficient is considered to be 1.0. If electrical space heating is done in a "superior" house, the demand coefficient is 1.6; for oil heating in an inferior house, the demand coefficient is .71.

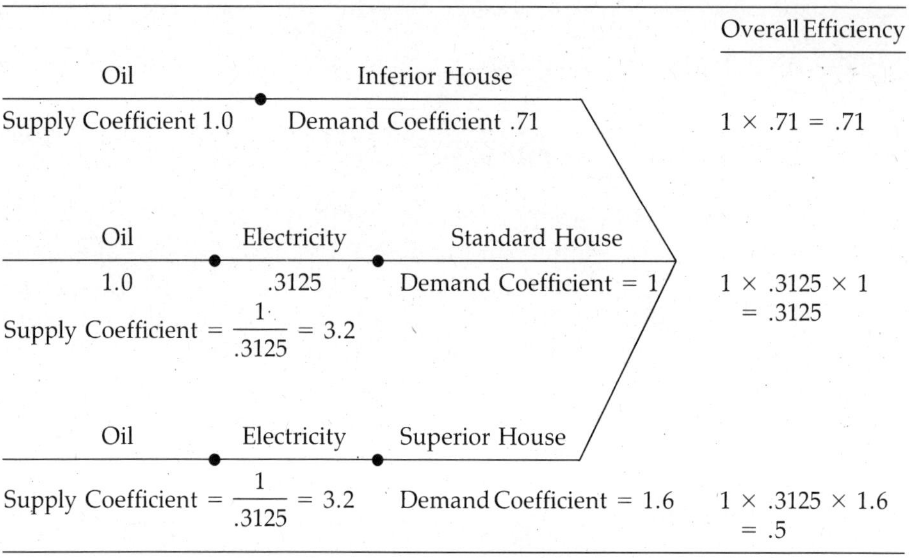

The overall efficiency (relative) is the product of e_i, which is the reciprocal of supply coefficient and d_k, the demand coefficient.

From the above, we can see that the overall efficiency of space heating by electricity in a superior house is lower than oil heating in an inferior house, in spite of better insulation, because of low conversion efficiency of oil electric generators. Thus it will be more efficient to use oil directly for space heating rather than through the medium of electricity.

MATHEMATICAL MODEL

The linear programming model is formulated in a manner similar to the classic transportation problem which determines the optimal routing of a product (in this case, intermediate forms of energy) from supply nodes to demand nodes. A cost and set of environmental impacts are identified for each unit of energy passing over each of the possible paths. The number of supply and demand nodes may vary according to the level of detail desired of the model. The U.S. energy system in the ERDA model includes 16 demands and 26 supplies. The example to be considered in this case is a simplification of the ERDA model and includes only 7 supply and 9 demand categories.

Between each supply category, $i\,(= 1, \ldots 7)$, and energy demand category, $j\,(= 1, \ldots 9)$, capable of using that energy resource, there exists a feasible path k (i.e., k is the set of feasible ij combinations). For each realization path k, the energy source (i) is converted (at an efficiency e_i) to an intermediate energy form (X_k). Hence, the maximum amount of intermediate energy available from a particular energy resource equals the energy resource's annual capacity of generation (S_i) multiplied by the associated conversion efficiency. In other words, to obtain an amount X_k of intermediate energy, X_k/e_i of energy resource is required. The intermediate energy is used to satisfy the demand (D_j) at a relative efficiency d_k. Wherever the energy resource is directly used for meeting the demand without intermediate conversion, for example, oil used for process heating, the conversion efficiency and hence supply coefficient is taken as 1, in this simplified model. The interpretation of the demand coefficients is explained under the relevant demand constraints.

A cost (c_k) and a set of environmental pollution coefficients (a_{kp}), are also defined per unit of intermediate energy form. The costs reflect extraction, refining, conversion, plant operations and capital cost, and transportation, as well as final conversion costs prior to utilization. The environmental and technical equations are constrained by placing upper limits (E_p) on the amounts of pollutant of type p.[1]

The mathematical formulation of the LP model is then as follows:

$$\text{Min:} \quad \sum_k c_k X_k$$

$$\text{Subject to:} \quad \sum_{k'} \frac{1}{e_i} X_k \leq S_i \qquad \text{for } i = 1, \ldots, 7$$

$$\sum_{k'} d_k X_k = D_j \qquad \text{for } j = 1, \ldots, 9$$

$$\sum_{k'} a_{kp} X_k - E_p = 0 \quad \text{for } p = 1,2,3$$

$$X_k \geq 0$$

Note: k' refers to a subset of k corresponding to those k's which are associated with the particular i (in the supply constraints), j (in the demand constraints), or p (in the pollution constraints).

It is now possible to enumerate the specific constraints that are required to solve the allocation problem for a 7 × 9 matrix of supplies and demands. The numbering of the X_k's, the costs (c_k) associated with each, and the values of S_i

[1]In the environmental equations, E_p is typically expressed in pounds per BTU (British Thermal Unit) of energy consumed and usually not used as a constraint value, because it is difficult to determine a national figure for the maximum level of environmental pollutants. It is, rather, moved to the left-hand side of the equation and incorporated as a variable in order to simply keep track of the emission level (as will be explicitly illustrated later in constraints 21, 22, and 23).

and D_j (the right-hand sides of the supply and demand constraint equations) are given in Table 1, Exhibit 4.[2] For example, in row 3 and column 1 (entry labelled as box 9 in Table 1, i.e., $k = 9$) natural gas used as fuel for space heating costs $1.25 per 10^6 BTUs and S_3, the maximum amount of natural gas that can be extracted and refined annually is 26.4×10^{15} BTUs. The demand coefficients, d_k, and the supply coefficients $1/e_i$, are given in Tables 2 and 3 of Exhibit 4.

Note that Table 1 does not specify the demand for peak load. The peak load demand depends on the total electrical energy used and hence on the variables which are determined internally in the model. In this simplified model, it is assumed, based on historical percentages, that specified fractions of total electrical energy used for meeting particular demands are to be supplied by peak load equipment, as indicated below.

End Use	*Percent*
Space Heat	Nil
Air-conditioning	20
Other Elec. Consumption	5
Process Heat	5
Public Transport	10

Other electric consumption refers to the requirement for electricity, not specificially included in the other eight categories, for end uses such as residential and commercial lighting, and so on. In this model, the peak load demand will be exclusively met by the turbines. A single supply limit is specified for oil-electric and turbines together, as all turbines are assumed to be oil fired.

The values of the decision variables X_k are in units of 10^{15} BTU. The constraints for the LP model can be derived as follows.

DEMAND CONSTRAINTS

The coefficients are obtained from Table 2 and the right-hand-side values from Table 1, both of Exhibit 4.

1. Space Heat

The demand of 10.6×10^{15} BTU for space heating is specified in terms of standard houses. If a house is built with better insulation, it requires less energy. In other words, one unit of energy can meet more than one unit of demand expressed in terms of requirement for a standard house. Similarly, if the house is poorly insulated, one unit of energy can meet less than one unit of demand. Generally, electrically heated houses are equipped with superior

[2]All values in Exhibit 4 are projected figures for 1985.

EXHIBIT 4

Table 1. Cost Coefficients and Constraining Values

Supply	Demand: Space Heat	Air-Conditioning	Other Electric Consumption	Peak Load	Process Heat	Petro-chemical	Air Transport	Public Transport	Private Transport	S_i (1985 values)
Coal					1 .53	2 1.14				21.6
Oil	3 1.84				4 1.45	5 1.21	6 1.64	7 2.17	8 2.51	52.0
Natural Gas	9 1.25	10 .90			11 .64	12 .64				26.4
Coal-Electric	13 9.32	14 11.14	15 5.87		16 6.13			17 8.17		11.5
Oil-Electric	18 9.66	19 11.12	20 5.86		21 7.06			22 8.70		7.5
Turbine	23 11.93	24 12.93	25 10.00	26 24.89	27 10.14			28 11.27		
Light Water Reactor (LWR)	29 9.69	30 11.73	31 5.84		32 6.12			33 8.39		16.2
D_j (1985 values)	10.6	2.2	7.2		19.0	8.0	3.2	1.6	2.4	

The cost coefficients c_k are in units of dollars per 10^6 BTU; D_j, S_i are in units of 10^{15} BTU.

Table 2. Demand Coefficients (d_k)

Supply	Demand: Space Heat	Air-Conditioning	Other Electric Consumption	Peak Load	Process Heat	Petro-chemical	Air Transport	Public Transport	Private Transport
Coal					1 .70	2 1.0			
Oil	3 .71				4 .68	5 1.0	6 .30	7 .20	8 .20
Natural Gas	9 .78	10 1.8			11 .64	12 1.0			
Coal-Electric	13 1.6	14 2.4	15 1.0		16 1.0			17 .27	
Oil-Electric	18 1.6	19 2.4	20 1.0		21 1.0			22 .27	
Turbine	23 1.6	24 2.4	25 1.0	26 1.0	27 1.0			28 .27	
Light Water Reactor (LWR)	29 1.6	30 2.4	31 1.0		32 1.0			33 .27	

(continued)

EXHIBIT 4 (Continued)

Table 3. Supply Coefficients ($1/ei$)

	Demand								
Supply	Space Heat	Air-Conditioning	Other Electric Consumption	Peak Load	Process Heat	Petro-chemical	Air Transport	Public Transport	Private Transport
Coal					1 1.0	2 1.0			
Oil	3 1.0				4 1.0	5 1.0	6 1.0	7 1.0	8 1.0
Natural Gas	9 1.0	10 1.0			11 1.0	12 1.0			
Coal-Electric	13 3.2	14 3.2	15 3.2		16 3.2			17 3.2	
Oil-Electric	18 3.2	19 3.2	20 3.2		21 3.2			22 3.2	
Turbine	23 5.0	24 5.0	25 5.0	26 5.0	27 5.0			28 5.0	
Light Water Reactor (LWR)	29 3.6	30 3.6	31 3.6		32 3.6			33 3.6	

insulation than standard houses and oil and gas heated houses are equipped with inferior insulation. The demand coefficients for space heating represent the product of the conversion efficiency of the equipment and the insulation efficiency (relative) of the house. Thus .71, .78, and 1.6 are the relative overall efficiencies or demand coefficients of oil heated, gas heated and electrically heated houses respectively.

The constraint can be specified as

$$.71X_3 + .78X_9 + 1.6(X_{13} + X_{18} + X_{23} + X_{29}) = 10.6$$

2. Air Conditioning

Air conditioning units are heat pumps, which transfer heat from a low to a high temperature reservoir. Under typical operating conditions, for each unit of energy used to run the unit, more than one unit of heat can be transferred. The demand of 2.2×10^{15} BTU for air conditioning represents the amount of heat energy to be transferred for this purpose. Each unit of energy derived from natural gas can transfer 1.8 units of heat and each unit of electrical energy consumed for air conditioning can transfer 2.4 units of heat.

Total heat energy to be transferred = 2.2

Heat transferred by gas air conditioners = $1.8X_{10}$

Heat to be transferred by electrical air conditioners $= 2.2 - 1.8X_{10}$

Electric energy required for transferring this heat $= \dfrac{2.2 - 1.8X_{10}}{2.4}$

20% of this energy has to be supplied by the peak-load devices—namely, turbines—and the remaining 80% by the base load equipment—namely, coal electric, oil electric, turbines (base load) and light water reactors. The constraint can thus be written as

$$.8\left(\frac{2.2 - 1.8X_{10}}{2.4}\right) = X_{14} + X_{19} + X_{24} + X_{30}$$

This can be rewritten as

$$1.8X_{10} + 3X_{14} + 3X_{19} + 3X_{24} + 3X_{30} = 2.2$$

The constraint is written in this form to facilitate interpretation of shadow prices.

3. Other Electric Consumption

As 5% of the electrical energy is to be supplied by the peak load turbines and the balance by the base load equipment, the constraint can be written as:

$$X_{15} + X_{20} + X_{25} + X_{31} = (1 - .05) \times 7.2$$

or

$$1.05X_{15} + 1.05X_{20} + 1.05X_{25} + 1.05X_{31} = 7.2$$

4. Process Heat

Electrical energy required for process heating

$$= 19 - .7X_1 - .68X_4 - .64X_{11}$$

95% of this energy is to be supplied by base load equipment. So

$$.95(19 - .7X_1 - .68X_4 - .64X_{11}) = 1.0(X_{16} + X_{21} + X_{27} + X_{32})$$

or

$$.7X_1 + .68X_4 + .64X_{11} + 1.05X_{16} + 1.05X_{21} + 1.05X_{27} + 1.05X_{32} = 19$$

5. Petrochemical Industries

$$1.0(X_2 + X_5 + X_{12}) = 2.0$$

6. Air Transport

$$.3X_6 = 3.2$$

7. Public Transport

Demand of energy for public transport to be met by electric generators = $1.6 - .2X_7$.

Energy to by supplied by electric generators = $(1.6 - .2X_7)/.27$ as .27 is the demand coefficient. Of this, 90% is to be supplied by base load equipment. Hence

$$\frac{.9(1.6 - .2X_7)}{.27} = X_{17} + X_{22} + X_{28} + X_{33}$$

This can be rewritten as

$$.2X_7 + .3X_{17} + .3X_{22} + .3X_{28} + .3X_{33} = 1.6$$

8. Private Transport

$$.20X_8 = 2.4$$

9. Peak Load

20%, 5%, 5%, and 10% of electrical energy used for air conditioning, other electric consumption, process heat, and public transport, respectively, is to be met by peak-load turbines. Hence:

The peak load energy requirement for air conditioning

$$= \frac{.2}{.8}(X_{14} + X_{19} + X_{24} + X_{30})$$

The peak load energy requirement for other electric consumption

$$= \frac{.05}{.95}(X_{15} + X_{20} + X_{25} + X_{31})$$

The peak load energy requirement for process heat

$$= \frac{.05}{.95}(X_{16} + X_{21} + X_{27} + X_{32})$$

The peak load energy requirement for public transport

$$= \frac{.10}{.90}(X_{17} + X_{22} + X_{28} + X_{33})$$

The total of these peak load energy requirements should be equal to energy output by turbines for meeting peak load (X_{26}). Hence the constraint can be written as

$$\begin{aligned} .25X_{14} + .053X_{15} + .053X_{16} + .111X_{17} + .25X_{19} \\ + .053X_{20} + .053X_{21} + .111X_{22} + .25X_{24} + .053X_{25} \\ - X_{26} + .053X_{27} + .111X_{28} + .25X_{30} + .053X_{31} \\ + .053X_{32} + .111X_{33} = 0 \end{aligned}$$

SUPPLY CONSTRAINTS

10. Coal

Coal can be used for process heat, petrochemical, and coal-electric equipment:

$$X_1 + X_2 + 3.2(X_{13} + X_{14} + X_{15} + X_{16} + X_{17}) \leq 21.6$$

11. Oil

Oil can be used for space heat, process heat, petrochemical, air, public and privâte transportation, oil-electric, and for running turbines.

$$\begin{aligned} X_3 + X_4 + X_5 + X_6 + X_7 + X_8 + 3.2(X_{18} + X_{19} + X_{20} + X_{21} + X_{22}) \\ + 5.0(X_{23} + X_{24} + X_{25} + X_{26} + X_{27} + X_{28}) \leq 52.0 \end{aligned}$$

12. Natural Gas

It is assumed that natural gas can be used for space heat, air conditioning, process heat, and petrochemicals only.

$$X_9 + X_{10} + X_{11} + X_{12} \leq 26.4$$

13. Coal-fired Electric Capacity

$$3.2(X_{13} + X_{14} + X_{15} + X_{16} + X_{17}) \leq 11.5$$

14. Oil-fired Electric Capacity (including turbines)

$$3.2(X_{18} + X_{19} + X_{20} + X_{21} + X_{22}) + 5.0(X_{23} + X_{24} + X_{25} + X_{26} + X_{27} + X_{28}) \leq 7.5$$

15. Nuclear (Light Water Reactor) Capacity

$$3.6(X_{29} + X_{30} + X_{31} + X_{32} + X_{33}) \leq 16.2$$

Special Constraints

The following special constraints are imposed considering the national usage figures.

16. Limit on the Amount of Electricity that Can Be Used for Space Heat

$$1.6(X_{13} + X_{18} + X_{23} + X_{29}) \leq 2.8$$

17. Amount of Electricity Required for Ground Transportation

This constraint explicitly recognizes that certain public transportation facilities use electricity, and their total requirement (including peak load) is at least 0.05.

$$\frac{.27}{.9}(X_{17} + X_{22} + X_{28} + X_{33}) \geq .05$$

or

$$.3X_{17} + .3X_{22} + .3X_{28} + .3X_{33} \geq .05$$

18. Oil/Gas Ratio for Petrochemical Industries

This constraint is imposed to ensure that the petrochemical industries are assigned the energy sources in the ratio required by them.

$$-X_5 + 5.6X_{12} = 0$$

19. Limit on the Amount of Coal Used by Petrochemical Industries

$$X_2 \leq .35$$

20. Limit on the Amount of Coal Used for Process Heat

Most of the existing equipment used to produce process heat is oil or gas fired. Extensive changes in this equipment will be necessary to use more coal for process heat. Hence this limit is imposed.

$$X_1 \leq 3.4$$

ENVIRONMENTAL CONSTRAINTS

Environmental considerations are included by defining three nonnegative variables (E_1, E_2, E_3), one for each of the three types of pollutants considered, and including constraints which set the sum of all a_{kp} X_k's equal to that variable. The three types of emissions considered are SO_2 ($p = 1$), hydrocarbons ($p = 2$), and thermal ($p = 3$). The a_{kp} values for these emissions are given in Tables 1, 2, and 3 of Exhibit 5.

These constraints are used only to determine the levels of E_1, E_2, E_3, rather than imposing constraints on the other variables.

21. SO_2

$$\begin{aligned} &1.2X_1 + .37(X_3 + X_4) + .07(X_6 + X_7 + X_8) \\ &\quad + 2.9(X_{13} + X_{14} + X_{15} + X_{16} + X_{17}) + 1.2(X_{18} + X_{19} \\ &\quad + X_{20} + X_{21} + X_{22}) + 1.8(X_{23} + X_{24} + X_{25} + X_{26} + X_{27} + X_{28}) \\ &\quad - E_1 = 0 \end{aligned}$$

22. Hydrocarbons

$$\begin{aligned} &.08X_1 + .22(X_3 + X_4) + 1.2(X_6 + X_7 + X_8) \\ &\quad + .25(X_{13} + X_{14} + X_{15} + X_{16} + X_{17}) \\ &\quad + 3.8(X_{18} + X_{19} + X_{20} + X_{21} + X_{22}) \\ &\quad + 6.0(X_{23} + X_{24} + X_{25} + X_{26} + X_{27} + X_{28}) - E_2 = 0 \end{aligned}$$

23. Thermal

$$1.5X_1 + 1.1(X_3 + X_4 + X_6 + X_7 + X_8) + 1.1(X_9 + X_{11})$$

$$+ 2.4(X_{13} + X_{14} + X_{15} + X_{16} + X_{17}) + 1.1(X_{18} + X_{19} + X_{20}$$
$$+ X_{21} + X_{22}) + 6.0(X_{23} + X_{24} + X_{25} + X_{26} + X_{27} + X_{28})$$
$$+ 3.3(X_{29} + X_{30} + X_{31} + X_{32} + X_{33}) - E_3 = 0$$

EXHIBIT 5

Table 1. Sulfur Dioxide (SO_2) Emission Factors (10^9 lb $SO_2/10^{15}$ BTU Fuel Consumed)

	Converted to Electricity	Turbine	Space Heat	Transportation	Process Heat
Coal	2.9	0	0	0	1.2
Oil (.33% Sulfur)	1.2	1.8	.37	.07	.37
Gas	0	0	0	0	0

Table 2. Hydrocarbon (HDC) Emission Factors (10^8 lb HDC/10^{15} BTU Fuel Consumed)

	Converted to Electricity	Turbine	Space Heat	Transportation	Process Heat
Coal	.25	0	0	0	.08
Oil (.33% Sulfur)	3.8	6.0	.22	1.2	.22
Gas	0	0	0	0	0

Table 3. Thermal Emission Factors (BTU of Heat Dissipated/BTU of Intermediate Energy Form Produced)

	Converted to Electricity	Turbine	Space Heat	Transportation	Process Heat
Coal	2.4	0	0	0	1.5
Oil	1.1	6.0	1.1	1.1	1.1
Natural Gas	0	0	1.1	0	1.1
Nuclear	3.3	0	3.3	3.3	3.3

RESULTS, LIMITATIONS, AND FINAL COMMENTS

The solution to the Energy Supply and Demand Model—the values of the energy flows and emission levels (decision variables), total cost (objective function value), and shadow prices (dual variables) for the supply and demand values—is given in Exhibit 6. At a national level, ERDA and other agencies use this information or information from revisions to this "base case" to determine the effects of a new strategy (whether it be an alternative energy source or a new end-use device) and to allocate its research and development funds accordingly.

The model ignores many factors and subtleties relevant to the complex U.S. energy system. Such things as the price elasticities of demand; technogolocal, economic, or political problems associated with expansion or substitution of fuels are not considered directly by the model. Obviously, effective energy policy cannot be established solely on the basis of analysis by models like this one. What the model does accomplish, however, is to permit evaluation of the impact of the substitution of fuels as a function of demand. These long-term changes are of considerable importance to agencies such as ERDA, with respect to establishing guidelines for new energy programs.

The structure of the model and type of analysis allowed requires information in a form that is somewhat different from that usually compiled for energy

EXHIBIT 6
Computer Results for Linear Program

	Demand											
Supply	Space Heat	Air-Conditioning	Other Electric Consumption	Peak Load	Process Heat	Petro-chemical	Air Transport	Public Transport	Private Transport	S_i (1985 values) 10^{15} BTU	Supply used 10^{15} BTU	Dual (\$/10^6 BTU)
Coal					1 3.4	2 .35				21.6	14.2	0
Oil	3 0				4 13.18	5 6.49	6 10.67	7 7.75	8 12.0	52.0	52.0	−2.729
Natural Gas	9 12.06	10 1.22			11 11.96	12 1.16				26.4	26.4	−3.294
Coal-Electric	13 .75	14 0	15 2.36		16 0			17 .17		11.5	10.5	0
Oil-Electric	18 0	19 0	20 0		21 0			22 0		7.5	1.9	0
Turbine	23 0	24 0	25 0	26 .38	27 0			28 0				
Light Water Reactor (LWR)	29 0	30 0	31 4.5		32 0			33 0		16.2	16.2	−.008
D_j (1985 values) 10^{15} BTU	10.6	2.2	7.2		19.0	8.0	3.2	1.6	2.4			
Dual Variable (\$/10^6 BTU)	5.825	2.330	7.536	−38.537	6.146	3.938	14.565	24.497	26.197			

Total Cost: \$176.119 × 10^9

Emissions
SO_2: 21.26 × 10^9 lb
HDC: 42.78 × 10^8 lb
Thermal: 104.47 × 10^{15} BTU

Entries in matrix are values of X_k in 10^{15} BTU.

planning and evaluation. The model addresses specific end-use categories, the technologies they employ, and the possibility of substitution between different energy forms. Detailed information on the various categories' load structure and other salient information is sometimes not directly available; therefore, reasonable approximations are made. In some instances, the constraints or coefficients specified contain an element of arbitrariness. Nonetheless, they are reasonable estimates and adequately serve as a standard set for analysis. More significantly, for actual long-range planning activities, extremely accurate estimates of demand are not necessarily required. In evaluating alternative options the specific levels of costs, emissions, and other parameters of the energy system are often of less importance than the relative difference between the various options which are available. A comparison of two policy approaches generally require only a knowledge of the relative change between them in order to anticipate the long-term significance of these alternatives. The linear programming model of the Reference Energy System is a tool for just that purpose.

Avis Rent a Car System, Inc.

COMPANY BACKGROUND

Avis was formed as "Avis Airlines Rent-A-Car System" in December 1946 when Warren E. Avis of Detroit, a Ford dealer, opened the country's first airport rental counter at Willow Run Airport. The operation proved successful and Mr. Avis proceeded to franchise independent rental operators throughout the country at airport locations. The word "Airline" was dropped from the System's name in 1948, after Avis outlets were opened at downtown locations.

In 1954, Mr. Avis sold his interest in the company to Richard S. Robie, owner of the U-Driv-It Corporation of Boston, Massachusetts. Mr. Robie was the largest franchisee of the Hertz System at the time. By 1956, Mr. Robie had increased the number of cities under corporate ownership, as opposed to franchised cities, from two to sixteen. The expansion of direct corporate activity caused a severe capital drain, and Mr. Robie found himself unable to finance Avis' continued growth. In September of 1956 he sold the Avis Rent-A-Car System to a group of investors headed by Amoskeag Co., a Boston investment firm. Mr. Winston V. Morrow, Jr., who had been an attorney on the acquisition for Amoskeag, came to Avis as Executive Vice President in 1957. As further expansion demanded additional capital, Avis made its first public offering for $7 million, in 1960. Control of Avis was acquired, in March 1962, by Lazard Freres & Co., international bankers, with headquarters in New York City. By

Source: Based on material prepared by Stephen P. Bradley.

this time annual volume was $25MM and fleet size was 7,500 vehicles. A number of other changes resulted at the time, including the hiring of Robert Townsend, as President, who later became widely known for his unique business views, the rehiring of Mr. Morrow who had left Avis in January and was planning to start work for Hertz on April 1, 1962, and the appointment of Doyle Dane Bernbach as the corporate advertising agency. The following January saw the introduction of the "We Try Harder" campaign, followed by spectacular growth. Corporate revenues (sales of corporate locations alone) grew from $35 million in 1962 to $219.2 million in 1970. As of the end of 1970, the Avis System (the corporation and its 305 licensees combined) operated a fleet of over 112,000 vehicles in 1,712 cities in 90 countries and 983 airports.

In February 1971, a new phase in the advertising program was introduced when Mr. Morrow announced that Avis was going to become number one in the car rental industry. Key to attainment of the goal is the planned installation of a real-time, computerized, rental and reservation system, named "The Wizard of Avis," and continued expansion of overseas business.

ENTER MANAGEMENT SCIENCE

Early in 1970 Mr. Morrow attended a computer concepts course where he learned of new management science techniques. The idea of using a management science model for fleet planning was discussed with various officers, including John F. Stolle, an Avis Vice-President experienced in operations research, who agreed with Mr. Morrow that fleet planning was a potentially fruitful area. He asked to be provided with two men, one experienced in operations and one in financial planning. Messrs. Jack McKinstry and Steve Foerder, both of Avis headquarters, were assigned to the project. Mr. McKinstry had started with Avis in January, 1955 as a counter agent. He worked his way up to District Manager, then was brought into Avis' New York headquarters as the Director of Fleet and Administration, which position he held at the time of this assignment. He was intimately acquainted with the problems of fleet planning and control from a practical point of view, and had worked with city managers in the past on these and other problems. Mr. Foerder joined the accounting staff at Avis in December 1965. He spent the next 3½ years in general accounting, taking night courses and receiving an MBA in controllership in the meantime. At the time of the assignment, he was the Manager of General Accounting. Mr. Stolle noted that the basic strategy of the project would be to take a first cut at the improvement potential by developing and testing tools in a relatively short time and then refining these tools over a longer period as they are actually used in the field. Making this practical required that one district be selected for the initial development and testing work, with nationwide application following successful development.

The fleet planning team began work under the direction of Mr. Stolle in September, 1969 by initiating an inventory study of the fleet in "Avisville," a large eastern city. This location was chosen because it was reasonably close to Avis headquarters, and because the city was large enough to provide a good incentive for cost savings (Avisville had a fleet of between 1,300 and 1,600 cars, one of Avis' larger fleets). Jack McKinstry and Steve Foerder were, at this point, concerned primarily with studying the distribution and utilization of Avisville's fleet rather than with management science techniques. The design of a management science model was entrusted to Mr. Jack Kent, a consultant from International Associates, a well-known consulting firm.

THE VIEW FROM AVISVILLE

Mr. George Evans was the city manager of Avisville, a large eastern seaboard city, and one of Avis' more profitable markets. Mr. Evans' fleet serviced the needs of a large airport, as well as the city itself. Like all Avis cities, Avisville was responsible for planning and purchasing its own fleet. Proper fleet planning was vital to good financial performance, since having too few cars in a fleet resulted in lost revenue, as well as loss of customer goodwill, and having too many cars resulted in needless expenses. Fleet planning was a complex process: it was complicated to forecast the desired fleet size throughout the coming year and equally complicated to devise an ordering plan to provide that fleet in a cost effective manner.

Avis replaces its fleet, generally speaking, twice a year. Since so-called "second cycle" ordering programs (manufacturers programs for cars ordered during the second half of the model year) are often restricted by purchases during the "first cycle" (first half of the model year), and since Avis requires revenue and acquisition forecasts on an annual basis for budgetary reasons, annual planning is done for an 18-month time period.

Construction of the ordering plan was a time consuming "cut and fit" process, with a bewildering number of ways of acquiring cars available. There were three basic choices, leasing, outright purchase, and purchase with an agreed buyback price at a particular time. Within each of these basic choices, a large number of specific plans were available from different manufacturers. Many of these plans contained fairly complex restrictions and provisions, for example limiting cars available during the "second cycle" to a certain percentage of those purchased or leased during the "first cycle."

Good fleet planning involved consideration not only of the purchase terms, but of the disposal terms as well. If cars were purchased outright, the city manager would have to consider what used car prices would be six months or so after delivery, to determine when to dispose of the cars, and how many cars the local market could absorb. The best disposal date would, however, depend not only on these factors, but on the number of leased and other cars in the fleet and the dates on which those cars were due to be returned, and upon the way the manager intended to replace those cars. Thus, the fleet planning task involved thinking well over a year in the future, and required a detailed analysis of the interactions among acquisition dates, disposal dates, and restrictions of all plans available. The city manager had to stay within fairly broad limits in ordering his fleet (for example, at least 51% of the fleet had to be Chrysler products, and Avis usually liked another 5–10% to be General Motors products), but he enjoyed a good deal of freedom in determining which cars to order, how many to order, and how to dispose of Avis-owned portions of his old fleet. However, constraints imposed by the need to dispose of the present fleet, operating restrictions, uneven demand patterns, and manufacturers restrictions which tie orders in one cycle to orders in another, made it difficult to identify feasible ordering plans that followed closely the goal fleet. Further complicating the process was the fact that today's decisions affected the decisions to be made in the next planning cycle.

City managers generally had their own methods for developing alternative ordering plans. Mr. Evans would first attempt to select an ordering plan, regardless of cost, such that the pattern of cars entering and leaving the fleet would match the forecasted fleet needs. Various graphical aids to clarify the effects of the different plans on the fleet size were available to the Avisville staff. He would then compute the cost per car-month for each possible acquisition plan.

An alternative was then examined to see if a lower cost acquisition plan could be substituted for a higher cost plan without disturbing the fleet size pattern. Then the new ordering plan was examined to see if further substitutions could be made, and so on until a final plan was derived. Working out each of these plans was a lengthy process, as changing even one month's ordering pattern frequently threw fleet size off in subsequent months, creating a need for further changes and so on.

Finally, once Mr. Evans and his staff had decided how many of each basic type of car to acquire, they had to decide how acquisitions were to be divided by types of cars, specific models and accessories.

Mr. Evans pointed out that in actual practice the acquisition plan was revised continually throughout the year in response to actual rental experience, late deliveries, etc. New plans were, as a rule, necessary every few weeks. The initial plan required considerable time and effort to construct, on the order of two weeks. The revisions took much less time per revision, but over the year they represented a great deal of effort.

THE DEVELOPMENT AND INTRODUCTION OF THE FLEET PLANNING MODEL

Until the spring of 1970, the McKinstry-Foerder inventory study and the Kent management science work proceeded somewhat independently, though the Kent work depended on the McKinstry-Foerder work for data and insight into key factors in fleet planning. There was some contact among the men involved in the projects, both through joint field work and through Jack Stolle. Mr. Kent on numerous occasions conducted in-depth discussions and strategy meetings with McKinstry and Foerder to gain information affecting his approach to the problem.

In December of 1969, Mr. Bob Kavee, of ITT Data Services, joined Jack Kent in structuring the management science phase of the work. By this time Messrs. Stolle and Kent had formally decided to concentrate on building a linear programming (LP) model to help develop fleet ordering plans. A number of standard computer programs were available for solving such problems, leaving the team free to concentrate on model structure and data acquisition problems. By March 1970, the following had been accomplished:

1. A mathematical model (set of equations) had been selected.
2. A consistent set of notation, primarily used for naming rows and columns, had been developed.
3. Data input formats were determined.
4. A large program to rearrange input data, check for inconsistencies and duplicates, and correct minor errors had been written.
5. A matrix generator, a program to arrange data into a form suitable for the standard LP problem solver had been written.
6. All necessary program- and job-cards to interface the output of the matrix generator with IBM's MPS package and control the computations were written.

When output from the model was first obtained for the Avisville data, significant differences were found between fleet plans being produced by the model and plans actually used in the field. In order to corroborate these findings it was decided to expand the study to five representative cities. Inventory studies either had been or were in the process of being done in most of the cities, so some of the necessary data had already been gathered. For other data necessary to run the model, the team tried having the city managers assess the data by themselves. This approach was later abandoned, however, because the city managers either did not understand what was wanted or they provided inconsistent data. For example, a manager might specify that a minimum of 10% of his fleet should be General Motors products at any time, yet the manager would, sometimes, when doing manual planning, allow the fleet to be as low as 3% GM. Similar misunderstandings came up in other areas, making it necessary for the team to work closely with each city in order to get trustworthy data. Messrs. McKinstry and Foerder were, by this time, becoming increasingly involved in the data gathering process and the entire LP project.

Data required from the field was extensive. As mentioned above, good fleet planning involved looking ahead for 18 months, this time being long enough to plan for orderly replacement of the fleet after six months and again after a year, yet short enough to keep the problem of manageable size. Thus, 18-month forecasts were required by month, for the following:

1. Total revenues expected.
2. Target revenue per unit.
3. Target fleet size, (which is just the quotient of the last two figures).
4. Used car prices and maximum number of cars saleable at that price.
5. Variable overhead (for gas, oil, insurance, etc.) expressed as a percent of total revenue.
6. Limits on the number of cars which could be processed in and out of the city fleet.

Other data required were:

1. Existing inventory of cars in the fleet, by manufacturer and specific purchase plan.
2. Existing commitments for future acquisitions, again by manufacturer and plan.
3. Acquisition and ownership costs, such as licenses and registration fees.
4. Limitations on fleet composition (for example, at least 51% of the fleet was usually required to be Chrysler products).
5. Acquisition costs of cars under each of the manufacturers plans (the plans, as announced by the manufacturers, usually applied nationwide, but local factors such as transportation and make ready charges would affect the ultimate price to the fleet).
6. Any special programs or needs of the cities.

A number of these figures were being computed by city managers for their manual fleet planning effort, but nonetheless, data collection and verification became one of the major concerns of the analysis team.

By April, computer runs for the five selected cities seemed to be reasonably accurate, and the managers of each city were invited to a review meeting at Avis headquarters. The purpose of the meeting was to present the concept of the LP model to the field, demonstrate results of the first runs, and solicit comments and suggestions. Up until this time the field personnel knew little about the model, as their primary contacts had been with McKinstry and Foerder in connection with the inventory study.

In the presentation, the general role and process of fleet planning was discussed, the concept of the LP model was presented, and results of comparisons made between LP produced fleets and fleets budgeted for the previous year and remaining months of 1970 were given (see Exhibits 1 and 2). Strong differences of opinion between headquarters and city operating personnel on one hand and the analysis team on the other, quickly became apparent. The reactions of operating personnel were summed up in the following terms:

> *This is a fast moving, fast growing industry requiring rapid decision making, often based on judgment. There was some question about the usefulness of a rigid system. Also, each city manager is somewhat of a local entrepreneur, and there was a concern that now a machine from outside would be runnng the city. In a sense, those behind the LP concept were viewed as strangers in the temple, making waves and impugning the abilities of the city managers.*

There was a good deal of spirited discussion at the meeting, and in the end, a number of specific objections to the model resulted:

1. The model was based on a monthly ordering cycle. This effectively ignored a two-week grace period allowed on Chrysler leased cars, which constituted the bulk of Avis' fleet. This grace period had the effect of turning a 6-month into a 6½-month lease for the same cost.
2. The model assumed cars ordered would arrive on the first of the month, and that cars disposed of would leave the fleet on the 30th of the month. In fact, cars would both enter and leave the fleet continually.

EXHIBIT 1
Avisville: Demand Versus Hard-planned Fleet

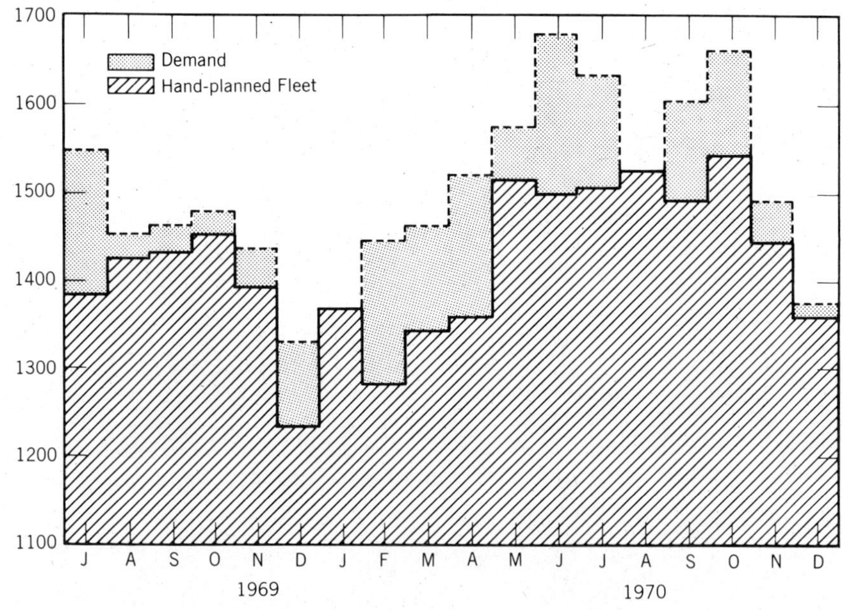

EXHIBIT 2
Avisville: Demand versus LP Fleet

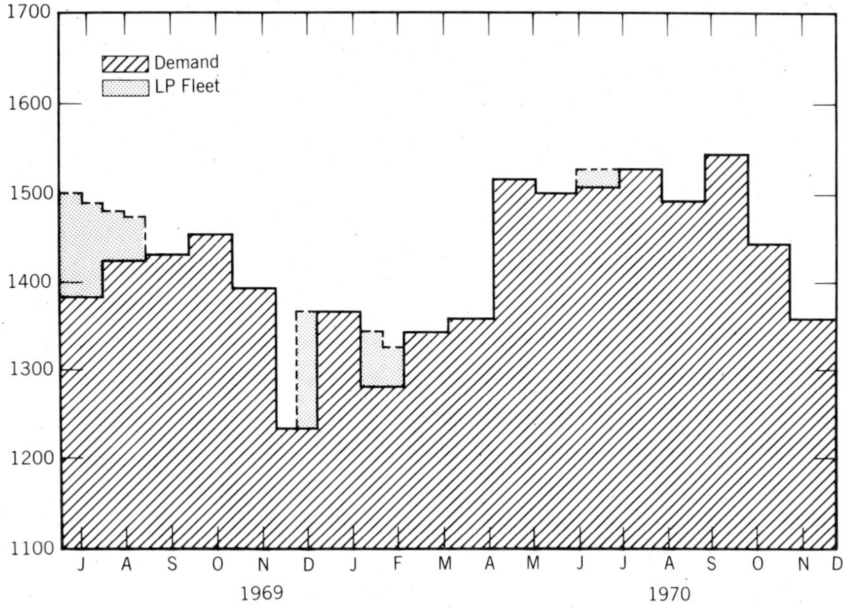

3. The model structure assumed one type of "basic" car was acquired from each of the manufacturers. In reality, a mixture of standard sized cars, intermediates, and compacts (and later subcompacts) were being acquired.

After the April meeting the LP team made a number of changes to overcome these objections. In order to allow for the grace period, the basic model cycle was changed from one month to a half month. The second objection was overcome by spreading the arrival and departure times evenly throughout the month. The team felt that the third objection could not be overcome by making specific allowance for each type of car, as this would make the LP model too large. Instead, the field would be asked to specify in advance the "mix" of, say, Chrysler cars desired. An "equivalent" car, whose costs were symbolic of the desired mix, would then be used in the model.

In June 1970, a presentation was made to Avis' top management, including Messrs. Morrow (President), Knight (General Manager of the Rent A Car Division), and Stepnowski (Vice President of Operation). The content of the presentation was similar to that of the April meeting (except that the suggestions discussed above had been incorporated), but the tone, in general, was that of a "soft" sell. In the words of Mr. Stepnowski, the group had "done their homework" between the April and June meetings, and the reception of the model concept was favorable. Mr. Knight, while not enthusiastic about the model, did not oppose it, and Mr. Morrow decided to go ahead with the application phase. He asked Jack Stolle how many cities should be brought into the model. Stolle committed himself and his team to analysis of Avis' 19 largest fleets, comprising about 85% of all cars owned by Avis in the U.S.

This represented an enormous problem to the LP team, as fleet plans would have to be prepared within about a month to affect September orders. Experience during the spring of 1970 indicated that the headquarters team would have to actively participate in the data gathering process. Also, there was no program written to translate results of the model into terms readily understood by management, so interpretation of the results was difficult and time consuming. Finally, Steve Foerder and Jack Kent left the analysis team at this time, so that Kavee and McKinstry were the only people, beside Stolle, who were intimately acquainted with the model. Steve Foerder had joined the team originally with the understanding that this was to be a one-year assignment. Since the year was nearly up and a new phase of the LP work was about to start, he felt that this was a good time to leave. Jack Kent left the project because his assignment, the structuring of the model, was essentially completed.

Just hours before the meeting at which the "go" decision was made, the auto manufacturers cancelled most discount plans for large fleet purchases. This somewhat simplified the fleet planning process by eliminating numerous alternatives, but also had the effect of raising ownership costs substantially. Since rental companies operate on low sales margins, and are reluctant to risk losing market share by raising rental rates, cost reduction became all the more important. On balance, Avis management decided to use the model, since even in the simplified planning environment some cost savings were anticipated.

Jack McKinstry built a 5-man team of some of the younger, more promising city personnel, spent the last half of June training them in the data gathering process, and went with them in the field in July. By the middle of July the first runs were being produced. Model results looked reasonable, except that the revenue per car figures resulting from the LP plans were poor. Upon examination, the main problems seemed to be:

1. City managers were imposing constraints which they did not follow themselves at all times.
2. The model required complete satisfaction of demand at all times, while city managers would allow for occasional turndowns during peak demand periods.
3. Some of the data were still inconsistent or unreasonable.

The first and last of these problems were attacked by discussing the issues directly with the field, a process made fairly difficult by the reluctance of many field managers to believe in the model. The second problem was solved by allowing the model to cut demand peaks in order to eliminate cars which would not pay for themselves. The team had planned, at some future date, to allow a city's fleet to be broken into segments. The first a% of the fleet, for example, would be specified to produce $A of revenue per car, the next b% to produce $B, where B is less than A, and so forth. This feature was incorporated into the model at this time to allow the peak cutting to be done in an optimum fashion. Mr. McKinstry spent a good deal of time discussing this structure with field personnel, and eventually derived cutoff points and revenue levels for each segment which seemed applicable to broad regions of the country. Using this structure, if the model were allowed to maximize profits with no limits on fleet size, then the less profitable upper segments of a city's fleet would be eliminated. The implication was, of course, that the marginal cars were not supporting themselves. Some peak cutting seemed justified and indeed, was done informally when city managers constructed manual fleet plans. Management, however, would have considered the fleet cutting done by the unconstrained model excessive, so Jack McKinstry put a limit, more or less arbitrarily established, on the degree of peak cutting done by the model. The next set of runs, incorporating these changes were "on the money" for revenue per unit, according to Mr. McKinstry. It soon became standard procedure to provide both a regular and a "cut peak" run for each city, with the "cut peak" plan being the recommended one.

By now it was too late to affect the September ordering process directly, but the runs were distributed to the field with the expectation of having them used for subsequent ordering periods. One result that management found particularly interesting was that the model requested the maximum number of Chrysler lease cars possible, while general feeling at Avis was that a large number of General Motors cars should be acquired by outright purchase. For this and other reasons, the level of acceptance of the LP plans by the field remained low. Field managers continued their manual planning as a double check on model results. At the direction of Mr. Stepnowski, the number of cars ordered by any city was the lesser of either the LP or the manual plans.

A major development came in September 1970, when in Avisville a number of outright purchase orders were recommended, mostly for GM cars, in direct opposition to the model's results. Those connected with the LP project objected to the recommendation, and Dick Knight stopped all ordering until the question could be resolved. Bob Kavee ran a new set of plans for Avisville, imposing the Avisville share of this order (530 cars) as a constraint on the model. The best plan the model could come up with, given that the 530 cars were on order, was $94,000 more expensive than the previous solution. This one order affected the ordering pattern for the entire year, resulting in excess cars for a number of months in the future. Only if the net cost of these cars to Avis were $250 less than forecasted (either through lower purchase cost or higher used car prices), and if the used car market would be able to absorb three times the maximum forecasted limit, would the new plan compare favorably with the old. Although the GM strike, by this time, had made the question of ordering any GM cars academic, this simulation convinced Dick Knight that the model was useful, and caused others with borderline opinions at Avis to favor the model. Field acceptance also improved after this time.

During the last few months of 1970, a program was written to make the model output much easier to read and understand. Continual improvements, particularly in this report package, continued to be made to the model.

GENERAL STRUCTURE OF THE MODEL

The formulation of the Avis LP model is summarized in Exhibit 3. The first row (Demand/Car Linkage) expressed the equation:

Beginning Inventory + ½ of Acquisitions − ½ of

Sales of Turnbacks = Average Fleet Size.

The factor "½" for the acquisitions and turnbacks reflects the assumption that cars are considered to enter and leave the fleet evenly throughout the acquisition and disposal periods. The second row (Segment Linkage) expresses the fact that when a city's fleet is broken into revenue bearing segments (the first a% producing $A, etc.) the sum of the number of cars in each segment must be the fleet size. The third row (Inventory Linkage) expresses the fact that the new inventory of cars must be equal to the old inventory plus acquisitions minus disposals. In the simplified structure given in Exhibit 3, both the new and the old inventories are in the "Inventory" column, and the ± 1.0 is used to remind one of this fact. The fourth row (Inequalities) uses a $\pm\%$ in the acquisitions column to allow for requirements such as "at least $x\%$ of the acquisitions must be car A" or "at most $y\%$ must be car B."

The "Bounds" section allows the user to keep the decision variables between certain bounds. For example, the inventory is required to be between the "Highest Inventory Permitted" and the "Lowest Inventory Permitted" at all times.

EXHIBIT 3
Simplified Structure of LP Formulation

			Columns				
Matrix	Fleet Size	Segments	Acquisitions	Inventory	Sale or Turnback		Right Hand Side
Demand/Car Linkage	−1.0		+.5	+1.0	−.5	=	0
Segment Linkage	−1.0	+1.0				=	0
Inventory Linkage			+1.0	±1.0	−1.0	=	0
Inequalities			±%			<	0

Rows

Bounds					
Upper Bound	None	Demand for Cars	Maximum Acquisitions	Highest Inventory Permitted	Maximum Exit
Lower Bound	Minimum Cars Supplied	0	Minimum Acquisitions	Lowest Inventory Permitted	Minimum Exit

Objective Function					
Profit	Periodic Fleet Charge (Taxes)	Marginal Profit Opportunity	Cost of Program	Cost of Holding Beyond Program Require-ments	Used Car Sales Price

Chapter 4

MULTIPLE CRITERIA DECISION MAKING

INTRODUCTION

Multiple criteria decision making has emerged as one of the most recent applications of management science. Traditional single criterion analytical decision models (e.g., linear programming) act as if a decision has its consequences on only a single dimension (e.g., profits) and does not affect other dimensions of the organization. This oversimplified assumption has been under attack for two reasons. First, it has been justly argued that a single criterion model does not represent the true characteristics of most real-world problems or the corporate decision-making process. In neoclassical economic theory, for instance, the firm has been viewed as a single goal organization, the goal being to maximize the owner's economic welfare. The ownership–management separation, along with the decentralization of stock ownership witnessed in the last few decades, has led to one criticism of the applicability of any single criterion model. Furthermore, studies have strongly indicated that, in practice, management decides upon allocation of scarce resources with reference to several (sometimes conflicting) goals rather than to a single one. In long-term (strategic) corporate planning these goals include: profit, market share, balanced business portfolio, long-range growth rate, risk, and so on. A multidimensional approach is also appropriate in short-term (tactical) planning. A production decision that takes into account only its possible effects on immediate profit and costs may have effects on other dimensions such as employment level, management-labor relations, product quality, and so on.

In addition, the recently recognized inadequacy of single criterion decision models stems from the growing pressure of social and environmental demands that the business community has experienced in the last decades. Management of most corporations have long recognized the need to respond to nonfinancial demands such as equal employment opportunities, pollution control, product safety, and work safety, to name only a few. Effective pressure has come from concerned private groups (environmentalist and minority groups, unions, consumer organizations, and the like) and governmental regulatory agencies. The effective answer to these demands is often an internal response that incorporates them in the company policy and obtains results by effective multidimensional planning, control, and performance evaluation.

As more management scientists recognized that satisfactory solutions to many real problems could not be obtained by modifying the problem formulation (for instance, by incorporating all the criteria but one in the constraint set of the problem), special methods have been developed. The most significant characteristic that distinguishes these methods from traditional solution procedures is the incorporation of the decision maker's preferences in the search for the optimal solution. Multiple objectives (often conflicting) that compete for the organization's scarce resources call for tradeoffs, and the decision maker plays a central role in the solution search. Thus, unlike traditional models, there is often no "objective" optimal solution to a multiple criteria problem. The purposes of this chapter are: (1) to demonstrate the multidimensional aspect of decision making; (2) to familiarize the reader with the formulation of the multiple criteria problem; (3) to introduce and illustrate the conceptual framework of the problem and the basic nature of its solution; and (4) to discuss some approaches to solve the problem in order to enable the nonmathematical reader to judge the adequacy or inadequacy of each approach. We make use of one particular example, selected mainly because of its simplicity; however, it contains the important aspects of the problem being discussed.

EXAMPLE: INVESTMENT PROBLEM

A firm has \$500,000 available to invest in one or both of two 1-year short-term investments. The first project has an estimated return of ten percent and is considered a low-risk venture, while the second is a high risk venture with an estimated return of 20%. The company can invest any amount (subject to the \$500,000 budget limitation) in either project and management wishes to obtain a balanced portfolio (in terms of profitabilty and risk involved). In particular, in view of the nature of its investments in other ventures, the company wants to invest at least \$100,000 in the low-risk project. How should the \$500,000 be budgeted?

Let the two decision variables x_1 and x_2 denote the amounts invested in the low- and high-risk projects, respectively. The constraints restricting these decision variables are:

$$(1)\quad x_1 + x_2 \leqslant 500$$

$$(2)\quad x_1 \qquad \geqslant 100$$

The two relevant criteria with which management is concerned are profitability and risk. Let c_p denote the profitability of the portfolio, which can be expressed as the total return:

$$c_p = .1x_1 + .2x_2$$

As an ad hoc way of looking at the criterion "risk," we define a new criterion—"security": the larger the amount invested in the low-risk project, the higher is the security of the portfolio. We denote security by c_s and measure it by the difference between the amounts invested in the low- and high-risk projects:

$$c_s = x_1 - x_2$$

Due to the problem's constraints, each criterion can assume values only in a certain range: if $x_1 = 500$, c_p assumes its lowest value, 50, and c_s assumes its highest value, 500. If, at the other extreme, $x_1 = 100$, c_p assumes its highest value, 90, and c_s assumes it lowest value, -300. To reflect that we would like each criterion to be as large as possible, the complete formulation of the problem can be written,

$$\begin{aligned}
&\text{Max} && c_p = .1x_1 + .2x_2 \\
&\text{Max} && c_s = x_1 - x_2 \\
&\text{Subject to} && \\
& && (1)\ x_1 + x_2 \leqslant 500 \\
& && (2)\ x_1 \qquad \geqslant 100
\end{aligned}$$

This simple example will be used for illustrative purposes throughout this chapter.

Geometry

Suppose that the decision maker already has decided on the relevant criteria for a problem at hand (the decision as to what criteria to include in the decision making process is a highly subjective one). Let $c_1, \ldots, c_m$ denote the relevant m criteria, and the m-dimensional space in which these criteria are defined will be referred to as the *criterion space*. Each feasible decision x_1, $x_2, \ldots, x_n$ has its consequences on these m criteria via the relations between the decision variables $x_1, \ldots, x_n$ and each of the criteria $c_1, \ldots, c_m$. In our example, the relationship between the c's and the x's is linear: $c_p = .1x_1 + .2x_2$ and $c_s = x_1 - x_2$. Thus, the consequences of the portfolio decision $(x_1, x_2) = (300, 200)$ is the criterion pair $(c_p, c_s) = (70,100)$. There is one, and only one, criterion pair associated with a specific decision. In general, a solution c_1, $c_2, \ldots, c_m$ (note that "solution" here is in terms of the *criteria*) is said to be *feasible* if, and only if, there exists at least one feasible decision (x's) that attains all the c_1 through c_m simultaneously. Thus, the solution $(c_p, c_s) = (90,0)$ is not feasible since there exists no feasible portfolio decision (x_1, x_2) that has a total return of $90,000 simultaneously invests the same amount in each project.

For our example, the feasible region or decision space, denoted X, in terms of x_1 and x_2, is shown by Figure 4-1. We can construct the set of all feasible solutions relative to c_p and c_s by calculating the criterion value associated with each feasible decision in the set X. We will denote this set of all feasible solutions by C. Figure 4-2 is the graphical representation of C for the example.[1]

Figure 4-2 will help us to clarify the two fundamental concepts of the multicriterion problem: *dominance* and *efficiency*. We say that one solution *dominates* another if the first has a higher value for at least one criterion and a lower value for no criteria. In Figure 4-2 we can see that $c_1 = (60, 50)$ is a *dominated* solution since $c_2 = (70, 100)$ is a solution with higher values for both criteria: c_2 dominates c_1. From studying the figure it also becomes clear that the solution $c_4 = (50, 500)$ is not dominated: there exists no other solution in C

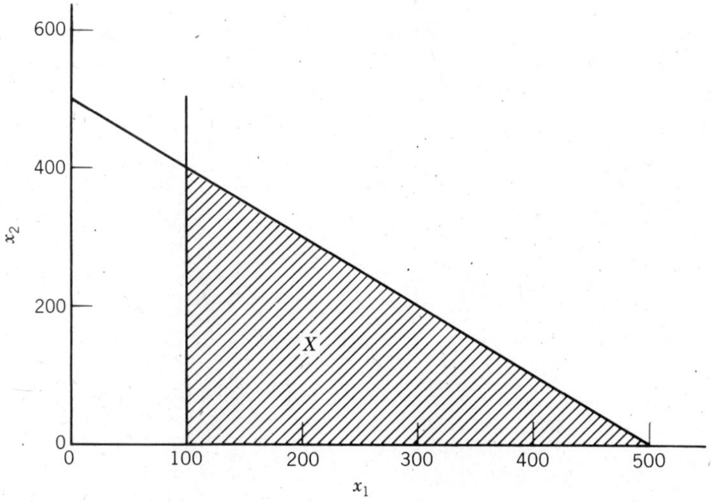

FIGURE 4-1. The decision space.

[1]For many problems, especially of constraints or criterion relationships are nonlinear, transforming X to C is not a trivial matter.

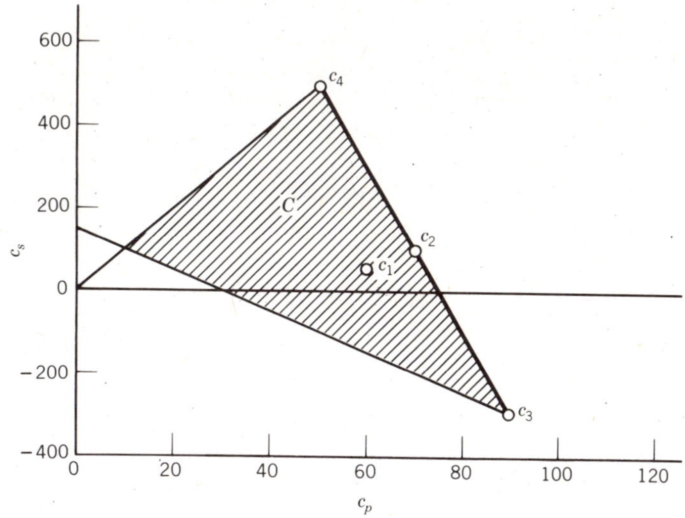

FIGURE 4-2. The criterion space.

having a higher value for at least one criterion. We notice that $c_2 = (70, 100)$ and $c_3 = (90, -300)$ are also undominated solutions. Further study of C reveals that all solutions along the heavy line are nondominated while *all* the other solutions in C are dominated. A nondominated solution is called an *efficient solution*, and the set of all efficient solutions (the heavy line) in C is called the *efficient frontier* of C. It should be noted that for an efficient solution there is no way to improve achievement on any criterion without worsening achievement on at least one other criterion. Thus, if the decision maker is not completely satisfied with the efficient solution $c_4 = (50, 500)$, which is the result of the portfolio decision $(x_1, x_2) = (500, 0)$, and wishes to increase the profitability of the portfolio, this can be achieved only at the expense of the security (i.e., increasing the risk).

PREFERENCES AND THE OPTIMAL SOLUTION TO THE MULTIPLE CRITERION PROBLEM

In this section we will define the concept of "optimal" for the multicriterion problem. As discussed earlier, there is no unique "objective" solution to this problem. The "best" solution depends a great deal on the decision maker's subjective preferences. The solution technique must also be subjective: the efficient frontier of the set C is explored by the decision maker who selects one efficient solution as the optimal solution to the problem.[2] The exploration of the efficient frontier is "subjective" since it is guided by the decision maker's preferences.

[2]Only an efficient solution can be considered truly optimal, since, for any nonefficient point, there is another that dominates it.

Preference Structure

If a sequence of efficient solutions are generated, the decision maker can select an optimal solution by *comparing* different ones. The classical approach to this problem is based on the fundamental notion of tradeoffs and preferences. For now forget about the set of attainable solutions in the criterion space (i.e., the set C). An indifference curve connects points at which the *value function* representing the decision maker's preferences is the same. Assessing an indifference curve that truly reflects preferences is, in practice, quite difficult. Any point on the curve represents a criterion pair (not necessarily feasible!) and if this is a true indifference curve, the decision maker is indifferent between any point on this curve (such as A) and any other that lies on the same curve (such as E or H). Suppose the decision maker is asked: "Consider the solution $A = (112, -225)$, which has high profitability but also a high degree of risk. Suppose the security criterion is increased by 100 (i.e., $100,000 is withdrawn from the high-risk project and invested in the low-risk one), how much would the profitability c_p have to decrease in order for you to remain indifferent?" The decision maker is very much concerned about the great deal of risk involved in solution A and after a deep reflection the answer is that the profitability has to be decreased by $20,000 in order for him to remain indifferent. From this we learn that he is indifferent between $A = (112, -225)$ and $B = (92, -125)$ and therefore these two points must lie on the *same* indifference curve, as shown in Figure 4-3. This is the essence of the tradeoff issue: the

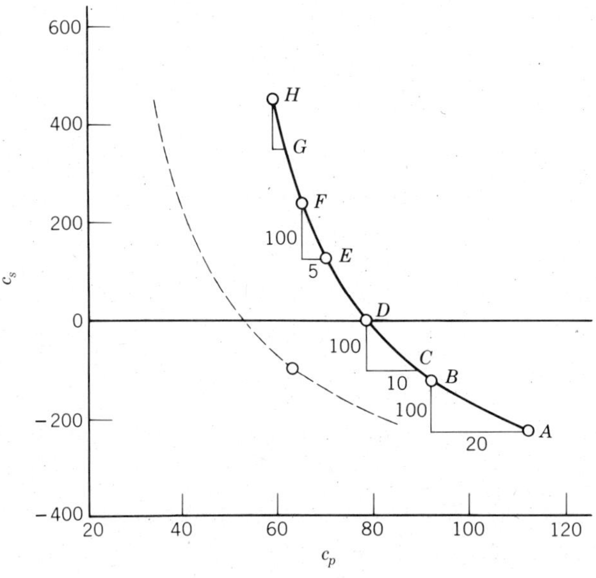

FIGURE 4-3. Indifference curves.

decision maker must give up some achievement in one criterion in order to improve achievement in some other. We continue to investigate the decision maker's preferences by asking the same question, but assuming now that $C = (88, -100)$ is the starting point. However, this time the decision maker is less concerned about the risk involved in solution C than when considering A, and therefore is willing to give up only \$10,000. This, again, implies that $C = (88, -100)$ and $D = (78, 0)$ lie on the same indifference curve. When the decision maker faces the same question at point $E = (70, 125)$, profitability is more important than risk and he will give up only \$5000 in order to decrease the risk. Finally, at point $G = (61,350)$, only \$2000 will be given up (see Figure 4-3). We notice that the answers of the decision maker to the tradeoffs questions *depend on the levels* of c_p and c_s at the relevant solution. We can quantify this concept by introducing the concept of *marginal rate of substitution* (MRS). If, at the point $A = (112, -225)$, the decision maker is willing to give up 20 units of c_p (i.e., \$20,000) in order to increase c_s by 100 units, then we say that the marginal rate of substitution of profitability for security at A is $20/100 = .20$. In other words, the decision maker is just willing to "pay" \$200 (.20 units of c_p) for one unit increase in c_s (i.e., withdrawing \$1,000 from the high-risk project and investing it in the low-risk project). At point C, however, the decision maker is willing to "pay" only \$100 for one unit increase in c_s (MRS(C) $= 10/100 = .10$); at point E only \$50 (MRS(E) $= 5/100 = .05$); and at point G only \$20 (MRS(G) $=$ $2/100 = .02$).

The indifference curves in the criterion space define the decision maker's *preference structure* in the sense that they enable the decision maker to compare *any* two solutions. The point A (Figure 4-3) is preferred to the point I (by the decision maker) since A lies on a high indifference curve than I would. This ability to compare in this way between any two solutions is necessary to obtain the optimal solution.

The Optimal Solution for a Multicriterion Problem

Figure 4-4 depicts the geometry of our multimaximization problem. The *optimal solution* is the efficient solution most preferred by the decision maker. The curves v_1, v_2, and v_3 represent three different indifference curves assessed by the decision maker. Any point on v_3 is preferred to any point on v_2; any point on v_2 is preferred to any point on v_1. Hence, the solution $c^* = (70, 100)$ is the optimal solution since any other efficient point (e.g., c_1) lies on a lower (or less preferred) indifference curve. Points c_2 and c_3, lying on v_3, would be preferred to c^*. Unfortunately, neither is feasible.

After determining the optimal solution (in criterion space), we must find the *optimal decision* that results in this solution. In the example, the optimal decision is (300, 200): to invest \$300,000 in the low-risk project and \$200,000 in the high-risk one.

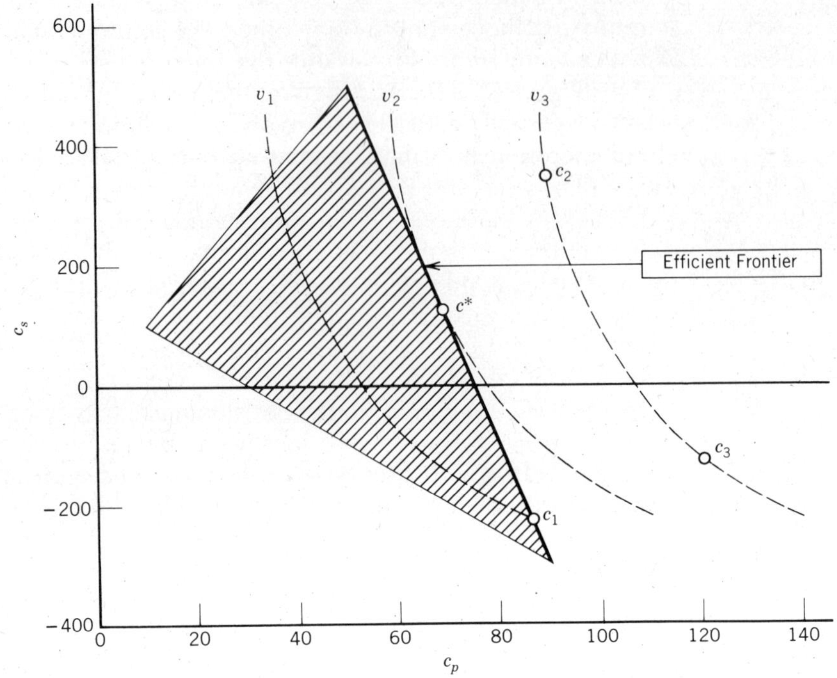

FIGURE 4-4. The optimal solution.

SOLUTION APPROACHES

In describing some of the various approaches to the multiple criteria problem, we pay special attention to the following points:

- Is an efficient solution obtained?
- Is the decision maker's preference structure taken into account?

We describe three general methodologies, for which many variations exist:

1. Satisficing approach.
2. Imposing an external and impersonal preference structure on the decision maker.
3. Interactive approach.

The section does not intend to be an exhaustive survey of the approaches and methods to the problem.

The Satisficing Approach

This approach does not look for the best achievable solution but for one which is "good enough" in the sense that it ensures the attainment of some pre-determined minimum *performance levels*. No objective function is defined here. One such approach is illustrated in Figure 4-5 using the example. The decision maker sets a minimum performance level for each criterion: for example, the portfolio must earn at least $45,000 and the amount invested in the high risk project cannot exceed the amount invested in the low-risk project by more than $50,000. These two lower bounds define the set of all solutions that are "good enough" since they ensure the attainment of the decision maker's pre-determined minimum goals. Though the set of "good enough" (or satisficing) solutions might include part of the efficient frontier (as it does in Figure 4-5) it is not necessary that an efficient solution be chosen. Instead, a solution such as the point in Figure 4-5 may be chosen. Another example is illustrated in Figure 4-6 where in addition to the minimum goals, the decision maker sets upper bounds on each criterion: $c_p \leqslant 75$ and $c_s \leqslant 50$, defining an acceptance interval for each criterion (the method, accordingly, is often called "interval programming").

The satisficing approach is not regarded by some as a serious attempt to solve the multiple criteria problem, since it avoids the two cornerstone issues of multiple criteria decision making: efficiency and tradeoffs. Satisficing meth-

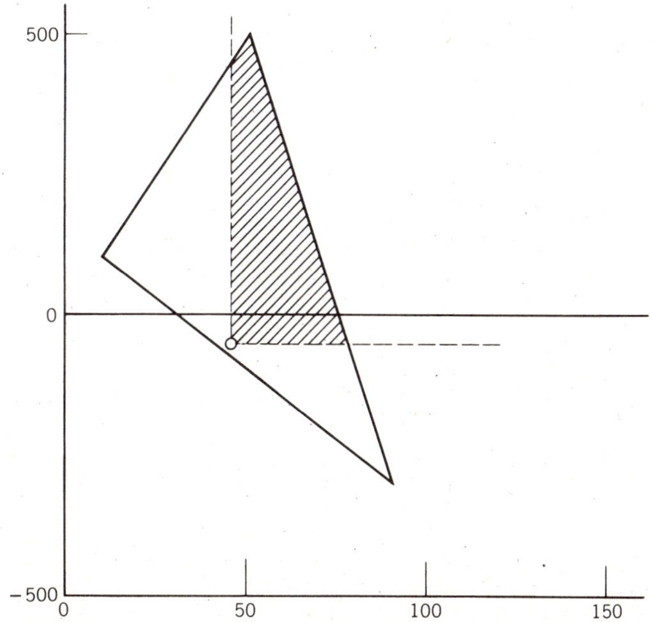

FIGURE 4-5. A satisficing solution.

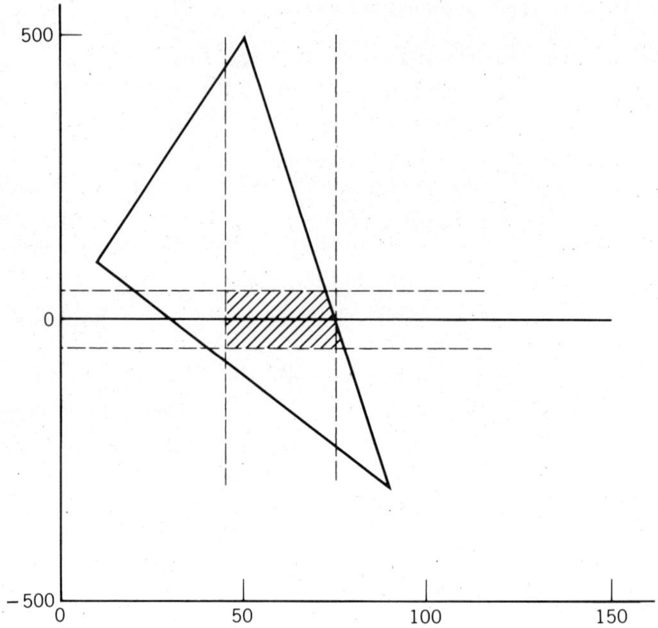

FIGURE 4-6. Interval programming.

ods have found acceptance, however, in practice, due to several factors, including:

1. They are simple, and give the decision maker control and a feeling of "hands on" choice.
2. Although the fact is obscured by a simple two-dimension example, generating the efficient frontier is often quite difficult.
3. In practice, assessment of indifference curves by specifying marginal rates of substitution is found by most decision makers to be nearly impossible, especially when three or more criteria are involved. Often, after an assessment has been made, fundamental inconsistencies are found—the decision maker has been unable to reconcile conflicts between the criteria through the specification of tradeoffs.

In this case, a "good enough" solution (though not efficient) is often acceptable.

Imposing External Value Functions

This approach, recognizing the difficulty expressed in factor (3) above, imposes a sometimes arbitrary, predetermined built-in-the-model value function on the decision maker. Though capable of producing efficient solutions, this methodology is not regarded by some as capable of generating truly optimal solutions since the decision maker's *own* preferences are not taken into consideration. Some of the more common methods of this type are:

Lexicographic Ordering. This method has the decision maker rank the criteria according to importance. The solution with the highest score on the most important criterion is selected as the best. The "second" criterion comes into consideration only if there is a tie on the first, and so on. Figure 4-7 illustrates the method for the example: if the decision maker decides that profitability (c_p) is more important to him than security (c_s), he will select solution *A* as the "best." This solution represents a portfolio with the highest possible profitability but which also involves the highest degree of risk. Three typical indifference curves that represent the value function implied by the lexicographical ordering are depicted in the Figure 4-7.

Converting Criteria into Constraints. The decision maker selects one criterion, determines minimum acceptable levels for the other criteria, and chooses the solution with the highest score on the most important criterion while ensuring the attainment of the minimum acceptable levels on the other criteria. Operationally this is done by incorporating all criteria but the most

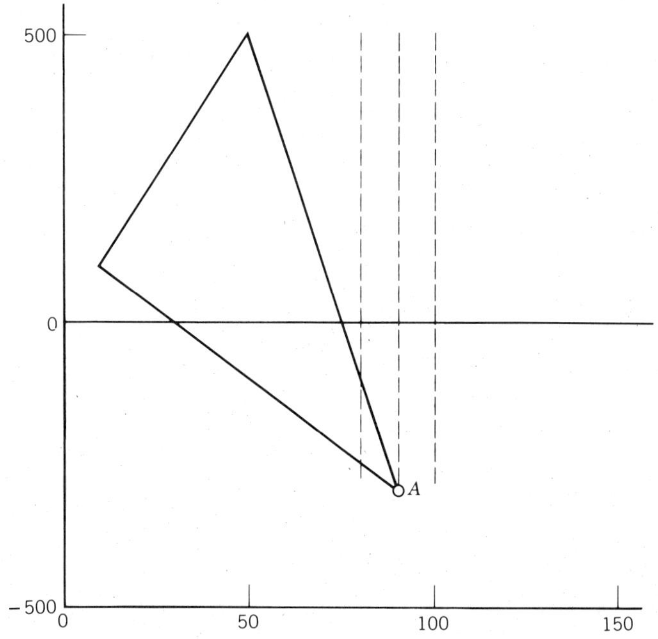

FIGURE 4-7. Lexicographical ordering.

important one in the constraint set of the problem. The method is illustrated in Figure 4-8. The profitability criterion is the most important and the decision maker decides that the amount invested in the high-risk project should not exceed the amount invested in the low-risk one by more than \$150,000 ($c_s \geqq -150$). Then the "best" point is A. Implied indifference curves are as shown. The method is similar to lexicographic ordering but gives some consideration to other criteria.

Goal Programming. This is perhaps the most commonly used of all multicriterion methods—a technique in which the decision maker is required to set a *goal* for each criterion. Since in most cases there exists no feasible decision that attains *all* these target goals simultaneously, the program finds a feasible decision that comes "as close as possible" to the simultaneous attainment of the target goals. Goal Programming methods differ from one another according to the way in which "closeness" is measured. The most widely used method will be discussed here.

For our example, assume that the decision maker sets the following target: a portfolio that will result in \$90,000 profit where the amount invested in the low-risk project exceeds the amount invested in the high-risk project by \$150,000. Examination of Figure 4-2 reveals that such a portfolio is *not* feasible. The decision maker may assign weights to the two criteria that reflect the

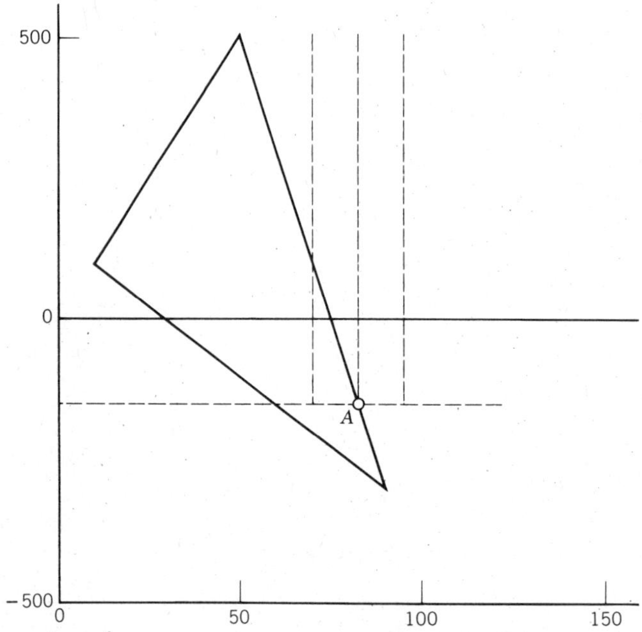

FIGURE 4-8. Converting a criterion into a constraint.

relative importance of each one. For instance, he might decide that the profitability criterion is twice as important than the security criterion.

Since any feasible solution will deviate from the goals (90, 150), we must establish criterion to measure these deviations. One possible measure of the deviation between a feasible solution (c_p, c_s) and the "target solution" (90, 150) is the weighted sum of the absolute values of the deviations. Mathematically, this weighted sum is expressed as

$$2|c_p - 90| + |c_s - 150|$$

The coefficient 2 reflects the fact that the decision maker has decided that profitability is twice as important as security. Since we want to come "as close as possible" to the attainment of the previously set goals, we wish to *minimize* this weighted sum. Since $c_p = .1x_1 + .2x_2$ and $c_s = x_1 - x_2$ in the example, we can formulate the goal programming model as follows:

$$\begin{aligned} &\text{Min} \quad 2|.1x_1 + .2x_2 - 90| + |x_1 - x_2 - 150| \\ &\text{Subject to} \\ &\qquad (1)\ x_1 + x_2 \leq 500 \\ &\qquad (2)\ x_1 \qquad \geqq 100 \end{aligned}$$

However, this is a nonlinear program, since the absolute value function is nonlinear. Fortunately, this problem can be reformulated as a linear program: Define the following variables:

d_p^+ = the *positive* deviation from the goal of $90,000 total return,

d_p^- = the *negative* deviation from the same goal,

d_s^+ = the positive deviation from the security goal of $150,000, and

d_s^- = the negative deviation from this goal.

Note that $d_p^+ > 0$ implies $d_p^- = 0$, since it is impossible to have *both* positive and negative deviation simultaneously. The same argument applies to d_s^+ and d_s^-. Then:

$$d_p^+ - d_p^- = 90 - (.1x_1 + .2x_2)$$

$$d_s^+ - d_s^- = 150 - (x_1 - x_2)$$

or

$$\begin{aligned} .1x_1 + .2x_2 - d_p^+ + d_p^- \qquad\qquad &= 90 \\ x_1 - \quad x_2 \qquad\qquad - d_s^+ + d_s^- &= 150 \end{aligned}$$

The goal programming model can then be reformulated as the following linear program:

$$\begin{array}{llr}
\text{Min} & 2(d_p^+ + d_p^-) + (d_s^+ + d_s^-) & \\
\text{Subject to} & & \\
& (1)\quad .1x_1 + .2x_2 - d_p^+ + d_p^- & = 90 \\
& (2)\quad x_1 - x_2 - d_s^+ + d_s^- & = 150 \\
& (3)\quad x_1 + x_2 & \leqq 500 \\
& (4)\quad x_1 & \geqq 100
\end{array}$$

The optimal solution is:

$$x_1^* = 325 \qquad d_p^+ = 0 \qquad d_s^+ = 0$$

$$x_2^* = 175 \qquad d_p^- = 22.5 \qquad d_s^- = 0$$

This portfolio has a total return of \$67,500, which is \$22,500 short of the profitability goal while the security goal is fully met.

To get some insight into this method we will examine the nature of this solution in the *criterion space*. First, observe that the objective function of the goal programming method, $2|c_p - 90| + |c_s - 150|$, is a *value function*. Furthermore, this value function is characterized by *constant* marginal rates of substitution. Implied indifference curves are shown in Figure 4-9. (Note that the scale of the horizontal axis is approximately 6.5 times that of the horizontal one.)

Several comments can be made at this point: (1) The model imposes on the decision maker an external value function that is naively simple: the indifference curves are linear. (2) The value structure imposed by the model is not necessarily "proper." For instance, it gives a higher value to $c^* = (c_p^* = 67.5, c_s^* = 150)$ than to $c_1 = (c_p = 80, c_s = 225)$, though it is clear that this latter point dominates c^*. (3) In this case, the model is not sensitive to changes in the relative weights. Though profitability goal was assigned a higher weight then the security goal, the *security goal* is the one fully met at the optimal solution. We would have to assign to the profitability goal a weight that is twenty one times the weight assigned to the security goal in order to change the optimal solution.

The Interactive Approach

The interactive approach to the multiple criteria problem stems, on one hand, from the idea that the particular decision maker's preferences must be taken into account in searching for the best solution and, on the other hand, from the recognition that the procedures for explicitly structuring this particular value function can be too demanding. Therefore, the decision maker is iteratively asked to provide information as to judgmental *local tradeoffs* in the neighborhood of a feasible point in the criterion space, and no attempt is made

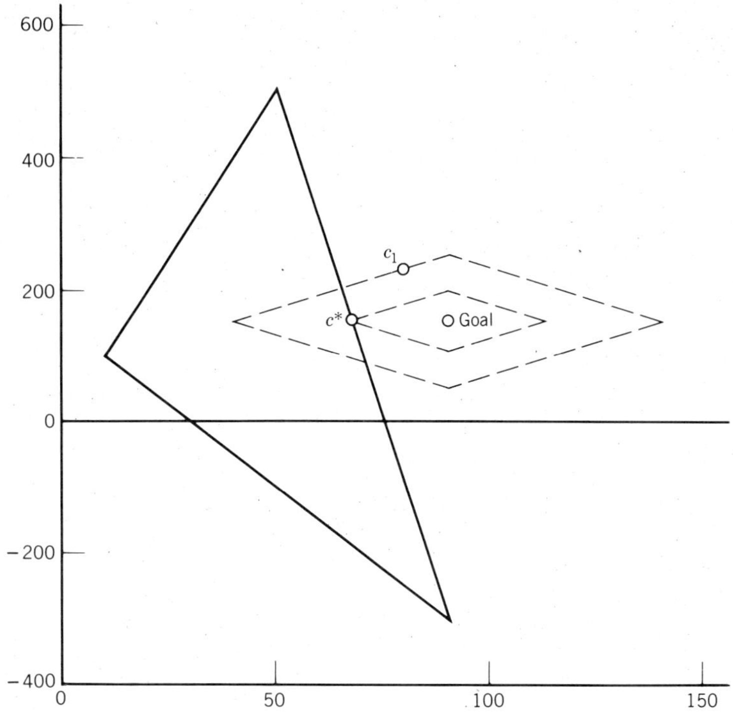

FIGURE 4-9. Goal programming.

to structure a complete preference order. The method is called "interactive" since each iteration is composed of two steps: a decision-making step and a computation step; that is, after each iteration, the decision maker is queried.

The search process can be described as follows. First, an initial feasible solution is selected. Then, a direction is determined in which the decision maker's preferences increase. To determine this direction, tradeoffs must be specified in the form of marginal rates of substitution. A new feasible solution is found along this direction such that it is preferred over the original point. This completes one iteration of the search process. This process is repeated until the "best" solution is found—that is, until a point is found such that any improving direction of movement, *in the decision maker's view,* lead to infeasibility. Figure 4-10 shows a typical search trajectory for this method. The nonlinear dashed curves are the contours of a hypothetical value function. (Notice that this value function is assumed only for illustration purposes; it is *not* constructed explicitly by the method.) The search trajectory shown is composed of the sequence (A, B, C^*), assuming that the initial chosen point is A.

It can be seen that in a two-dimensional problem, the convergence to the optimal solution is very fast. The initial solution A is not efficient and it has been chosen for illustrative purposes only. In an actual situation the method

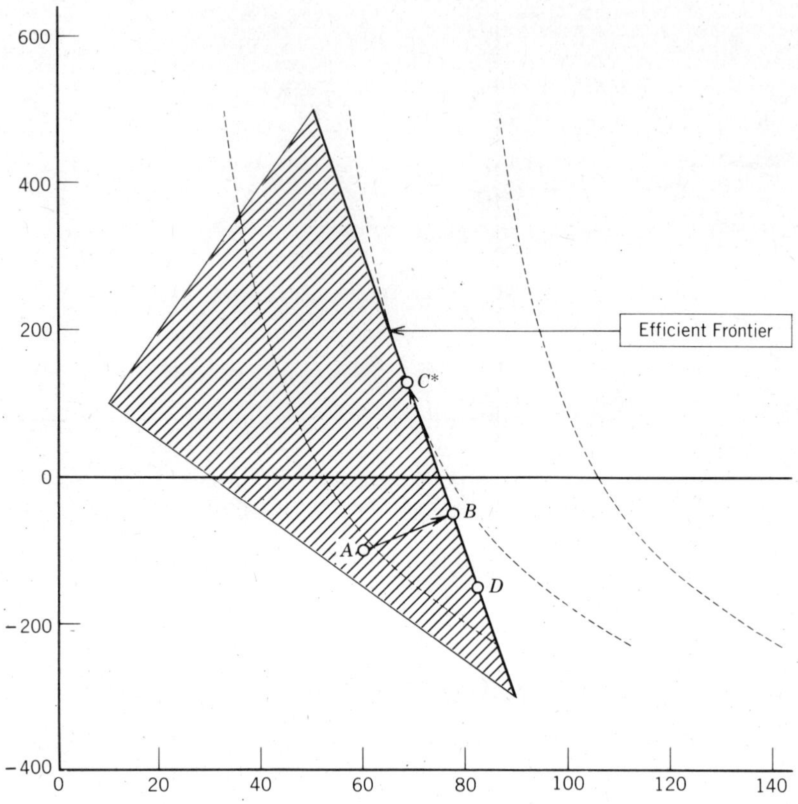

FIGURE 4-10. The trajectory followed by an interactive approach.

can always start with an initial solution that is efficient (such as point *D* in Figure 4-10). Then, *all* the solutions along the search trajectory are efficient.

Interactive approaches work by either (1) presenting a number of solutions to the decision maker and asking which he prefers or (2) presenting a solution and asking how much of each criterion he is willing to sacrifice for a unit increase in another particular criterion. Following is sample output from two interactive multicriterion programs.

The decision problem with which the following computer ouput[3] is concerned is the determination of an aggregate production plan for which the decision variables are the amounts of various products to be produced subject to constraints on demand capacity, manpower, investment capital, and so on. Rather than merely minimize cost, the user is concerned with three criteria: cost, stock-outs (expressed as a percent of unit sales), and layoffs (expressed as a percent of the man-days worked). The following two methods allow the user to "search" for a satisfactory solution.

[3]Adapted from *Interactive Multiple Criteria Decision Methods: An Investigation and an Approach* by Jyrki Wallenius, published by The Helsinki School of Economics.

Method 1

```
GIVE THE STARTING SOLUTION IN
CRITERION FUNCTION VALUES

1200000.,10.,15.

THE COMBINATION 1200000.00 10.00 15.00
IS FEASIBLE

DO YOU WANT TO SKIP THE TRADEOFF-ROUTINE
YES (1) OR NO(0)?

0

SELECT THE CHANGE IN THE FIRST CRITERION

-20000.

SELECT THE CHANGE IN THE SECOND CRITERION

-1.

DO YOU PREFER THE FIRST (1) OR THE SECOND (2)
CRITERION VECTOR OR ARE YOU INDIFFERENT (0)?

          1200000.00   10.00   15.00
          1180000.00   11.00   15.00

1

DO YOU PREFER THE FIRST (1) OR THE SECOND (2)
CRITERION VECTOR OR ARE YOU INDIFFERENT (0)?

          1200000.00   10.00   15.00
          1180000.00   10.00   15.00

0

SELECT THE CHANGE IN THE THIRD CRITERION

-1

DO YOU PREFER THE FIRST (1) OR THE SECOND (2)
CRITERION VECTOR OR ARE YOU INDIFFERENT (0)?

          1200000.00   10.00   15.00
          1180000.00   10.00   16.00

0

YOUR TRADEOFFS ARE 1.    40000.    20000
THE NEW OPERATING POINT IS

          1375000.000   6.355   1.747

ENTER NUMBER OF POINTS TO SEE IN STEP-SIZE PROBLEM

5

          1200000.00   10.00   15.00
          1243750.00    9.09   11.69
          1287500.00    8.18    8.37
          1330250.00    7.27    5.06
          1375000.00    6.36    1.75

WHICH DO YOU PREFER?

4

THE VALUES FOR THE CRITERIA ARE 1331250.00  7.27 5.06

IS THE SOLUTION SATISFACTORY?

0

DO YOU WANT TO SKIP THE THE TRADEOFF-ROUTINE
YES (1) OR (0)?

0

.
.
.
```

The decision maker has chosen a starting point of $1.2 million, 10% stock-outs, and 15% layoffs.

Here, the interactive method gives the decision maker a choice of two "solutions." The first is chosen.

Again, the decision maker is given a choice. Now, he is indifferent. This defines two points on an indifference curve.

Here, the choice forces the decision maker to tradeoff between cost and level of layoffs.

The program has moved to a new point (cost=$1,375 million, stock-outs = 6.355% and layoffs = 1.747%) and gives the decision maker a choice of three other points between this new point and the original one.

The decision maker chooses to continue the search.

continuing...

```
YOUR TRADEOFFS ARE 1.  40000.  40000.
THE NEW OPERATING POINT IS

          1375000.000  6.355   1.747

ENTER NUMBER OF POINTS TO SEE IN STEP-SIZE PROBLEM

3

          1331250.00   7.27    5.06
          1353120.00   6.81    3.40
          1375000.00   6.36    1.75

WHICH DO YOU PREFER?

1

THE VALUES FOR THE CRITERIA ARE 1331250.00  7.27  5.06

IS THE SOLUTION SATISFACTORY?

1

OPTIMAL SOLUTION IS 1331250.00  7.27  5.06

*** JOB COMPLETED ***
```

After a few more iterations, the decision maker sees that no feasible point is better than the current one, and the program stops.

Method 2

```
*** THE PAYOFF MATRIX ***

          1021610.00  22.93  19.45
          1076110.00  10.48  19.45
          1180000.00  22.93   5.33

FUNCTION   IDEAL        COMPROMISE
           SOLUTION     SOLUTION

  1        1021610.00   1180000.00
  2             10.48        11.92
  3              5.33         9.02

ARE ANY OF THE COMPONENTS SATISFACTORY (YES/NO)?

YES

ARE ALL THE COMPONENTS SATISFACTORY?

NO

HOW MANY COMPONENTS DO YOU WISH TO RELAX?

1

CHOOSE THE COMPONENT TO BE RELAXED (GIVE THE NUMBER)

2

GIVE THE AMOUNT OF RELAXATION (WITH A DECIMAL POINT)

1.

FUNCTION   IDEAL        COMPROMISE
           SOLUTION     SOLUTION

  1        1021610.00   1180000.00
  2             10.48        12.92
  3              5.33         8.51

CHOOSE THE COMPONENT TO BE RELAXED (GIVE THE NUMBER)

2

GIVE THE AMOUNT OF RELAXATION (WITH A DECIMAL POINT)

1.
```

Method 2 allows the decision maker to "relax" one criterion at a time, and shows the impact on another pre-specified criterion.

Here the decision maker has said, "Allow the stock-out level to increase from 11.92% to 12.92%. Remaining on the efficient frontier, what happens to layoffs if cost stays the same?" Answer: layoffs can decrease to 8.51%.

```
FUNCTION    IDEAL          COMPROMISE
            SOLUTION       SOLUTION

   1        1021610.00     1180000.00
   2             10.48          13.92
   3              5.33           8.04
```

> *Again.*

```
.

.

.

CHOOSE THE COMPONENT TO BE RELAXED (GIVE THE NUMBER )

3

GIVE THE AMOUNT OF RELAXATION (WITH A DECIMAL POINT)

2.

FUNCTION    IDEAL          COMPROMISE
            SOLUTION       SOLUTION

   1        1021610.00     1155460.00
   2             10.48          13.92
   3              5.33           6.02
```

> *Final Solution-- this decision maker is quite a bit more cost-conscious.*

```
ARE ANY OF THE COMPONENTS SATISFACTORY (YES/NO)?

YES

ARE ALL THE COMPONENTS SATISFACTORY?

YES

*** JOB COMPLETED ***
```

EXERCISES

1. A large electronics firm is about to launch a new home computer, relying on a national advertising campaign to stimulate interest. Each magazine insert is expected to cost \$20,000 and reach 200,000 people. Each TV ad will cost \$50,000 and reach 700,000 people. No more than \$500,000 can be spent on TV ads.

a. Formulate a two-criterion problem to reach as many people as possible and spend as little money as possible.
b. Graph the problem in criterion space and indicate the efficient frontier.
c. If the firm's *goals* are to reach six million people and to not exceed its \$500,000 budget, find the optimal efficient solution.

2. A chemical firm produces two organic solvents that compete for the same resources: labor, machine capacity, and raw materials as follows:

	Labor hours required (per lb of solvent)	Machine capacity (in production units) required	Raw Material (lb) required
Solvent 1	.2	3	1
Solvent 2	.2	1	2

For the coming week, 140 hours of labor, 1800 units of machine capacity and 1100 pounds of raw material are available. The sales manager feels that the firm could sell as many product 1's as are produced, but that the weekly production of product 2 should be limited to 500 units. Typically, in scheduling production, the plant manager has been concerned with maximizing contribution ($5.60/lb for solvent 1; $8.20/lb for solvent 2). Recently, however, the firm has come under increasing scrutiny from the EPA due to the water pollution levels (measured in mg. chlorinated ethanes per liter of water) in the effluent from the plant. The production of one pound of Solvent 1 creates .0237 mg./liter of chlorinated ethane pollutants; the manufacture of one pound of Solvent 2, .0109 mg./liter. It would benefit the firm to have the production schedule set to keep the total effluent pollution level as low as possible.

a. Formulate the two-criterion problem. Graph the feasible region in criterion space and indicate the efficient frontier.
b. If the firm's weekly *goals* were $7500 in contribution and *no* pollution, what efficient solution is optimal?

3. Example 3 of Chapter 2 was a bond portfolio problem where criteria concerning average quality and average time to maturity were included as constraints. Reformulate this problem as a goal program with the following goals:

a. After-tax return at 7%.
b. $3 million in municipal bonds.
c. An average quality rating of 1.4.
d. An average maturity of 5 years.
e. No borrowing ($10 million is available without additional borrowing).

4. A hospital administrator is reviewing departmental requests prior to the design of a new emergency room. At issue is the number of beds for each department. The current plans call for a 15,000 square foot facility. The hospital board has established the following goals:

	No. of Beds Requested	Cost per Bed (incl. equipment)	Area per Bed (sq ft)	Peak requirements (max. no. of patients at one time)
Major Trauma	5	$12,600	474	3
Surgical	20	5,400	542	18
Medical	20	8,600	438	15

a. Avoid overspending the $300,000 budget.
b. Avoid a plan requiring more than 15,000 square feet.
c. Meet peak requirements (Surgical and Medical can be interchanged, but Major Trauma patients cannot use other facilities. Hence, peak Major Trauma requirements *must* be fulfilled.)
d. Meet the departmental requests. This goal is viewed as twice as important as any of the others.

Formulate a goal program to help the hospital administrator make this decision.

5. An airline is considering three possible route expansions under airline deregulation. As indicated by the table below, the choice of route impacts a number of financial and performance measures:

	Route		
	1	2	3
Demand (per week)	7200	4600	2500
Profit/seat ($)	40	55	80
Required cabin attendants per trip	4	7	10
Flight crew hours per trip	6	9	13.5
Aircraft capacity (seats/trip)	100	140	250
Average Load Factor[a]	.75	.60	.52

[a]Fraction of seats filled on an average flight. For example, route 1 flights typically carried 75 passengers (.75 × 100), route 2 flights typically carried 84 passengers (.6 × 140), route 3 flights, 130.

The airline has 75 cabin attendants and 75 flight crew members available for these routes. By law, flight crew members cannot spend more than 16 hours per week each. By contrast, a cabin attendant can make as many as 16 *trips* per week.

Due to maintenance and networking requirements for other routes, aircraft availability for route 1 is limited to 90 trips per week, for route 2 to 60 trips per week, 3 to 15 trips per week.

Formulate a goal program to determine the optimal number of trips to schedule on each of the three routes given the following goals:

a. Do not divert additional aircraft to any of the three routes.
b. Avoid the use of more than the 75 flight crew members available.
c. Profit should be at least $600,000 per week.
d. As many passengers as possible should be carried but demand figures should not be exceeded.
e. Avoid the use of more than 75 cabin attendants.

Goal (a) is viewed as important as the other goals combined. Goals (b), (c), and (d) are equally important. The three goals in (d) are to be weighted in proportion to the size of the market on each route. Goal (e) is only half as important as (b).

*6. The text alluded to the difficulty that arises in trying to explicitly incorporate decision maker preferences into a value function. Suppose a firm faces a decision problem with two criteria, profitability and unemployment. The decision maker has given us the following statements about his preferences:

a. ($10,000; 0%) is better than ($0; 0%).
b. ($7,000; 8%) is better than ($120,000; 10%).
c. ($130,000; 15%) is better than (−$5,000; 10%).
d. (−$5,000; 10%) is better than ($25,000; 25%).

Suppose we wish to approximate the decision maker's value function with the quadratic polynomial

$$V(x,y) = ax + by + cxy + dx^2 + ey^2$$

How might an appropriate V be determined from the above preference statements?

*Adapted from D. Bell, "Defining and Quantifying Goals for Measuring the State of an Ecosystem," IIASA Research Report RR-74-26, December 1974.

Mitchell Enterprises (C)

Mr. Gordon Mitchell, President of Mitchell Enterprises, called a meeting of the firm's Investment Review Committee in December, 1976. The committee consisted of Mr. Charles Gilbert, the Treasurer, Ms. Roberta Phillips, the Controller, and Mr. Paul Chesler, Special Assistant to Mr. Mitchell. Mr. Mitchell had called the meeting because he was very disturbed about the latest results of the Linear Program Mr. Chesler had been using to determine the firm's investment portfolio (see Exhibits 1, 2, and 3). In the past 2 years, the firm had done a thorough review of different methods of evaluating investment projects and had finally decided that linear programming was the best method. The quantity to be maximized was pre-tax cash flow at the end of an appropriate planning horizon. The firm had successfully used this method to make their investment decisions for the previous year, and the Investment Review Committee had agreed that this was the appropriate method and had been quite satisfied with the results.

EXHIBIT 1
Memorandum

12/1/76
TO: The Investment Review Committee
FROM: Paul Chesler
RE: Investment Possibilities for 1977–79

The table below describes five projects which compete for our investment dollar. The table shows the cash flow that will result from investing one dollar. Project A is a 2-year investment available at the beginning of 1977 which pays 30 cents per dollar invested at the end of the first year and returns an additional dollar per dollar invested at the end of the second. At most $600,000 can be invested in A. Project B is identical to A except that it is available a year later and is limited to $500,000. C is a 1-year investment available only at the beginning of 1977 which pays $1.10 per dollar invested at the end of that year, with an upper limit of $250,000. Project D is a 3-year investment available at the beginning of 1977 which pays $1.75 per dollar invested at the end of 1979. E will become available at the beginning of 1979 and will, after a year, pay $1.40 per dollar invested. Project E is limited to a maximum investment of $1,000,000. Of course, the cash we receive from any of these projects may be reinvested in others which are available at the time. In addition, we could obtain 6% via short-term bank accounts for any money not invested in a given year. We have $1,000,000 to invest by the first of the year.

Cash Flow per Dollar Invested

Project:	A	B	C	D	E
1/1/77	−1.00	0	−1.00	−1.00	0
1/1/78	+ .30	−1.00	+1.10	0	0
1/1/79	+1.00	+ .30	0	0	−1.00
12/31/79	0	+1.00	0	+1.75	+1.40
Limit	$600,000	$500,000	$250,000	None	$1,000,000

EXHIBIT 2
LP Formulated with 1979 Payout as the Objective

```
DATA SET NAME? MTCH1

TITLE: MITCHELL PAYOUT 79
PROCEED, DISPLAY, OR REJECT? D

OBJECTIVES:

              A          B          C          D          E          BANK77
              BANK78     BANK79

PAYOUT79      .0000      1.000      0000       1.750      1.400      .0000
              .0000      1.060
```

(continued)

EXHIBIT 2 (Continued)

CONSTRAINTS:

	A	B	C	D	E	BANK77
	BANK78	BANK79	RELATION	CASH		
77INVEST	-1. 000	. 0000	-1. 000	-1. 000	. 0000	-1. 000
	. 0000	. 0000	EQ	-1000.		
78INVEST	. 3000	-1. 000	1. 100	. 0000	. 0000	1. 060
	-1. 000	. 0000	EQ	. 0000		
79INVEST	1. 000	. 3000	. 0000	. 0000	-1. 000	. 0000
	1. 060	-1. 000	EQ	. 0000		
LIMITA	1. 000	. 0000	. 0000	. 0000	. 0000	. 0000
	. 0000	. 0000	LE	600. 0		
LIMITB	. 0000	1. 000	. 0000	. 0000	. 0000	. 0000
	. 0000	. 0000	LE	500. 0		
LIMITC	. 0000	. 0000	1. 000	. 0000	. 0000	. 0000
	. 0000	. 0000	LE	250. 0		
LIMITE	. 0000	. 0000	. 0000	. 0000	1. 000	. 0000
	. 0000	. 0000	LE	1000.		

PROCEED OR REJECT? P

EXHIBIT 3
Solution to LP Formulated in Exhibit 2

```
                    MTCH1

TITLE: MITCHELL PAYOUT 79
PROCEED, DISPLAY, OR REJECT? P

MAXIMIZE OR MINIMIZE? MAX

OPTIMAL SOLUTION FOUND.
     PAYOUT79      1807. 12

OUTPUT OPTION? E

ALL ITEMS NOT LISTED IN SECTIONS 1 - 4 HAVE THE VALUE ZERO.

*1* DECISION VARIABLES
  1. A           600. 000
  4. D           400. 000
  5. E           790. 800
  7. BANK78      180. 000

*2* SLACK(+) AND SURPLUS(-) IN CONSTRAINTS
  5. +LIMITB    500. 000
  6. +LIMITC    250. 000
  7. +LIMITE    209. 200

*3* SHADOW PRICES FOR CONSTRAINTS
  1. 77INVEST -1. 75000
  2. 78INVEST -1. 48400
  3. 79INVEST -1. 40000
  4. LIMITA     . 952000E-01

*4* REDUCED COSTS FOR DECISION VARIABLES
  2. B          -. 640000E-01
  3. C          -. 117600
  6. BANK77     -. 176960
  8. BANK79     -. 340000

*5* RANGES ON COEFFICIENTS OF OBJECTIVE PAYOUT79
     VARIABLE  LOWER BOUND  CURRENT VALUE  UPPER BOUND
  1. A         -. 95200E-01    . 00000      UNBOUNDED
  2. B         UNBOUNDED      1. 0000        1. 0640
  3. C         UNBOUNDED       . 00000        . 11760
  4. D           1. 6324      1. 7500        1. 8452
  5. E           1. 3278      1. 4000        1. 5009
  6. BANK77    UNBOUNDED       . 00000        . 17696
  7. BANK78    -. 64000E-01    . 00000        . 10691
  8. BANK79    UNBOUNDED      1. 0600        1. 4000
```

(continued)

EXHIBIT 3 (Continued)

```
*6* RANGES ON VALUES OF RIGHT-HAND-SIDE CASH
      CONSTRNT   LOWER BOUND   CURRENT VALUE   UPPER BOUND
 1.   77INVEST   UNBOUNDED       -1000.0         -600.00
 2.   78INVEST   -197.36          .00000          180.00
 3.   79INVEST   -209.20          .00000          790.80
 4.   LIMITA      .00000         600.00          758.73
 5.   LIMITB      .00000         500.00          UNBOUNDED
 6.   LIMITC      .00000         250.00          UNBOUNDED
 7.   LIMITE     790.80          1000.0          UNBOUNDED
```

Mr. Mitchell opened the meeting by reminding the committee that the firm was planning to make a public stock offering in early 1978. The following dialogue ensued:

Mitchell: *I realize that the committee agreed last year that maximizing cash flow at the end of our planning horizon was the best way for us to do investment planning, but now that we are about to go public, we have to be concerned with profits as well. We need to have our 1977 earnings as high as possible, so we can get a good price for our stock. It looks like following the results of this LP will cause our earnings for 1977 to be lower than they were for 1976.*

Gilbert: *How did you figure the 1977 earnings?*

Mitchell: *Well, it's not explicit in the output, but you can look at the decision variables and calculate the cash inflow in the first year—30% for Project A, 10% for Project C, and 6% for banked money. Since there are no expenses, the pretax earnings will be equal to this cash flow—$180,000.*

Chesler: *I thought we all agreed last year that maximizing the cash flow at the 3-year horizon was the best way to plan our portfolio.*

Mitchell: *Well, I don't see how it can be best if it makes our profits look so bad. We'd better do some more thinking about this whole linear programming idea. The trouble is that you have to pick only one thing to maximize.*

Chesler: *Well, I could change the program so it will maximize 1977 earnings if that's what you really want. But what about taxes?*

Phillips: *Since we'll have credits from other parts of our business, which should be enough to shield this income from taxes, why don't you just look at pretax earnings now. That should give us a good enough idea of what will happen and we can worry about the details of taxes later, if necessary.*

Gilbert: *Well, I would certainly like to see the results with first-year earnings as the objective. I've been saying all along that profits are what really matter.*

Phillips: *Now, wait a minute, Charlie. Profits in the coming year won't be worth a damn to us if we have to sacrifice long run cash flow in order to get them.*

Mitchell: *Well, Paul, you'd better rerun the program, and we'll see what you came up with. Can you have the results by tomorrow afternoon? We'll meet again then.*

When the committee reconvened the next afternoon, Mr. Chesler passed around copies of the results of the linear program with 1977 profit as the objective function. He said, "Now your 1977 profits are up, but you may not like what that does to your cash flow for 1979." As the committee members pored over the computer output (Exhibits 4 and 5), Ms. Phillips did some calculations and said, "I couldn't possibly vote approval for that portfolio. It looks like the payout drops to a little over $1.4 million. That's almost $400,000 lower than yesterday's plan." Mr. Gilbert replied that he, on the other hand, would refuse to approve the original plan.

"Wait a minute," said Mitchell, "I can't understand what this program is doing. It seems to me that if, instead of putting money in the bank in 1979 we invested it in Project E, we'd improve our 1979 payout without decreasing our 1977 profits. Why doesn't the program do that?"

EXHIBIT 4
LP Formulated with 1977 Profit as the Objective

```
DATA SET NAME? MTCH2

TITLE: MITCHELL PROFIT 77
PROCEED, DISPLAY, OR REJECT? D

OBJECTIVES:

              A          B          C          D          E          BANK77
              BANK78     BANK79

PROFIT77      .3000      .0000      .1000      .0000      .0000      .6000E-01
              .0000      .0000

CONSTRAINTS:

              A          B          C          D          E          BANK77
              BANK78     BANK79     RELATION   CASH

77INVEST     -1.000      .0000     -1.000     -1.000      .0000     -1.000
              .0000      .0000      EQ        -1000.
78INVEST      .3000     -1.000      1.100      .0000      .0000      1.060
             -1.000      .0000      EQ         .0000
79INVEST      1.000      .3000      .0000      .0000     -1.000      .0000
              1.060     -1.000      EQ         .0000
LIMITA        1.000      .0000      .0000      .0000      .0000      .0000
              .0000      .0000      LE         600.0
LIMITB        .0000      1.000      .0000      .0000      .0000      .0000
              .0000      .0000      LE         500.0
LIMITC        .0000      .0000      1.000      .0000      .0000      .0000
              .0000      .0000      LE         250.0
LIMITE        .0000      .0000      .0000      .0000      1.000      .0000
              .0000      .0000      LE         1000.

PROCEED OR REJECT?
```

Phillips remarked, "Investing in E up to the $1,000,000 limit would even be better. Why doesn't the program do that?" "Well, Paul," said Mitchell, "You're the quantitative expert. First, you'd better explain what this program is doing, and then try to figure out a way to help us make an efficient compromise."

EXHIBIT 5
Solution to LP Formulated in Exhibit 4

```
OPTION? OPT

DATA SET NAME? MTCH2

TITLE: MITCHELL PROFIT 77
PROCEED, DISPLAY, OR REJECT? P

MAXIMIZE OR MINIMIZE? MAX

OPTIMAL SOLUTION FOUND.
     PROFIT77      214.000

OUTPUT OPTION? E

ALL ITEMS NOT LISTED IN SECTIONS 1 - 4 HAVE THE VALUE ZERO.

*1* DECISION VARIABLES
  1. A           600.000
  2. B           500.000
  3. C           250.000
  6. BANK77      150.000
  7. BANK78      114.000
  8. BANK79      870.840

*2* SLACK(+) AND SURPLUS(-) IN CONSTRAINTS
  7. +LIMITE    1000.00

*3* SHADOW PRICES FOR CONSTRAINTS
  1. 77INVEST -.600000E-01
  2. 78INVEST  .000000
  3. 79INVEST  .000000
  4. LIMITA    .240000
  5. LIMITB    .000000
  6. LIMITC    .400000E-01

*4* REDUCED COSTS FOR DECISION VARIABLES
  4. D          -.600000E-01
  5. E           .000000

*5* RANGES ON COEFFICIENTS OF OBJECTIVE PROFIT77
     VARIABLE  LOWER BOUND  CURRENT VALUE  UPPER BOUND
  1. A          .60000E-01    .30000        UNBOUNDED
  2. B          .00000        .00000        UNBOUNDED
  3. C          .60000E-01    .10000        UNBOUNDED
  4. D         UNBOUNDED      .00000         .60000E-01
  5. E         UNBOUNDED      .00000         .00000
  6. BANK77     .00000        .60000E-01     .10000
  7. BANK78   -.56604E-01     .00000         .00000
  8. BANK79     .00000        .00000         .00000

*6* RANGES ON VALUES OF RIGHT-HAND-SIDE CASH
     CONSTRNT  LOWER BOUND  CURRENT VALUE  UPPER BOUND
 1.  77INVEST  UNBOUNDED     -1000.0        -892.45
 2.  78INVEST  UNBOUNDED      .00000         114.00
 3.  79INVEST  UNBOUNDED      .00000         870.84
 4.  LIMITA     .00000       600.00          750.00
 5.  LIMITB   -.38147E-05    500.00          614.00
 6.  LIMITC     .00000       250.00          400.00
 7.  LIMITE     .00000       1000.0         UNBOUNDED

OUTPUT OPTION? N
```

Quadraphonic Systems, Inc.

In the fall of 1976, the Managing Board of Quadraphonic Systems, Inc. met to develop an aggregate production plan for the coming calendar year. The Board was comprised of Scott Isaacson, Vice President of Marketing and Sales; Jeff Gault, Vice President of Production; and Jim Stottlemeyer, Director of Employee Relations.

Quad Systems was a moderate sized producer of quadraphonic, high-fidelity audio equipment located in Chehalis, Washington, a small town about 40 miles south of Tacoma. Although sales had been increasing steadily through the 1970s, the company had suffered a net loss of $210,000 on 1975 sales of $17.1 million. Problems in the production process and quality control in the growing company had been identified as the prime reasons. Major efforts to deal with these problems met with success. Financial results for the first half of 1976 indicated a net profit of about 2% on sales of 9.2 million, with even brighter prospects for the second half of the year. The company was anxious to show a respectable profit for 1976 and hoped for even better profitability for 1977 so they could more easily finance their expansion in this growing market.

THE QUADRAPHONIC AUDIO MARKET

Within the past several years, the major innovation in high-fidelity audio equipment has been the development and introduction of quadraphonic, or four-channel systems. In the late sixties, two separate methods for reproducing quadraphonic sound were developed, known as the "discrete" and "matrix" systems. The matrix method synthesizes, or encodes, four separate sound channels into two recorded channels which are cut into a record. The record can be played with a conventional stereo stylus and cartridge with the audio signal being sent to a "decoder" where the four channels are restored into two dominant channels and two other, less audible ones.

The developers of the discrete technology claim that the matrix method does not produce a truly four-channel sound as does the discrete method. The discrete system actually records four distinct channels on the record vinyl which can then be reproduced individually using a special quadraphonic cartridge. Due to the quantity of information encoded on the record, however, the playing time is shortened, a purer vinyl is required, and the frequency response of the system is slightly narrowed. Initially, the discrete systems were much more expensive, but advancing technologies have now made them more competitive.

In 1976, recordings for both types of quadraphonic music systems were being produced; most large companies choosing one type or the other exclusively, although a few did produce both types. As the availability of discrete systems had increased, so had sales. The proponents of the discrete technology claimed that discrete provided a richer, more encompassing quadraphonic effect, although the clipping of the highs and lows was sometimes noticeable. The matrix was marketed earlier and remained less expensive. The deciding factor in the market battle will probably be the FCC's decision as to which type of quadraphonic FM broadcasting to authorize. The Commission is said to be leaning toward matrix, but it could conceivably switch or even authorize both.

COMPANY COSTS, DEMAND PROJECTIONS, AND POLICIES

Although the 1976 market was quite competitve, Quad Systems had successfully straddled the market, having produced both types of systems for several years. In 1976, they had a substantial share of the quadraphonic market, one which was youth-oriented and growing. 1977 demand projections for Quad Systems represented an approximately 20% increase over projected 1976 demand. Exhibit 1 gives what marketing personnel considered reliable estimates for 1977. The seasonal fluctuations are due primarily to the high volume of student sales and holiday sales in the third and fourth quarters, respectively. Each system is comprised of 4 speakers, a turntable equipped with cartridge and stylus, and a receiver-amplifier. Most of the tuner section and integrated circuitry of the receiver-amplifier is purchased prepackaged from a Japanese electronics firm. Likewise, most turntable components are purchased ready-made from one of the major stereo producers and assembled at the plant. The two-way speakers are assembled from pre-formed cabinets and standard woofers and tweeters. Thus, the manufacturing process has high costs for the purchase of many component parts and a relatively lower labor cost.

EXHIBIT 1
Projected Demand for 1977

	1st Quarter	2d Quarter	3d Quarter	4th Quarter
Matrix System Units	2500	3100	4000	4000
Discrete System Units	2500	2700	3200	5000

The matrix system costs $400 to produce, exclusive of labor, and is sold to wholesalers for $600 per unit. The discrete system with non-labor production costs of $450, is sold for $750 per unit. The greater margin on the discrete system is due to its "snob" appeal.

Management was worried about the potential turnover in employees due to the increasingly seasonal nature of demand. They feared falling significantly short of demand if labor stability were to be enforced. All production workers are skilled or semi-skilled and average $16.00 per hour including benefits and $25.00 (slightly over time-and-a-half) for overtime. Due to the many ready-made components, only 4 hours of labor input per matrix system, and 6 hours per discrete system are required.

The employees of the Quad-system factory were non-unionized. This was primarily because of the small town setting, good wages, and a history of good relationship with management. Management wanted to maintain a non-union shop. The increasing cyclicality of hiring practices, the recent depression of the area's defense industries, and the union presence in a few other, nearby factories had, however, increased the chances of unionization in Quad Systems' factory.

Quad Systems had more than adequate inventory storage space, with storage and holding costs per quarter averaging $20 per unit for a matrix system and $30 per unit for a discrete system. A labor force which could provide 20,000 hours per quarter was projected to be on hand as of January 1977. Strict company policy restricted an employee's overtime to no more than 25% of his regular time. Because of relatively high unemployment in the area, additional labor, even of the more specialized kind, is easily obtained and essentially no costs are associated with an increase or decrease in the working force.

THE MEETING OF THE MANAGING BOARD

Excerpts from the Managing Board's discussion are presented below:

Isaacson: *Jim, I realize you want to maintain an efficient, cost conscious production operation. Remember, however, we're in a competitive, growing market, and our customers aren't going to wait for delayed deliveries. We have to be "eager beavers." If orders can't be filled quickly, the customers will go elsewhere. As a result we need to have an adequate inventory to meet immediate orders and to beef up our labor force to meet the seasonal demand. We want to show a solid profit in 1977, and eliminating shortages is the way to do it.*

Gault: *We've also got to remember that if we hire and then lay off too many employees, we can expect problems with employee morale. That'd be likely to cause a reduction in production efficiency and quality which could seriously effect our profits. The layoffs could also increase the pressures for unionization and I bet you'd start having problems recruiting competent people soon. I think it's wise to keep a slim but stable workforce.*

Stottlemeyer: *Yes, it's a real problem, but I think we can handle it. True, we want increased sales and profits, but we also need to keep control of costs, efficiency, and quality. Carrying a large inventory or keeping employees when they're not needed puts us in an untenable financial position. Hiring and firing a lot of employees will also create problems which will affect our profits. These issues involve trade-offs and we need to find some compromise position. I think we need to do some more thinking about how to analyze these problems.*

Gartland Steel

Thursday, October 11, 1979. Dan Crossan, Director of Environmental Engineering at Gartland Steel, returned from lunch and found a memo left for him by Jay Peeler, Senior Project Engineer. Crossan had spend a good deal of time in the past year promoting the new "bubble" method of determining air pollution emissions compliance for industrial plants. Rather than restricting pollution from each individual point source (smokestacks, vents, etc.) within the plant, the bubble method restricted the total pollution output of a plant—as if it were under a bubble. This new method promised both capital spending reductions and energy savings to industry while giving the same pollution emissions performance, since industry would be left to meet the goals in the most cost-effective way. The day before, Crossan had sent Peeler a memo asking for some information on the production processes at Gartland's Salem, Ohio steel mill. Crossan, who was attempting to convince the EPA to approve a bubble plan for the Salem mill, felt this information would be very helpful.

GARTLAND STEEL

Gartland Steel was a fully integrated steel producer. In 1978, it had 9 plants in the U.S. and was the country's 4th largest steelmaker in tonnage and 3rd largest in sales. Gartland, since its founding in 1901, had led the industry in method and product innovations, starting with commercial development of taconite ores, large-scale direct reduction furnaces, continuous flat rolling, development of electrical steel products and many grades of stainless steel, among others.

Gartland had also led the industry in both sensitivity to environmental questions and in actions to alleviate pollution. By the end of 1978, Gartland had spent over $300 million on pollution control; 85% of this was spent before being required by state or federal authorities. A 1977 report, by the Council on Economic Priorities, concerning pollution in the steel industry placed Gartland's performance first, with the other six major firms "tied for last."

Source: This case was written by Richard E. Stone. Harvard Business School Case 0-183-133.

THE CLEAN AIR ACT AND THE EPA

The Clean Air Act (CAA) of 1970 created the Environmental Protection Agency (EPA) to establish and enforce air quality goals. (The EPA was also responsible for control of water pollution and solid wastes.) The EPA set up National Ambient Air Quality Standards (NAAQS) for six airborne pollutants: sulfur oxides, nitrogen dioxide, carbon monoxide, photochemicals, hydrocarbons, and particulates. These standards postulate "threshold levels" beyond which ambient pollutants are damaging to health and welfare. States were given the primary responsibility to develop and implement programs to achieve these standards. In the attempt to bring regions or firms into compliance, emission rate standards were also set "stack-by-stack" for most industrial processes. In most cases, the state specified what equipment had to be installed and where. In addition, 247 "air quality control regions" were established and classed either as "clean" or "nonattainment" regions. "Nonattainment" meant that the NAAQS for at least one pollutant was not met. (Air Quality Control Region boundaries were set according to the geographic size of the regions and the availability of personnel to monitor pollution levels. The dispersal characteristics of the pollutants were not considered.)

The EPA's first task was to bring nonattainment regions into compliance. The EPA brought pressure on the companies within the regions to clean up or shut down existing pollution sources and threatened to deny permits for further new industrial plants producing pollution until all NAAQS were attained. Despite these actions, by the 1975 deadline set in the Clean Air Act, 160 of the 247 regions were classed as "nonattainment." Often, emissions improvements as stipulated stack-by-stack seemed to be more costly than industry could bear and survive. Industries such as steel, however, were often crucial to local economies. Under conflicting pressure from industry and local governments as well as the environmental movement, the EPA had by 1979 evolved compromise policies.

THE "OFFSET" POLICY

The EPA's first step was the creation of the "offset" policy in late 1976. "Offsets" allow firms to trade pollution "rights" within nonattainment regions. A firm may add a pollution source if it reduces pollution by a greater amount somewhere within the air quality control region, by buying controls for another firm, by adding controls to its own operations, or by shutting down one of its present sources to name a few possibilities. One category of pollution could not be traded for another, for example, sulfur oxides could only be traded for sulfur oxides. By 1977, firms could "bank" excess pollution reductions and a market in pollution "rights" developed, though it was far from clear how a price per ton of pollutant could be set in such a thin market without some idea of the value of an offset to a firm.

THE BUBBLE POLICY

The "bubble" policy, which the EPA had officially proposed on January 18, 1979, is a conceptual relative of the offset policy. Rather than installing mandated stack-by-stack equipment, a firm would be free to decide how to bring the net level of pollution for the entire plant within the regulated standards for each type of pollutant. However, approval procedures of federal, state, and local officials for a firm's compliance plans, prepared under the bubble policy, were rather complex. Industry faced the risk that a bubble plan prepared in good faith would be rejected. Rejection involved retroactive non-compliance fines as well as having to go back to the stack-by-stack controls. For this reason, industry had made few commitments to bubble policy compliance plans.

THE SALEM PLANT

Crossan was currently concerned with bringing Gartland's Salem plant into compliance with EPA standards for particulate pollution. The week before, he had spoken with Peeler about this. Peeler said he knew that a local rubber company, in the same quality control region, was planning to shut down permanently one of its plants. He felt that they would sell it to Gartland for $750,000. Gartland could then shut it down and receive a 250,000 pounds/year pollution "offset." Crossan felt that this might alienate the EPA and asked Peeler what could be done within the production process.

Peeler referred to a memo he had sent Crossan as background for their meeting. (See Exhibit 1.) Off-hand, the only thing Peeler could think of doing would be to change the mix in some of the open hearth furnaces. They would then consume .25 tons of pig iron and .95 tons of scrap to produce 1 ton of steel and .02 tons of scrap. The operating cost per open hearth would fall by $5/ton and pollution from each furnace would decline to .795 pounds/ton. Peeler didn't know if this was worth it. Then again, you could just cut back production somewhere along the line, but the question was where and what would be the cost?

EXHIBIT 1
Memorandum

October 1, 1979

TO: Daniel Crossan, Director of Environmental Engineering
FROM: Jay Peeler, Senior Project Engineer
SUBJECT: Linear Programming Model of the Salem Plant

As background to our meeting coming up this Thursday, I am sending you material describing the linear program used to help set the levels of production at the Salem plant.

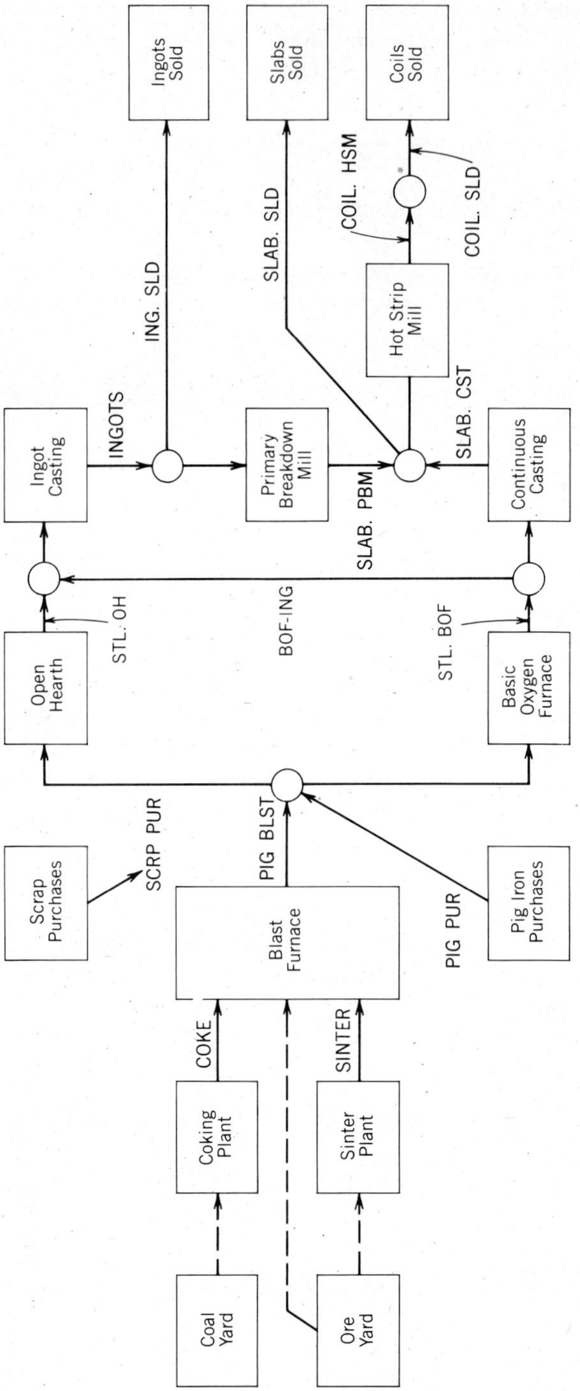

*Flows of particulate pollution and scrap not shown

FIGURE A. Process flow diagram.

Briefly, Figure A is a process flow diagram of the activities in the plant. Figure B is a chart of required amounts of input per unit of output for each activity. This includes expected 1980 unit operating costs. Figure C is the expected 1980 market price for our raw materials and final products. Figure D is a description of the decision variables and constraints used in the linear programming model. Figure E is the tableau of the linear program, and Figure F is the computer output.

This will help us determine the consequences of any plans we develop on Thursday to deal with pollution, and it might suggest some strategies or directions to follow.

	Coking Plant	Sinter Plant	Blast Furnace	Open Hearth Furnace	Basic Oxygen Furnace	Ingot Caster	Primary Breakdown Mill	Continuous Caster	Hot Strip Mill
1 ton of Outputs	Coke	Sinter	Pig Iron	Steel	Steel	Ingots	Slabs	Slabs	Coils
Produces Scrap (tons)	—	—	.01	.02	.02	.02	.18	.04	.04
Produces Particulate Pollution (pounds)	1.254	.777	.474	.861	.505	—	.156	—	—
Requires (tons)	1.43 Coal	.52 Ore	1.12 Ore .65 Coke .47 Sinter	1.08 Scrap .14 Pig Iron	.32 Scrap .81 Pig Iron	1.02 Steel	1.20 Ingots	1.06 Steel	1.04 Slabs
Operating Costs	$36.21	$18.77	$57.82	$52.88	$28.10	$4.56	$19.76	$32.30	$53.72
Cost of Coal and Ore Used	$68.05	$18.82	$40.54	—	—	—	—	—	—
Energy Credits	$1.86	$0.00	$0.16	$0.06	$0.13	$0.34	$3.45	$0.44	$0.41
Net Costs	$102.40	$37.59	$98.20	$52.82	$27.97	$4.22	$16.31	$31.86	$53.31

FIGURE B. Process technologies.

Coal	$47.59
Hot Rolled Coils	$350.00
Ingots	$200.00
Ore	$36.20
Pig Iron	$191.10
Scrap	$98.48
Slabs	$250.00

FIGURE C. Expected 1980 market prices (per ton).

Figure E is a linear program based on the process flow diagram. The decision variables (all measured in thousands of tons per year) are marked on the process flow diagram in capital letters:

1. COKE—coke from coke ovens.
2. SINTER—sinter from sinter plant.
3. PIG. BLST—pig iron from blast furnace.
4. SCRP. PUR—scrap purchased from outside.
5. PIG. PUR—pig iron purchased from outside.
6. STL. OH—steel from open hearth furnaces.
7. STL. BOF—steel from basic oxygen furnace (BOF).
8. BOF. ING—steel from BOF sent to ingot caster.
9. INGOTS—ingots produced by ingot caster.
10. SLAB. PBM—slabs produced by primary breakdown mill.
11. SLAB. CST—slabs produced by continuous caster.
12. COIL. HSM—hot rolled coils produced by hot strip mill.
13. COIL. SLD—hot rolled coils sold outside.
14. SLAB. SLD—slabs sold outside.
15. ING. SLD—ingots sold outside.

The constraints are of two types: nine balance constraints and six process capacity constraints. The exact coefficients for the balance equations are taken from Figure B. The constraints are:

1. COKE. BAL—all coke goes to the blast furnace (.65 tons of coke per ton of pig iron produced).
2. SINT. BAL—all sinter goes to blast furnace.

Objectives Contrib Constraints	COKE (1) −102.4	SINTER (2) −37.59	PIG. BLST (3) −98.2	SCRP. PUR (4) −98.48	PIG. PUR (5) −191.1	STL. OH (6) −52.82	STL. BOF (7) −27.97	BOF-ING (8) 0
1. COKE. BAL	−1	0	.65	0	0	0	0	0
2. SINT. BAL	0	−1	.47	0	0	0	0	0
3. PIG. BAL	0	0	−1	0	−1	.14	.81	0
4. STL1. BAL	0	0	0	0	0	−1	0	−1
5. STL2. BAL	0	0	0	0	0	0	−1	1
6. ING. BAL	0	0	0	0	0	0	0	0
7. SLAB. BAL	0	0	0	0	0	0	0	0
8. COIL. BAL	0	0	0	0	0	0	0	0
9. SCRP. BAL	0	0	−.01	−1	0	1.06	.3	0
10. BLST. CAP	0	0	1	0	0	0	0	0
11. OH. CAP	0	0	0	0	0	1	0	0
12. BOF. CAP	0	0	0	0	0	0	1	0
13. ING. CAP	0	0	0	0	0	0	0	0
14. CONT. CAP	0	0	0	0	0	0	0	0
15. HSM. CAP	0	0	0	0	0	0	0	0

3. PIG. BAL—pig iron from the blast furnace and pig iron purchases go to the open hearths and/or the BOF.

4. STL1. BAL—steel from the ingot caster comes from either the open hearths or the BOF.

5. STL2. BAL—steel from the BOF goes either to the continuous caster or the ingot caster.

6. ING. BAL—ingots from the ingot caster go the the primary breakdown mill or are sold.

7. SLAB. BAL—slabs from the continous caster and primary breakdown mill go to the hot strip mill or are sold.

8. COIL. BAL—hot rolled coils from the hot strip mill are sold.

9. SCRP. BAL—scrap purchased plus "home" scrap generated by various processes goes to either the open hearth furnace or the BOF.

10. BLST. CAP—limits the capacity of the blast furnace.

11. OH. CAP—limits the capacity of the open heart furnaces.

12. BOF. CAP—limits the capacity of the BOF.

13. ING. CAP—limits the capacity of the ingot caster.

14. CONT. CAP—limits the capacity of the continuous caster.

15. HSM. CAP—limits the capacity of the hot strip mill.

The coefficients of the objective function come from the net costs in Figure B and the market prices of various raw materials in Figure C (which Gartland uses for all transfer pricing).

Notice that separate variables for the amount of coal and ore used are not included (since these yards are the responsibility of another division). Rather, the cost of coal and ore have been added as a cost of operating the coking plant, sinter plant, and blast furnace.

The output of the linear program appears in Figure F.

FIGURE D. Definition of variables and constraints.

	INGOTS (9) −4.22	SLAB. PBM (10) −16.31	SLAB. CST (11) −31.86	COIL. HSM (12) −53.31	COIL. SLD (13) 350	SLAB. SLD (14) 250	ING. SLD (15) 200		BAL/LIM
1.	0	0	0	0	0	0	0	=	0
2.	0	0	0	0	0	0	0	=	0
3.	0	0	0	0	0	0	0	=	0
4.	1.02	0	0	0	0	0	0	=	0
5.	0	0	1.06	0	0	0	0	=	0
6.	−1	1.2	0	0	0	0	1	=	0
7.	0	−1	−1	1.04	0	1	0	=	0
8.	0	0	0	−1	1	0	0	=	0
9.	−.02	−.18	−.04	−.04	0	0	0	=	0
10.	0	0	0	0	0	0	0	≤	1404
11.	0	0	0	0	0	0	0	≤	1740
12.	0	0	0	0	0	0	0	≤	2700
13.	1	0	0	0	0	0	0	≤	3408
14.	0	0	1	0	0	0	0	≤	720
15.	0	0	0	1	0	0	0	≤	3480

FIGURE E. LP formulation.

```
MAXIMIZE OR MINIMIZE? max

OPTIMAL SOLUTION FOUND.
     CONTRIB       168164.


ALL ITEMS NOT LISTED IN SECTIONS 1 - 4 HAVE THE VALUE ZERO.

*1* DECISION VARIABLES
  1. COKE       912.600
  2. SINTER     659.880
  3. PIG.BLST   1404.00
  4. SCRP.PUR   1835.08
  5. PIG.PUR    864.082
  6. STL.OH     1740.00
  7. STL.BOF    2499.36
  8. BOF-ING    1736.16
  9. INGOTS     3408.00
 10. SLAB.PBM   2840.00
 11. SLAB.CST   720.000
 12. COIL.HSM   3423.08
 13. COIL.SLD   3423.08


*2* SLACK(+) AND SURPLUS(-) IN CONSTRAINTS
 12. +BOF.CAP   200.640
 15. +HSM.CAP   56.9231


*3* SHADOW PRICES FOR CONSTRAINTS
  1. COKE.BAL   102.400
  2. SINT.BAL   37.5900
  3. PIG.BAL    191.100
  4. STL1.BAL   212.305
  5. STL2.BAL   212.305
  6. ING.BAL    242.069
  7. SLAB.BAL   289.067
  8. COIL.BAL   350.000
  9. SCRP.BAL   98.4800
 10. BLST.CAP   9.65750
 11. OH.CAP     28.3422
 13. ING.CAP    23.2676
 14. CONT.CAP   36.1024


*4* REDUCED COSTS FOR DECISION VARIABLES
 14. SLAB.SLD -39.0665
 15. ING.SLD  -42.0691


*5* RANGES ON COEFFICIENTS OF OBJECTIVE CONTRIB
    VARIABLE   LOWER BOUND   CURRENT VALUE  UPPER BOUND
  1. COKE      -117.26         -102.40       UNBOUNDED
  2. SINTER    -58.138         -37.590       UNBOUNDED
  3. PIG.BLST  -107.86         -98.200       UNBOUNDED
  4. SCRP.PUR  -135.77         -98.480        132.60
  5. PIG.PUR   -219.26         -191.10       -181.44
  6. STL.OH    -81.162         -52.820       UNBOUNDED
  7. STL.BOF   -50.781         -27.970        .37220
  8. BOF-ING   -22.811          .00000        28.342
  9. INGOTS    -27.488         -4.2200       UNBOUNDED
 10. SLAB.PBM  -44.231         -16.310       UNBOUNDED
```

(continued)

```
11.  SLAB.CST   -67.962           -31.860          UNBOUNDED
12.  COIL.HSM   -82.348           -53.310          UNBOUNDED
13.  COIL.SLD    320.96            350.00          UNBOUNDED
14.  SLAB.SLD   UNBOUNDED          250.00           289.07
15.  ING.SLD    UNBOUNDED          200.00           242.07

*6* RANGES ON VALUES OF RIGHT-HAND-SIDE BAL/LIM
     CONSTRNT  LOWER BOUND  CURRENT VALUE  UPPER BOUND
  1.   COKE.BAL  UNBOUNDED       .00000        912.60
  2.   SINT.BAL  UNBOUNDED       .00000        659.88
  3.   PIG.BAL   UNBOUNDED       .00000        864.08
  4.   STL1.BAL  -200.64         .00000        1066.8
  5.   STL2.BAL  -200.64         .00000        1066.8
  6.   ING.BAL   -3408.0         .00000        71.040
  7.   SLAB.BAL  -3560.0         .00000        59.200
  8.   COIL.BAL  -3423.1         .00000       UNBOUNDED
  9.   SCRP.BAL  UNBOUNDED       .00000        1835.1
 10.   BLST.CAP    .00000        1404.0        2268.1
 11.   OH.CAP     1539.4         1740.0        3029.7
 12.   BOF.CAP    2499.4         2700.0       UNBOUNDED
 13.   ING.CAP    2362.1         3408.0        3479.0
 14.   CONT.CAP    .00000        720.00        779.20
 15.   HSM.CAP    3423.1         3480.0       UNBOUNDED
```

FIGURE F. Linear program output without pollution constraint.

They decided to rerun the linear program described in the memo with a pollution constraint that would require pollution to stay under 5163 thousand pounds per year. Current pollution levels were 5525 thousand pounds per year. The EPA wanted Gartland to reduce particulate pollution by 6.6%, hence the restriction to 5163 thousand pounds per year. The stack-by-stack approach, which Crossan was hoping to avoid being forced to use, would cost $9 million per year (including operating expenses and amortization of capital expenses).

Last Monday, October 8, Crossan got a memo from Peeler containing the new linear program and its results. (See Exhibits 2 and 3.) At the time the memo arrived, Crossan was writing down some notes on a "bubble" plan for the Salem plant. Much of the plant's pollution came from the coal and ore yards—.04 pounds per ton of coal and .589 pounds per ton of ore. This would be relatively inexpensive to clean up. If Crossan could convince the EPA that this additional 1181 thousand pounds/year of particulate pollution should be included "under the bubble," then reducing total pollution 6.6% would be much cheaper. However, convincing the EPA to go along with it would not be that easy, and Crossan wanted more information before finalizing any plans. Yesterday, he sent Peeler a memo asking him to include the pollution from the coal and ore yards into the linear program. Today's memo from Peeler was the result. Included was a chart showing the maximum level of contribution that could be achieved for each level of pollution restriction. (See Exhibit 4.)

Crossan felt it would be a major advance if Gartland could get the EPA to approve some sort of bubble plan. It would cause other firms and industries to look into getting similar plans and would help convince the EPA that the bubble policy was good for everyone.

EXHIBIT 2
LP Formulation with Pollution Constraint

Objectives Contrib Constraints	COKE (1) −102.4	SINTER (2) −37.59	PIG. BLST (3) −98.2	SCRP. PUR (4) −98.48	PIG. PUR (5) −191.1	STL. OH (6) −52.82	STL. BOF (7) −27.97	BOF-ING (8) 0
1. COKE. BAL	−1	0	.65	0	0	0	0	0
2. SINT. BAL	0	−1	.47	0	0	0	0	0
3. PIG. BAL	0	0	−1	0	−1	.14	.81	0
4. STL1. BAL	0	0	0	0	0	−1	0	−1
5. STL2. BAL	0	0	0	0	0	0	−1	1
6. ING. BAL	0	0	0	0	0	0	0	0
7. SLAB. BAL	0	0	0	0	0	0	0	0
8. COIL. BAL	0	0	0	0	0	0	0	0
9. SCRP. BAL	0	0	−.01	−1	0	1.06	.3	0
10. BLST. CAP	0	0	1	0	0	0	0	0
11. OH. CAP	0	0	0	0	0	1	0	0
12. BOF. CAP	0	0	0	0	0	0	1	0
13. ING. CAP	0	0	0	0	0	0	0	0
14. CONT. CAP	0	0	0	0	0	0	0	0
15. HSM. CAP	0	0	0	0	0	0	0	0
16. POLLUT'N	1.254	.777	.474	0	0	.861	.505	0

EXHIBIT 3
LP Output with Pollution Constraint

```
                Exhibit 3

Linear Program Output With Pollution Constraint


MAXIMIZE OR MINIMIZE? max

OPTIMAL SOLUTION FOUND.
     CONTRIB         166045.

OUTPUT OPTION? 1 2 3 4


ALL ITEMS NOT LISTED IN SECTIONS 1 - 4 HAVE THE VALUE ZERO.

*1* DECISION VARIABLES
  1. COKE        769.979
  2. SINTER      556.754
  3. PIG.BLST    1184.58
  4. SCRP.PUR    1837.28
  5. PIG.PUR     1083.50
  6. STL.OH      1740.00
  7. STL.BOF     2499.36
  8. BOF-ING     1736.16
  9. INGOTS      3408.00
 10. SLAB.PBM    2840.00
 11. SLAB.CST    720.000
 12. COIL.HSM    3423.08
 13. COIL.SLD    3423.08

*2* SLACK(+) AND SURPLUS(-) IN CONSTRAINTS
 10. +BLST.CAP  219.417
 12. +BOF.CAP   200.640
 15. +HSM.CAP   56.9231

*3* SHADOW PRICES FOR CONSTRAINTS
  1. COKE.BAL    109.721
  2. SINT.BAL    42.1260
  3. PIG.BAL     191.100
  4. STL1.BAL    215.253
  5. STL2.BAL    215.253
  6. ING.BAL     241.310
  7. SLAB.BAL    289.067
  8. COIL.BAL    350.000
```

	INGOTS (9) −4.22	SLAB. PBM (10) −16.31	SLAB. CST (11) −31.86	COIL. HSM (12) −53.31	COIL. SLD (13) 350	SLAB. SLD (14) 250	ING. SLD (15) 200		BAL/LIM
1.	0	0	0	0	0	0	0	=	0
2.	0	0	0	0	0	0	0	=	0
3.	0	0	0	0	0	0	0	=	0
4.	1.02	0	0	0	0	0	0	=	0
5.	0	0	1.06	0	0	0	0	=	0
6.	−1	1.2	0	0	0	0	1	=	0
7.	0	−1	−1	1.04	0	1	0	=	0
8.	0	0	0	−1	1	0	0	=	0
9.	−.02	−.18	−.04	−.04	0	0	0	=	0
10.	0	0	0	0	0	0	0	≤	1404
11.	0	0	0	0	0	0	0	≤	1740
12.	0	0	0	0	0	0	0	≤	2700
13.	1	0	0	0	0	0	0	≤	3408
14.	0	0	1	0	0	0	0	≤	720
15.	0	0	0	1	0	0	0	≤	3480
16.	0	.156	0	0	0	0	0	≤	5163

```
  9. SCRP.BAL   98.4800
 11. OH.CAP     26.2639
 13. ING.CAP    19.5016
 14. CONT.CAP   32.9774
 16. POLLUT'N   5.83785

*4* REDUCED COSTS FOR DECISION VARIABLES
 14. SLAB.SLD -39.0665
 15. ING.SLD  -41.3102

OUTPUT OPTION?
                5 6
*5* RANGES ON COEFFICIENTS OF OBJECTIVE CONTRIB
     VARIABLE    LOWER BOUND   CURRENT VALUE   UPPER BOUND
  1. COKE        -117.26       -102.40         -25.462
  2. SINTER      -58.138       -37.590          68.814
  3. PIG.BLST    -107.86       -98.200         -48.190
  4. SCRP.PUR    -132.94       -98.480          129.42
  5. PIG.PUR     -207.14       -191.10         -181.44
  6. STL.OH      -79.084       -52.820         UNBOUNDED
  7. STL.BOF     -47.089       -27.970         -1.7061
  8. BOF-ING     -19.119       .00000           26.264
  9. INGOTS      -23.722       -4.2200         UNBOUNDED
 10. SLAB.PBM    -39.712       -16.310         UNBOUNDED
 11. SLAB.CST    -64.837       -31.860         UNBOUNDED
 12. COIL.HSM    -77.648       -53.310         UNBOUNDED
 13. COIL.SLD     325.66        350.00         UNBOUNDED
 14. SLAB.SLD    UNBOUNDED      250.00          289.07
 15. ING.SLD     UNBOUNDED      200.00          241.31

*6* RANGES ON VALUES OF RIGHT-HAND-SIDE BAL/LIM
     CONSTRNT     LOWER BOUND   CURRENT VALUE   UPPER BOUND
  1.  COKE.BAL    -1562.7        .00000          289.46
  2.  SINT.BAL    -2522.1        .00000          467.16
  3.  PIG.BAL     UNBOUNDED      .00000          1083.5
  4.  STL1.BAL    -200.64        .00000          718.77
  5.  STL2.BAL    -200.64        .00000          718.77
  6.  ING.BAL     -2792.2        .00000          71.040
  7.  SLAB.BAL    -3560.0        .00000          59.200
  8.  COIL.BAL    -3423.1        .00000          UNBOUNDED
  9.  SCRP.BAL    UNBOUNDED      .00000          1837.3
 10.  BLST.CAP     1184.6        1404.0          UNBOUNDED
 11.  OH.CAP       1539.4        1740.0          3476.2
 12.  BOF.CAP      2499.4        2700.0          UNBOUNDED
 13.  ING.CAP      2845.3        3408.0          3479.0
 14.  CONT.CAP     41.913        720.00          779.20
 15.  HSM.CAP      3423.1        3480.0          UNBOUNDED
 16.  POLLUT'N     3203.4        5163.0          5526.0
```

EXHIBIT 4
Contribution Versus Pollution

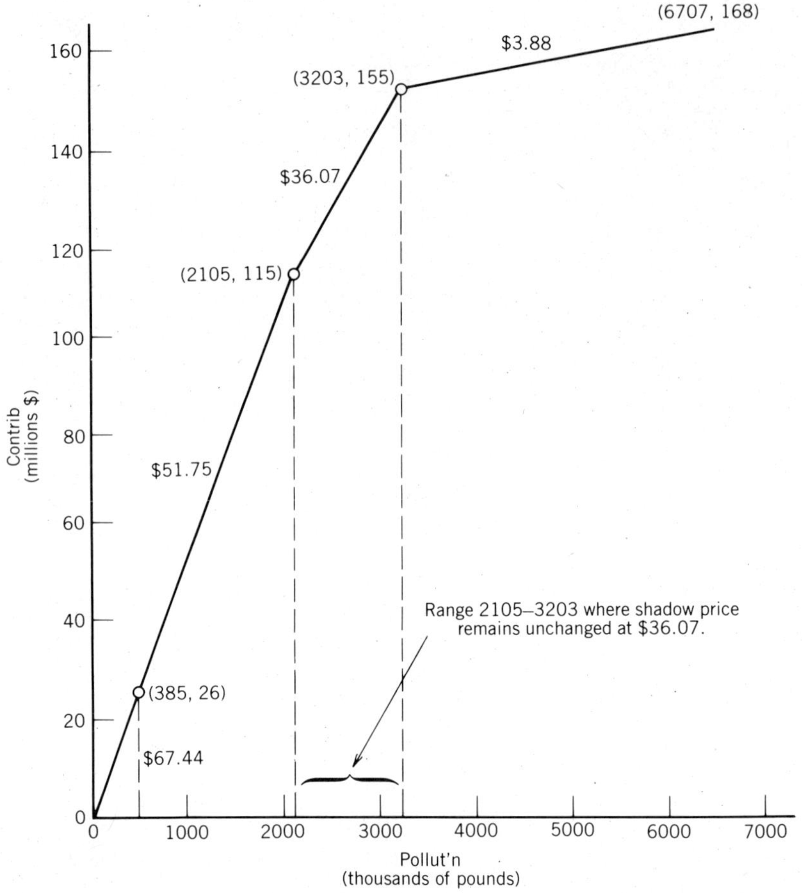

Wentworth Pacific Telephone Company (A)

Early in the winter of 1976, Mr. Richard Karney, Director of the Manpower Planning Group within the Wentworth Pacific Telephone Company, was wondering how he might integrate the new Equal Employment Opportunity (EEO) goals for WPTC into his projections for future personnel needs. Since 1972 when the U.S. Congress passed the Equal Employment Opportunity Act, WPTC had been actively recruiting qualified women and minorities for mana-

Source: This case was prepared by John Mulvey. Copyright © 1978 by the President and Fellows of Harvard College. Harvard Business School Case 9-178-146.

gerial positions. However, a sizeable imbalance still existed in 1976 because of the intense competition for people in these categories. On several occasions, Mr. Karney had been challenged by individuals in the company to produce concrete evidence that WPTC was fulfilling its moral obligation by doing *all it could* to change the complexion of the workforce. Mr. Karney's typical response was to rely upon historical statistics showing the WPTC had doubled the percentage of women middle-level managers, from 3½% to 7%, in 4 years. Furthermore, one out of every five newly hired managers was female. Similar evidence was available concerning the number of minorities in the workforce.

Although Mr. Karney felt that WPTC was doing "all that it could" to meet EEO objectives, he was particularly worried about the new EEO goal of 25% women managers by December 31, 1980. In addition, the EEO commission could single out WPTC by filing a suit charging employment discrimination. It would then be WPTC's responsibility to show that any discrimination was job related. Since Mr. Karney wanted to be able to anticipate any eventuality, he was attempting to put together a systematic and thorough plan for setting *and* meeting realistic EEO objectives. He had called a meeting for January 8, 1976 of the three administrators who would influence the development of such a plan. Mr. Karney intended initially to concentrate on the operator division as a test of the plan's feasibility.

COMPANY BACKGROUND

Wentworth Pacific Telephone Company was found in 1903 and quickly became one of the largest independent telephone companies in the U.S. Although the company was still considered a regional firm, the total workforce was substantial (almost 9,000 employees), and company policies affected the total employment picture within the region. Despite the decline in the number of independent telephone companies from fifty-five hundred in 1950 to sixteen hundred in 1976, Wentworth Pacific had slowly grown during this period.

The operator division employed the largest number of people, approximately 3,211, within WPTC. There were three primary tiers of jobs—3,030 operators, 151 supervisors, and 30 managers. Ninety-five percent of the operators in 1976 were women. This division was organized into teams of twenty operators who were coordinated by a supervisor. These teams (or work units) were assigned schedules as a block, including holidays and vacations. (Any newly hired operator was assigned to one of these teams). The team supervisor mainly dealt with scheduling problems—seeing that part timers replaced employees who were sick or on vacation—and was required to help "man" the switchboard whenever her[1] or his scheduling tasks were finished or whenever a peak period occurred. The team approach worked well since WPTC rewarded superior performance by incentives such as prizes and extra vacations. They recognized that cooperation was a key ingredient to productivity.

[1]Traditionally, a supervisor was female. In 1976 the percentage of women supervisors was 85%.

It was a longstanding policy to have a manager cover five supervisors. A manager had responsibility to hire personnel and to recommend promotions for supervisor openings. In 1976, 93% of the managers were men.

The historical promotional rates are shown in Exhibit 1. This table is called a Markov Matrix; the number in any row show the fraction of people moving from a given position to any other position over a 5-year period. For example, 62.5% of the operators who were currently employed would remain as operators 5 years later, whereas 2% of the operators would be promoted to supervisors, and 35.5% would leave the organization. No operators would become managers in five years. The company adhered to a policy of no layoffs; therefore, attrition was the sole method for reducing the workforce.

EXHIBIT 1

A Markov Matrix for Employees on January 1, 1976

Now \ 5 Years Later	Operators	Supervisors	Managers	Left the Division
Operators	.625	.02	0	.355
Supervisors	.007	.535	.0094	.4486
Managers	0	.002	.904	.094

THE MEETING

Richard Karney called the meeting on January 8, 1976 to decide upon the proper course of action for achieving the EEO objectives. Besides himself, the following individuals were in attendance: Harriet Narod (Karney's assistant); a management scientist who was familiar with formal mathematical models for manpower planning; Michael Barr, the manager of the economic-forecasting department of WPTC; and Fred Ross, the head of the personnel department.

Excerpts from that meeting are provided below:

R. Karney: *As we all know, the EEO commission has recently reviewed us and set up a list of numerical goals, really ratios, for different job categories to be achieved by January 1, 1981. For example, in the operator division we are expected to reduce the percentage of women operators to 80%, the percent-*

age of women supervisors to 80%, and the percentage of men managers to 75%.

H. Narod: *What happens if we don't reach these quotas?*

R. Karney: *From what I understand, we must demonstrate to the EEO commission's satisfaction that the attainment of these goals was impossible, or else face a possible charge of discrimination.*

M. Barr: *Say Fred, what is the prospects for growth in the O-division over the 5-year horizon?*

F. Ross: *I really don't know. Since 1959, we have experienced growth over a long period, but the question of automation is still unresolved.*[2] *Give me a few days to put together a forecast.*

H. Narod: *Don't forget to provide the possible spread in the forecast in addition to your usual point estimates.*

R. Karney: *I view the problem from several perspectives. First of all, we must meet the day-to-day operating needs of the company by providing acceptable service. Secondly, the costs of providing that service must be kept low enough so that a fair profit is maintained. Thirdly, we have a responsibility to the community to put together a workforce which reflects needs within the community.*

F. Ross: *But we also must realize the limitations within that community. The pool of eligible candidates is not unlimited. Besides, the costs of hiring must be taken into account.*

H. Narod: *How about citing some figures for those costs?*

F. Ross: *According to our estimates, we calculated $2,000 per supervisor, $1,500 per male operator, $1,000 per female operator, $3,000 per male manager and $7,000 per female manager hired. Although these costs represent several amounts, they are fairly accurate and could be used for future projections.*

M. Barr: *Will the new internal training and promotion policies affect those amounts?*

F. Ross: *No. But the costs of training are high, about $3,000 for training an operator to become a supervisor and $7,500 for training a supervisor to be a manager. The former requires four courses, and the latter requires sixteen. Traditionally, a small segment of the employees will pay for*[3] *these courses on their own, and any promotions based on these extra courses are reflected in the historical Markov matrix. We don't expect the new policies to affect*

[2]See Exhibit 2.

[3]WPTC had recently decided to give a limited number of scholarships to employees who wanted to increase their chances of promotion within WPTC.

EXHIBIT 2
Historical Data Pertaining to the Demand for Operators

Year	# Operator (hundreds)	Regional Population (millions)	# Telephones (millions)	# Completed Local Calls per day (millions)	% Of Operator-Assisted Local Calls	# Completed Toll Calls per day (millions)
59	23.3	6.84	2.72	7.86	20	.367
60	22.9	6.95	2.86	8.28	19	.387
61	22.8	7.06	2.98	8.55	18	.405
62	22.3	7.17	3.10	9.15	17	.429
63	22.2	7.28	3.25	9.47	16	.453
64	22.5	7.38	3.41	9.87	16	.492
65	24.1	7.47	3.60	10.50	16	.538
66	25.6	7.56	3.80	11.07	14	.592
67	25.9	7.64	3.99	11.48	14	.642
68	26.8	7.72	4.20	12.00	13	.708
69	29.0	7.79	4.43	13.00	14	.796
70	30.5	7.88	4.62	13.70	13	.884
71	30.5	7.96	4.81	14.40	12	.923
72	30.7	8.03	5.08	15.20	12	1.038
73	31.5	8.09	5.31	15.90	12	1.150
74	31.2	8.15	5.54	16.80	11	1.230
75	30.3	8.21	5.73	17.20	10	1.270

EXHIBIT 3
A Markov Matrix for Period January 1, 1976 to January 1, 1981

Hired during period Jan. 1, 1976 to Jan. 1, 1981. \ January 1, 1981	Operators	Supervisors	Managers	Left the Division
Operators	.83	0	0	.17
Supervisors	0	.905	0	.095
Managers	0	0	.97	.03

these transition rates but anyone selected for internal training will be automatically promoted to supervisor (from operator) or to manager (from supervisor) at the completion of their training. For instance, I can't see how we can hire more than 500 male operators in the next 5 years. In addition, not all the people whom we hire during the next 5 years will be with the company in 1981.[4] *We might also have difficulty finding enough qualified*

[4]See Exhibit 3 for a Markov matrix, which gives the average transition rates affecting people hired during the 5-year period.

Year	% Of Operator-Assisted Toll Calls	GNP (billions)	Price Deflator (1972=100)	Telephone Price Index (1972=100)	Total Personal Consumption (billions)	Personal Consumption for Telephones (billions)
59	85%	475	67.6	71.9	161	4.2
60	77%	506	68.7	72.8	173	4.5
61	73%	520	69.3	73.8	185	4.8
62	65%	560	70.6	74.6	196	5.1
63	61%	590	71.6	75.8	211	5.5
64	59%	632	72.7	76.8	227	5.9
65	58%	688	74.3	77.6	250	6.5
66	55%	750	76.8	78.6	265	6.9
67	51%	794	79.0	80.3	288	7.5
68	47%	864	82.6	82.9	315	8.2
69	44%	930	86.7	85.6	330	9.0
70	40%	982	91.4	89.2	350	10.1
71	36%	1063	96.0	95.3	375	11.0
72	31%	1171	100.0	100.0	411	12.4
73	27%	1307	105.9	104.2	458	14.0
74	25%	1413	116.2	116.4	498	16.0
75	21%	1516	126.4	128.0	615	19.0

female managers, but at least we can provide more scholarships for female supervisors and promote them from within.

R. Karney: *Company policy says that we should promote internally at least one person to manager for every four people hired as manager.*

H. Narod: *Perhaps I can integrate all of this information into a formal mathematical model. Let's get back together in a few days.*

APPENDIX **Background on Equal Employment Opportunity**

The history of equal employment opportunity law starts with the Civil Right Acts of 1866 and 1871. For nearly 100 years the courts read those Acts as being almost without application. So, it was really not until 1941 that the development of modern EEO law began, when President Roosevelt's Executive Order 8802 (1) established a Fair Employment Practices Committee, and (2) required a non-discrimination clause to be included in all defense contracts.

President Roosevelt strengthened the power of the Committee in 1943, when he issued Executive Order 9346 requiring a non-discrimination clause in

Source: Reproduced from K. Lewis, "Manpower Planning in Equal Employment Opportunities: As Applied to the U.S. Navy Civilian Work Force," Ph.D. Dissertation, Carnegie-Mellon University, 1977.

all government contracts, rather than just in defense contracts. Since government contracts were widespread and important throughout industry, Executive Order 9346 could have potentially produced an important impact. However, an amendment to an appropriations bill that same year killed the Executive Order and the Committee as well.

Nothing further in the form of official documents—such as Executive Orders—occurred until President Truman, in 1948, ordered the Armed Forces desegregated. In 1951, he also issued Executive Order 10308, which again required non-discrimination in Federal contracts, and set up a committee to oversee compliance. Success was at best limited in terms of the effects of these orders.

President Kennedy, in 1961, issued the first effective order in ten years to strengthen the non-discrimination clause. Executive Order 10925 applied not only to the particular plant holding a Federal contract, but also to all the employment practices carried on by Federal contractors. It also contained, for the first time, references to affirmative action. In 1963, Executive Order 10925 was further extended in the form of Executive Order 11114 to include Federally assisted construction contracts. This development put Federal pressure on the construction unions and skilled trades to accept minorities for the first time. That same year, Congress passed the Equal Pay Act, endorsing the basic concept of equal pay for equal work, regardless of sex.

Then, in 1964, the Civil Rights Act was passed, establishing what later became the Equal Employment Opportunity Commission. This act has become the mainstay of EEO legislation. Title VII of the act places broad prohibitions on employment discrimination on the basis of race, sex, or national origin. In 1967, the Department of Justice put a high priority on the enforcement of Title VII, with the specific objective of forcing case decisions in the law in order to (1) provide Federal agencies with specific guidelines implied by the law, and (2) provide private employers and unions with an explicit understanding of their obligations.

This Department of Justice program passed initiative to the judiciary, as cases addressing the meaning of Title VII were pressed between 1968 and 1971. Two interpretations of Title VII were possible. The narrower one saw Title VII as prohibiting only those employment acts which were consciously motivated by social prejudice. The broader interpretation, however, was that Title VII prohibited any kind of employment practice that had a discriminatory impact, unless that practice was required by some business need. This second interpretation was upheld by the Supreme Court in Griggs vs. Duke Power Co. (401 U.S. 424, 1971). The Supreme Court ruling in that case stated that a test or high school education requirement which has a discriminatory impact on a group that was previously the object of discrimination is unlawful, unless the practice is required by business necessity.That is, seemingly neutral practices—practices that on their face do not discriminate with respect to race or sex, but which have a discriminatory impact—are unlawful unless required by business need. Notice, therefore, the two-fold character of this decision which may be interpreted as follows. It does not say that there must be no discrimination. It says rather that any discrimination must be shown to be job related, and thus the responsibility for showing this is on the organization that discriminates. If there is no evidence of discrimination, however, the law has no further interest in the reasons for this state of affairs. It is this Supreme Court ruling that has defined to a large extent the basic underpinnings of the EEO considerations that are important to managers today.

Even while this was going on a new Executive Order 11246 was issued, further strengthening equal employment opportunity by applying numerical goals to workers on federally assisted construction contracts. This additional move was challenged over several years, but finally, in Contractors Association of Eastern Pennsylvania vs. Schultz (442 F. 2d 159, cert. denied, 404 U.S. 854), the Third U.S. Circuit Court of Appeals sustained the power of the Executive Branch to impose such plans, and specifically endorsed the use of numerical goals as objectives in Federally assisted construction employment. Since that time, the use of numerical goals and timetables has become a regular part of relief in Title VII cases.

More recently, the U.S. Congress passed the Equal Employment Opportunity Act in 1972, thereby strengthening the Equal Opportunity Commission. The new Act expanded the Commission's jurisidiction to cover the employees of private educational institutions, state and local governments, and all companies and all unions with at least 15 workers, thereby extending the impact of equal employment opportunity compliance to the private sector as well as the public sector. The Act empowered the Commission to seek Federal court action against employment discrimination—although a filibuster prevented passage of an additional provision which would allow the Commission to issue binding cease-and-desist orders against discriminators. Later that year, the Department of Health, Education and Welfare issued 17 pages of guidelines for minority employment directed at 2,500 colleges and universities with Federal contracts. No quotas were set in these guidelines, but the development of numerical goals and timetables was required.

Throughout 1973, the Equal Employment Opportunity Commission filed suits charging employment discrimination against such firms as General Motors, Ford, General Electric, and Xerox, as well as locals of such unions as the United Auto Workers and the International Longshoremen. The Commission also joined the Department of Labor to obtain a settlement with the American Telephone and Telegraph Company under which the company paid women and minority employees compensation for past hiring and promotion restrictions. A Federal judge in Birmingham, Alabama, ordered both the United States Steel Company and the United Steelworkers Union to alter discriminatory hiring and seniority systems at the Fairfield Works while also requiring back-pay compensation by the company, a uniform seniority system, and specific hiring timetables. Most recently, the Civil Service Commission, working with the Equal Employment Opportunity Commission, issued a new Federal personnel hiring and promotion directive that describes standards for hiring and promotion testing which are intended to apply to *all* sectors of employment.

Wentworth Pacific Telephone Company (B)

On January 15, one week after Richard Karney had called a meeting to discuss plans for meeting EEO objectives, he received a short report from Mike Barr containing his department's estimate of the number of operators needed by January 1, 1981.[1]

Analysts in WPTC's economic-forecasting department viewed the requirements for operators to be dependent on three factors: first, the total number of calls requiring operator assistance; second, the number of telephones in service; and third, the impact of differential inflation. This view is depicted graphically in an excerpt from Barr's report, shown in Exhibit 1. The impact of inflation on telephone prices was captured in the ratio of the telephone price index to the index of all prices (i.e., the price deflator).

EXHIBIT 1

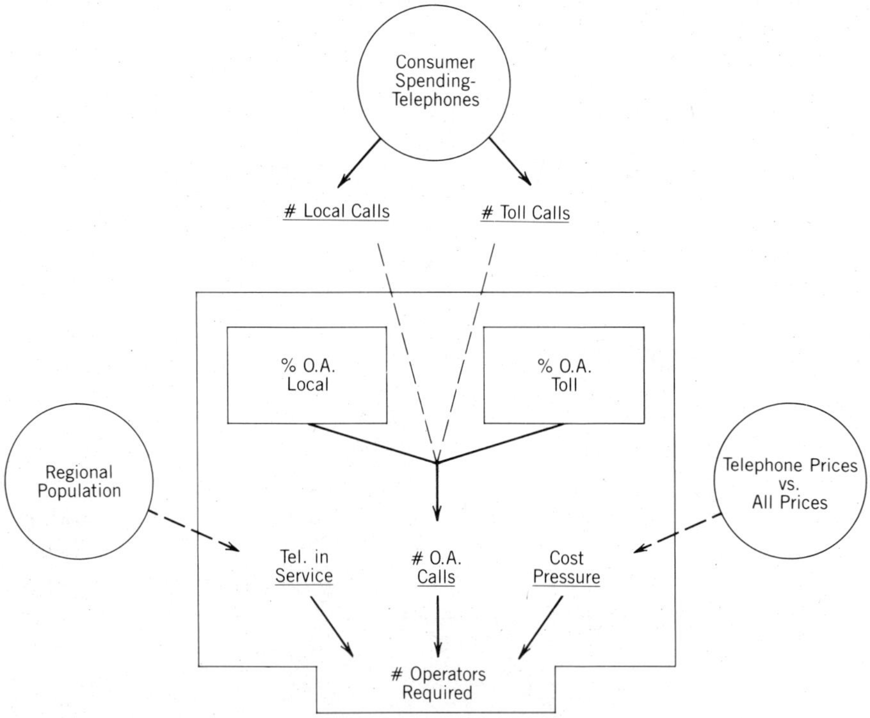

[1]See Wentworth Pacific Telephone Company (A).

A regression model was constructed to measure the effect of these factors on operator requirements. Since, according to Barr, the three independent variables could be predicted quite accurately, the model could project the number of operators needed through 1980. Exhibits 2 and 3, also excerpts from Barr's report, show the relevant data and the forecasts through 1980. Interestingly enough, the number of operators expected to be needed in 1980 was 3030, the same number as employed at present.

EXHIBIT 2

RELEVANT DATA, WPTC ECONOMIC FORECASTING DEPARTMENT

	1960	*1961*	*1962*	*1963*	*1964*	*1965*	*1966*	*1967*	*1968*	*1969*	*1970*	*1971*	*1972*	*1973*	*1974*	*1975*
OPERATORS REQUIRED (IN HUNDREDS)	22.90	22.80	22.30	22.20	22.50	24.10	25.60	25.90	26.80	29.00	30.50	30.50	30.70	31.50	31.20	30.30
%CH	−1.7	−0.4	−2.2	−0.4	1.4	7.1	6.2	1.2	3.5	8.2	5.2	0.0	0.7	2.6	−1.0	−2.9
TOTAL OPERATOR ASSISTED CALLS (IN MILLIONS)	1.87	1.83	1.83	1.79	1.87	1.99	1.88	1.93	1.89	2.17	2.13	2.06	2.15	2.22	2.16	1.99
%CH	−0.7	−2.0	0.0	−2.3	4.4	6.6	−5.9	3.2	−2.2	14.7	−1.6	−3.5	4.1	3.4	−2.8	−7.8
TOLL CALLS	0.39	0.41	0.43	0.45	0.49	0.54	0.59	0.64	0.71	0.80	0.88	0.92	1.04	1.15	1.23	1.27
%CH	5.4	4.7	5.9	5.6	8.6	9.3	10.0	8.4	10.3	12.4	11.1	4.4	12.5	10.8	7.0	3.3
%O.A.*	77	73	65	61	59	58	55	51	47	44	40	36	31	27	25	21
%CH	−9.4	−5.2	−11.0	−6.2	−3.3	−1.7	−5.2	−7.3	−7.8	−6.4	−9.1	−10.0	−13.9	−12.9	−7.4	−16.0
LOCAL CALLS	8.28	8.55	9.15	9.47	9.87	10.50	11.07	11.48	12.00	13.00	13.70	14.40	15.20	15.90	16.80	17.20
%CH	5.3	3.3	7.0	3.5	4.2	6.4	5.4	3.7	4.5	8.3	5.4	5.1	5.6	4.6	5.7	2.4
%O.A.*	19	18	17	16	16	16	14	14	13	14	13	12	12	12	11	10
%CH	−5.0	−5.3	−5.6	−5.9	0.0	0.0	−12.5	0.0	−7.1	7.7	−7.1	−7.7	0.0	0.0	−8.3	−9.1
TELEPHONES IN SERVICE (IN MILLIONS)	2.86	2.98	3.10	3.25	3.41	3.60	3.80	3.99	4.20	4.43	4.62	4.81	5.08	5.31	5.54	5.73
%CH	5.1	4.2	4.0	4.8	4.9	5.6	5.6	5.0	5.3	5.5	4.3	4.1	5.6	4.5	4.3	3.4
COST SQUEEZE INDEX**	105.9	106.6	105.8	105.9	105.6	104.4	102.4	101.6	100.4	98.7	97.7	99.3	100.0	98.5	100.3	100.6
%CH	1.9	0.6	−0.7	0.1	−0.3	−1.1	−1.9	−0.8	−1.1	−1.7	−1.1	1.6	0.8	−1.5	1.8	0.3

ECONOMIC ASSUMPTIONS

	1960	*1961*	*1962*	*1963*	*1964*	*1965*	*1966*	*1967*	*1968*	*1969*	*1970*	*1971*	*1972*	*1973*	*1974*	*1975*
REAL GNP (BIL. $ 1972)	737	755	799	831	874	926	981	1008	1052	1079	1075	1108	1171	1235	1218	1202
%CH	2.3	2.5	5.8	4.0	5.3	5.9	6.0	2.7	4.4	2.6	−0.3	3.0	5.7	5.5	−1.4	−1.1
REAL PERSONAL EXPEND. TELEPHONES (BIL. $ 1972)	5.0	5.4	5.7	6.1	6.6	7.3	8.1	8.7	9.5	10.4	11.1	11.6	12.4	13.7	14.4	15.7
%CH	6.4	8.0	5.6	7.0	8.2	10.6	11.0	7.4	9.2	9.5	6.7	4.5	6.9	10.5	5.1	9.0

(continued)

EXHIBIT 2 (Continued)

TELEPHONE PRICES (1972=1)	0.728	0.738	0.746	0.758	0.768	0.776	0.786	0.803	0.829	0.856	0.892	0.953	1.000	1.042	1.164	1.280
%CH	3.6	1.5	1.1	1.6	1.3	1.1	1.3	2.1	3.3	3.3	4.2	6.8	4.9	4.2	11.7	9.9
ALL PRICES (1972=1)	0.687	0.693	0.706	0.716	0.727	0.743	0.768	0.790	0.826	0.867	0.914	0.960	1.000	1.058	1.160	1.271
%CH	1.7	0.9	1.8	1.5	1.6	2.2	3.3	3.0	4.5	5.0	5.4	5.1	4.1	5.8	9.7	9.6
POPULATION (IN MILLIONS)	6.95	7.06	7.17	7.28	7.48	7.47	7.56	7.64	7.72	7.79	7.80	7.96	8.03	8.09	8.15	8.21
%CH	1.6	1.6	1.5	1.5	1.4	1.2	1.2	1.1	1.0	0.9	1.2	1.0	0.9	0.7	0.7	0.7

*PERCENTAGE OF CALLS REQUIRING OPERATOR ASSISTANCE.
**INDEX = TELEPHONE PRICES/ALL PRICES.

EXHIBIT 3

THE OUTLOOK TO 1980
(PREPARED ON 1/13/76 BY WPTC ECONOMIC FORECASTING DEPARTMENT)

	1975	1976	1977	1978	1979	1980'
OPERATORS REQUIRED (IN HUNDREDS)	30.30	29.86	30.00	30.37	30.31	30.30
% CH	−2.9	−1.4	0.5	1.2	−0.2	0.0
TOTAL OPERATOR ASSISTED CALLS (IN MILLIONS)	1.99	2.05	2.05	2.06	1.98	1.92
% CH	−7.8	3.2	1.4	−0.9	−3.8	−3.2
TOLL CALLS	1.27	1.38	1.52	1.63	1.70	1.79
% CH	3.3	9.0	9.7	7.0	4.6	5.0
% O.A.*	21	20	19	18	16	15
% CH	−16.0	−4.8	−5.0	−5.3	−11.1	−6.3
LOCAL CALLS	17.20	18.61	19.95	21.01	21.76	22.61
% CH	2.4	8.2	7.2	5.3	3.6	3.9
% O.A.*	10	10	9	8	8	7
% CH	−9.1	−4.8	−5.8	−6.2	−6.6	−7.1
TELEPHONES IN SERVICE (IN MILLIONS)	5.73	5.34	6.18	6.43	6.68	6.95
% CH	3.4	3.6	4.1	4.1	3.9	3.9
COST SQUEEZE INDEX**	100.6	103.3	104.4	104.3	104.6	105.0
% CH	0.3	2.7	1.0	0.0	0.3	0.3

ECONOMIC ASSUMPTIONS

	1975	1976	1977	1978	1979	1980'
REAL GNP (BIL. $ 1972)	1202	1275	1338	1397	1452	1519
% CH	−1.3	6.0	4.9	4.5	3.9	4.7

(continued)

EXHIBIT 3 (Continued)

REAL PERSONAL EXPEND. TELEPHONES (BIL. $ 1972)	15.7	16.8	18.3	19.5	20.4	21.3
% CH	9.0	4.7	9.2	6.6	4.4	4.8
TELEPHONE PRICES (1972=1)	1.280	1.383	1.475	1.563	1.658	1.761
% CH	9.9	8.1	6.6	6.0	6.1	6.2
ALL PRICES (1972=1)	1.271	1.339	1.413	1.498	1.584	1.678
% CH	9.6	5.3	5.6	6.0	5.8	5.9
POPULATION (IN MILLIONS)	8.21	8.28	8.34	8.41	8.48	8.55
% CH	0.7	0.8	0.8	0.8	0.8	0.9

*Percentage of calls requiring operator assistance.
**Index = (Telephone Prices/All Prices) × 100.

Meanwhile, Karney and his assistant, Harriet Narod, had been discussing how they could determine the number of operators, supervisors, and managers of each sex to be hired over the next 5 years, as well as how many should be selected for internal training. This hiring and training had to be planned in such a way that enough personnel would be available to meet operating needs while, also, EEO goals were met. Narod suggested that a linear programming model might aid in this decision and formulated a model (see Exhibit 4) that would minimize the costs of hiring and training while satisfying company and EEO constraints. Karney agreed that the LP would do just what they wanted. The only other input needed was Barr's forecast.

EXHIBIT 4
Narod's Linear Programming Formulation

Decision Variables

o_m; o_f	Operators (male, female) employed on 1/1/81
s_m; s_f	Supervisors (male, female) employed on 1/1/81
m_m; m_f	Managers (male, female) employed on 1/1/81
ho_m; ho_f	Males, females *hired* into operator ranks during period 1/1/76–1/1/81
hs_m; hs_f	Males, females *hired* into supervisor ranks during period 1/1/76–1/1/81
hm_m; hm_f	Males, females *hired* into manager ranks during period 1/1/76–1/1/81
ts_m; ts_f	Males, females *trained* to be supervisors (from employees in operator ranks)
tm_m; tm_f	Males, females *trained* to be managers (from employees in supervisor ranks)

(continued)

EXHIBIT 4 (Continued)

Objective Function—Cost

$$\text{Min} \quad ho_f + 1.5ho_m + 2hs_f + 2hs_m + 7hm_f + 3hm_m + 3ts_f + 3ts_m + 7.5tm_f + 7.5tm_m$$

Constraints

1. *Transitions*

o_m = .625 × (current male operators) + .007 × (current male supervisors) + $.8265ho_m - ts_m$

o_f = .625 × (current female operators) + .007 × (current female supervisors) + $.8265ho_f - ts_f$

s_m = .02 × (current male operators) + .535 × (current male supervisors) + .002 × (current male managers) + $.0035ho_m + .905hs_m + ts_m - tm_m$

s_f = .02 × (current female operators) + .535 × (current female supervisos) + .002 × (current female managers) + $.0035ho_f + .905hs_f + ts_f - tm_f$

m_m = .0095 × (current male supervisors) + .904 × (current male managers) + $.97hm_m + tm_m$

m_f = .0095 × (current female supervisors) + .904 × (current female managers) + $.97hm_f + tm_f$

2. *WPTC Personnel Requirements*

$$o_m + o_f \geqslant 3030$$

$$s_m + s_f \geqslant 151$$

$$m_m + m_f \geqslant 30$$

3. *EEO Goals*

$$\frac{o_f}{o_f + o_m} \leqslant .8$$

$$\frac{s_f}{s_f + s_m} \leqslant .8$$

$$\frac{m_m}{m_f + m_m} \leqslant .75$$

(continued)

EXHIBIT 4 (Continued)

4. *Training Policy*

$$\frac{.0094 \times (\text{current supervisors}) + tm_m + tm_f}{hm_m + hm_f} \geqslant .25$$

5. *Supply of Male Candidates for Operator Position*

$$ho_m \leqslant 500$$

After reading Barr's report, Narod completed the LP tableau (Exhibit 5), and submitted it to Computer Operations. She was pleasantly surprised to receive the output that very afternoon; pleased, that is, until she turned to the results of the computer run and saw:

NO FEASIBLE SOLUTION EXISTS

Stepping into Karney's office, she showed him the printout. "What now?" she asked him glumly.

EXHIBIT 5

LINEAR: NAROD'S MODEL

OBJECTIVES	OM	OF	SM	SF	MM	MF	HOM	HOF	HSM	HSF	HMM	HMF	TSM	TSF	TMM	TMF		
COST	0	0	0	0	0	0	1.5	1	2	2	3	7	3	3	7.5	7.5		
CONSTRAINTS	OM	OF	SM	SF	MM	MF	HOM	HOF	HSM	HSF	HMM	HMF	TSM	TSF	TMM	TMF	RELATION	RHS
TRANS-OM	-1	0	0	0	0	0	.8265	0	0	0	0	0	-1	0	0	0	EQ	-95.16
TRANS-OF	0	-1	0	0	0	0	0	.8265	0	0	0	0	0	-1	0	0	EQ	-1800
TRANS-SM	0	0	-1	0	0	0	3.5E-3	0	.905	0	0	0	1	0	-1	0	EQ	-15.41
TRANS-SF	0	0	0	-1	0	0	0	3.5E-3	0	.905	0	0	0	1	0	-1	EQ	-126
TRANS-MM	0	0	0	0	-1	0	0	0	0	0	.97	0	0	0	1	0	EQ	-25.53
TRANS-MF	0	0	0	0	0	-1	0	0	0	0	0	.97	0	0	0	1	EQ	-3.024
REG-OP	1	1	0	0	0	0	0	0	0	0	0	0	0	0	0	0	GE	3030
REQ-SUP	0	0	1	1	0	0	0	0	0	0	0	0	0	0	0	0	GE	151
REG-MGR	0	0	0	0	1	1	0	0	0	0	0	0	0	0	0	0	GE	30
EEO-OP	-.8	.2	0	0	0	0	0	0	0	0	0	0	0	0	0	0	LE	0
EEO-SUP	0	0	-.8	.2	0	0	0	0	0	0	0	0	0	0	0	0	LE	0
EEO-MGR	0	0	0	0	.25	-.75	0	0	0	0	0	0	0	0	0	0	LE	0
TRAINING	0	0	0	0	0	0	0	0	0	0	-.25	-.25	0	0	1	1	GE	-1.418
OM-SUPPL	0	0	0	0	0	0	1	0	0	0	0	0	0	0	0	0	LE	500

Chapter 5

NETWORK OPTIMIZATION MODELS

There are a number of linear programming models whose special structure permits the use of more efficient solution techniques than would otherwise be possible. The foremost of these is *network optimization models*. The algorithms developed to exploit the special network structure enable the solution of models far too large for conventional solution techniques. As an illustrative example, the U.S. Treasury Department is now in the process of implementing a network model with 63 million variables!

Originally applied to electrical engineering, network analysis has been used successfully over the past few decades in a wide range of applications. Some of the most important include information theory and communications systems, production and employment scheduling, transportation and distribution systems, and project planning and control. Very generally, a network model is any linear program that can be represented pictorially as a network ("links" between various "locations"), whose variables are the *flows* occurring on the links of that network, and whose constraints are the material balance equations on those flows.

The purpose of this chapter is to familiarize the reader with some of the basic concepts of network theory and to formulate a few of the most commonly applied network models that have been successfully implemented. We will not be concerned here with solution methodologies. Due, however, to the historical significance as well as the simplicity of the transportation method, we illustrate that solution procedure in Appendix B. It gives the interested reader the "flavor" of techniques for optimizing flows in networks; we will not deal with any other specific solution procedures.

NOTATION

A network consists of a number of *nodes* with certain pairs of nodes joined by one or more *arcs*. The nodes are usually denoted by numbers—that is, 1, 2, ... An arc joining nodes i and j is denoted (i, j). An example of a network with 8 nodes and 13 arcs is shown in Figure 5-1. An arc is called *directed* if the flow along this arc is restricted to one direction only. All arcs in Figure 5-1 are directed. If flow along a given arc (i, j) can only be from i to j, the directed arc is denoted $i \to j$. If directions are assigned to all arcs, the resulting network is called *directed.*

A *network* is a graph with *flow* of some type over its arcs. Examples include:

- A network of roads (arcs), intersections (nodes), and vehicles (flow).
- A network of airports (nodes), airlanes (arcs), and aircraft (flow).
- A network of pipelines (arcs), pumping stations (nodes), and oil (flow).

A network need not be directed; it may be feasible to have flow in either direction along an arc. If a network is directed, the orientation of the arc is the feasible direction of flow.

In most practical instances, there are capacity restrictions on the flow; there are upper limits on the amount that can flow along each arc. For example, oil pipelines have flow capacity determined by the pipe diameter. Most applications, in addition, involve costs per unit of flow, assigned to each arc. The network of Figure 5-1 is redrawn with costs and capacities in Figure 5-2. For

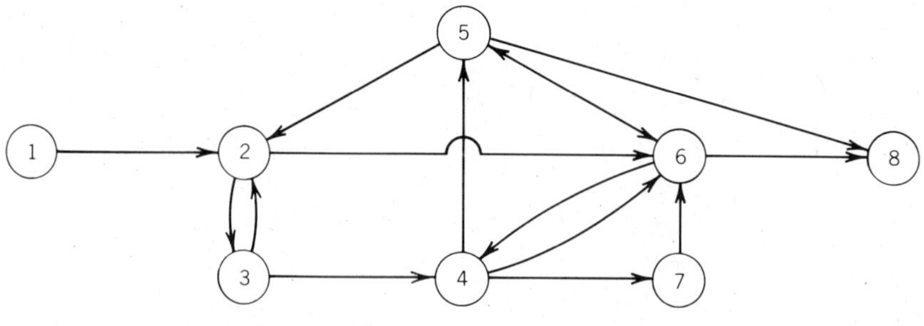

FIGURE 5-1. A directed network.

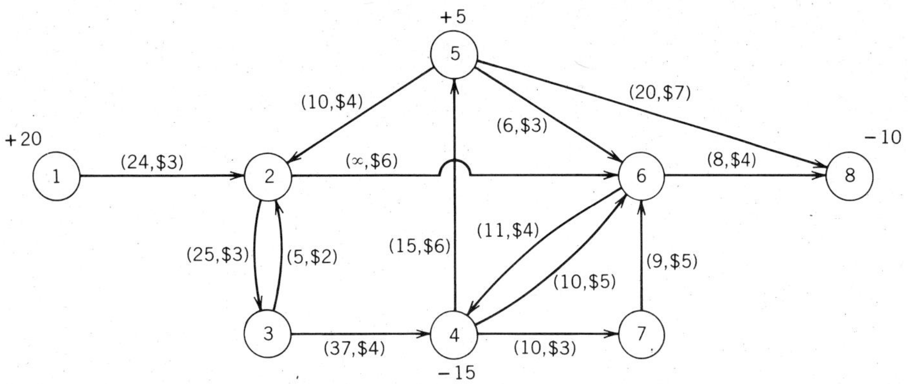

FIGURE 5-2. A directed network with costs and capacities.

example, (24, \$3) on arc (1→2) denotes that a maximum of 24 units can flow along this arc and the cost of shipping one unit is \$3. In general, the cost of shipping one unit from node i to node j is not necessarily the same as the cost of shipping one unit from node j to node i.

We are also often interested in the flow originating (supplied) at some nodes and being consumed (demanded) at others. In the context of distribution systems, one unit of "flow" may represent one unit of product traveling in the system from factory to warehouse or customer. These latter nodes consume flow. In Figure 5-2, the quantities "+20" and "+5" at nodes 1 and 5, respectively, are the units available at those *sources* (origins). The quantities "−15" and "−10" at nodes 4 and 8 are the amounts demanded at those nodes, called *sinks (destinations)*. The remaining nodes (2, 3, 6, and 7) have no net supply or demand: the flow into each of these nodes must be equal to the flow out; that is, there is a conservation of flow at any node other than the source and sink nodes. These nodes are sometimes called *transshipment* nodes.

THE NODE ARC INCIDENCE MATRIX

In the network of Figure 5-2 there are many possible routes over which the 25 units from source nodes 1 and 5 may be shipped to sink nodes 4 and 8. We wish to find the minimum cost route to fulfill this demand. This problem can be formulated as a linear program. Let

$$x_{ij} = \text{amount of flow over arc } i \to j$$

The objective function to be minimized is the total shipping cost; that is,

$$3x_{12} + 3x_{23} + 6x_{26} + 2x_{32} + 4x_{34} + 6x_{45} + 5x_{46} + 3x_{47} + 4x_{52} + 7x_{58}$$
$$+ 4x_{64} + 3x_{65} + 4x_{68} + 5x_{76}$$

There is a constraint associated with each node in the network that expresses the conservation of flow at the node:

$$\text{Flow out of a node} - \text{Flow into a node} = \text{Net flow at node}$$

By convention, at each of the source nodes, the net flow is positive; at each sink node the net flow is negative; and the net flow at each transshipment node is zero. The eight *flow conservation* or *material balance* equations are given below.

$$\begin{array}{ll}
\text{Node 1:} & x_{12} = 20 \\
\text{Node 2:} & -x_{12} + x_{23} + x_{26} - x_{32} - x_{52} = 0 \\
\text{Node 3:} & -x_{23} + x_{32} + x_{34} = 0 \\
\text{Node 4:} & -x_{34} + x_{45} + x_{46} + x_{47} - x_{64} = -15 \\
\text{Node 5:} & -x_{45} + x_{52} + x_{58} - x_{65} = 5 \\
\text{Node 6:} & -x_{26} - x_{46} + x_{64} + x_{65} + x_{68} - x_{76} = 0 \\
\text{Node 7:} & -x_{47} + x_{76} = 0 \\
\text{Node 8:} & -x_{58} - x_{68} = -10
\end{array}$$

In addition to the material balance constraints there are the flow capacity constraints for each arc:

$$\begin{array}{lll}
0 \leq x_{12} \leq 24 & 0 \leq x_{46} \leq 10 & 0 \leq x_{68} \leq 8 \\
0 \leq x_{23} \leq 25 & 0 \leq x_{47} \leq 10 & 0 \leq x_{76} \leq 9 \\
0 \leq x_{26} \leq \infty & 0 \leq x_{52} \leq 10 & \\
0 \leq x_{32} \leq 5 & 0 \leq x_{58} \leq 20 & \\
0 \leq x_{34} \leq 37 & 0 \leq x_{64} \leq 11 & \\
0 \leq x_{45} \leq 15 & 0 \leq x_{65} \leq 6 &
\end{array}$$

The optimal flow consists of that set of x_{ij} which minimizes the total costs and satisfies the balance and capacity constraints. In addition to determining the optimal amount to be shipped over each arc, the optimal shipping route is determined: each nonzero x_{ij} corresponds to arc $(i \rightarrow j)$ being included in the optimal route.

As was discussed above, the Simplex method could be used to solve this linear program, but the network model has a structure that can be exploited by more efficient solution procedures. This structure can best be seen from the tableau of material balance equations in Table 5-1. This tableau is called the *node arc incidence matrix* of the original network; it completely describes the network's physical characteristics—which nodes are connected to which others by which arcs. Note that (1) the variables x_{ij} have only 0, $+1$, and -1 coefficients

TABLE 5-1

Nodes	x_{12}	x_{23}	x_{26}	x_{32}	x_{34}	x_{45}	x_{46}	x_{47}	x_{52}	x_{58}	x_{64}	x_{65}	x_{68}	x_{76}	Net Flow
1	1														20
2	−1	1	1	−1					−1						0
3		−1		1	1										0
4					−1	1	1	1			−1				−15
5						−1			1	1		−1			5
6			−1				−1				1	1	1	−1	0
7								−1						1	0
8										−1			−1		−10
Capacity	24	25	∞	5	37	15	10	10	10	20	11	6	8	9	
Cost	3	3	6	2	4	6	5	3	4	7	4	3	4	5	

and (2) each variable appears in only two balance constraints, once with a +1 coefficient and once with a −1 coefficient. We have already seen that each balance constraint (a *row* in the incidence matrix) corresponds to a particular *node* in the network and constrains the flow from and into that node. We note now that each column of the incidence matrix corresponds to a particular directed *arc* of the network: the +1 coefficient in each column corresponds to the node from which this arc emanates and the −1 coefficient corresponds to the node into which the arc is incident. Thus, given the physical representation of a network, its node arc incidence matrix is easy to construct.

Sometimes a linear program's original form must be algebraically manipulated, such as by combining equations, changing the signs of coefficients, multiplying constraints by a constant, or by adding or eliminating a redundant relation in order for it to have the form above. Once it is in this form, however, it can be solved easily by available network codes.

CAPACITATED NODES

We have seen that capacity restrictions on the arcs can be handled simply by upper-bounding the x_{ij} variable which corresponds to that arc. What about capacity restrictions on nodes? For example, Figure 5-3*a* shows part of a network depicting a firm's distribution plans from plants to warehouses to customers. Node Z, shown in Figure 5-3*a*, represents a warehouse that has a throughput limitation of 100,000 units. Unfortunately, the capacities on the arcs leading *into* Z are not sufficiently restrictive to guarantee that less than 100,000 units will flow into Z. One way of explicitly incorporating this node capacity is to split node Z into two nodes Z′ and Z″ as shown on Figure 5-3*b*. One arc, with zero cost and capacity 100,000, links Z′ and Z″. This effectively

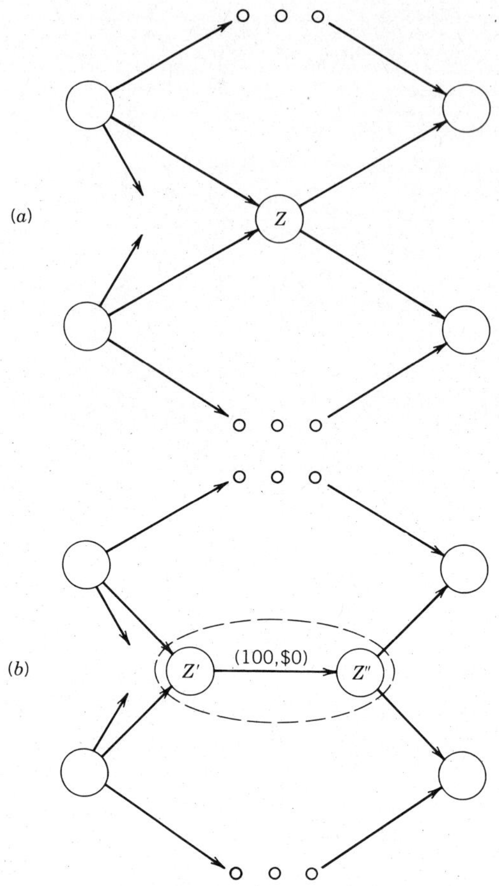

FIGURE 5-3. A capacitated node.

guarantees that since no more than 100,000 units can leave Z', no more than 100,000 units can enter Z'. Similarly, since no more than 100,000 units can flow into Z'', no more than 100,000 units can flow out of Z''.[1] Hence, the "expanded node" represented by the dotted lines in Figure 5-3*b* has capacity 100,000.

THE GENERAL FORMULATION OF THE NETWORK FLOW PROBLEM

Let u_{ij} and l_{ij} be the upper and lower capacity bounds on the total flow on arc $(i \rightarrow j)$. Also, let P_k be the net flow at node k. A source node has a positive P_k (net supply), a sink node has a negative P_k (net demand), and a transshipment

[1]You might think of Z' as Z's receiving dock and Z'' as Z's shipping dock. The arc representing flow between this "front door" and this "back door" has a capacity of 100,000 units.

node is characterized by $P_k = 0$. Let c_{ij} be the cost per one unit of flow shipped from node i to node j. Then a general network flow problem with n nodes is

$$\text{Min} \qquad \sum_i \sum_j c_{ij} x_{ij}$$

$$\text{Subject to} \quad \sum_j x_{kj} - \sum_i x_{ik} = P_k \quad (k = 1, 2, \ldots, n)$$

$$l_{ij} \leq x_{ij} \leq u_{ij}$$

The summations are taken only over existing arcs; that is, if any arc ($i \rightarrow j$) is not in the network, x_{ij} is undefined.

The optimum solution to a network problem has an important characteristic. If all the P_k, u_{ij}, and l_{ij} in the general minimum cost network model are integers, the optimal flow x_{ij} will be integer valued.[2] Thus if variables have an integer-value restriction, there is no need to worry about special procedures such as integer programming. In addition, since integer programming problems are so much harder to solve (as we will discuss later in the book), if an integer program is seen to have the special network structure above, it may be solved without the computationally unwelcome integer restriction.

SPECIAL NETWORK MODELS

The Transportation Problem

Consider the following problem. A certain rent-a-car company has to fulfill the excess demand for cars in four different cities on a specific future day. Two cars are requested in city A, three in city B, five in city C, and seven cars in city D. The company has offices in three cities E, F, and G, each of which has excess supply. Six cars are available in city E on this specific day, one in city F, and 10 cars in city G. The distances between each of the company's offices and any destination city are given in Table 5-2 (in hundreds of miles). What is in the optimal distribution policy, that minimizes total traveling distance and meets the demand at each city?

The network depicting this situation is shown in Figure 5-4. Note that it is a special case of the type of network we saw earlier. There are two major differences: (1) there are no transshipment nodes—nodes E, F, and G are sources, nodes A, B, C, and D are destinations, and (2) there are no capacity restrictions on the flow variables—any amount may be shipped from each source to each destination. Twelve flow variables, one for each directed arc, are necessary. Let x_{ij} = number of cars sent from origin i to destination j, where

[2] To be strictly correct, since there may be multiple optima, we should say that for at least one optimum (but not necessarily all), the x_{ij}'s will be integers.

TABLE 5-2

	Destinations			
Origins	A	B	C	D
E	7	11	3	2
F	7	6	10	1
G	9	15	8	5

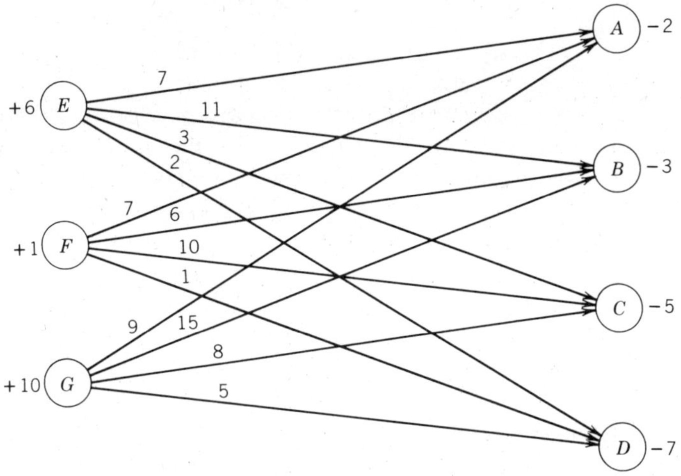

FIGURE 5-4. A transportation network.

$i = E, F, G$ and $j = A, B, C, D$. Then, the linear program corresponding to Figure 5-4 is:

$$\begin{aligned} \text{Min} \quad & 7x_{EA} + 11x_{EB} + 3x_{EC} + 2x_{ED} + 7x_{FA} + 6x_{FB} + 10x_{FC} + 1x_{FD} + \\ & 9x_{GA} + 15x_{GB} + 8x_{GC} + 5x_{GD} \end{aligned}$$

Subject to

$$\begin{array}{lllllllllllllll}
(1) & x_{EA} + x_{EB} + x_{EC} + x_{ED} & & & & & & & & & & & & = 6 \\
(2) & & x_{FA} + x_{FB} + x_{FC} + x_{FD} & & & & & & & & & & & = 1 \\
(3) & & & x_{GA} + x_{GB} + x_{GC} + x_{GD} & & & & & & & & & & = 10 \\
(4) & -x_{EA} & -x_{FA} & -x_{GA} & & & & & & & & & & = -2 \\
(5) & -x_{EB} & -x_{FB} & -x_{GB} & & & & & & & & & & = -3 \\
(6) & -x_{EC} & -x_{FC} & -x_{GC} & & & & & & & & & & = -5 \\
(7) & -x_{ED} & -x_{FD} & -x_{GD} & & & & & & & & & & = -7
\end{array}$$

Constraints (1), (2), and (3) are *supply constraints* and correspond to the origin nodes E, F, and G, respectively. For instance, the first equation states that the total shipment out of source E must equal the supply at this source, six units. Constraints (4), (5), (6), and (7) are *demand constraints* and correspond to the destination nodes A, B, C, and D, respectively. For instance, equation (4) states that the total shipment into destination A must equal the demand at this destination, two units. The objective to be minimized is the total traveling distance. Note the similarity between the structure of this problem and that of the general network: each column of the tableau contains one +1 and one −1. (The last equations would usually be written without the minus signs for ease of notation.)

This problem is an example of the *transportation* (or *distribution*) model, which is of a special importance among network models. The model has been used successfully by hundreds of firms with several plants, warehouses, sales districts, and distribution outlets. In its standard form, the model assumes that there are m supply points and n demand points (or destinations). The supply of a certain item available at origin i is S_i and the demand for this item at destination j is D_j. The cost of shipping each unit from origin i to destination j is c_{ij}. The objective is to find the optimal quantities x_{ij} ($I = 1, \ldots, m$; $j = 1, \ldots, n$) to be shipped from origin i to destination j that result in the minimum shipping costs. In general, the problem may be formulated as follows:

$$\text{Min} \quad \sum_{i=1}^{m} \sum_{j=1}^{n} c_{ij} x_{ij}$$

Subject to

$$\text{(a)} \quad \sum_{j=1}^{n} x_{ij} = S_i, \qquad i = 1, \ldots, m \ \text{(supply equations)}$$

$$\text{(b)} \quad \sum_{i=1}^{m} (-x_{ij}) = -D_j, \qquad j = 1, \ldots, n \ \text{(demand equations)}$$

To have a feasible solution of the transportation problem as formulated here we must assume that the total quantity required at all destinations is precisely the same as the total amount available at the origins:

$$\sum_{i=1}^{m} S_i = \sum_{j=1}^{n} D_j$$

Suppose that instead we consider the more general case where more units may be available at the origins than are required at the destinations. In this case

$$\sum_{i=1}^{m} S_i \geq \sum_{j=1}^{n} D_j$$

and the model is

$$\text{Min} \quad \sum_{i=1}^{m} \sum_{j=1}^{n} c_{ij} x_{ij}$$

Subject to

$$\text{(a)} \quad \sum_{j=1}^{n} x_{ij} \leq S_i, \qquad i = 1, \ldots, m$$

$$\text{(b)} \quad \sum_{i=1}^{m} x_{ij} \geq D_j, \qquad j = 1, \ldots, n$$

This model assumes, in fact, that not all the amount produced in plant i, say, is shipped and therefore it allows production for inventory in each origin. Indeed, this idea is used to convert the inequalities into equalities and therefore to reformulate the model as a standard transportation model. To do it, we create a fictitious (often called *slack* or *dummy*) destination and number it $n + 1$. In addition, we add one arc from each origin i to this fictitious destination and assign a cost of 0 to this arc: $c_{i,n+1} = 0$ for all i. In this way, m flow variables are added: $x_{1,n+1}, \ldots, x_{m,n+1}$. We interpret $x_{i,n+1}$ as *surplus* at origin i. Furthermore, we assume that the "demand" at the fictitious destination is

$$D_{n+1} = \sum_{i=1}^{m} S_i - \sum_{j=1}^{n} D_j$$

Figure 5-5 illustrates these modifications, assuming $m = 2$ and $n = 3$.

If the shipping routes have capacity restrictions, the model may easily be altered by adding upper limits on the flow variables.

The transportation model is, without a doubt, the single most successful of all linear programming models. It fills the need for solution of a very common industrial problem where the previous ad hoc procedures do not suffice. The solution procedures developed for this problem are fast, efficient, and permit the solution of very large problems—a problem with a thousand sources and a thousand sinks, corresponding to a tableau with one million entries (most zero, of course), could be solved quite easily.

In fact, solving the transportation problem is so easy that it has a special historical significance. In the late 1940s and early 1950s, after the Simplex method had been constructed but at a time when the few existing computers were too slow and unreliable to be any more than experimental gadgets, transportation problems with as many as thirty sources and thirty sinks (900 variables!) were being solved routinely in many companies by hand! No one would have thought of using the Simplex method by hand, routinely, for problems of even one-tenth that size. Hence, partly in the historical tradition, Appendix B illustrates a method for solving the transportation problem.

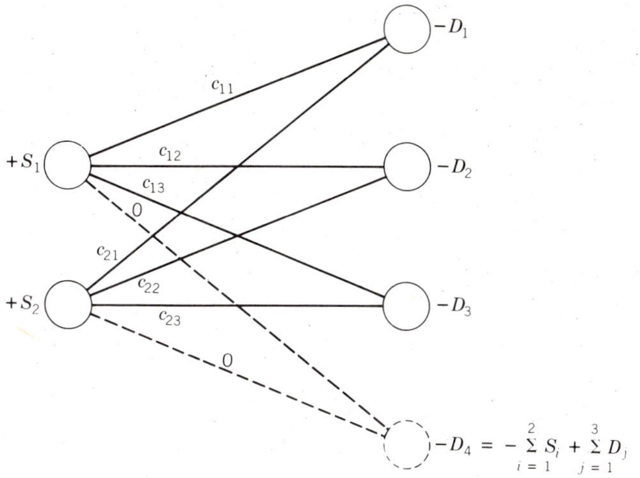

FIGURE 5-5. A dummy sink.

The Shortest-Route Model

Consider the modified version of the network of Figure 5-1 given below in Figure 5-6. This network has a single source (node 1) and a single sink (node 8). The problem is: given the distances c_{ij} (which may be cost, travel time, etc.), what is the shortest route from the source to the sink? In flow terms we can imagine that a unit of flow (person, for instance) wants to travel from node 1 to node 8 taking the shortest route. This is again a special case of the network model discussed earlier, where $p_1 = 1$, $P_8 = -1$, and $P_k = 0$ for any other node in the network. The formulation of this model follows:

$$\text{Min} \quad \sum_i \sum_j c_{ij} x_{ij}$$

Subject to

$$\sum_j x_{kj} - \sum_i x_{ik} = \begin{cases} 1 \text{ for } k = \text{source} \\ 0 \text{ for all other } k \\ -1 \text{ for } k = \text{sink} \end{cases}$$

In addition to finding the least-cost route for transshipment problems, there are other examples of large-scale network models that have the shortest route problem as a subproblem. The model also has important application outside the strict "shortest route" setting. For example, consider a machine shop that uses an expensive extrusion press which must be maintained and replaced periodically as it wears out. The purchase of a new model has been authorized, and the production foreman has been asked to determine the optimal replacement plan for the next 6 years after which the press will no longer be needed. Table 5-3 contains the necessary information. For example, if the press were purchased at the beginning of year 1, used for 3 years, and

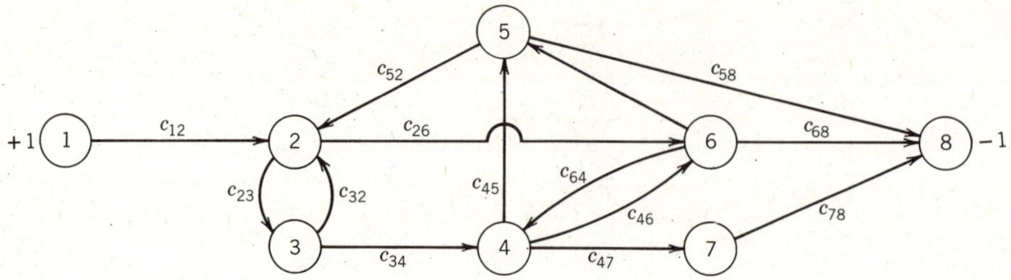

FIGURE 5-6. A shortest-route network.

TABLE 5-3

k	Purchase Price at the Beginning of Year k	Salvage Value at End of k Years of Use	Operating Cost of Equip. During kth Year of Use
1	100	50	30
2	110	25	40
3	105	10	45
4	90	5	50
5	90	2	70
6	90	0	100

another purchase made at the beginning of year 4, the total cost would be the discounted sum of the costs shown in Table 5-4. We might represent this problem as the network shown in Figure 5-7.

The unit of flow is the extrusion press; node i represents year i, arc $(i \rightarrow j)$ represents the policy of purchasing a press in year i and selling it in year j. Node 1 is the source, node 7, the sink. To arc $(i \rightarrow j)$ corresponds a cost c_{ij}, where c_{ij} is the discounted cost of purchase, operation, and salvage. In the example illustrated above, the cost referred to would be $c_{14} + c_{47}$ and if the corresponding policy was optimal, the solution would be $x_{14} = 1$, $x_{47} = 1$, all other $x_{ij} = 0$. Our objective is to find the "shortest route"—that is, minimum cost route—from node 1 to node 7.

The networks discussed in this section are members of a special class of directed networks, called *acyclic* because there are no "loops"; that is, there is no path that allows a return to a previously visited node. In practice, dynamic

TABLE 5-4

Beginning of year:	1	2	3	4	5	6	7
Purchase:	100			90			
Salvage:				−10			−10
Operation:	30	40	45	30	40	45	
Total annual cost:	130	40	45	110	40	45	−10

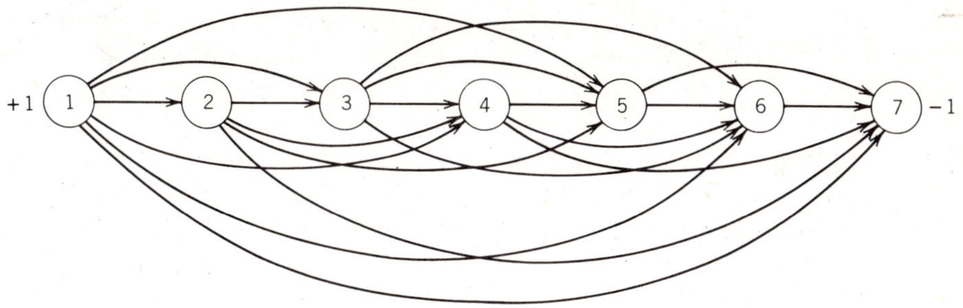

FIGURE 5-7. An equipment replacement network.

programming algorithms, discussed in Chapter 9, have been found to be extremely effective in solving acyclic network optimization problems.

An application of a "longest-route" model is seen in connection with *critical path scheduling*.

CRITICAL PATH SCHEDULING

The critical path method (CPM) is a technique applied to schedule large-scale projects, such as construction projects, research and development programs, maintenance projects, introduction of new systems, or new products, and so on. Each of such projects consists of a well-specified set of tasks (or jobs) that is to be executed in a certain prescribed sequence. That sequence is a series of *precedence relationships* that describe what tasks must precede what other tasks.

CPM analysis provides the project's management with the earliest time at which the project can be completed, allowing management to identify critical tasks that are most likely to be bottlenecks, and indicates where special effort should be made in order to meet specified deadlines. It also assists in evaluating the effect of changes in the schedule and therefore in optimal allocation of resources and efforts among different activities.

The special network structure of a CPM problem is not obvious from the outset, but a simple transformation reveals this structure.

For example, consider a construction firm planning to build a house whose design it is particularly familiar with. Based on previous experience, the project manager has detailed by way of Table 5-5 the tasks required, their immediate predecessors, and an estimate of how long (in days) each task will take. It is clear that there is no need to indicate that the siding must be put up before the outside painting can begin, since putting up the siding precedes installing the windows, which precedes the outside painting. It is always convenient to identify a "finish" task, which has, as immediate predecessors, all tasks that in actuality have no successors.

If we assume that our objective is to minimize the elapsed time of the project, we can formulate a linear programming problem. First, we define the decision variables t_i for i = 1,2, . . . 6, as the earliest starting times for each of the tasks. Table 5-5 gives the earliest starting times where the same earliest starting time is assigned to tasks with the same immediate predecessors. For instance, tasks 4 and 5 have task 3 as their immediate predecessor. Obviously, they cannot start until task 3 is finished; therefore, they should have the same earliest starting time. The earliest starting time of the fictitious "finish" activity, t_6, is the earliest completion time of the entire project, so our objective is to minimize the project duration, given by $t_6 - t_1$, subject to the precedence constraints among tasks. For example, consider roofing. Its earliest starting time is t_2 and its immediate predecessor is framing. The earliest starting time for framing is t_1, and framing is estimated to take 12 days. Hence the earliest time at which roofing may start is $t_1 + 12$. That is,

$$t_2 \geqslant t_1 + 12$$

Since siding has the same immediate predecessor as roofing as well as the same earliest starting time, its precedence constraint is identical and thus need not be repeated. Now, consider wiring (task 6). Its earliest starting time is t_4 and its immediate predecessors are tasks 2 (roofing) and 4 (windows). Their earliest starting times are t_2 and t_3, respectively, while their durations are 6 and 13 days, respectively. Hence, the earliest starting time of task 6 must satisfy:

$$t_4 \geqslant t_2 + 6$$

$$t_4 \geqslant t_3 + 13$$

In general, if t_j is the earliest starting time of a task, t_i is the earliest starting time of an immediate predecessor to task j, and d_{ij} is the duration of that immediate predecessor, then

$$t_j \geqslant t_i + d_{ij}$$

TABLE 5-5

Task	Immediate Predecessors	Duration (days)	Earliest Starting Times
1. Framing	0	12	t_1
2. Roofing	1	6	t_2
3. Siding	1	16	t_2
4. Windows	3	13	t_3
5. Plumbing	3	8	t_3
6. Wiring	2,4	11	t_4
7. Inside finishing	5,6	23	t_5
8. Outside painting	2,4	15	t_4
9. Finish	7,8	0	t_6

The complete linear program is thus:

$$
\begin{array}{llllllll}
\text{Min} & -t_1 & & & & & +t_6 & \\
\text{Subject to} & & & & & & & \\
\text{Roofing/siding} & -t_1 & +t_2 & & & & & \geqslant 12 \\
\text{Windows/plumbing} & & -t_2 & +t_3 & & & & \geqslant 16 \\
\text{Wiring/outside painting} & & -t_2 & & +t_4 & & & \geqslant 6 \\
 & & & -t_3 & +t_4 & & & \geqslant 13 \\
\text{Inside finishing} & & & -t_3 & & +t_5 & & \geqslant 8 \\
 & & & & -t_4 & +t_5 & & \geqslant 11 \\
\text{Finish} & & & & & -t_5 & +t_6 & \geqslant 23 \\
 & & & & -t_4 & & +t_6 & \geqslant 15
\end{array}
$$

The tableau for this linear program is shown by Table 5-6.

Table 5-6, unfortunately, does not possess the structure of a network flow model (i.e., it does not have the property that each column has two nonzero coefficients—one +1, one −1). However, note that the *rows* of Table 5-6 do have this property. Let us, therefore, formulate the *dual* of this problem:

$$
\begin{array}{l}
\text{Max} \quad 12x_{12} + 6x_{23} + 16x_{24} + 13x_{34} + 8x_{35} + 11x_{95} + 15x_{16} + 23x_{56} \\
\text{Subject to}
\end{array}
$$

$$
\begin{array}{ccccccccccc}
-x_{12} & & & & & & & & & = & -1 \\
x_{12} & -x_{23} & -x_{24} & & & & & & & = & 0 \\
 & x_{23} & & -x_{34} & -x_{35} & & & & & = & 0 \\
 & & x_{24} & +x_{34} & & -x_{45} & -x_{46} & & & = & 0 \\
 & & & & x_{35} & x_{45} & & -x_{56} & & = & 0 \\
 & & & & & & x_{46} & +x_{56} & & = & 1
\end{array}
$$

The equalities in the dual correspond to the t_i being unrestricted in sign. If we multiply each equation by (-1), Table 5-7 gives the tableau for the dual. This tableau now demonstrates the special network structure we have been looking for. The network is shown in Figure 5-8.

Note that, in this network, the arcs represent the tasks, while the nodes

TABLE 5-6

t_1	t_2	t_3	t_4	t_5	t_6	Relation	RHS
−1	1					≥	12
	−1	1				≥	6
	−1		1			≥	16
		−1	1			≥	13
		−1		1		≥	8
			−1	1		≥	11
			−1		1	≥	15
				−1	1	≥	23
1					1		Min

TABLE 5-7

x_{12}	x_{23}	x_{24}	x_{34}	x_{35}	x_{45}	x_{46}	x_{56}	Relation	RHS
1								=	1
−1	1	1						=	0
	−1		1	1				=	0
		−1	−1		1	1		=	0
				−1	−1		1	=	0
						−1	−1	=	−1
12	6	16	13	8	11	15	23		Max

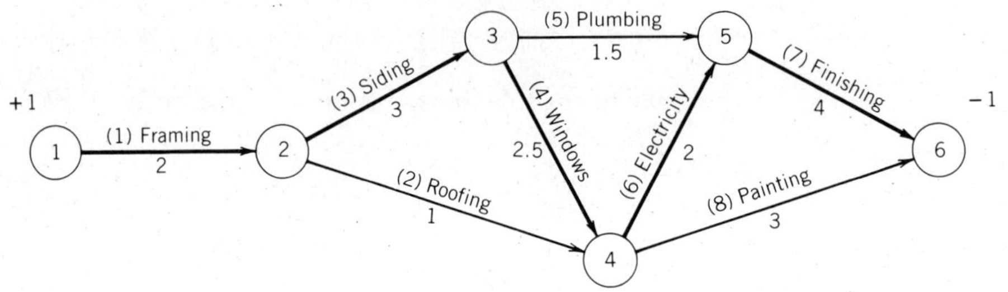

FIGURE 5-8. A critical path network.

describe the precedence relationships among tasks. Node 1 is the source, node 7 is the sink. The quantities on the arcs represent task durations. Since the objective in Table 5-7 is to be *maximized*, this is a "longest route" problem. A longest route formulation is intuitively reasonable since the overall project will be defined (in terms of time required) by the longest sequence of jobs. If cuts are to be made, they should be made along that "route." The solution in this case is $1 \rightarrow 2 \rightarrow 3 \rightarrow 4 \rightarrow 5 \rightarrow 6$, shown by the heavy lines in Figure 5-8, with a path length of 75 days. The path is the *critical path*. This corresponds to a primal solution of $t_1 = 0$, $t_2 = 12$, $t_3 = 28$, $t_4 = 41$, $t_5 = 52$, $t_6 = 75$ with the objective $t_6 - t_1 = 75$. The noncritical activities—roofing, plumbing, and painting—are ones for which shortening their durations will not improve the needed 75-day time for the complete job. Note that the constraints (second, third, fifth, and seventh) relating to these noncritical activities in Table 5-6 are nonbinding. All others are binding.

In contrast to the preceding examples, CPM involved a network maximization problem. So does the next commonly applied network example.

THE MAXIMAL FLOW MODEL

Consider an oil refinery with pumping stations and pipelines as depicted in Figure 5-9. The symbol u_{ij} represents the capacity (in gallons of oil per day) of the pipeline represented by arc ($i \rightarrow j$). The firm is interested in determining the maximum daily quantitiy of oil that can be pumped from station 1 to station 7.

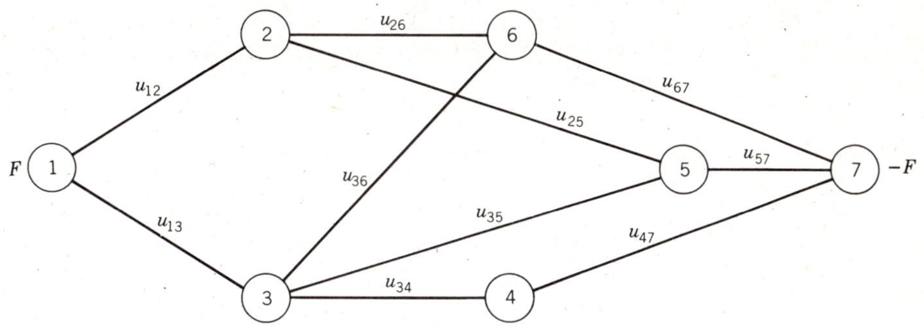

FIGURE 5-9. A maximal flow network.

Let x_{ij} be the number of gallons of oil flowing per day along arc ($i \rightarrow j$). We introduce an additional variable F to represent the maximal flow and consider it a supply at the source (node 1) and a demand at the sink (node 7). The constraints of the network model will "transport" this amount F from source to sink, and the objective will cause F to be as large as possible; that is,

$$\text{Max } F$$

$$\text{Subject to} \quad \sum_j x_{kj} - \sum_i x_{ik} = \begin{cases} F \text{ if } k = 1 \\ 0 \text{ if } k = 2, 3, \ldots, 6 \\ -F \text{ if } k = 7 \end{cases}$$

$$x_{ij} \leq u_{ij}$$

CONCLUDING REMARKS

Network optimization models are the most important, certainly the most used, special linear programming structures. Distribution models alone have allowed firms to cut billions of dollars from their costs. Their usefulness stems from three characteristics:

1. If the model data is integral, so will be the solution.
2. Solution computation is many times more efficient than that necessary to solve a general linear program.
3. The "picture" is, by itself, a *complete* description of the problem.

From a managerial standpoint, this last characteristic is, perhaps, the most important. Since there is a one-to-one correspondence between the network and the corresponding linear program (via the node arc incidence matrix), *the network diagram is the model formulation.* This fact makes communication between the manager and analyst nearly trivial, especially in today's world of multi-color computer graphics. Network optimization models are living proof of the variant of the well-known saying, "A picture is worth a thousand equations."

EXERCISES

1. A gas company owns a pipeline network to pump natural gas from its main field to a distribution center. It has two intermediate pumping stations *A* and *B* connected to the main field and to the distribution center. Each of these pumping stations can pump gas from the main field to the distribution center or from the main field to the other pumping station. At present the company nets 1200 mcf (million cubic feet) of gas per month from its main field and must transport that entire amount to the distribution center. The capacity of the pipelines and the current transport costs ($/mcf) are given below:

Pipeline			
From	To	Capacity	Cost
Main Field	*A*	500	20
Main Field	*B*	900	25
A	*B*	700	10
B	*A*	400	15
A	Distribution Center	600	20
B	Distribution Center	1,000	40

Each of the pumping stations is capable of handling the entire 1200 mcf. Formulate a network optimization model that will determine the least cost distribution pattern and draw the corresponding network.

2. An agricultural cooperative grows wheat on farms at locations *A* and *B*, ships it to mills at locations *C* and *D* where it is processed and sends it to distribution centers *E*, *F*, and *G* where it is sold to customers. The maximum possible amount that can be harvested at locations *A* and *B* is 400 and 500 tons, respectively. (Any wheat not sent to be milled is sold at a price equal to the cost of growing and harvesting.) The demand at the distribution centers *E*, *F*, and *G* is 250, 325, and 300 tons, respectively. Mill *C* can process up to 1000 tons; mill *D* can process up to 600 tons. The cost of growing and harvesting at farm *A* is $2/ton; at farm *B*, $2.25/ton. Cost of milling is $1/ton at mill *C* and $1.10/ton at mill *D*. The shipping costs are given in the tables below.

	A	*B*
C	.40	.30
D	.33	.25

	E	*F*	*G*
C	.10	.20	.15
D	.12	.18	.15

Formulate a network model that will plan the most cost effective way to harvest and mill in order to meet demand, and draw the corresponding network.

3. A manufacturing firm produces diesel engines in its plants in Los Angeles, Cleveland, Louisville and Detroit. Production costs and capacities are given below:

	Production Cost (per units)	Capacity (units per month)
Los Angeles	\$730	25
Cleveland	\$680	125
Louisville	\$660	100
Detroit	\$690	125

For the coming month, orders have been received from firms in three cities:

Cincinnati	110 engines
Denver	150 engines
Atlanta	95 engines

Due to past problems the Cincinnati firm will not accept engines from the Cleveland facility.

a. Formulate a model to schedule next month's production and transportation given the transport rates (\$ per unit) below:

From/To	Cincinnati	Denver	Atlanta
Los Angeles	22	15	28
Cleveland	X	26	18
Louisville	18	22	14
Detroit	19	23	20

Draw the corresponding network

b. Find the optimal solution

4. The Greater New York Little League Association provides umpires for scheduled Little League baseball games. Two umpires are assigned to each game. For the upcoming Saturday schedule, the Association wants to assign six umpires to three games so that total travel time is minimized. Given the table of travel times below for the ten officials who are available for the coming Saturday, find the appropriate assignment.

	Game Location		
Umpire	Bronx	Queens	Brooklyn
1	53	15	22
2	18	45	38

(continued)

Umpire	Bronx	Queens	Brooklyn
	Game Location		
3	5	44	42
4	10	37	50
5	20	25	40
6	30	60	50
7	18	50	44
8	36	5	28
9	42	16	18
10	32	18	35

Travel time (minutes)

5. A firm in a very seasonal business leases its delivery vans. Its requirements, monthly, for the next three quarters is shown below:

	April	May	June	July	Aug.	Sept.	Oct.	Nov.	Dec.
Vans Required	47	41	32	20	18	28	38	50	44

On April 1, there will be 22 vans on lease. Ten of these go off lease May 1; the other twelve are leased until June 1. Delivery vans can be leased on any of three plans, with costs and lease lengths as follows:

	Cost	Length
Plan 1	$680	3 months
Plan 2	$880	4 months
Plan 3	$1040	5 months

Formulate a minimum cost network model to help this firm do its fleet planning. Draw the corresponding network.

6. In the midst of its R&D effort to design a new product, a company has learned that a competitor is about to introduce a similar product which could capture a major part of the market. In view of this information, corporate headquarters has agreed to allocate $8 million to rush the product to the market place. The five remaining phases can be speeded up at various additional costs, as follows:

Remaining Research: Time (wks)	Remaining Research: Add'l Cost ($ mill)	Development: Time (wks)	Development: Add'l Cost ($ mill)	Manufacturing Design: Time (wks)	Manufacturing Design: Add'l Cost ($ mill)	Supplier Analysis: Time (wks)	Supplier Analysis: Cost ($ mill)	Initiation of Production: Time (wks)	Initiation of Production: Cost ($ mill)
4	0	3	0	5	0	3	0	2	0
3	1	2	2	4	1	2	1	1	2
2	3	1	3	2	4				

Formulate a network that will help the R&D department decide how to best allocate the $8 million.

*7. A firm's financial staff is responsible for the management of it's monthly cash balances to cover cash needs and yet obtain the maximum return. For the coming quarter, accounts receivable and payable have been forecast as summarized below:

Month	Inflow from Accounts Receivable	Accounts Payable
October	$150,000	$600,000
November	300,000	900,000
December	600,000	750,000

Payables can be delayed until the next month, with a 2% penalty. In order to retain its excellent credit rating, the firm's policy prohibits delaying payables more than one month.

The firm has two marketable securities that are about to mature: the first, with a face value of $300,000, on November 1; the second, with a face value of $750,000 on December 1. These securities could be sold early, if needed, but early sale would reduce the cash inflow by ½% of the face value if sold one month early and by 1½% if sold two months early.

The firm also has a credit line with a local bank, large enough to finance any of the firm's anticipated cash needs. Funds borrowed in October carry an interest charge of 3%; in November or December, the interest charge is 2% (monthly).

Lastly, excess cash may be invested in short-term financial instruments that mature in either 30 days, paying 0.7%; or 60 days, paying 1.5%.

Supposing the firm has no cash on hand on October 1, that accounts receivable are received on the first of the month, and that accounts payable are due on the first of the month. Formulate a model to help this firm's financial staff plan its cash balances. Draw the corresponding network. What assumptions, implicit in this model, do you feel are unrealistic? How might you alter the model to make it more realistic? Would you?

8. An architectural firm has been selected as one of two firms asked to prepare blueprints for a suburban mall. The tasks required for this proposal are shown below:

Task Description	Immediate Predecessors	Time (days)
1. Preliminary Sketches		2
2. Specification Outline		1

(continued)

*Based on Srinivasan, "A Transshipment Model for Cash Decisions," *Management Science* Vol. 20, No. 10, June 1974.

Task Description	Immediate Predecessors	Time (days)
3. Drawings	1	3
4. Detailed Specs	1, 2	2
5. Duplicate Drawings	3	1
6. Print Specs	4	3
7. Assemble Proposal	5, 6	1

Formulate a model to plan starting times for each task so as to complete the project in as little time as possible. Draw a network that could be solved to specify these times.

9. An independent air freight company knows that over the next four days it will require for its operation 25, 35, 25, and 20 planes, respectively. Maintenance is required on each plane after each day of use. The maintenance can be done over a 24-hour period (requiring that the plane be out of service for one day), or it can be done overnight at twice the cost. Planes can be held unused without maintenance indefinitely at no cost. The total number of planes owned by the firm is 40, of which 20 are currently on hand with maintenance completed. Formulate a network model to plan the maintenance over the next four days and draw the corresponding network.

10. A shoe manufacturer is planning its production of running shoes for July, August, and September, during which time much of the workforce will be on summer vacation. Each pair of shoes requires one person-hour of labor that costs the company \$4 for straight-time hours and \$6 at overtime rates. Demand for shoes is expected to be 1000 in July, 1750 in August, and 1400 in September. Cost of carrying a pair of shoes in finished goods inventory for one month is \$1.50. Regular hours available are 1100 in July, 700 in August, and 1100 in September. The company has a firm policy of never allowing overtime hours to exceed 40% of regular hours during each month.

Formulate a network model that will allow the company to plan production to meet as much demand as possible at minimum cost, and draw the corresponding network.

11. *An investment problem.* You're concerned about covering the college educational expenses for your only child, now 10 years old. You've looked into a number of investments, and you seek to maximize your return over an 8-year planning horizon. Contrary to what you may have learned about financial theory, your set of investments are somewhat unusual:

One-year investment paying 8%.

Two-year investment paying 6% per annum.

Three-year investment paying 9% per annum.

Four-year investment paying 7% per annum.

These returns hold now and for any time during the next 8 years. Formulate a network optimization model that will maximize your return after 8 years.

Bay Area Bakery Company

PROPOSAL TO BUILD A NEW BAKERY

Bay Area Bakery Company was a regional baker and distributor of bread, and operated six bakeries shown on the map in Exhibit 1. The manager of Transportation and Customer Service for Bay Area Bakery Company was asked to prepare a report outlining the effect of the proposed San Jose, California, baking facility on the Company's physical distribution system. The prob-

EXHIBIT 1
Map of Bakeries and Marketing Territories Served by Each Bay Area Bakery

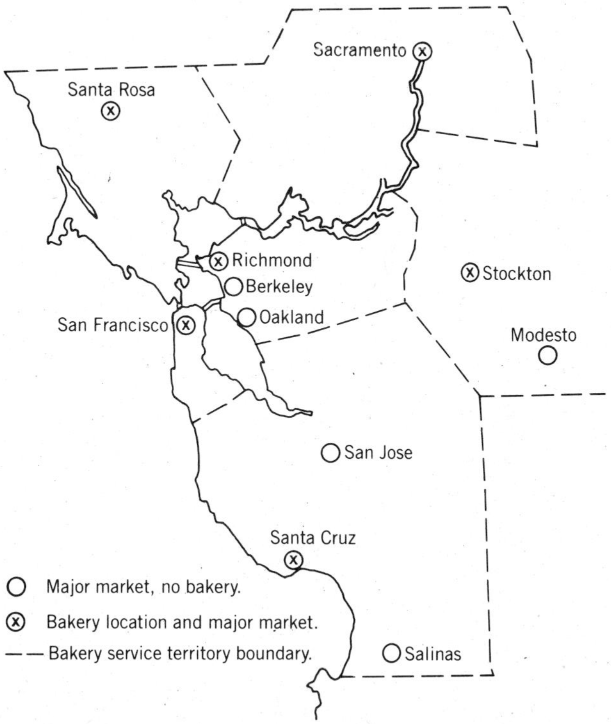

Source: This case has been taken from *Case Problems in Business Logistics* by J. L. Heskett, L. M. Schneider, R. M. Ivie, and N. A. Glaskowsky, Jr. Wiley, New York, 1973.

lem centered around the fact that marketing territories served by the Santa Cruz and Stockton bakeries had increased rapidly in population, necessitating consideration for locating and equipping a new baking plant at San Jose.

Varying costs of labor, ingredients, and operation affected the total cost of baking a standard quantity of bread at each of the locations. This average cost for each of the bakeries is shown in Exhibit 2, along with the daily capacity at each location. The full line of the Company's goods was baked at each of the six locations.

Bay Area sold its products at an average delivered price of $23 per cwt. and maintained uniform delivered prices on comparable quantity orders throughout its marketing territory. Average daily sales from each bakery for each major center of consumption are shown in Exhibit 3.

EXHIBIT 2

Baking Costs and Capacities-Bay Area Bakery Company Bakeries

Location	Baking Cost (per cwt)	Daily Baking Capacity (in cwt)[a]
Santa Rosa	$19	500
Sacramento	$17	1,000
Richmond	$16	2,700
San Francisco	$17	2,000
Stockton	$18	500
Santa Cruz	$21	800

[a]Current value of each plant was roughly estimated at $300 per unit of daily capacity (cwt.).

EXHIBIT 3

Current Average Daily Sales, Production Costs, and Transportation Costs, Bay Area Bakery Company (by bakery and major market)[a]

Major Market	Bakery of Origin	Quantity (in cwt)	Production Costs	Transportation Costs
Santa Rosa	Santa Rosa	300	$ 5,700	$ 600
Sacramento	Sacramento	500	8,500	750
Richmond	Richmond	600	9,600	600
Berkeley	Richmond	400	6,400	400
Oakland	Richmond	1,100	17,600	1,320
San Francisco	San Francisco	1,300	22,100	1,300
San Jose	Santa Cruz	600	12,600	1,260
Santa Cruz	Santa Cruz	100	2,100	200
Salinas	Santa Cruz	100	2,100	280
Stockton	Stockton	400	7,200	600
Modesto	Stockton	100	1,800	260
Total		5,500	$95,700	$7570

[a]Bakeries operated approximately 300 days per year.

To control physical distribution costs closely and to maintain tight control over the freshness of baked items, Bay Area operated its own fleet of privately owned local delivery trucks and over-the-road units. It was the Transportation and Customer Service Manager's job to minimize transportation costs from each of the baking plants to market areas. Within the limitations of the Company's operations, he had achieved levels of costs shown in Exhibit 3. He had also prepared an analysis of projected delivery costs between all Company baking facilities and market areas, as shown in Exhibit 4. Included in this projection were those costs of delivery which could be expected between the proposed plant at San Jose and other market areas.

A major reason for the suggested new baking plant at San Jose was the especially rapidly growing market which the area represented. It was expected that this market would double in the next five years, compared to an expected 10% increase in other markets (except less than 10% in San Francisco) during the same period. Also, the inefficiences involved in serving San Jose from the Santa Cruz bakery could be relieved by the construction of a more efficient bakery in the growing San Jose market.

A bakery with daily capacity of 1,200 cwt. was planned. It was estimated that a facility of this size would cost $4,000,000, fully equipped, and would be able to turn out bakery products at an average cost of $17 per hundred pounds. With the expected 5-year increase in demand, the supply pattern shown in Exhibit 5 was projected for 5 years into the future.

EXHIBIT 4

Estimated Delivery Costs Between Bakeries and Customers, Bay Area Bakery Company[a] (per cwt)

	Plants						
Markets	Santa Rosa	Sacramento	Richmond	San Francisco	Stockton	Santa Cruz	San Jose[b]
Santa Rosa	$2.00	$4.40	$3.20	$3.20	$4.20	$4.80	$4.20
Sacramento	3.90	1.50	2.90	3.30	2.50	4.50	3.90
Richmond	2.00	2.40	1.00	1.40	2.60	2.60	2.00
Berkeley	2.00	2.40	1.00	1.40	2.60	2.60	2.00
Oakland	2.20	2.60	1.20	1.20	2.60	2.40	1.80
San Francisco	2.20	2.80	1.40	1.00	2.80	2.60	2.00
San Jose	3.70	3.90	2.50	2.50	2.90	2.10	1.50
Santa Cruz	4.80	5.00	3.60	3.60	4.00	2.00	2.60
Salinas	5.60	5.60	4.20	4.40	4.60	2.80	3.20
Stockton	3.70	2.50	3.10	3.10	1.50	3.50	2.90
Modesto	4.80	3.60	4.00	4.00	2.60	4.20	3.60

[a]Delivery costs are influenced primarily by (1) the distance of the market area from the bakery, and (2) the distance between deliveries in the market area.

[b]Proposed plant.

EXHIBIT 5
Five-Year Projected Supply Pattern, Bay Area Bakery Company

Major Market	Bakery of Origin	Daily Quantity (in cwt)
Santa Rosa	Santa Rosa	330
Sacramento	Sacramento	550
Richmond	Richmond	660
Berkeley	Richmond	440
Oakland	Richmond	1210
San Francisco	San Francisco	1300
San Jose	San Jose	1200
Santa Cruz	Santa Cruz	110
Salinas	Santa Cruz	110
Stockton	Stockton	390
Stockton	Sacramento	50
Modesto	Stockton	110

Chrysler Corporation Plant Loading Model

Gordon Olson was pleased. Chrysler's Plant Loading Analysis System, which he had developed in the early 1960s, had recently performed well in a new role. Since its development, it had been used successfully by marketing and manufacturing groups to set production levels and the subsequent distribution patterns for the various Chrysler models as well as by the Source and Facilities Planning Group as a tool to explore new plant sites. Now, it was being used to tackle the important problem of assigning individual dealer orders to assembly plants. In this task, the system was saving, by mid-1978, at the rate of $1 million a year for car order allocation and an additional $1.5 million a year for truck orders. Olson leafed through a report on his desk dated July 28, 1961 and reflected on the system's 17-year history.

Here was a system which had been developed when, to some extent, computers were still a novelty, had gained support from top management, and was still in use 17 years later. Although there had been some changes over the years, the underlying model was the same. Olson, Director of Operations Planning at Chrysler, had had experience with the development and implementation of other models. This one was different:

> *I'm in charge of all Chrysler's corporate systems, and I don't know of any computer system, be it model or be it an application program, that has held together as long as*

this one. This has much of the same code in it that was written in early 1960—the same code, the same statements, the same output format. I've been involved with many other programs that have long since disappeared—our press loading model that determined how to load our press plants, another one to determine what the optimal line balance is in assembly plants, and I could go on and on. And I think most people that have worked in operations research could tell you about numerous cases of models that have disappeared from the face of the earth. Yet, here's a model that has survived.

COMPANY BACKGROUND

The automobile industry is characterized by the vast resources necessary to produce cars and trucks in large volume at a price within the means of the average consumer. Vehicle design and production requires nearly every known science, craft and skill, plus millions of square feet of manufacturing area.[1]

Chrysler Corporation is the world's third largest automotive enterprise and one of the ten largest industrial companies with manufacturing operations in 24 countries, 245,000 employees and 220,000 shareholders. Manufacturing facilities occupy over 100 million square feet and are capable of production which exceeds in value the entire industrial output of many nations.

Chrysler Corporation was incorporated in Delaware on June 6, 1925, assuming the business and assets of Maxwell Motors Corporation. The company's founding was a direct result of the success of the "Chrysler Six," a completely new car introduced to the public by Walter P. Chrysler in New York in 1924. The new Chrysler combined in a mass-produced car such features as four-wheel hydraulic brakes, a high compression engine, shock absorbers, and many other innovations. Nearly 32,000 new Chryslers were sold in 1925, a first-year record for the introduction of a new passenger car.

In 1928, Chrysler Corporation acquired Dodge Brothers, Inc. In that same year, the Plymouth was introduced and the firm assumed its position as one of the world's three largest automobile manufacturers. Over the past half-century, Chrysler has maintained its reputation for engineering excellence as the direct result of a large number of automotive innovations, as well as contributions to the design, research, and manufacture of military and aerospace products.

During the years 1970–1976 Chrysler Corporation's financial position (see Exhibit 1) reflected the pressures of inflation, rising labor and material costs not fully recovered in selling prices, and an extreme fluctuation in the automotive market. After record years in 1972 and 1973, sales fell abruptly as the country entered its most severe post-war economic recession. In 1975, sales of Chrysler products began to recover gradually, although Chrysler took only 14.9% of the

[1]*Financial and General Fact Book*, Chrysler Corporation, June 1973.

domestic automobile market. This loss of market share resulted primarily from a drop in the company's share of the compact market where Valiants and Darts competed against new entries from GM and Ford. The Plymouth Volaré and

EXHIBIT 1
Financial Statistics

Operating Data (in millions of dollars)	1976	1975	1974	1973	1972	1971	1970
Passenger cars and trucks sold (in thousands of units)	3,130	2,476	2,763	3,402	3,028	2,663	2,434
Share of Domestic Market							
Passenger cars (%)	15.9	14.9	16.6	16.2	16.6	18.5	17.5
Trucks(%)	14.1	12.7	11.6	11.1	11.4	8.4	8.3
Net sales	$ 15,538	11,598	10,860	11,667	9,641	7,893	6,887
Interest expense	$ 181	205	176	114	96	96	107
Maintenance and repairs	$ 449	302	338	403	307	247	216
Research and development	$ 280	199	239	245	185	143	137
Taxes other than on income	$ 279	214	203	201	166	450	385
Taxes on income (credit)	$ 212	26	(78)	203	192	66	(14)
Earnings (loss) from continuing operations	$ 328	(207)	(41)	266	226	92	—
A share (in dollars)	$ 5.45	(3.46)	(0.73)	4.99	4.38	1.82	—
Net earnings (loss)	$ 423	(260)	(52)	255	220	84	(8)
A share (in dollars)	$ 7.02	(4.33)	(0.92)	4.80	4.27	1.67	(0.16)
Average shares outstanding (in thousands)	60,205	59,942	56,421	53,182	51,643	50,214	48,693
Cash dividends paid	$ 18	—	79	69	47	30	29
A share (in dollars)	$ 0.30	—	1.40	1.30	0.90	0.60	0.60
Net earnings as a percent of sales	2.7%	(2.2%)	(0.5%)	2.2%	2.3%	1.1%	(0.1%)
Capital expenditures	$ 424	384	468	629	335	250	416
Financial Position-Year End (in millions of dollars)							
Current assets	$ 3,876	3,116	3,697	3,238	2,896	2,411	2,167
Current liabilities	$ 2,826	2,462	2,709	2,094	1,941	1,648	1,548
Net current assets	$ 1,052	654	988	1,144	955	763	619
Net property, plant, equipment & tools	$ 2,087	2,115	2,062	1,926	1,680	1,729	1,803
Total Assets	$ 7,074	6,267	6,733	6,105	5,497	5,000	4,816
Long-term debt	$ 2,815	2,409	2,660	2,728	2,489	2,269	2,156
Shareholders' investment	$ 2,815	2,409	2,660	2,728	2,489	2,269	2,156
Shares outstanding (in thousands)	$ 60,247	60,102	59,266	54,442	52,359	50,950	49,499
Shareholders' investment a share (in dollars)	$ 46.73	40.09	44.89	50.10	47.54	44.53	43.55
Numbers of shareholders	217,359	233,915	232,435	208,771	188,914	201,945	204,945

Dodge Aspen, introduced in the fall of 1975, greatly strengthened Chrysler's competitive position, and helped the firm increase its share of the compact market in the early selling periods of 1976. In that year, as total passenger car sales by all U.S. manufacturers increased 22% over 1975, Chrysler's U.S. car sales increased 31%. In the domestic small car market, Chrysler took 23.2% compared to 16.7% in 1975.

In 1976, Chrysler began to implement a long-range plan to strengthen operations by controlling costs, improving operating efficiencies, and introducing new, resized, fuel-efficient models in all major markets. The programs implemented include:

- Parts standardization and simplification for lower unit costs.
- Action to recognize and then strengthen or eliminate operations which threaten to be a drain on profits.
- Improved production scheduling to reflect market changes and avoid inventory buildup.
- Implementation of supplier suggestions to reduce the cost of purchased components.

New products to come on the market in 1976–1978 included:

- Successful new compact cars—the Dodge Aspen and Plymouth Volaré.
- Mid-size speciality cars—the Chrysler Cordoba and Dodge Charger S.E.
- Luxury intermediates—the Dodge Diplomat and Chrysler LeBaron.
- The first domestic small cars with front wheel drive—the Dodge Omni and Plymouth Horizon, introduced early in 1978.

HISTORY OF THE MODEL

Gordon Olson joined Chrysler Corporation in the mid-1950s and began work in the Dodge Division as part of a group involved in Chrysler's "Forward Planning" project specifically in the area of facilities location. As he describes it:

> *In those days, all plans were evaluated manually. It was a long, tedious process. I didn't even know there was such a thing as linear programming, but I figured there must be a better way to do what we were doing.*

Olson's first step was to develop a routine for early electronic data processing machines which mimicked the manual calculations. He would choose a new assembly plant location, input relevant costs and the routine would generate estimated dollar savings for different levels of plant volume. Olson would use this output to generate curves such as that below. These would then go to management.

This was a step forward, but it was a difficult process—computers were, at that time, notoriously slow, unreliable, and noisy, and input/output was tedious at best. As Olson describes the early machines on which his program ran: "It sat out on the floor and clunked away like an old threshing machine."

In the late 1950s, Olson learned about linear programming and programmed an IBM 650 with vacuum tube memory to solve a linear programming formulation which determined the optimal distribution of automobile models from assembly plants to Chrysler's dealers. Input data would then be punched onto computer cards for all regions. These would be fed into the 650, and after the appropriate calculations, another set of computer cards would be output. This was one "iteration." The output cards would then be rearranged and become input from another iteration. Olson describes his experience:

It took a full weekend using the 650 (30 or 40 computer hours) to get the solution to one problem. And even though this answer was correct, it just wasn't practical.

In the early sixties an IBM 7090 was acquired by Chrysler,[2] and Olson, then Manager of Operations Analysis, was able to program the 7090 to run the program in under eight hours without the input/output problems he had with the 650. Although it was slow, this program began to be used for plant loading and operating decisions. Olson relates:

Even with the 7090 computer, which was fairly fast and had 32K words in core, it took many hours for one solution. Of course, the time depended on the complexity of the problem. It would vary from a low of three to four hours up to six or eight for one solution. At that time, my job had evolved to one that included scheduling the assembly plants, but it was awfully difficult to get the computer for that length of time because it was a production computer.

Then came the important breakthrough:

I, and a couple of my people, developed a radically new algorithm for linear programming. We were attempting to improve on the computer time of the

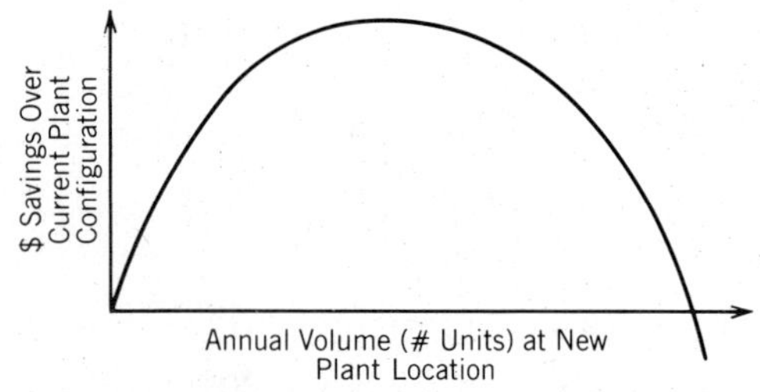

[2]The IBM 7090 was one of the first of the second generation computers. It was an order of magnitude more powerful (in size and speed) than first generation computers such as the 650 whose computing power and ease of use was comparable to a good programmable handheld calculator.

traditional method by getting a better initial solution. So we started with a solution which was good from a cost standpoint except that it didn't meet the constraints, and developed a technique to iterate to produce an initial solution which was feasible. All we were doing was changing the initial solution algorithm. And one day we ran the model and it came to the end of its initial solution, met all its constraints, went onto the traditional transportation linear programming algorithm and took no iterations to reach optimum. In fact, we ran for several months in this mode and for many months there were no iterations in the transportation model. It took a fair amount of work to mathematically prove that in fact what we were using as an initial solution was in all cases an optimal solution. So, on the same computer, a six-hour problem became a six-minute problem.

At first there was some hesitation to believe this "computer" model, but Olson was able to demonstrate that it was able to process and analyze more information than the traditional approaches. On one hand, manufacturing personnel had a tendency to choose plant loading configurations which were easiest to operate, often with little attention to transportation costs—thus generating plant-model configuration slanted toward one model per plant. On the other hand, marketing personnel would like to see all assembly plants capable of simultaneously producing all models—disregarding operating costs. The plant loading model could make optimal tradeoffs between these two extremes, and as such, was accepted as an important decision aid in plant loading plans.

The output format was particularly well thought out. As Olson says:

As early as 1961, the computer output from this model was very commonly used, just by xeroxing the summary management report that came out of this system, including maps of the United States printed out by computer, summary data that showed not only allocation of production between plants, but also the transportation and manufacturing costs. This summary output went directly to management without even going through a typewriter.

After it was printed out by the computer, I had transparent overlays[3] *prepared which had the same outline of the United States and had state outlines and our sales zones outlined on it. Then you could immediately get a computer printout and lay this grid over the printout (of course, we had no computer graphics in those days), and you could not only see the picture of the United States, but, superimposed over that, you could see the state lines and the sales zones.*

In 1964, Olson became Manager of Facilities Planning for the corporation.

As I moved from job to job, the model moved with me, and I used it. My responsibilities always included doing facilities studies, plant loading studies, and so forth. In the process I trained several people that understood the model technically and could use it.

[3]See Exhibit A-7 for examples.

In 1968, Olson was transferred to Chrysler's International Division, and during the next 5 years held positions in Mexico and South America. During that period, the cadre of people from Olson's staff who understood the inner workings of the model became dispersed throughout the corporation. Without a strong backer, the model, over the next 5 years, gradually fell into disuse. As Olson says:

> *Even though top management accepted the model and used it, it still takes somebody that really understands the insides of it.*

In November of 1975, Gordon Olson became Director of Operations Planning, overseeing Manufacturing Group staff departments including Resource Planning as well as Production Control, Traffic, Industrial Engineering, Computer Installation—approximately 500 professionals. At his request, Jim Haselwood (HBS '69), Manager of Resource Planning, along with Herm Cabello, an expert programmer newly transferred from the Parts Division, began work to bring the plant loading model back into use and acceptance.

There were two major changes made. The first was to reconfigure the program for time-shared rather than batch use.

> *Today we're doing it on a time-shared computer through terminals. That puts the terminals in offices where they belong instead of putting users in the computer room with a deck of cards.*

The next change was to the program's input requirements so that, rather than needing copious data input manually, the system would access tariff and cost data which existed on other computer files.

By 1978, the system was being used regularly by top management to aid in annual plant loading decisions, evaluating new expansion options, and, most recently, to further decrease costs by optimally assigning individual dealer orders to assembly plants.

THE CURRENT PLANT LOADING MODEL

Perhaps Chrysler's plant loading model might be better designated a plant loading and distribution model. Theoretically, it has the capability of assigning car lines to assembly plants, determining the number of units on each line to be produced in each plant, and the number of units to be shipped to each Chrysler dealer. In reality, however, there is a good deal of preselection and aggregation which is done judgmentally prior to a model run to conform to a number of customer service, quality, and corporate policy constraints as well as program size restrictions. Chrysler's many models of passenger cars are aggregated for most runs into 11 model lines according to size, "shell" (basic body shape), number of doors, and other marketing characteristics.[4] There are 10 assembly

[4]The plant loading program is capable of handling up to 50 "lines," and for special runs, the 11 usual lines may be disaggregated into as many as 30 classes, some containing a single model.

plants, listed in Exhibit 2. In addition, the Hamtramck plant has two assembly lines which run independently and are considered by the program to be two separate plants (HAMT1 and HAMT2) at the same location. For most runs the program considers only a few of the many carline-plant combinations to be feasible. Usually, for example, several of the 11 lines are restricted to a single plant each. This preselection is done in accordance with changeover costs, quality considerations, and other corporate policy. Lastly, the almost 5000 Chrysler dealers through-out the U.S. and Canada are aggregated into 550 "trading areas." The program then determines which carlines will be made at which plants and the distribution from plants to trading zones which minimizes production and transportation cost subject to demand and plant capacity constraints.

Gathering input data for the plant loading model requires substantial effort but is expedited by a "satellite" program, the Manufacturing Cost Program, run by the Industrial Engineering Department, which calculates labor cost per unit as a function of the mix of assembly linespeeds.[5] A given linespeed for a given model implies a certain number of direct, indirect, and salaried hours. These labor hours, together with performance variances, appropriate "complexity factors," and other variable costs all input by IE personnel allow the calculation of the cost per unit for a given model at a given plant. One time tooling and changeover costs are not considered explicitly by the model; instead, these are part of the judgmental preselection process which determines the appropriate assignment of model lines to plants. For each plant, overtime costs are incorporated by having, for each real plant, an additional plant in the model with correspondingly higher unit cost. Since the program minimizes costs, it will never allow production in an "overtime plant" until the "standard time plant" is at capacity. The model also includes a "dummy" plant, i.e., one with virtually infinity capacity and very high produc-

EXHIBIT 2
Chrysler Corporation's Vehicle Assembly Plants

Belvidere Assembly	Belvidere, Illinois
Hamtramck Assembly[a]	Hamtramck, Michigan
Jefferson Assembly	Detroit, Michigan
Lynch Road Assembly	Detroit, Michigan
Missouri Truck Assembly	Fenton, Missouri
Newark Assembly	Newark, Delaware
Pillette Truck Assembly	Windsor, Ontario
St. Louis Assembly	Fenton, Missouri
Warren Truck Assembly	Warren, Michigan
Windsor Assembly	Windsor, Ontario

[a]The Hamtramck Assembly Plant has two independent lines considered to be separate plants by the plant loading model.

[5]The linespeed at each assembly plant is usually set once per model year. The cost of assembly line changeover is approximately $250,000.

tion costs. This plant can assemble all models and ship to all trading zones, but will be used by a cost-minimizing model only when demand cannot be satisfied by any other feasible means. The inclusion of this dummy source alerts management to the need for capacity expansion or demand-cutting whenever the dummy plant is included in the optimal solution.

The cost of shipping from each plant to each trading zone is also necessary input. Chrysler ships its cars and trucks by rail and haul-away truck. Whether rail or truck is used for shipping from a particular plant to a particular trading zone is an economic decision which incorporates loading/unloading costs, cost of transporting vehicles to the railheads, and railcar loading factors,[6] as well as actual point-to-point truck and rail rates. One of the tasks for which Olson's staff is responsible is auditing the freight bills presented to Chrysler by its carriers. In order to do this, Traffic has, on file, the approved tariffs from each assembly plant to each trading area. These are updated annually. The file, denoted the Freight File, is accessed by programs which extract the desired data and put it into the format required to run the plant loading model. These rates are then combined with "inbound" rates (from supplier to assembly plant), as well as other relevant factors (e.g., vehicle weight), input by the user, to produce total transportation costs.

Finally, the plant loading model requires demand projections for each of the 550 trading zones. Rather than project each zone separately, historical data is used to construct a Distribution File which specifies the estimated *percent* of the total demand for each line which will emanate from each trading zone. On any given run, the user may choose any of several such percentage breakdowns (called distribution "patterns") for each model line.[7] The Distribution File is updated annually. Total U.S. and Canadian volumes for each model line must then be input for each run. These are the subject of much discussion and often disagreement among marketing, sales, and manufacturing personnel. However, the use of the model has eased this source of conflict. As Haselwood says:

> *We can separate the argument over volumes from the discussion of the plant loading decisions. We argue the volumes in marketing and sales meetings. Once we agree, then we use those numbers.*

This input, along with plant capacities (in terms of total available labor hours and number of shifts) and output format specifications, defines a given run. Most of the input is stored on disk so that multiple runs may be made with various changes without respecifying all the data.

The output available from the plant loading model consists of:

- A concise table specifying the optimal solution and the capacities required.

[6]Rail shipping policy dictates that cars be held over at a railhead until railcars can be filled (approximately 15 cars/railcar).

[7]In fact, the program has the flexibility to allow the user to specify a weighted average of up to eight distribution patterns.

- Several more detailed tables specifying costs and requirements by line and plant.
- A lengthy (60–70 pages) list specifying for each trading zone and model line, the optimal and next best assembly plant source along with the resultant volumes and costs.
- The Plant Loading Analysis Report which gives the cost/benefit and optimal action caused by decreasing/increasing capacities for each model line at each assembly plant.
- The Plant Loading Analysis Plot, a computer generated graph of savings versus volume at each assembly plant for each model line.
- Computer-generated maps of the regions serviced by each plant for each model line which can be distributed from more than one plant.

Appendix A contains sample data input and detailed program output for a typical run of the plant loading model.

IMPLEMENTATION

The crucial policy decision made by the users of the plant loading model is the choice of configuration alternatives to be run, i.e., what is the appropriate set of plant-model combinations to be allowed? This is argued at length by marketing and manufacturing personnel before the model is run, and the choice of approximately a dozen alternatives is made.[8] Running these requires 5 to 15 minutes of central processor time on Chrysler's Burroughs Dual Processor 7600. The best three or four are then chosen for presentation to management.

For the annual plant loading decision, the model is run on a single period basis with the annual data. Since adjustments are necessary when actual demand differs from the forecast, the model has recently been run on a monthly basis with actual orders received input as trading zone demand. As such, the model becomes a tool to aid in deciding which order to fill from which plant.

The plant loading model is also used regularly to answer a wide range of "what if" questions. In this way, management has been able to explore the consequences of:

- A change in a plant's assembly linespeed.
- Capacity expansion in an existing plant versus new facility acquisition or construction.

[8]On occasion, the full set of assignments (every line at every plant) is run. This helps to eliminate certain configurations purely from a cost-standpoint. Corporate policy then eliminates others for quality control reasons.

- Potential new plant locations.
- Changes in demand.
- New model introduction.

As an example, the plant loading model was an integral part of Chrysler management's recent decision to sell its idle and partially completed assembly plant in New Stanton, Pennsylvania.

APPENDIX A

The following exhibits represent a plant loading configuration for six carlines and four assembly plants:

Lines

F-29: Aspen/Volaré two-door coupe

F-41: Aspen/Volaré four-door sedan

F-45: Aspen/Volaré wagon

M-22: LeBaron/Diplomat two-door

M-41: LeBaron/Diplomat four-door sedan

M-45: LeBaron/Diplomat wagon

Assembly Plants[1]

HAMT1: Hamtramck Assembly Plant (line 1), Hamtramck, Michigan

HAMT2: Hamtramck Assembly Plant (line 2), Hamtramck, Michigan

NEWARK: Newark Assembly Plant, Newark, Delaware

ST LOU: St. Louis Assembly Plant, Fenton, Missouri

Allowed Assignments

	HAMT1	HAMT2	NEWARK	ST LOU
F-29	√		√	
F-41	√	√	√	
F-45		√	√	√
M-22				√
M-41				√
M-45				√

[1]The model considers a fifth unnamed plant (dummy plant) that can take on excess demand.

The following exhibits give samples of the model's output.[2]

Exhibit A-1: *Problem Input Report* gives tabular recap of input specifications, most importantly, volumes (U.S. and Canada) of each carline; assembly plant standard and overtime labor days, variable cost, and capacity characteristics; and inbound transportation (IBT) charges by carline and assembly plant.

Exhibit A-2: Sample of *Cost Matrix Report* specifies optimal annual shipments by carline from assembly plant to trading zones; gives "next best plant" for each zone; and specifies the freight cost/vehicle for the optimal and "next best" plant-zone link.

Exhibit A-3: *Plant Operating Pattern* gives a summary of the optimal plant loading, capacity utilization and costs; number of units to be assembled per day by carline and plant; standard time and overtime required; and annual costs.

Exhibit A-4: *Plant Operating Pattern* gives a more detailed breakdown of costs and volumes by carline and assembly plant—straight time, overtime, and total.

Exhibit A-5: Sample of *Plant Loading Analysis Report* for carline F-29 at the Newark Assembly Plant (there is one such report for each feasible carline—plant combination if the line can be assembled elsewhere)—specifies the action to be taken if the actual volume of F-29s at Newark deviates from the optimal volume.

Exhibit A-6: *Plant Loading Analysis Plot* for carline F-29 at the Newark Assembly Plant— a graphical summary of the detailed cost information given by the Plant Loading Analysis Report; in particular, a graph of the increase in cost which accompanies a deviation from the optimal volume.

Exhibit A-7: *Actual Shipping Area Map* (one for each carline which is more than single-sourced) presents a computer-generated map of the trading zones serviced by each assembly plant.

EXHIBIT A-1
Problem Input Report

ASSEMBLY PLANT LOADING — ALTERNATE IV

CAR	LINE	MODEL LOGIC	DIST CODE	WEIGHT	DIST US	VOLUMES CANADA		TOTAL VOLUME
1	F-29	CAR	313	3250.	180213.	27432.	3649.	211294.
2	F-41	CAR	313	3265.	216744.	32993.	4389.	254126.

(continued)

[2]The input data that resulted in this output has been disguised to guarantee confidentiality.

EXHIBIT A-1 (Continued)

ASSEMBLY PLANT LOADING ALTERNATE IV

CAR	LINE	MODEL LOGIC	DIST CODE	WEIGHT	DIST US	VOLUMES CANADA		TOTAL VOLUME
3	F-45	CAR	313	3565.	220901.	33627.	4475.	259003.
4	M-22	CAR	313	3250.	75795.	15976.	2338.	94109.
5	M-41	CAR	313	3265.	85961.	14320.	2096.	102377.
6	M-45	CAR	313	3565.	26597.	4384.	643.	31624.
					806211.	128732.	17590.	952533.

ASSEMBLY PLANT					
NAME	HAMT1	HAMT2	NEWARK	ST LOU	
POSITION	1	2	3	4	5
DAYS	243.0	242.0	242.0	239.0	1000.0
OVERTIME					
. DAYS	39.0	39.0	39.0	39.0	36.0
SHIFTS	2	2	2	2	2
FIXED COST	0.	0.	0.	0.	0.
VARIABLE MFG COST					
STRAIGHT TIME	380.73	383.49	326.95	344.49	1000.00
OVERTIME	144.17	146.77	137.28	144.17	0.00
PLANT CAPACITY					
STRAIGHT TIME	880.00	960.00	880.00	960.00	1000.00
OVERTIME	880.00	960.00	880.00	960.00	1000.00

PLANT LINE UP					
NAME	HAMT1	HAMT2	NEWARK	ST LOU	
OVERTIME					
ALLOWED	1.	1.	1.	1.	1.
1 F-29 IBT	41.63	......	100.12	......	......
RAILCAR	DET/TR	......	NEW/TR	......	......
LOAD FACTOR	15.00	......	15.00	......	......
2 F-41 IBT	41.82	41.82	100.59	......	......
RAILCAR	DET/TR	DET/TR	NEW/TR	......	......
LOAD FACTOR	15.00	15.00	15.00	......	......
3 F-45 IBT	......	45.66	109.84	92.27	......
RAILCAR	......	DET/TR	NEW/TR	STL/TR	......
LOAD FACTOR	......	15.00	15.00	15.00	......
4 M-22 IBT	......	......	......	84.12	......
RAILCAR	......	......	......	STL/TR	......
LOAD FACTOR	......	......	......	15.00	......
5 M-41 IBT	......	......	......	84.51	......
RAILCAR	......	......	......	STL/TR	......
LOAD FACTOR	......	......	......	15.00	.....
6 M-45 IBT	......	......	......	92.27	......
RAILCAR	......	......	......	STL/TR	......
LOAD FACTOR	......	......	......	15.00	......

EXHIBIT A-2
Cost Matrix Report

EXCERPT FROM COST MATRIX REPORT

COST MATRIX REPORT 08/17/77 PROBLEM NUMBER 78414
ASSEMBLY PLANT LOADING ALTERNATE IV

DISTRIBUTION ZONE	CARLINE	ANNUAL VOLUME	IDEAL PLANT	NEXT BEST* PLANT	IDEAL FREIGHT	NEXT BEST FREIGHT
1	F-29	74	HAMT1	NEWARK	621.	692.
1	F-41	89	HAMT1	HAMT2	621.	624.
1	F-45	91	ST LOU	HAMT2	616.	628.
1	M-22	31	ST LOU		608.	752.
1	M-41	35	ST LOU		608.	752.
1	M-45	11	ST LOU		615.	760.
2	F-29	31	HAMT1	NEWARK	757.	835.
2	F-41	37	HAMT1	HAMT2	757.	760.
2	F-45	38	ST LOU	HAMT2	752.	764.
2	M-22	13	ST LOU		744.	888.
2	M-41	14	ST LOU		745.	889.
2	M-45	4	ST LOU		752.	897.
3	F-29	83	HAMT1	NEWARK	638.	684.
3	F-41	101	HAMT1	HAMT2	638.	641.
3	F-45	102	ST LOU	HAMT2	583.	645.
3	M-22	35	ST LOU		574.	719.
3	M-41	40	ST LOU		575.	719.
3	M-45	12	ST LOU		583.	727.
4	F-29	25	HAMT1	NEWARK	587.	653.
4	F-41	31	HAMT1	HAMT2	587.	590.
4	F-45	31	ST LOU	HAMT2	540.	594.
4	M-22	10	ST LOU		532.	676.
4	M-41	12	ST LOU		532.	676.
4	M-45	3	ST LOU		540.	684.
5	F-29	84	HAMT1	NEWARK	456.	511.
5	F-41	101	HAMT1	HAMT2	456.	459.
5	F-45	103	HAMT2	NEWARK	463.	521.
5	M-22	35	ST LOU		537.	681.
5	M-41	40	ST LOU		537.	681.
5	M-45	12	ST LOU		545.	689.
6	F-29	1193	HAMT1	NEWARK	483.	524.
6	F-41	1435	HAMT1	HAMT2	483.	486.
6	F-45	1463	HAMT2	NEWARK	490.	534.
6	M-22	502	ST LOU		548.	692.
6	M-41	569	ST LOU		548.	693.
6	M-45	176	ST LOU		556.	700.
7	F-29	55	NEWARK	HAMT1	562.	566.
7	F-41	66	NEWARK	HAMT1	563.	567.
7	F-45	67	ST LOU	NEWARK	558.	572.
7	M-22	23	ST LOU		549.	694.
7	M-41	26	ST LOU		550.	694.
7	M-45	8	ST LOU		558.	702.
8	F-29	2015	NEWARK	HAMT1	515.	545.
8	F-41	2424	NEWARK	HAMT1	515.	545.
8	F-45	2471	NEWARK	HAMT2	524.	552.
8	M-22	847	ST LOU		604.	748.
8	M-41	961	ST LOU		604.	748.

(continued)

EXHIBIT A-2 (Continued)

DISTRIBUTION ZONE	CARLINE	ANNUAL VOLUME	IDEAL PLANT	NEXT BEST PLANT	IDEAL FREIGHT	NEXT BEST FREIGHT
8	M-45	297	ST LOU		612.	756.
9	F-29	356	HAMT1	NEWARK	649.	724.
9	F-41	428	HAMT1	HAMT2	650.	652.
9	F-45	437	ST LOU	HAMT2	617.	656.
9	M-22	150	ST LOU		609.	753.
9	M-41	170	ST LOU		610.	754.
9	M-45	52	ST LOU		617.	761.
10	F-29	76	HAMT1	NEWARK	565.	602.
10	F-41	92	HAMT1	HAMT2	565.	568.
10	F-45	94	ST LOU	HAMT2	542.	572.
10	M-22	32	ST LOU		534.	678.
10	M-41	36	ST LOU		535.	679.
10	M-45	11	ST LOU		542.	687.
11	F-29	947	NEWARK	HAMT1	478.	517.
11	F-41	1139	NEWARK	HAMT1	478.	517.
11	F-45	1161	NEWARK	HAMT2	487.	524.
11	M-22	398	ST LOU		609.	753.
11	M-41	451	ST LOU		610.	754.
11	M-45	139	ST LOU		617.	761.
12	F-29	37	HAMT1	NEWARK	499.	554.
12	F-41	44	HAMT1	HAMT2	499.	501.
12	F-45	45	HAMT2	NEWARK	505.	564.
12	M-22	15	ST LOU		580.	724.
12	M-41	17	ST LOU		580.	724.
12	M-45	5	ST LOU		588.	732.
13	F-29	116	NEWARK	HAMT1	496.	515.
13	F-41	555	NEWARK	HAMT1	496.	515.
13	F-45	566	NEWARK	HAMT2	505.	522.
13	M-22	194	ST LOU		563.	707.
13	M-41	220	ST LOU		563.	707.
13	M-45	68	ST LOU		571.	715.
14	F-29	117	HAMT1	NEWARK	583.	649.
14	F-41	141	HAMT1	HAMT2	583.	586.
14	F-45	144	ST LOU	HAMT2	536.	590.
14	M-22	49	ST LOU		528.	672.
14	M-41	56	ST LOU		528.	672.
14	M-45	17	ST LOU		536.	680.
15	F-29	187	HAMT1	NEWARK	951.	0.
15	F-41	225	HAMT1	HAMT2	951.	954.
15	F-45	230	ST LOU	HAMT2	946.	957.
15	M-22	78	ST LOU		938.	0.
15	M-41	89	ST LOU		938.	0.
15	M-45	27	ST LOU		946.	0.
16	F-29	21	HAMT1	NEWARK	497.	553.
16	F-41	26	HAMT1	HAMT2	497.	500.
16	F-45	26	HAMT2	ST LOU	504.	528.
16	M-22	9	ST LOU		520.	664.

EXHIBIT A-3
Plant Operating Pattern

PROBLEM NO 78414

DAILY UNITS	HAMT1	HAMT2	NEWARK	ST LOU		TOTAL
F-29	607.3	0.0	263.3	0.0	0.0	870.6
F-41	272.7	461.6	314.7	0.0	0.0	1049.0
F-45	0.0	498.4	411.7	162.2	0.0	1072.3
M-22	0.0	0.0	0.0	393.8	0.0	393.8

(continued)

EXHIBIT A-3 (Continued)

PROBLEM NO. 78414

DAILY UNITS	HAMT1	HAMT2	NEWARK	ST LOU		TOTAL
M-41	0.0	0.0	0.0	428.4	0.0	428.4
M-45	0.0	0.0	0.0	132.3	0.0	132.3
TOTAL REQ.	880.0	960.0	989.6	1116.7	0.0	3946.3
TOTAL CAP.	880.0	960.0	880.0	960.0	1000.0	4680.0
TOTAL OVER.	0.0	0.0	109.6	156.7	0.0	266.3
ANNUAL UNITS	213840.	232320.	239493.	266880.	0.0	952533.
NO. SHIFTS	2	2	2	2	2	10
WORK DAYS	243.0	242.0	242.0	239.0	1000.0	
OVERTIME DAYS	0.0	0.0	30.2	39.0	0.0	
ANNUAL COST-THOUS DOLLARS						
FREIGHT	32895.	35295.	42918.	65017.	0.	176126.
VAR. MFG.	81415.	89092.	78302.	91937.	0.	340747.
OVERTIME	0.	0.	3642.	5398.	0.	9040.
TOTAL	114310.	124388.	124863.	162353.	0.	525913.

NOTES-

1 ASSEMBLY PLANT LOADING ALTERNATE IV
2 ANNUAL COST EXPRESSED IN THOUSANDS OF DOLLARS
3 ANNUAL UNIT CHANGE 0.00
4 MAXIMUM ERROR 54.91

EXHIBIT A-4
Plant Operating Pattern

ANNUAL COSTS	HAMT1	HAMT2	NEWARK	ST LOU		TOTAL
F-29	80097.8	0.0	26891.1	0.0	0.0	106988.9
F-41	34212.1	62727.3	34855.1	0.0	0.0	131794.6
F-45	0.0	61660.5	45710.8	22455.2	0.0	129826.4
M-22	0.0	0.0	0.0	46570.7	0.0	46570.7
M-41	0.0	0.0	0.0	50764.6	0.0	50764.6
M-45	0.0	0.0	0.0	15858.8	0.0	15858.8
TOTAL	114309.9	124387.8	107457.0	135649.3	0.0	481804.0
ANNUAL REQUIREMENTS						
F-29	147582.5	0.0	54155.0	0.0	0.0	201737.5
F-41	66257.5	111716.4	69944.7	0.0	0.0	247918.8
F-45	0.0	120603.6	88860.3	35619.1	0.0	245082.9
M-22	0.0	0.0	0.0	80029.7	0.0	80029.7
M-41	0.0	0.0	0.0	86982.5	0.0	86982.5
M-45	0.0	0.0	0.0	26808.7	0.0	26808.7
TOTAL	213840.0	232320.0	212960.0	229440.0	0.0	888560.0
MARKET SHARE PERCENT						
F-29	69.8	0.0	30.2	0.0	0.0	100.0
F-41	26.0	44.0	30.0	0.0	0.0	100.0
F-45	0.0	46.5	38.4	15.1	0.0	100.0
M-22	0.0	0.0	0.0	100.0	0.0	100.0
M-41	0.0	0.0	0.0	100.0	0.0	100.0
M-45	0.0	0.0	0.0	100.0	0.0	100.0
	22.3	24.3	25.1	28.3	0.0	100.0

(continued)

EXHIBIT A-4 (Continued)

COSTS BY TYPE AND PER UNIT

	INBOUND		OUTBOUND		VARIABLE MFG		TOTAL COST	
	TOTAL	PER/UNIT	TOTAL	PER/UNIT	TOTAL	PER/UNIT	TOTAL	PER/UNIT
F-29	11565.9	57.33	21528.0	106.71	73895.1	366.29	106988.9	530.34
F-41	14478.6	58.40	26379.2	106.40	90936.8	366.80	131794.6	531.60
F-45	18553.7	75.70	23699.1	96.70	87573.6	357.32	129826.4	529.72
M-22	6732.1	84.12	12269.1	153.31	27569.4	344.49	46570.7	581.92
M-41	7350.9	84.51	13449.1	154.62	29964.6	344.49	50764.6	583.62
M-45	2473.6	92.27	4149.9	154.80	9235.3	344.49	15858.8	591.56
TOTAL	61154.8	68.82	101464.4	111.20	319174.8	358.20	481804.0	542.23

OVERTIME OPERATION PROBLEM NO. 78414

ASSEMBLY PLANT LOADING ALTERNATE IV

ANNUAL COSTS	HAMT1	HAMT2	NEWARK	ST LOU		
F-29	0.0	0.0	6232.9	0.0	0.0	6232.9
F-41	0.0	0.0	4040.3	0.0	0.0	4040.3
F-45	0.0	0.0	7132.8	2278.2	0.0	9411.0
M-22	0.0	0.0	0.0	9998.2	0.0	9998.2
M-41	0.0	0.0	0.0	10962.5	0.0	10962.5
M-45	0.0	0.0	0.0	3464.4	0.0	3464.4
TOTAL	0.0	0.0	17406.0	26703.2	0.0	44109.2

ANNUAL REQUIREMENTS

F-29	0.0	0.0	9556.5	0.0	0.0	9556.5
F-41	0.0	0.0	6207.4	0.0	0.0	6207.4
F-45	0.0	0.0	10769.2	3150.9	0.0	13920.1
M-22	0.0	0.0	0.0	14079.3	0.0	14079.3
M-41	0.0	0.0	0.0	15394.5	0.0	15394.5
M-45	0.0	0.0	0.0	4815.3	0.0	4815.3
TOTAL	0.0	0.0	26533.0	37440.0	0.0	63973.0

MARKET SHARE PERCENT

F-29	69.8	0.0	30.2	0.0	0.0	100.0
F-41	26.0	44.0	30.0	0.0	0.0	100.0
F-45	0.0	46.5	38.4	15.1	0.0	100.0
M-22	0.0	0.0	0.0	100.0	0.0	100.0
M-41	0.0	0.0	0.0	100.0	0.0	100.0
M-45	0.0	0.0	0.0	100.0	0.0	100.0
	22.3	24.3	25.1	28.3	0.0	100.0

COSTS BY TYPE AND PER UNIT

	INBOUND		OUTBOUND		VARIABLE MFG		TOTAL COST	
	TOTAL	PER/UNIT	TOTAL	PER/UNIT	TOTAL	PER/UNIT	TOTAL	PER/UNIT
F-29	956.8	100.12	839.7	87.86	4436.4	464.23	6232.9	652.21
F-41	624.4	100.59	534.3	86.07	2881.6	464.23	4040.3	650.89
F-45	1473.6	105.86	1398.3	100.45	6539.1	469.76	9411.0	676.08
M-22	1184.4	84.12	1933.8	137.35	6880.0	488.66	9998.2	710.13
M-41	1301.0	84.51	2138.8	138.93	7522.7	488.66	10962.5	712.10
M-45	444.3	92.27	667.0	138.51	2353.1	488.66	3464.4	719.44
TOTAL	5984.5	93.55	7511.9	117.42	30612.8	478.53	44109.2	689.50

(continued)

EXHIBIT A-4 (Continued)

TOTAL OPERATION PROBLEM NO. 78414.
ASSEMBLY PLANT LOADING ALTERNATE IV

ANNUAL COSTS	HAMT1	HAMT2	NEWARK	ST LOU		TOTAL
F-29	80097.8	0.0	33124.0	0.0	0.0	113221.8
F-41	34212.1	62727.3	38895.5	0.0	0.0	135834.9
F-45	0.0	61660.5	52843.6	24733.4	0.0	139237.4
M-22	0.0	0.0	0.0	56568.9	0.0	56568.9
M-41	0.0	0.0	0.0	61727.1	0.0	61727.1
M-45	0.0	0.0	0.0	19323.2	0.0	19323.2
TOTAL	114309.9	124387.8	124863.0	162352.5	0.0	525913.3
ANNUAL REQUIREMENTS						
F-29	147582.5	0.0	63411.5	0.0	0.0	211294.0
F-41	66257.5	111716.4	76152.1	0.0	0.0	254126.0
F-45	0.0	120603.6	99629.4	38770.0	0.0	259003.0
M-22	0.0	0.0	0.0	94109.0	0.0	94109.0
M-41	0.0	0.0	0.0	102377.0	0.0	102377.0
M-45	0.0	0.0	0.0	31624.0	0.0	31624.0
TOTAL	213840.0	232320.0	239493.0	266880.0	0.0	952533.0
MARKET SHARE PERCENT						
F-29	69.8	0.0	30.2	0.0	0.0	100.0
F-41	26.0	44.0	30.0	0.0	0.0	100.0
F-45	0.0	46.5	38.4	15.1	0.0	100.0
M-22	0.0	0.0	0.0	100.0	0.0	100.0
M-41	0.0	0.0	0.0	100.0	0.0	100.0
M-45	0.0	0.0	0.0	100.0	0.0	100.0
	22.3	24.3	25.1	28.3	0.0	100.0

COSTS BY TYPE AND PER UNITS

	INBOUND		OUTBOUND		VARIABLE MFG		TOTAL COST	
	TOTAL	PER/UNIT	TOTAL	PER/UNIT	TOTAL	PER/UNIT	TOTAL	PER/UNIT
F-29	12522.7	59.27	22367.7	105.86	78331.5	370.72	113221.8	535.85
F-41	15103.0	59.43	26917.5	105.91	93818.4	369.18	135834.9	534.52
F-45	20027.4	77.32	25097.4	96.90	94112.6	363.37	139237.4	537.59
M-22	7916.4	84.12	14203.0	150.92	34449.4	366.06	56568.9	601.10
M-41	8651.9	84.51	15588.0	152.26	37487.3	366.17	61727.1	602.94
M-45	2917.9	92.27	4816.8	152.32	11588.4	366.44	19323.2	611.03
TOTAL	67139.3	73.49	108986.4	114.42	349787.6	367.22	525913.3	552.12

EXHIBIT A-5
Plant Loading Analysis Report

The Plant Loading Analysis Report specifies the increase in cost and the action to be taken if the optimal production level is not reached or is exceeded. The column labelled "ACCUM. VOLUME" gives successive limits (in units per *day*) for deviations (positive and negative) from the optimal volume.

For example, from Exhibit A-3, we have determined that the optimal daily production of line F-29 at the Newark plant is 263.3 units. Line 29 of the Plant Loading Analysis Report below specifies that if this volume falls short by .14 units/day, the optimal action would be to transfer the responsibility for serving trading zone #27 from Newark to Hamtramck (1), at an additional cost of $0.09/unit or $3.00/year. If the optimal volume falls short by .522 units/day (see line 28 below), the best action would be to transfer

(continued)

EXHIBIT A-5 (Continued)

zones 27 and 107 to Hamtramck, with a concurrent cost increase of $65.00/year, as calculated below:

	Annual Units Transferred	Penalty Per Unit	Total Penalty Per Year
Zone 27	33.9	$ .09	$ 3.00
Zone 107	92.4	.67	62.00
			$65.00

or as indicated in the column labelled AC(cumulated) PEN(alty).

As a somewhat more complex example, suppose the Newark plants production on line F-29 were to fall short by 20 units, we look for the "accumulated volume" limits between which 20 falls, in this case 19.565 on line 15 below and 20.922 on line 14 below. The optimal action then is to "unload" trading zones 27, 107, 436, 217, 7, 470, 455, 438, 379, 264, 236, 28, 25, 23, and 120 and service these areas from Hamtramck (1). In addition, .435 units/day (20 − 19.565) of zone 466's 1.81 units/day volume should be transferred to Hamtramck (1). The cost increase which accompanies this transfer is $17,507/year ($16,272 + $5.11 × .435 units/day × 242 days).

If the optimal production level is exceeded, the plant "expands" its service region. Sensitivity analysis of this type is performed in the same way as "unloading" trading zones. Say, for example, the Newark plant were to exceed the optimal F-29 production level of 263.3 units/day by 20.33 units/day. Then from line 34 below, the best action is to have the Newark plant service trading zones 35, 214, 347, 323, and 298 instead of having the Hamtramck (1) plant do so. The increase in annual cost which accompanies this transfer is $16,971.

(continued)

EXHIBIT A-5 (Continued)

PAGE 3 ASSEMBLY PLANT LOADING ALTERNATE IV CARLINE F-29
PROB. NO. 78414 PLANT LOADING ANALYSIS REPORT - IDEAL SOLUTION

OVERTIME NOTE - EFFECTIVE UNIT COST PENALTY FOR OVERTIME OPERATION AT NEWARK = 137.28

LINE NO.	ZONE NO.	ANN. ZONE VOLUME	ZONE VOLUME	ACCUM. VOLUME	NEWARK ACTION	UNIT COST	ZONE COST	ALT. PLANT	UNIT PENALTY	ZONE PEN.	AC PEN
1	77	310.5	1.28	36.660	UNLOAD	536.66	166625.	HAMT1	13.79	4281.	63184.
2	100	71.7	0.30	35.377	UNLOAD	543.41	38954.	HAMT1	13.79	988.	58903.
3	147	78.7	0.33	35.081	UNLOAD	534.05	42041.	HAMT1	13.79	1085.	57915.
4	177	203.2	0.84	34.756	UNLOAD	535.47	108796.	HAMT1	13.79	2801.	56830.
5	408	166.2	0.69	33.916	UNLOAD	545.22	90636.	HAMT1	13.79	2292.	54028.
6	423	120.5	0.50	33.229	UNLOAD	536.42	64639.	HAMT1	13.79	1661.	51736.
7	434	24.2	0.10	32.731	UNLOAD	527.07	12749.	HAMT1	13.79	333.	50075.
8	150	177.7	0.73	32.631	UNLOAD	577.90	102677.	HAMT1	13.18	2341.	49741.
9	388	636.8	2.63	31.897	UNLOAD	557.00	354701.	HAMT1	13.18	8392.	47400.
10	406	213.3	0.88	29.266	UNLOAD	562.88	120060.	HAMT1	13.13	2811.	39008.
11	439	595.9	2.46	28.384	UNLOAD	560.45	333976.	HAMT1	13.18	7853.	36197.
12	325	1100.3	4.55	25.922	UNLOAD	507.02	557894.	HAMT1	8.69	9562.	28343.
13	208	109.5	0.45	21.375	UNLOAD	525.22	57515.	HAMT1	7.59	831.	18781.
14	466	328.5	1.36	20.922	UNLOAD	552.25	181424.	HAMT1	5.11	1679.	17950.
15	120	437.6	1.81	19.565	UNLOAD	504.57	220792.	HAMT1	4.69	2052.	16272.
16	23	70.8	0.29	17.757	UNLOAD	538.38	38120.	HAMT1	4.14	293.	14219.
17	25	1391.0	5.75	17.464	UNLOAD	531.27	739019.	HAMT1	4.14	5764.	13926.
18	28	208.5	0.86	11.716	UNLOAD	542.54	113097.	HAMT1	4.14	864.	8161.
19	236	37.4	0.15	10.855	UNLOAD	538.90	20145.	HAMT1	4.14	155.	7298.
20	264	178.6	0.74	10.700	UNLOAD	547.65	97785.	HAMT1	4.14	740.	7143.
21	379	10.6	0.04	9.962	UNLOAD	538.38	5683.	HAMT1	4.14	44.	6403.
22	438	190.0	0.79	9.919	UNLOAD	576.67	109560.	HAMT1	4.14	787.	6359.
23	455	78.3	0.32	9.134	UNLOAD	570.55	44664.	HAMT1	4.14	324.	5572.
24	470	17.2	0.07	8.810	UNLOAD	568.91	9758.	HAMT1	4.14	71.	5247.
25	7	55.4	0.23	8.739	UNLOAD	562.18	31152.	HAMT1	4.14	230.	5176.
26	217	579.2	2.39	8.510	UNLOAD	503.02	291347.	HAMT1	3.94	2282.	4947.
27	436	1354.1	5.60	6.117	UNLOAD	537.18	727393.	HAMT1	1.92	2600.	2665.
28	107	92.4	0.38	0.522	UNLOAD	532.58	49186.	HAMT1	0.67	62.	65.
29	27	33.9	0.14	0.140	UNLOAD	532.15	18620.	HAMT1	0.09	3.	3.
30	35	120.9	0.50	0.500	EXPAND	517.14	62543.	HAMT1	1.97	238.	238.
31	214	297.3	1.23	1.728	EXPAND	514.77	153038.	HAMT1	2.94	874.	1112.
32	347	4281.7	17.69	19.421	EXPAND	500.79	2144252.	HAMT1	3.42	14644.	15756.
33	323	90.2	0.37	19.794	EXPAND	512.40	46196.	HAMT1	5.52	498.	16254.
34	298	129.7	0.54	20.330	EXPAND	515.06	66822.	HAMT1	5.53	717.	16971
35	87	226.5	0.94	21.266	EXPAND	525.23	118959.	HAMT1	6.45	1463.	18434.
36	137	94.6	0.39	21.657	EXPAND	516.83	48868.	HAMT1	6.97	659.	19093.
37	203	99.4	0.41	22.067	EXPAND	510.40	50729.	HAMT1	7.95	790.	19883.
38	68	211.5	0.87	22.941	EXPAND	508.47	107560.	HAMT1	9.12	1929.	21812.
39	18	56.3	0.23	23.174	EXPAND	554.14	31194.	HAMT1	9.12	513.	22326.
40	49	591.1	2.44	25.617	EXPAND	549.36	324712.	HAMT1	9.12	5391.	27717.
41	96	131.5	0.54	26.160	EXPAND	562.04	73906.	HAMT1	9.12	1199.	28917.
42	118	75.2	0.31	26.471	EXPAND	582.91	43637.	HAMT1	9.12	686.	29603.
43	160	67.3	0.28	26.749	EXPAND	553.16	37221.	HAMT1	9.12	614.	30216.
44	297	179.4	0.74	27.490	EXPAND	564.86	101354.	HAMT1	9.12	1637.	31853.
45	339	329.4	1.36	28.851	EXPAND	584.94	192679.	HAMT1	9.12	3005.	34858.

EXHIBIT A-6
Plant Loading Analysis Plot

```
              ASSEMBLY PLANT LOADING        ALTERNATE IV                                                08/17/77

PROB. NO.78414                    PLANT LOADING ANALYSIS  PLOT -        IDEAL SOLUTION                  NEWARK ASSEMBLY PLANT
              ACCUMULATED PENALTY (AND UNIT PENALTY) VERSUS ACCUMULATED VOLUME              CAPLI      F-29

              ------------------------------------------------------------------------------------      -----------------
  7246436.                                                                                               (-53.88)
  7101507.                                                                                               -53.35)
  6956578.                                                                                          *    52.88)
  6811649.                                                                                          *    52.49)
  6666721.                                                                                         *(    2.12)
  6521792.                                                                                         *(-   .74)
  6376863.                                                                                         *(-   .35)
  6231935.                                                                                        *(-5   96 )
  6087006.                                                                                       *(-50   9)
  5942077.                                                                                      *(-50.   )
  5797148.                                                                                      *(-49.   )
  5652220.                                                                                     *(-49.4
  5507291.                                                                                    *(-49.06
  5362362.                                                                                   *(-48.64)
  5217434.                                                                                   *(-48.22)
  5072505.                                                                                  *(-47.78)
  4927576.                                                                                 *(-47.34)
  4782647.                                                                                *(-46.92)
  4637719.                                                                               *(-46.50)
  4492790.                                                                              *(-46.07)
  4347861.                                                                              *(-45.61)
  4202933.                                                                             *(-45.13)
  4058004.                                                                            *(-44.64)
  3913075.                                                                           *(-44.11)
  3768146.                                                                          *(-43.56)
  3623218.                                                                         *(-42.98)
  3478289.       *( 46.01)                                                        *(-42.37)
  3333360.       *( 45.07)                                                       *(-41.73)
  3188432.        *( 44.33)                                                     *(-41.05)
  3043503.         *( 43.55)                                                   *(-40.32)
  2898574.          *( 42.73)                                                 *(-39.57)
  2753646.          *( 41.86)                                                *(-39.81)
  2608717.           *( 40.94)                                              *(-38.02)
  2463788.            *( 39.95)                                             *(-37.21)
  2318859.             *( 38.89)                                           *(-36.37)
  2173931.              *( 37.76)                                         *(-35.56)
  2029002.              *( 36.59)                                        *(-34.68)
  1884073.               *( 35.48)                                      *(-33.71)
  1739145.                *( 34.32)                                    *(-32.65)
  1594216.                 *( 33.12)                                  *(-31.49)
  1449287.                  *( 31.86)                                *(-30.20)
  1304358.                   *( 30.44)                              *(-28.82)
  1159430.                    *( 29.00)                            *(-27.27)
  1014501.                     *( 27.50)                          *(-25.56)
   869572.                       *( 25.75)                      *(-23.65)
   724644.                        *( 23.69)                    *(-21.54)
   579715.                         *( 21.30)                  *(-19.11)
   434786.                           *( 18.49)               *(-16.37)
   289857.                            *( 15.10)            *(-13.24)
   144929.                              *( 10.62)  *( -9.89)
        0.                                  *(  0.00)

              ------------------------------------------------------------------------------------      -----------------
              V=  315.       V=  158.      V=    0.                  V= -278.                           = -556.

                             ACCUMULATED DAILY VOLUME

              ------------------------------------------------------------------------------------      -----------------
```

EXHIBIT A-7
Actual Shipping Area Map

The Federal Power Commission Transport of Alaskan Natural Gas

On October 22, 1976, the Alaska Natural Gas Transportation Act (Public Law 94-586) was signed into law. This Act established a special decision-making process for the selection of a system to transport natural gas from the Prudhoe Bay area of Alaska's North Slope to United States markets. The Act was designed to provide for the participation of the President and Congress in the final decision. Ordinarily the Federal Power Commission (FPC) has the final authority to approve proposed natural gas transmission systems.

As mandated by the Act, Richard L. Dunham, Chairman of the FPC, had until May 2, 1977 to present the recommendations of the FPC to the President. Three competitors, the Alcan Pipeline Company, the Alaskan Arctic Gas Pipeline Company, and El Paso Alaska Company had submitted proposals to the FPC which detailed the location, costs, and timing of their delivery systems and the amounts of natural gas to be delivered annually to the "lower 48." Feasibility studies and some prior analysis had been conducted by the FPC and Department of Interior (DOI) in accordance with the Trans-Alaska Oil Pipeline Authorization Act of 1974. It was now up to the FPC to do the detailed analysis which would determine which of these three consortia would be awarded the right to transport the gas.

BACKGROUND

Natural gas is a clean burning, easily transportable fuel which supplies almost one third of the nation's energy needs. Proven reserves in the continental United States have been declining since 1968. The decreasing supply of natural gas to consumers had resulted in the denial of gas service to industries, homeowners, and commercial businesses. The 1968 discovery of large deposits of natural gas in the Prudhoe Bay area of Northern Alaska presented a potential new source. Based on a reservoir simulation run of the Prudhoe Bay reserves, a DOI study concluded that *salable* reserves of 18.2 trillion cubic feet (TCF) exist, approximately one tenth of the lower 48's proven reserves. This gas could be produced and sold over a 20 year period at the rate of 2.4 billion cubic feet per day (BCFD), approximately 5% of the 1975 U.S. daily consumption of natural gas.

The Alcan Pipeline Company (Alcan) and the Alaskan Arctic Gas Pipeline Company (Arctic), both consortia of U.S. and Canadian companies, had submitted two quite similar proposals; both proposed overland pipelines traveling through Alaska, Canada, and dividing into a western leg which would terminate in Antioch, California (near San Francisco) and an eastern leg which would terminate in Dwight, Illinois (near Chicago). The El Paso Alaska Company (El Paso) recommended a combination of pipeline and tanker. For this proposed system, the gas would flow, via pipeline, to Gravina Point in Southern Alaska where it would be liquified. The liquid natural gas (LNG) would then be transported, via cryogenic tankers, to Point Conception, California, where it would be unloaded, regasified, and fed into existing natural gas pipelines. Exhibits 1, 2, and 3 provide more detailed descriptions of these three proposals.

THE FPC

The Federal Power Commission regulates the interstate aspects of the nation's electrical power and natural gas industries. The FPC is an independent agency established under the Federal Water Power Act of 1920, whose charter is to insure that the Nation's consumers have adequate supplies of gas and electricity at the lowest reasonable rates. Its powers and responsibilities include issuing permits for non-Federal hydroelectric power projects; regulating the rates and other aspects of interstate wholesale transactions in electric power and natural gas; issuing licenses for interstate gas sales and for construction and operation of interstate pipeline facilities and the importation and exportation of gas; and investigating electric power and natural gas industries to ascertain if their programs are in accordance with national conservation and efficient utilization of resources objectives. In addition, the FPC prescribes and enforces a uniform system of accounts for regulated electric utilities and natural gas pipeline companies. The FPC collects data and prepares public reports on national gas supply and demand, and conducts special studies of gas reported to be shut in or uncommitted.

The Commission also allocates available supplies of interstate natural gas, on the basis of end use, during periods of shortage to assure the best use of available supplies. It has established a set of service guidelines for pipelines to follow when it is necessary to curtail deliveries of gas to their wholesale or industrial customers. When these matters are contested, or the public interest otherwise so requires, the FPC holds public hearings to resolve the issues.

Under the Alaska Natural Gas Transportation Act, the FPC was to recommend a route for the natural gas pipeline to President Carter by May 2, 1977. Mr. Carter would have until December 1, 1977 to make a decision on the pipeline route. Unless both houses of Congress were to reject the presidential choice, it would be final and construction would begin.

THE EVALUATION METHODOLOGY

The major economic criterion which was selected by the FPC for judging the alternative systems was one traditionally used by Federal agencies in

EXHIBIT 1
Alcan Pipeline Company Proposal—Proposed Route

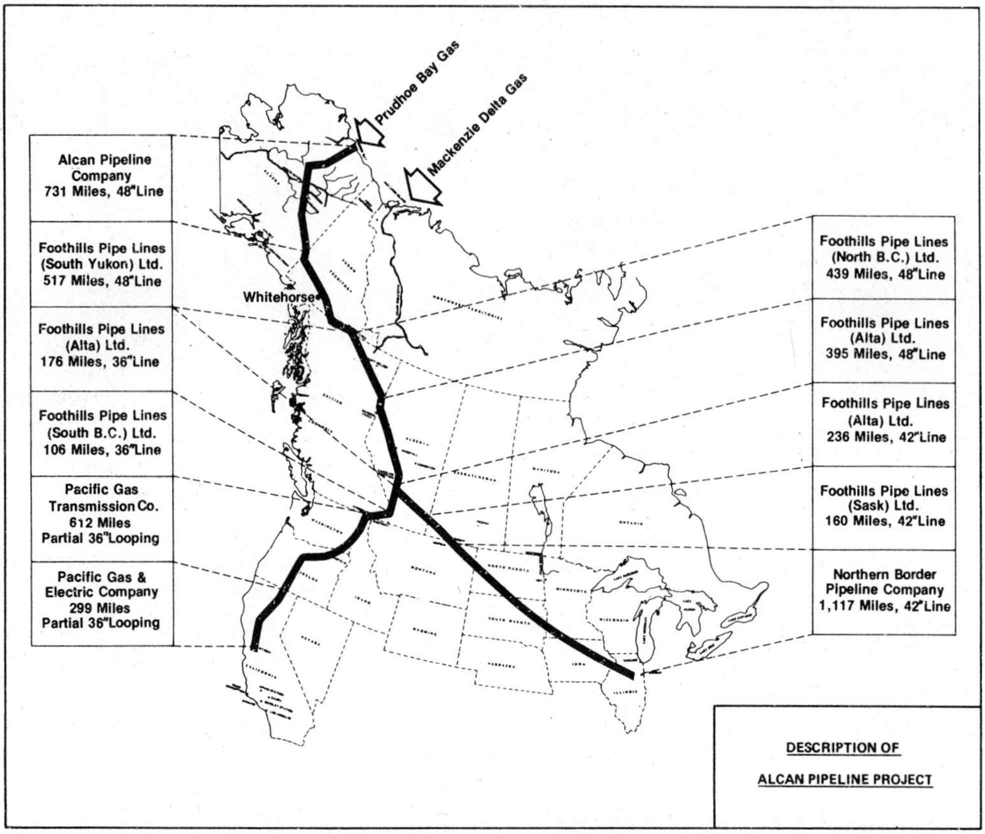

Total Pipeline Length: 4797 miles

Total Shrinkage: 6.32% for flow of 2.4 BCFD[a]

Projected Completion Date: June 1982

Estimated Project Cost (1975 dollars): $6.76 Billion

Projected Flow Rate: 2.4 BCFD

Maximum Flow: 3.2 BCFD (without major systems additions or expenditures)

[a]Shrinkage is the loss of natural gas due to (1) transportation losses or (2) use of gas for pipeline compressor and refrigeration system along the route.

Source: Decision and Report to Congress on the Alaska Natural Gas Transportation System, Publication No. 95-56, U.S. Government Printing Office, Washington, October 1977.

EXHIBIT 2
Alaskan Arctic Gas Pipeline Company Proposal—Proposed Route

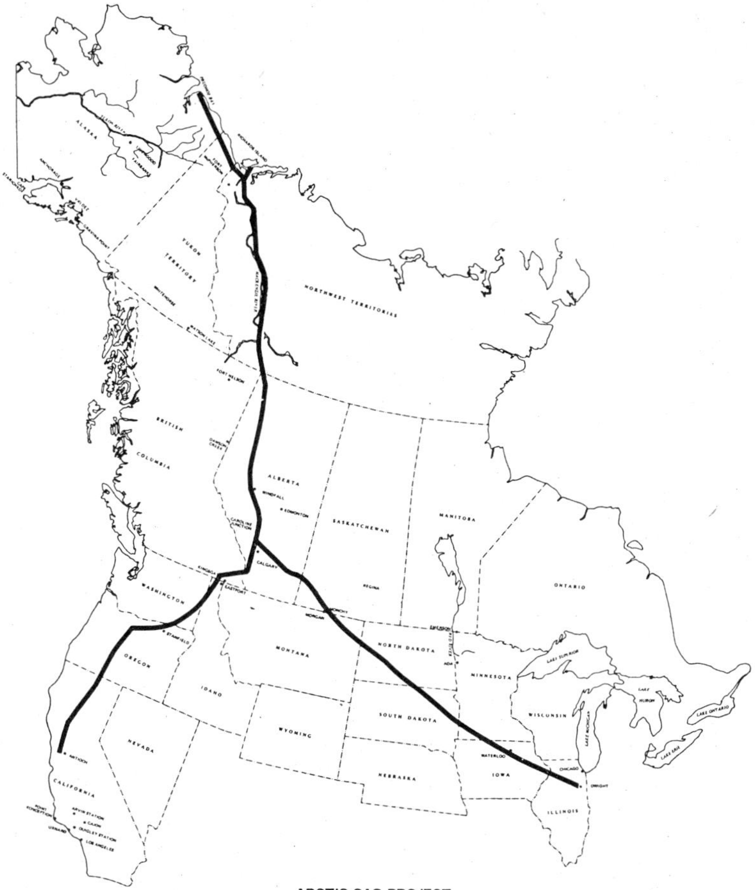

ARCTIC GAS PROJECT

Total Pipeline Length: 4500 miles

Total Shrinkage: 6.31% for flow of 2.4 BCFD

Projected Completion Date: June 1983

Estimated Project Cost (1975 dollars): $6.73 Billion

Projected Flow Rate: 2.4 BCFD

Maximum Flow: 3.5 BCFD (without major system additions or expenditures)

Source: Decision and Report to Congress on the Alaska Natural Gas Transportation System, Publication No. 95-56, U.S. Government Printing Office, Washington, October 1977.

EXHIBIT 3
El Paso Alaska Company Proposal—Proposed Route

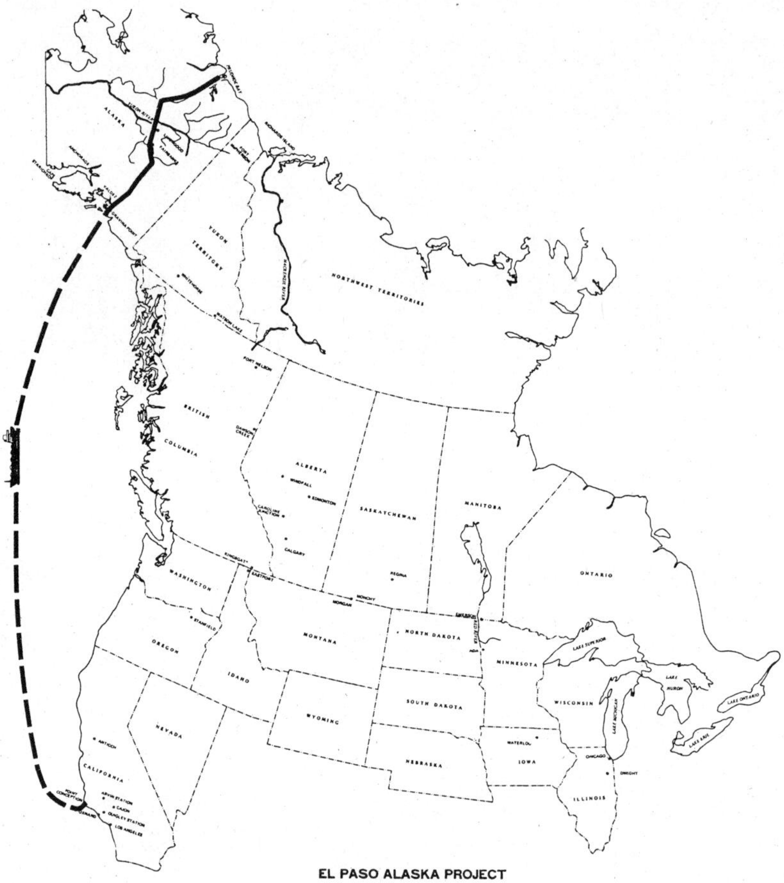

Total System Length: 809 miles of 42″ pipeline and 1900 nautical miles for 8 cryogenic tankers

Total Shrinkage: 10.9% for flow of 2.36 BCFD

Projected Completion Date: June 1983

Estimated Project Cost (1975 dollars): $6.57 Billion

Projected Flow Rate: 2.36 BCFD

Maximum Flow: 2.36 BCFD (flow increases require major expenditures and system additions)

Source: Decision and Report to Congress on the Alaska Natural Gas Transportation System, Publication No.95-56, U.S. Government Printing Office, Washington, October 1977.

evaluating government projects or policies—Net National Economic Benefit (NNEB). This is a measure in dollar terms, of the benefits and costs to the entire nation of the construction and operation of the natural gas transportation system. The FPC defined NNEB as the present value of the future stream of project benefits minus the present value of the future stream of project costs.

Other factors also enter the analysis of the three proposals, including environmental impacts, socio-economic impacts (local and regional), international relations, national security implications, and financial support for the project. However, if the proposals met acceptable standards on such qualitative criteria NNEB would be the deciding factor.

For each year from 1977 until 25 years after the project's initial gas delivery, the annual project costs were subtracted from the dollar value of the annual project "benefits," discounted to January 1, 1977, and summed to give the total project NNEB. All costs and benefits were measured in 1975 dollars. Most of the preliminary work done by DOI and the consortia themselves assumed a 10% discount rate based upon their assessment that this figure approximated the real rate of return on private investment experienced in the U.S. economy in the past. The FPC, however, felt 10% to be too high for what they termed "a real social rate of discount" which should be used for NNEB, and as such discounted at 6%. They did, nonetheless, do the NNEB analysis and present their results at both discount rates. Calculating NNEB using two discount factors also provide a rough analysis of the sensitivity of any ranking to the discount rate chosen.

The types of costs included in the NNEB analysis are as follows:

1. Incremental field cost of gas—the expense of gathering and conditioning gas for delivery (the gas would be produced in conjunction with oil which was being produced in any event). This cost included the initial capital outlay, over a 4-year period, for facilities construction, and an annual operating and maintenance charge.

2. The construction of the transportation system.

3. The annual operating and maintenance expense of the transportation system. Fuel costs are not specifically included, but are reflected in lower volumes of delivered gas (i.e., shrinkage). Minor working capital requirements are also included.

4. Costs associated with taxes. U.S. income taxes are excluded from an NNEB cost on the grounds that they are transfer payments rather than true resource costs. Taxes other than income (e.g., property taxes) are included as proxies for the costs of governmental services (e.g., roads, health systems and schools) required as a result of a system's construction. Canadian taxes, both income and other, are included.

All costs for the three projects are shown in Exhibits 4-1 through 4-3. A cost summary, discounting at both 6% and 10%, is given for all three proposals in Exhibit 5.

EXHIBIT 4-1
Arctic NNEB Components

(1) YEAR	(2) DELIVERED GAS (TRILLIONS BTU'S)	(3) FIELD GATHERING & CONDITIONING ($MILLION)	(4) FIELD G & C O & M ($MILLION)	(5) TRANSPORT FACILITIES ($MILLION)	(6) WORKING CAPITAL ($MILLION)	(7) O & M ($MILLION)	(8) U. S. OTHER TAXES ($MILLION)	(9) CANADIAN TAXES OTHER ($MILLION)	(10) CANADIAN TAXES INCOME ($MILLION)
1977	0.0	0.0	0.0	0.0	0.0	0.0	0.0	0.0	0.0
1978	0.0	0.0	0.0	18.2	0.0	0.0	0.0	0.0	0.0
1979	0.0	200.0	0.0	623.7	0.0	0.0	0.0	0.0	0.0
1980	0.0	344.0	0.0	1345.5	0.0	0.0	0.0	0.0	0.0
1981	0.0	400.0	0.0	1921.4	0.0	0.0	0.0	0.0	0.0
1982	0.0	500.0	0.0	1208.5	9.7	0.0	0.0	0.0	0.0
1983	471.1	0.0	4.0	407.2	8.4	42.5	24.3	0.5	0.0
1984	942.2	0.0	8.0	0.1	8.9	61.7	48.0	12.3	0.0
1985	942.2	0.0	8.0	8.7	0.7	60.8	46.1	13.0	0.0
1986	942.2	0.0	8.0	3.4	-0.1	59.8	44.3	13.2	56.4
1987	942.2	0.0	8.0	0.1	0.1	58.8	42.5	13.5	215.1
1988	942.2	0.0	8.0	6.2	-0.1	57.9	40.5	13.8	233.0
1989	942.2	0.0	8.0	3.7	0.0	57.1	38.7	13.9	254.0
1990	942.2	0.0	8.0	0.9	0.1	56.4	36.9	14.2	268.0
1991	942.2	0.0	8.0	22.7	-0.1	57.1	34.9	14.4	265.7
1992	942.2	0.0	8.0	18.4	-0.1	56.6	33.3	14.7	262.3
1993	942.2	0.0	8.0	17.7	0.4	56.5	31.7	14.9	257.7
1994	942.2	0.0	8.0	14.1	0.4	56.7	30.5	15.2	256.5
1995	942.2	0.0	8.0	0.0	0.4	56.8	29.5	15.5	255.4
1996	942.2	0.0	8.0	0.0	0.4	57.0	27.4	15.7	254.5
1997	942.2	0.0	8.0	0.0	0.0	57.0	25.3	16.0	254.2
1998	942.2	0.0	8.0	0.0	0.0	57.0	23.6	16.3	253.5
1999	942.2	0.0	8.0	0.0	0.0	57.0	21.2	16.6	252.9
2000	942.2	0.0	8.0	0.0	0.0	57.0	19.1	16.9	252.5
2001	942.2	0.0	8.0	0.0	0.0	57.0	17.0	17.2	252.2
2002	942.2	0.0	8.0	0.0	0.0	57.0	14.9	18.7	252.1
2003	942.2	0.0	8.0	0.0	0.0	57.0	12.9	19.1	252.9
2004	942.2	0.0	8.0	0.0	0.0	57.0	10.8	19.4	250.4
2005	942.2	0.0	8.0	0.0	0.0	57.0	9.0	19.8	232.4
2006	942.2	0.0	8.0	0.0	0.0	57.0	7.2	20.1	220.4
2007	942.2	0.0	8.0	0.0	0.0	57.0	5.4	20.5	205.0
TOTAL	23083.9	1444.0	196.0	5620.5	29.1	1422.7	675.0	385.4	5255.1

Prudhoe Bay Flow Rate: 2.4 Bcfd @ 1148 Btu beginning 7/1/83.
Total Energy Input Through 2007: 24,638 Trillion Btus.
Fuel Consumption: 6.31%.

Sources. Columns (3) and (4): See text.
Column (5): See Chapter VIII.
Columns (6), (8), (9), and (10): Arctic Gas Cost of Service, 4/15/77.
Column (7): FPC Staff, based upon Arctic's cost of service computer model.

All 1975 dollars except Columns (8), (9), and (10) which are nominal dollars.

All 1975 dollars except columns (8), (9), (10), which are nominal dollars.
Source: FPC Recommendation to the President: Alaska Natural Gas Transportation Systems, May 1977.

EXHIBIT 4-2
El Paso NNEB Components

(1) YEAR	(2) DELIVERED GAS (TRILLIONS BTU'S)	(3) FIELD GATHERING & CONDITIONING ($MILLION)	(4) FIELD G & C O & M ($MILLION)	(5) TRANSPORT FACILITIES ($MILLION)	(6) WORKING CAPITAL ($MILLION)	(7) O & M ($MILLION)	(8) U. S. OTHER TAXES ($MILLION)
1977	0.0	0.0	0.0	0.0	0.0	0.0	0.0
1978	0.0	0.0	0.0	35.6	0.0	0.0	0.0
1979	0.0	200.0	0.0	139.9	0.0	0.0	0.0
1980	0.0	344.0	0.0	1629.2	0.0	0.0	0.0
1981	0.0	400.0	0.0	2081.5	0.0	0.0	0.0
1982	0.0	500.0	0.0	1361.5	0.0	0.0	0.0
1983	433.9	0.0	4.0	339.8	55.4	86.4	41.7
1984	867.8	0.0	8.0	0.0	7.7	172.9	83.4
1985	867.8	0.0	8.0	0.0	0.0	172.9	80.1
1986	867.8	0.0	8.0	0.0	0.0	172.9	76.9
1987	867.8	0.0	8.0	0.0	0.0	172.9	73.8
1988	867.8	0.0	8.0	0.0	0.0	172.9	70.9
1989	867.8	0.0	8.0	0.0	0.0	172.9	68.1
1990	867.8	0.0	8.0	0.0	0.0	172.9	65.3
1991	867.8	0.0	8.0	0.0	0.0	172.9	62.7
1992	867.8	0.0	8.0	0.0	0.0	172.9	60.2
1993	867.8	0.0	8.0	0.0	0.0	172.9	57.8
1994	867.8	0.0	8.0	0.0	0.0	172.9	55.5
1995	867.8	0.0	8.0	0.0	0.0	172.9	53.3
1996	867.8	0.0	8.0	0.0	0.0	172.9	51.1
1997	867.8	0.0	8.0	0.0	0.0	172.9	49.1
1998	867.8	0.0	8.0	0.0	0.0	172.9	47.1
1999	867.8	0.0	8.0	0.0	0.0	172.9	45.2
2000	867.8	0.0	8.0	0.0	0.0	172.9	43.4
2001	867.8	0.0	8.0	0.0	0.0	172.9	41.7
2002	867.8	0.0	8.0	0.0	0.0	172.9	40.0
2003	867.8	0.0	8.0	0.0	0.0	172.9	38.4
2004	867.8	0.0	8.0	0.0	0.0	172.9	36.9
2005	867.8	0.0	8.0	0.0	0.0	172.9	35.4
2006	867.8	0.0	8.0	0.0	0.0	172.9	34.0
2007	867.8	0.0	8.0	0.0	0.0	172.9	32.6
TOTAL	21261.1	1444.0	196.0	5587.5	63.1	4236.0	1344.6

Prudhoe Bay Flow Rate: 2.3614 Bcfd @ 1130 Btus beginning 7/1/83.
Total Energy Input Through 2007: 23,862 Trillion Btus.
Fuel Consumption: 10.90%

Sources. Columns (3) and (4): See text.
Column (5): See Chapter VIII.
Column (6): AA-127, Schedule 6.
Column (7): EP-202, EP-228, EP-215, WL-46, EP-229, EP-265 and 4/13/77 data request response.
Column (8): AA-127. This does not agree with El Paso's 4/13/77 data request response which erroneously escalates "Other Taxes". Beginning figures in AA-127 are close to El Paso estimates contained in exhibits listed under Column (7) Sources; hence, AA-127 representation is adopted here.

All 1975 dollars.

All 1975 dollars.
Source: FPC Recommendation to the President: Alaska Natural Gas Transportation System, May 1977.

EXHIBIT 4-3
Alcan NNEB Components

(1) YEAR	(2) DELIVERED GAS (TRILLIONS BTU'S)	(3) FIELD GATHERING & CONDITIONING ($MILLION)	(4) FIELD G & C O & M ($MILLION)	(5) TRANSPORT FACILITIES ($MILLION)	(6) WORKING CAPITAL ($MILLION)	(7) O & M ($MILLION)	(8) U. S. OTHER TAXES ($MILLION)	(9) CANADIAN TAXES OTHER ($MILLION)	(10) CANADIAN TAXES INCOME ($MILLION)
1977	0.0	0.0	0.0	4.6	0.0	0.0	0.0	0.0	0.0
1978	0.0	200.0	0.0	57.2	0.0	0.0	0.0	0.0	0.0
1979	0.0	344.0	0.0	437.5	0.0	0.0	0.0	0.0	0.0
1980	0.0	400.0	0.0	1527.1	0.0	0.0	0.0	0.0	0.0
1981	0.0	500.0	0.0	2245.3	0.0	0.0	0.0	0.0	0.0
1982	321.4	0.0	3.0	1413.4	27.3	17.9	41.6	13.1	56.6
1983	933.5	0.0	8.0	3.2	12.4	53.8	98.4	27.2	109.9
1984	933.5	0.0	8.0	8.7	2.0	54.7	92.4	28.3	114.9
1985	933.5	0.0	8.0	3.4	0.5	55.6	87.7	29.5	120.2
1986	933.5	0.0	8.0	0.2	1.0	55.6	81.5	30.8	122.3
1987	933.5	0.0	8.0	6.1	0.7	55.6	77.7	32.0	125.3
1988	933.5	0.0	8.0	0.6	0.8	55.6	73.2	33.4	123.4
1989	933.5	0.0	8.0	1.0	0.8	55.6	68.8	34.8	119.9
1990	933.5	0.0	8.0	22.3	0.7	55.8	64.1	36.4	116.4
1991	933.5	0.0	8.0	18.4	0.7	56.0	60.0	37.5	111.8
1992	933.5	0.0	8.0	17.8	1.5	56.2	56.5	39.6	109.7
1993	933.5	0.0	8.0	14.1	1.5	56.2	53.2	41.3	106.6
1994	933.5	0.0	8.0	0.0	1.3	56.2	50.1	43.1	104.4
1995	933.5	0.0	8.0	0.0	1.5	56.2	46.3	45.1	101.6
1996	933.5	0.0	8.0	0.0	0.8	56.2	42.5	47.1	98.0
1997	933.5	0.0	8.0	0.0	0.8	56.2	39.4	49.3	94.6
1998	933.5	0.0	8.0	0.0	1.0	56.2	35.6	51.4	92.5
1999	933.5	0.0	8.0	0.0	1.1	56.2	32.3	53.8	83.9
2000	933.5	0.0	8.0	0.0	1.0	56.2	29.2	56.4	138.0
2001	933.5	0.0	8.0	0.0	1.1	56.2	25.8	58.9	128.8
2002	933.5	0.0	8.0	0.0	1.0	56.2	22.2	62.0	119.6
2003	933.5	0.0	8.0	0.0	0.0	56.2	18.7	64.9	110.3
2004	933.5	0.0	8.0	0.0	0.0	56.2	15.4	68.0	101.2
2005	933.5	0.0	8.0	0.0	0.0	56.2	12.1	71.4	92.4
2006	933.5	0.0	8.0	0.0	0.0	56.2	8.8	75.0	88.4
TOTAL	22725.4	1444.0	195.0	5780.9	59.5	1359.2	1233.5	1130.3	2690.7

Prudhoe Bay Flow Rate: 1.6 Bcfd @ 1138 Btus beginning 7/1/82 (3.1 TBtus to Fairbanks in 1982).
2.4 Bcfd @ 1138 Btus beginning 1/1/83 (12.4 TBtus in 1983 and 18.7 TBtus to Fairbanks in 1984 and thereafter).

Total Energy Input Through 2006: 24,258 Trillion Btus.
Fuel Consumption: 6.32%.

Sources. Columns (3) and (4): See text.
Column (5): See Chapter VIII.
Column (6): Alcan Project Cost of Service, 4/15/77; Arctic Gas Cost of Service Filing, 4/15/77 (for lower U.S. facilities); Canadian portion estimated.
Column (7): Alcan's Answer to Interrogatories, 4/8/77, pp. 362-373.
Columns (8), (9), and (10): Alcan's Project Cost of Service, 4/15/77.

All 1975 dollars except Columns (8), (9), and (10) which are nominal dollars.

All 1975 dollars except columns (8), (9), and (10), which are nominal dollars.
Source: FPC Recommendation to the President: Alaska Natural Gas Transportation Systems, May 1977.

EXHIBIT 5
Cost Summary: Net National Economic Benefit (billions of 1975 dollars)

	Arctic		El Paso		Alcan	
	10%	6%	10%	6%	10%	6%
Field Gathering & Conditioning	0.961	1.124	0.961	1.124	1.057	1.192
Field O & M	0.041	0.071	0.041	0.071	0.044	0.075
Transportation Facilities	3.682	4.329	3.653	4.304	3.820	4.478
Working Capital	0.015	0.019	0.034	0.043	0.028	0.036
System O & M	0.304	0.527	0.883	1.545	0.306	0.520
U.S. Other Taxes	0.090	0.139	0.238	0.373	0.198	0.292
Canadian Income Taxes	0.355	0.648	—	—	0.309	0.484
Canadian Other Taxes	0.031	0.053	—	—	0.094	0.157
Total Costs	5.479	6.910	5.808	7.461	5.856	7.235

Notes: All dollar flows assumed to be mid year.
Present value at 1/1/77.
All taxes were deflated by 5% to 1975 before discounting.

Source: *FPC Recommendation to the President—Alaska Natural Gas Transportation Systems,* May 1, 1977.

The yearly benefit was taken by the FPC to be the sum of the revenue obtained from the sale of the transported natural gas plus the *additional* consumer surplus[1] which accrues to U.S. buyers due to the lower natural gas prices. That is, if the natural gas price/demand relationship is as shown below, the rectangle *A* is revenue lost to the supplier due to the decrease in price, but is surplus gained by the consumer for a net change of zero; the triangle *B* is a gain

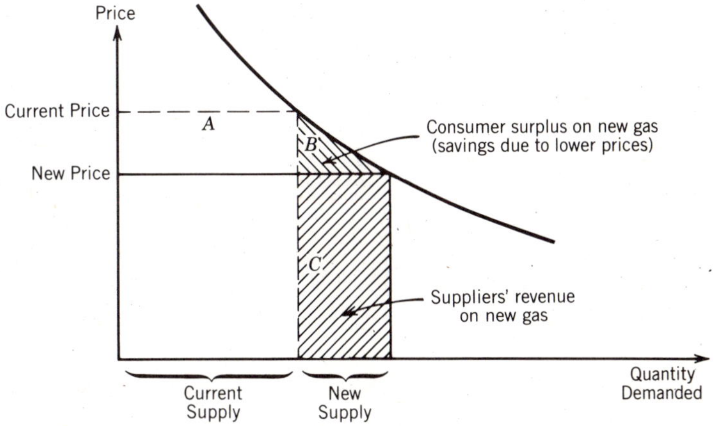

[1]See the Appendix for a brief discussion of the concept of consumer surplus.

in consumer surplus; and rectangle C represents additional revenue to the supplier. Thus the net change is the shaded area $B + C$.

The development of the appropriate price-demand relationships was a difficult task for the FPC and required complex estimation procedures. Not only does this relationship change over time, but it varies drastically from region to region due to differences in dependence on oil, coal, and so on. For example, New England is less dependent on natural gas but more dependent on oil than Midwest regions. The FPC divided the U.S. into 9 demand regions, as shown in Exhibit 6, and for *each region*, developed a price-demand curve for the years 1980–2005. From these curves, the gross benefit can be calculated as a function of the amount delivered to that region in the given year. As an example, Exhibit 7 gives the nine curves for 1980.[2]

A few other issues were raised during the course of the study and suggested for inclusion on the benefit side of NNEB. The first of these was additional employment provided by the project. The El Paso company argued that investment expenditures create jobs and income which multiply through

EXHIBIT 6
The FPC-defined Natural Gas Demand Regions

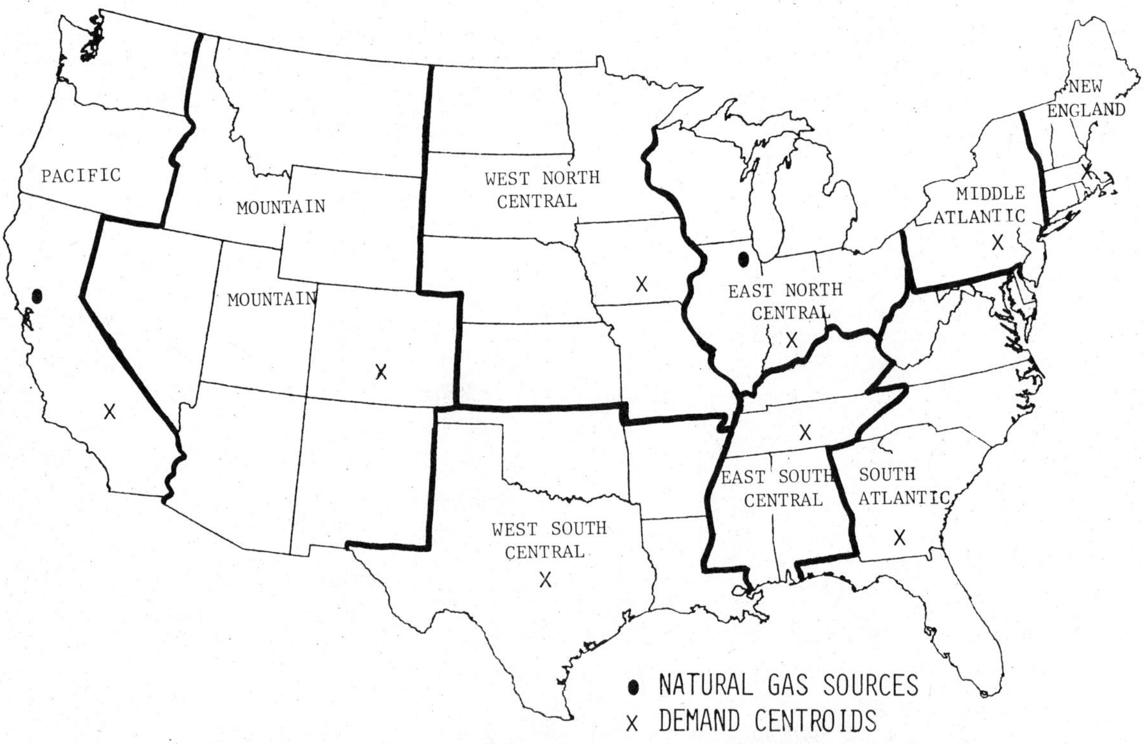

[2]It can be demonstrated that these curves will always have the concave shape exemplified in Exhibit 7.

EXHIBIT 7
Regional Gross Benefits for Natural Gas (1980)

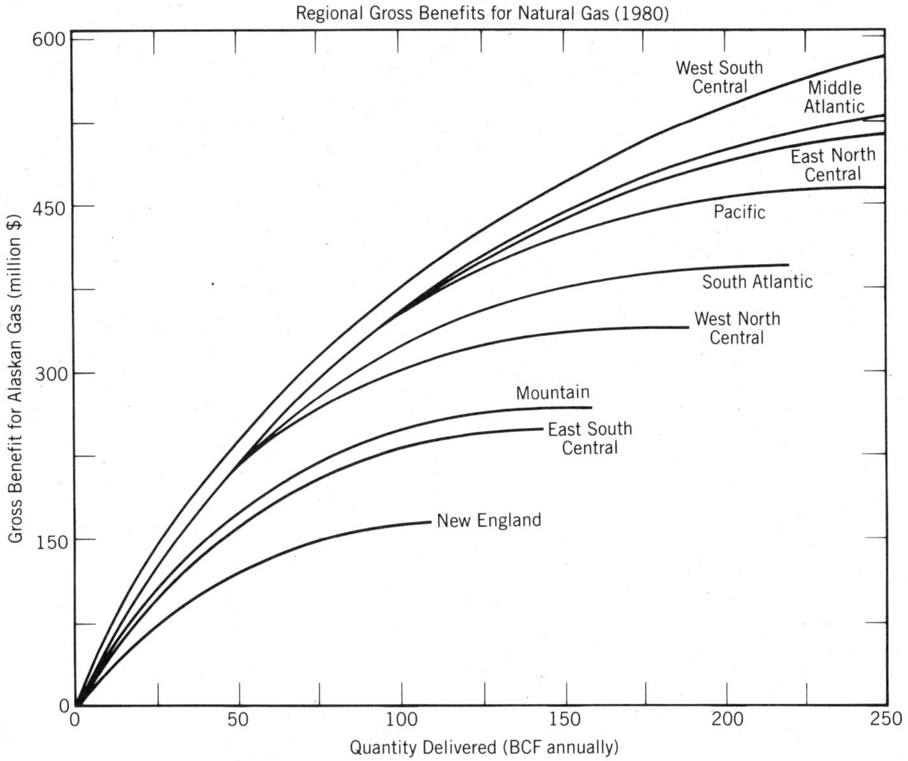

Source: Derived from FPC 1980 demand-curves (assuming the availability of $12/barrel oil).

the economy. Since El Paso planned to spend more in the U.S. than did the other applicants, they contended that this "multiple-employment" benefit would be greater for their proposed system. Others opposed the inclusion of this benefit, stating that other private capital expenditures, or monetary or fiscal stimulation, could provide the same effect as El Paso's expenditures. The FPC analysts concluded, however:

> *While this argument is theoretically correct, the government may not apply a comparable stimulus, in spite of high projected unemployment. Thus we find that some weight should be given to this factor, and the El Paso project credited with a small, subjective advantage in this regard.*

Arctic suggested that the maintenance of Canadian gas exports to the U.S. was a benefit. The Arctic proposal included a pipeline connection to the Mackenzie Delta natural gas field in Northern Canada. The Mackenzie Delta gas would be mixed with that from Prudhoe Bay and a comparable amount (.5–.9 BCFD) extracted from the pipeline in Southern Canada.

In the face of growing Canadian domestic needs, Arctic argued, this additional supply would allow the Canadians to continue exporting gas to the U.S. Because this exported gas is presently priced well below the average U.S. cost of domestic gas, this would represent an economic benefit. The FPC, however, rejected this contention on the ground that it was unlikely that the price of this export gas would stay below the average price of the U.S. gas, even if Arctic's system were to make more Canadian gas available for export. The National Energy Board of Canada had already proposed substantial price increases on this export gas.

A third benefit suggested for inclusion was that resulting from reduced requirements for an oil storage program as a result of the additional domestic gas supply. The FPC decided not to quantify and incorporate this factor in their analysis, although they did admit that there were benefits.

> *The availability of natural gas may reduce our necessary strategic oil reserves, giving us greater bargaining power in world energy markets, and add important energy diversity to the nation's portfolio of energy supplies. These benefits are, however, too elusive to quantify accurately for a cost-benefit analysis. Additionally, each of the proposed systems would reduce dependence on oil approximately equally, so non-inclusion of this consideration penalizes no applicant.*

THE FPC'S ANALYSIS

The initial phase of the study was spent evaluating the engineering designs of both proposals and determining the criteria and methodology to be used. FPC analysts agreed that the major uncertainty to which the NNEB evaluation was most sensitive was the time of start-up. A year or more of delay would cause a significant drop in this measure; hence, significant effort was devoted to verifying the various schedule components, and analyzing uncertainties involved.

The next phase of the analysis was a cross comparison of the three proposals and the subsequent elimination of the El Paso project from consideration, for several reasons. First, a comparison of project costs (see Exhibit 5) was unfavorable to El Paso at the 6% rate which the FPC felt was more appropriate. These costs, coupled with El Paso's lower gas flow rate and higher gas shrinkage, rendered it uncompetitive with respect to NNEB as well as from an efficiency viewpoint. Additionally there were potentially serious environmental safety considerations. Point Conception, California, the selected site for the El Paso Project's marine terminal and regasification plant, is located in an area that has had a history of seismic activity. El Paso's LNG system was also considered to be more prone to interruptions of flow because of the difficult weather conditions for shipping, possible seismic damage to the liquefication/

regasification system, and technical difficulties in the complex liquefication process. Finally, with respect to expansion and flexibility, the El Paso LNG system's cost for additional capacity would increase in rough proportion to the original unit cost. For Arctic and Alcan, however, additional capacity could be provided at lower per-unit costs than for the initial capacity.

The next phase of the evaluation was concerned with the comparison of the Arctic and Alcan proposals. Since these proposals had similar qualitative characteristics, an accurate NNEB analysis was necessary. Particular attention was given to ascertaining the benefits, or value, of the additional gas under each of the two systems.

Since the value of new gas provided would depend on the quantity of gas delivered to each of the nine regions shown in Exhibit 6, FPC analysts formulated a deterministic network optimization model to ascertain the most advantageous allocation of the new Alaskan gas within the U.S., and thus derived the maximum NNEB from each project. Alcan and Arctic had proposed very different allocations of their 2.4 BCFD gas flows between the Eastern and Western legs of the U.S. pipeline. Optimal allocation of the Alaskan gas from Antioch, California, and Dwight, Illinois to the nine U.S. regions would, therefore, result in different benefits for the two proposals due to the differing regional delivery-benefit relationships illustrated in Exhibit 7. The transportation costs to each region was calculated by the FPC by measuring the distance between the source and the demand centroid[3] of the region and assigning a cost of 2¢ per thousand cubic feet per 100 miles.

RESULTS OF THE EVALUATION

The analysis done by the FPC on the proposals highlighted several points of comparison.

1. NNEB analysis revealed the following results:

	Arctic		Alcan	
	10%	6%	10%	6%
Value of Gas	12.601	22.059	13.508	22.890
Total Costs	5.479	6.910	5.856	7.235
NNEB	7.122	15.149	7.652	15.655

Note: All values represent present value on 1/1/77 in billions of 1975 dollars.
Source: FPC Recommendation to the President—Alaska Natural Gas Transportation System, May 1, 1977.

[3]A demand centroid is a point which represents the aggregate demand of the entire region. It is determined by weighting each population center in proportion to its demand and calculating the resultant center of mass.

2. Alcan's proposed plan had a more favorable completion time and Arctic's proposed plan had more risk of construction delays and cost overruns.

3. Each system would have some adverse environmental impacts. These impacts were considered acceptable, given proper precautions or measures. Arctic's pipeline would involve crossing the Arctic National Wildlife Range, and other lands now little used by man. Alcan would generally follow existing utility corridors—a distinct environmental advantage.

4. Arctic's proposal would have the advantage of passing through the Mackenzie Delta region of Northern Canada. This route would facilitate the development of gas reserves in that area by providing a transportation system at a reduced cost to the Canadian government.

5. Should additional gas be found in the vicinity of the transportation system, expansion capability could become very important. Arctic could expand to deliver 3.5 BCFD from Prudhoe Bay, at a small cost. Alcan, however, could only expand to 3.2 BCFD at cost levels comparable to Arctic's.

The final plan will require endorsement of Canada's National Energy Board as well as presidential approval. Additionally, the U.S. and Canada have not yet concluded a treaty which would protect gas owned by one nation as it traveled through the other. Final disposition and recommendation to President Carter was still to be made by the FPC.

APPENDIX **Consumer Surplus**

Consumer surplus is a measure of the benefit to the community of being able to purchase goods at lower prices than each individual would have to pay if he were the only purchaser. For example, Figure A-1 depicts the price-demand curve for sugar in the U.S. during 1975. Hypothetically, if few sugar consumers existed so that demand was only one million tons annually, the price that they would pay would be high—\$1.10 per pound. In actuality 1975 demand was 11 million tons, so consumers paid 40¢ per pound. Those first consumers *benefit* from the existence of the others. The amount of their benefit (their *surplus*) is 70¢ per pound. Their consumer surplus can be calculated as the shaded area in Figure A-1*b*.

This concept can be extended to the total consumer surplus, defined as the area under the price-demand curve above the current price level. In 1975, the consumer surplus for sugar was given by the shaded area in Figure A-2.

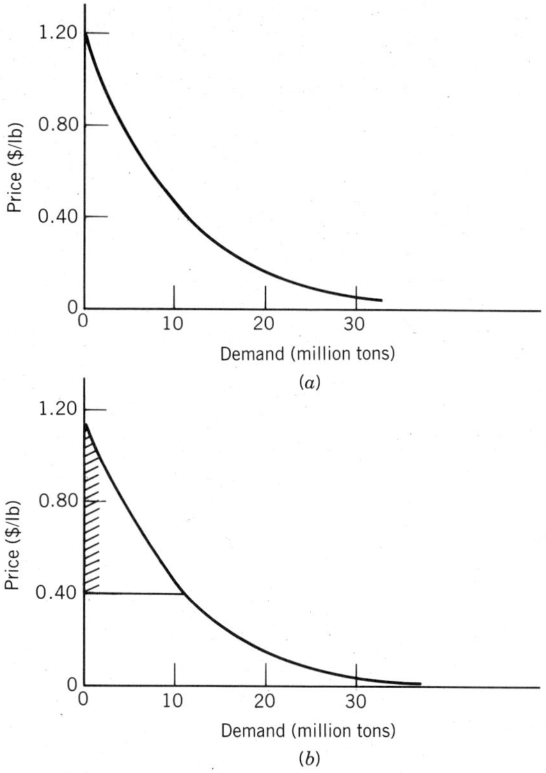

FIGURE A-1. 1975 U.S. price-demand relationship for sugar.

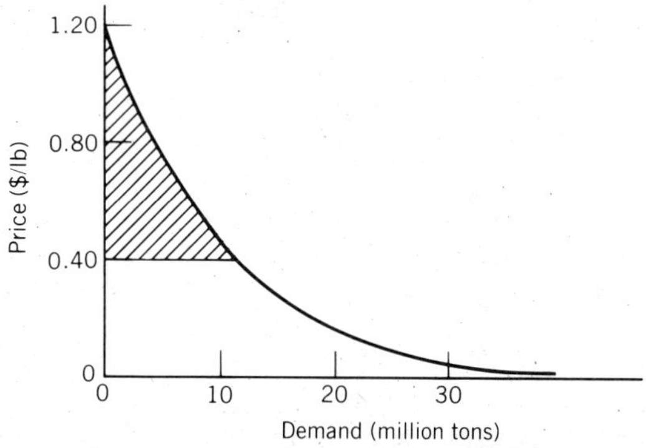

FIGURE A-2. 1975 U.S. consumer surplus for sugar.

BIBLIOGRAPHY

Alaskan Natural Gas Transportation Systems: Economic and Risk Analysis, prepared for the Department of the Interior by The Aerospace Corp., June 1975.

Alaskan Natural Gas Transportation Systems, A Report to Congress pursuant to Public Law 93-153, United States Department of the Interior, December 1975.

Hearings Before the Subcommittee on Energy and Power of the Committee on Interstate and Foreign Commerce, House of Representatives, Serial No. 94-133, May-August 1976.

Hearings Before the Committee on Energy and Natural Resources, United States Senate, Publication No. 95-73, September-October 1977.

Decision and Report to Congress on the Alaska Natural Gas Transportation System, Executive Office of the President, Energy Policy and Planning, Stock No. 060-000-00084-1, September 1977.

Aluminum Products International

In January of 1968 Paul Fraser, head of the recently formed Logistics Division of ALPRO (the acronym for Aluminum Products International), was considering how he should approach the problem of developing an information and control system to be used for both planning and day-to-day smelter operations. Top management had expressed repeatedly its concern about the lack of effective support given to all the medium and long range planning decisions. Andrew Gibson, the company's treasurer, presented the views of most of the high executives in ALPRO when he told Fraser that a planning system should be developed to address the problems of:

1. Evaluating different options for increasing capacity at the existing aluminum smelters of the company.
2. Measuring the economic consequences of installing a new smelter, whose size and location had yet to be determined.
3. Defining the desired quantity and the price to be paid for various purchasing and swapping contracts.[1]
4. Setting general guidelines as to the levels of operation for the company's smelters.
5. Assessing the attractiveness of each of the present market areas and defining a strategy for potential growth in these areas.

At the same time, Fraser was under considerable pressure from Howard Rogers, the smelter operations manager, to give him a system that would, as Howard put it, "allow me to plan and organize, in a rational way, what I'm doing on a day-to-day basis." Since Fraser had just completed the staffing of the logistics group, this assignment was going to be the first major project that his five-man department would be undertaking. Consequently, he was eager to produce fast and useful results for the company, so as to establish the kind of reputation that would allow his group to make significant contributions throughout the organization.

ALPRO AND THE SMELTER DIVISION

In 1968 ALPRO was one of the largest fully integrated aluminum companies in the world, supplying its products to approximately a hundred different countries. In 1968, ALPRO had six smelters, all operating almost at capacity, and demand for ingot aluminum was expected to continue to grow. Of the six smelters, two were located in the Michigan City, Indiana area; two were located in the Austin, Texas area; one was located in Trondheim, Norway; and one was located in Aberdeen, Scotland. In addition to its own facilities, ALPRO also was able to procure aluminum from external sources through swapping contracts.

In view of the present level of operations and the forecasted growth in demand, ALPRO was considering various alternatives for smelter capacity expansion. One of the alternatives consisted in the construction of a new smelter, either in the United States or in a foreign country. A major concern of top management was the decision affecting the location, capacity, and starting construction date of the new smelter. In addition, it was realized that a capacity expansion program was required in the existing facilities. It was imperative to organize this effort in a well balanced form, allocating the resources of the company in those areas which offered the highest potential payoffs.

---

[1] "Swapping" is the term used for the common practice in the aluminum industry of having a competitor produce one of *your* orders for a particular type of aluminum alloy and delivery it directly to *your* end use customer under your label. The reasons for a company wanting to do this may vary, but the most common one is to save on transportation costs of the finished product by selecting a competitor's smelter producing closer to your customer area. These agreements are generally reciprocal in nature and are negotiated prior to the start of a year. The terms of the contract generally state that Company A will "buy" a fixed amount of certain aluminum products from Company B during the coming year and that during that same year Company B will "buy" a certain amount of aluminum products from Company A. This is actually a trade of material since usually no money is exchanged and the major part of the negotiation concerns the amount of material that one company will produce for the other. All warrantees, complaints, etc. will fall back to the company doing the contracting and not the company producing and delivering the aluminum. Such swapping agreements are also a common practice in industries such as petroleum and fertilizer.

Source: This case was prepared by Arnoldo C. Hax and Edward G. Niblock. Harvard Business School Case 9-172-184.

Top management had also expressed great concern about the quality of the present order assignment procedures and the process of production scheduling for the smelter division. As a result, an internal study team had been assembled and had examined in depth the entire process. The conclusions of the team were that the people currently in the logistics management branch were doing a satisfactory job within the operational constraints of the task and the present organizational structure. Top management was aware of the fact that a smooth functioning logistics group should be responsible for transportation and traffic, inventory control and overall production planning, order assignment administration, and logistics system research and development. Top management also knew about the recent use of mathematical model building for problem solving within industries where distribution costs were a major cost factor. They were impressed with the reported results that had been achieved in these industries and were anxious to examine the possible application of these techniques within ALPRO. As a result, they had formed the Logistics Department.

THE LOGISTICS DEPARTMENT

The Logistics Department was established directly under the Vice President in charge of the Smelter Division. This placed the group on an organizational level with Howard Rogers, the Smelter Operations Department Manager. The group was established at this level because of the variety of tasks required and the significant way in which the decisions of this group impacted on corporate operations. The second major reason for the high status of this group was to ensure that the organizational structure encouraged cross functional information flow. Hence, the high interdependent departments of the Smelter Division are organized formally to be on the same hierarchical level. Exhibit 1 provides the higher echelons of ALPRO's organizational chart.

It was felt that in order for the Logistics Department to be successful, it would have to be staffed with individuals who have the variety of capabilities necessary to undertake the tasks required of the group. To accomplish this, top management selected Paul Fraser and gave him a free hand to choose the other members of the group. Fraser was a graduate of the School of Business Administration at the University of Michigan, where he had also obtained a Master's degree in computer science. Since graduating from the business school, he worked in the corporate planning division of ALPRO. To assist him, Fraser selected four individuals who, as a group, possessed the following capabilities:

1. A good background in mathematics, in general, and quantitative model building in particular.
2. An understanding of the aluminum industry and the flexibility that exists in its various process areas.

EXHIBIT 1
Partial ALPRO Organizational Chart

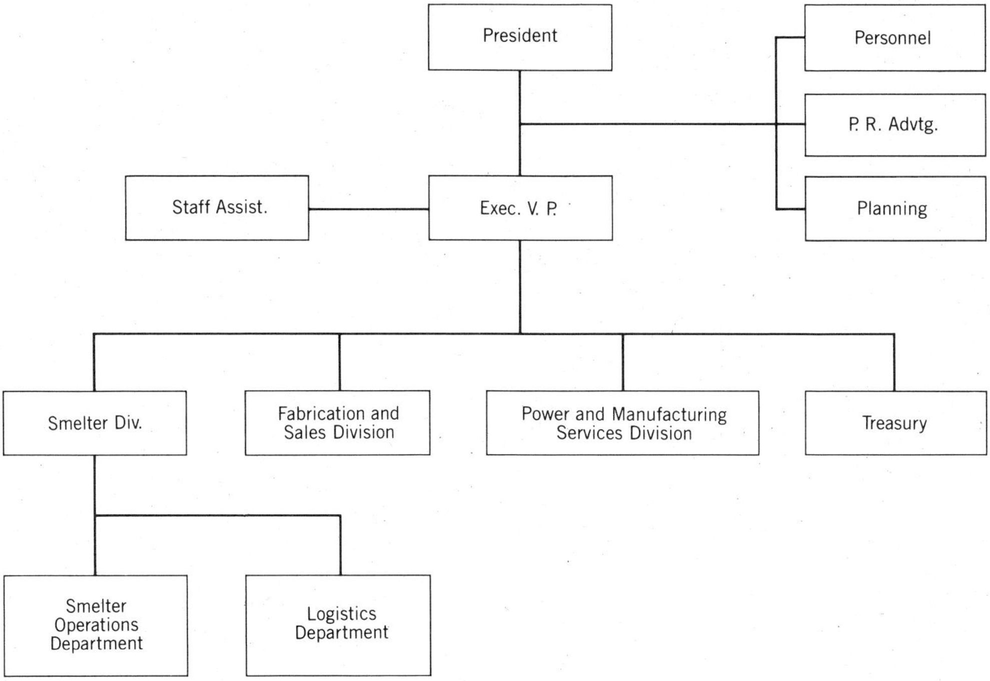

3. An understanding of traffic operations and rate structures.
4. An understanding of production scheduling and inventory trade-off techniques and procedures.
5. An excellent communications capability, both oral and written.

THE NATURE OF THE ALUMINUM INDUSTRY AND SMELTER OPERATIONS

The production of aluminum from raw materials to finished goods is a fairly involved process, but the basic concept of the smelting part of the process is straightforward. (See Exhibit 2.) Calcined alumina is reduced to aluminum metal in electrolytic cells or "pots." The passing of the electric current through the molten electrolyte causes an electrolytic action that reduces the alumina into two materials; molten aluminum and carbon dioxide. Periodically the aluminum is drawn off through a siphon into large crucibles. The crucible then is used to transfer the molten aluminum to a holding furnace where the

EXHIBIT 2
The Aluminum Production Process

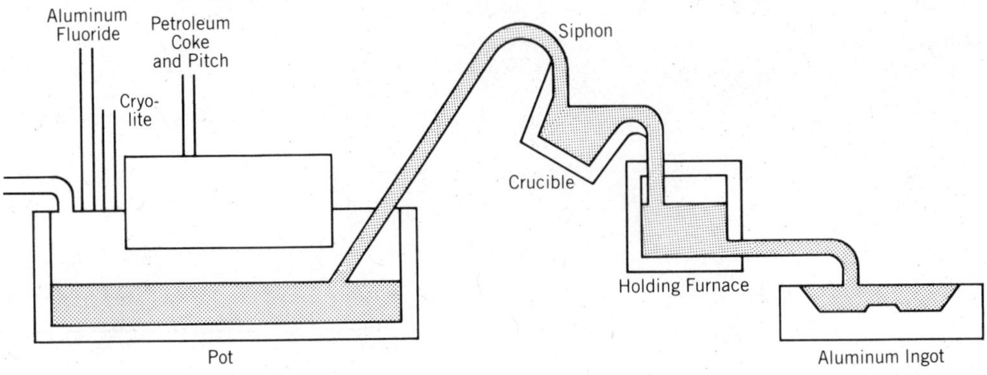

blending or alloying of the metal takes place. Next the aluminum is cast into ingots of various sizes and shapes ranging from 5 lb to 20,000 lb depending on the type of equipment which will be used in processing the metal. This is the major part of the smelting process that is of concern to Paul Fraser and the logistics division.

When considering an aluminum company's capacity for planning purposes, the main limitations can be viewed as the aggregate casting and hot metal capacity at each of the smelters. However, if the company is concerned about what its capacity is for day to day operations, it must consider many more details such as peculiar equipment requirements for a given item within a product type, or a particular order that must be filled only from one smelter primarily because of quality control requirements.

The large amounts of electricity consumed (approximately 8 KWH/lb) is one of the most significant features of the reduction process. In fact, the single most important consideration in determining a smelter's location is the cost of electricity, and thus electricity costs limit the choice of a location for a new smelter to a very small number of different geographic areas.

A second major cost factor is the cost of blending and casting the aluminum. However, the large number of different types of aluminum alloys and casting shapes make it impractical if not impossible to compare blending and casting costs between smelters. In addition, while the absolute magnitude of the cost of blending and casting is substantial, the differential variable blending and casting costs between smelters is not. Hence the question of blending and casting costs for many aluminum companies, including ALPRO may be ignored for order assignment considerations.

Another major relevant cost consideration for decisions concerning either a new smelter location or order allocations among existing smelters is transportation. Transportation cost is the total cost involved for both shipping the raw materials to a smelter and shipping the finished goods to a given customer. Because of the nature of the material (weight and volume) and the wide dispersion of ALPRO'S customers, transportation represents a large portion of the controllable variable cost of aluminum ingot. The problem is simplified for ALPRO in that the company owns its own ship line for delivering raw materials to the smelters. Since variable costs for a ship line are minimal, the company can exclude the cost of transporting raw materials to the smelter since this is essentially a fixed cost.

The final two major cost items to be considered when making logistic decisions are in transit inventory and customs duties. While both of these costs are relatively small when compared with the other variable costs, they are, nonetheless, still large enough to warrant consideration.

THE PLANNING MODEL

Several factors prompted Fraser to work on the planning model before the operating model. The timing was of primary importance since in a short time management would be negotiating the next year's swapping contracts and a new way to approach these decisions was desperately needed. Another reason was that since the company was currently operating close to capacity a decision concerning a new smelter was extremely urgent. A third reason concerned mechanics, in that some of the output from the planning model would be required by the operating model, but the planning model was not dependent on the operating model. The final reason involved internal politics in that the planning model was far less complicated than the operating model and Fraser felt that it was essential, in order to ensure that his department would gain in status, that the group first work on a project he was sure to be successful before taking on more complex projects.

The planning model was to be a tool that would be used by top management for market evaluation, the development of corporate policy capacity facilities planning, and external contract negotiation (swapping). It should deal with information at the aggregate level without getting into the unnecessary detail needed only for operating decisions.

After considerable discussion with the other members of the logistic division, Fraser decided that the most desirable way to approach the problems would be to develop a linear programming model. Fraser wanted to be able to consider nine different metal sources (four existing smelter locations, one new smelter, three swapping sources, and one other possible source). He felt the use of nine sources would build a flexibility into the model that would make it (1) feasible to evaluate precisely the value to be attached to swapping agreements, (2) possible to evaluate the marginal economic worth of the various alternate locations for a new smelter, and (3) possible to allow for an unanticipated source of metal. At the same time, the model would also provide management with sufficient detailed information to answer questions regarding the operation of existing smelters (capacity planning, the level of operation desired, etc.).

Customer areas and swapping destinations were initially broken down into forty different market groups. These forty different locations provided a sufficient segmentation of the market to allow for different demand centers that would result in different costs for transportation, customs duties, or in-transit inventory charges.

Fraser's analysis of the product line of ALPRO revealed that while the variety was extensive, an aggregation into only eight major categories would be sufficient detail for the planning model.

The quality of the results of the model was to a large measure dependent on the quality of the forecast subsystem. The forecast subsystem provided a forecast of one year's demand for each of the eight product types in each of the forty customer areas (one year was selected as the time horizon since it would be long enough to analyze the kinds of questions under consideration but still short enough to make a linear programming model practical). This subsystem was already operational. Fraser planned to analyze the critical nature of the forecast accuracy through the use of sensitivity analysis for he realized that no matter how well the model was able to forecast he would still have to face uncertainties in the projected demand.

There were several additional items that Fraser had to consider when building the model. He knew that he wanted the model to tell him the amount of each product to be produced at each smelter and how much of it would be shipped to each of the customer areas. In addition he wanted to be able to identify in-transit inventory costs for each product and what the company paid for customs duties. This would be useful information to have in addressing the major policy questions facing the company.

Fraser was also aware of the many limitations within which ALPRO had to work. The company was physically limited in both hot metal and casting capacities. Company policy dictated that no order would be refused and hence the planners must fill all orders. It was also company policy to have twice as much spare hot metal capacity in Indiana as was available in Texas. This policy was a result of the fact that the labor force in Texas was relatively static and that it would be very costly to expand or contract the size of the labor force in Texas.

Fraser also realized that pricing for the same types of aluminum alloys varies considerably for ALPRO'S different customers. Thus, it would be virtually impossible to obtain ALPRO'S selling prices.

Exhibit 3 represents some of the notations that Fraser felt he would use in formulating the linear programming model.

EXHIBIT 3
Notation to be Used by Paul Fraser

Smelter and Purchase Source Locations

Small "s" will represent the location of smelters or purchase sources according to the following convention:

$s = 1$	Texas
$s = 2$	Indiana
$s = 3$	Norway
$s = 4$	Scotland
$s = 5$	new smelter
$s = 6,7,8$	swapping
$s = 9$	other

Customer Areas

Small "a" will represent the 40 locations of the various customer areas and swapping recipients.

Product Types

Small "p" will represent the 8 different product types.

Readily Available

The following is the notation to be used for the data for the planning model:

r_{sp} = the reduction cost of product "p" at source "s," in \$/tons

c_{sp} = the casting cost of product "p" at source "s," in \$/ton (while this cost may not be relevant for ALPRO now, inclusion would provide flexibility in the future)

t_{sa} = the transportation cost from source "s" to customer area "a" in \$/ton (the cost/ton is the same for all product types)

o_{sa} = the customs duty charge for the shipments from source "s" to customer area "a" in \$/ton (the cost is the same for all product types)

l_{sa} = the lead time and in-transit time required to ship from source "s" to customer area "a" (the time is the same for all product types)

h_p = inventory holding cost for product "p," in \$/ton/day

d_{ap} = the forecasted demand for product "p" at customer location "a," in tons

$\underline{m_s}, \overline{m_s}$ = the lower and upper bounds respectively, for the hot metal capacity at source "s," in tons

$\underline{e_{sp}}, \overline{e_{sp}}$ = the lower and upper bounds respectively, for the equipment casting capacity at source "s" for product "p," in tons

(continued)

EXHIBIT 3 (Continued)

Possible Decision Variables

The following is the notation to be used for some of the variables to be determined by the planning model:

Q_{sap} = *the quantity of product "p" to be shipped from source "s" to customer area "a," in tons*

M_s = *the total hot metal output at source "s," in tons*

O = *the total overseas customs duty for all products shipped from all sources to all foreign customer areas, in dollars*

I_p = *the total in-transit inventory costs for product "p," in dollars*

E_{sp} = *the total amount of product "p" to be casted at source "s," in tons*

Chapter 6

INTEGER PROGRAMMING

Frequently, planners face circumstances that require models having integer-valued variables. In addition to the obvious situations that necessitate decisions about indivisible products/resources—for example, workers, trucks, machines, and so on—integer variables are essential in the formulation of any problem with "yes-no" variables or "either-or" constraints. In practice, the former situation is often handled by allowing the variable to take on noninteger values and then rounding the results to the appropriate integer. This procedure usually is sufficient unless the numbers to be rounded are small. In that case, serious difficulties can arise unless the variables are restricted to integer values. The latter group of models involved *binary* variables—that is, variables restricted to either 0 or 1. Examples include capital budgeting, sequencing, scheduling, routing, plant location, and a host of others.

Most people would guess that problems with integer-valued variables are easier to solve than similar problems with variables that are not so restricted. One might argue that $x_1 = 0$ or $x_1 = 1$ contains considerably fewer alternatives than the continuum $0 \leq x_1 \leq 1$. In fact, it is far more difficult to optimize medium to large-scale problems involving binary alternatives than problems involving continuous variables. For this reason, constructing new, faster algorithms for solving these problems is currently an important area of research among technical specialists, and several significant improvements have been made recently.

This note describes the basics of *integer programming*—that is, linear programming with the further restriction that some[1] or all of the decision variables must take on only integer values. The emphasis here, as has been our practice, will be on recognition and formulation; but we will discuss briefly the most commonly used methods of solving integer programs.

FORMULATION POSSIBILITIES WITH BINARY VARIABLES

This section discusses several examples of the ways in which 0−1 variables may be useful in formulation. For ease of exposition, consider a specific example, a manufacturer faced with a product-mix selection problem: four products, each of which uses three resources as shown in Table 6-1—steel, electrical power, and labor. There is sufficient demand but only a limited quantity of each of the resources, and the manufacturer must decide which products to produce in order to maximize contribution. The manufacturer can formulate a simple linear program to aid in this decision:

$$\begin{array}{ll} \text{Max} & x_1 + 7x_2 + 8x_3 + 4x_4 \\ \text{Subject to} & x_1 + 10x_2 + 3x_3 \leq 5000 \\ & 2x_1 + 6x_2 + 4x_3 + x_4 \leq 3000 \\ & 2x_1 + 2x_3 + 5x_4 \leq 3000 \end{array}$$

where x_i = the amount of product i to be produced. Solution shows that the maximal profit may be obtained by producing 666 units of product 3 and 333 units of product 4, for a total contribution of $6660. Let us change this example in several ways.

TABLE 6-1
Use of Resources by Four Products

Resource	Products	1	2	3	4	Amount of Resource Available
Steel (lbs.)		1	10	3	0	5000
Labor (hrs.)		2	6	4	1	3000
Power (Kwatts)		2	0	2	5	3000
Contribution		$1	$7	$8	$4	

[1]Problems in which some of the variables are integer-valued and the others are not so restricted are called *mixed integer programs*. Problems in which all variables are restricted to be binary are sometimes called *zero-one programs*.

Fixed Costs

Suppose that product 3 has associated with its manufacture a fixed cost of \$3000 and that products 2 and 4 are produced on a machine which requires set up costing \$1000. In this case, it makes little sense to maximize contribution, since these fixed costs are not incurred if none of these products is made. This fact certainly must affect the manufacturer's decision. In order to include these fixed costs in the model, define two 0–1 variables $y_{2,4}$ and y_3 as follows:

$$y_3 = \begin{cases} 1 \text{ if } x_3 \text{ is nonzero—that is, if any product 3 is made} \\ 0 \text{ if } x_3 \text{ is zero—that is, if no product 3 is made} \end{cases}$$

$$y_{2,4} = \begin{cases} 1 \text{ if either } x_2 \text{ or } x_4 \text{ is nonzero—that is, if either product} \\ \quad 2 \text{ or 4 (or both) is produced in any quantity} \\ 0 \text{ if both } x_2 \text{ and } x_4 \text{ are zero—that is, if neither product} \\ \quad 2 \text{ nor 4 are made} \end{cases}$$

For product 3, then, contribution less the associated fixed costs can be represented as

$$8x_3 - 3000y_3$$

Only if x_3 is nonzero will the \$3000 be subtracted out. Similarly, for products 2 and 4, we can write the contribution less the fixed product costs as

$$7x_2 + 4x_4 - 1000y_{2,4}$$

If any of products 2 or 4 are made, $y_{2,4} = 1$, and the set-up costs will be subtracted. In addition to these changes to the objective function, these new integer variables must be constrained so that they will correspond to x_2, x_3, and x_4 in the intended way. For example, because y_3 appears in the objective function with a negative coefficient, the program will try to keep it at a zero level. However, if we add

$$y_3 \geqslant x_3$$

then y_3 cannot take on a zero value unless x_3 does also. This constraint is inconsistent, however, with $y_3 \leqslant 1$ since x_3 may be in the thousands. If we change it to

$$y_3 \geqslant \frac{x_3}{5000}$$

then y_3 will be zero only when x_3 is, and in addition, y_3 will be less than or equal to 1. The number 5000 was chosen *arbitrarily* to be large enough so that it would be bigger than x_3 (so $x_3/5000$ is smaller than 1). Similarly, we can write

$$y_{2,4} \geqslant (x_2 + x_4)/5000$$

and, after rearranging terms, the full model can be formulated as

$$
\begin{array}{lllllllll}
\text{Max} & x_1 + & 7x_2 + & 8x_3 + & 4x_4 & - 3000y_3 & - 1000y_{2,4} & \\
\text{Subject to}^2 & x_1 + & 10x_2 + & 3x_3 & & & & \leq 5000 \\
& 2x_1 + & 6x_2 + & 4x_3 + & x_4 & & & \leq 3000 \\
& 2x_1 & & + 2x_3 + & 5x_4 & & & \leq 3000 \\
& & & - x_3 & & + 5000y_3 & & \geq 0 \\
& - & x_2 & & - x_4 & & + 5000y_{2,4} & \geq 0 \\
& & & y_3 \text{ and } y_{2,4} = 0 \text{ or } 1 & & & &
\end{array}
$$

With fixed costs taken into account the optimal production plan changes. Product 3 becomes too expensive to produce, and, instead, 400 units of product 2 are produced along with 600 units of product 4.

Batch Sizes

If product 4 can be produced only in a batch containing at least 500—that is, either $x_4 = 0$ or $x_4 \geq 500$—we can use a 0–1 variable to incorporate this "either-or" constraint. We add one binary variable, y_4, and two constraints:

$$x_4 \leq 5000y_4$$

$$x_4 \geq 500y_4$$

where, again, "5000" is chosen arbitrarily as an upper bound for x_4. When $y_4 = 0$, the first constraint guarantees that there is no production of product 4. When $y_4 = 1$, product 4 is constrained to be produced in quantities greater than 500, the required lot size. No change is necessary in the objective function.

In this example, the lot size requirement causes no product 4 to be made ($y_4^* = 0$). Instead, manufacture of product 3 is increased to 750 units with a resultant contribution of \$6000. In the next section, we extend this idea.

Product Line Policies

Suppose the Marketing Department specifies that a minimum of 100 units of at least three of the four products be produced in order to guarantee a reasonably complete product line. In principle, we would like to add the following four constraints in such a way that only three need be satisfied:

$$x_1 \geq 100$$

$$x_2 \geq 100$$

$$x_3 \geq 100$$

$$x_4 \geq 100$$

If we define

$$y_i = \begin{cases} 1 \text{ if } x_i \geqslant 100 \\ 0 \text{ otherwise} \end{cases}$$

then adding to our model

$$x_1 \geqslant 100y_1$$

$$x_2 \geqslant 100y_2$$

$$x_3 \geqslant 100y_3$$

$$x_4 \geqslant 100y_4$$

in addition to

$$y_1 + y_2 + y_3 + y_4 \geqslant 3$$

accomplishes this goal. The solution of this model specifies the manufacture of products 1, 3, and 4; product 1 at its minimum level of 100 units; products 3 and 4 at levels of 622 units and 311 units, respectively.

Economies of Scale

Often, total cost decreases as production volume increases. Suppose that product 3's costs exhibit this characteristic as shown in Figure 6-1. This is a nonlinear cost function and no linear (or linear integer) program can deal with it exactly. However, 0−1 variables make available a good approximation via three linear segments (dashed lines in Figure 6-2). Let

$x_{3,1}$ = number of product 3s, if less than 200 are to be produced

$x_{3,2}$ = number of product 3s, if between 200 and 500 are to be produced

$x_{3,3}$ = number of product 3s, if more than 500 are to be produced

Note that, as defined, the three regions are mutually exclusive, so that at most one of the above variables may be nonzero. For example, if the program specifies that 400 product 3s are to be produced,

$$x_{3,1} = 0, \quad x_{3,2} = 400, \quad x_{3,3} = 0$$

[2]We will typically indicate, in formulating a program, which, if any, variables are restricted to be integer (binary). This is a message to the analyst that this program must be solved with a special-purpose algorithm equipped to handle integer restrictions.

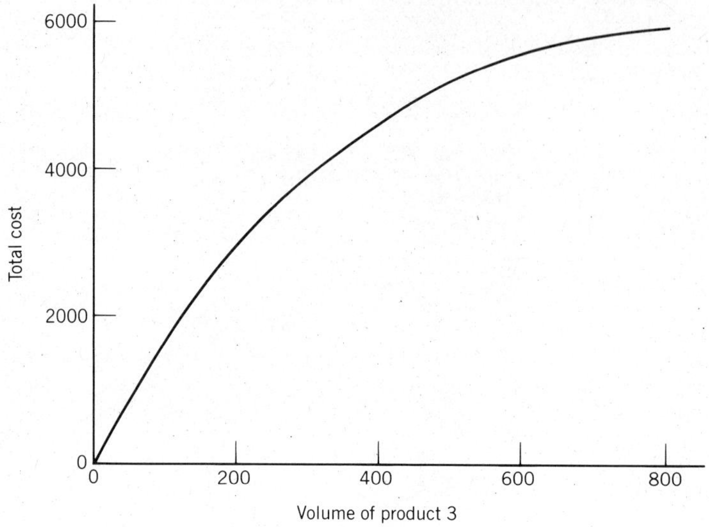

FIGURE 6-1. A nonlinear function.

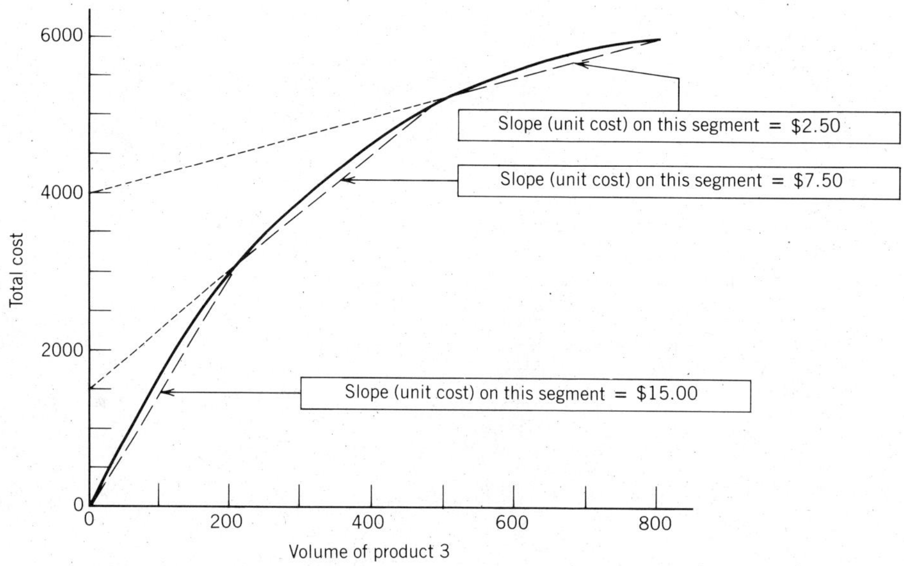

FIGURE 6-2. A piecewise linear approximation.

From Figure 6-2, note that if 400 units of product 3 are to be produced, then we might as well think of the cost as a per unit cost of \$7.50 plus a fixed cost of \$1500 (an extension of dashed line in Figure 6-2 toward the vertical axis intercepts the cost axis at \$1500). Similarly, if 600 are to be produced, the cost is as if the variable cost were \$2.50 and the fixed cost were \$4000. If less than 200 are produced, there is no fixed cost, only a variable cost of \$15 per unit. We can represent the total cost as

$\$15.00\ x_{3,1}$	if production is less than 200 units
$\$\ 7.50\ x_{3,2} + \$1500$	if production is between 200 and 500 units
$\$\ 2.50\ x_{3,3} + \$4000$	if production is greater than 500 units

To implement this idea, define three 0–1 variables y_1, y_2, y_3 to accompany $x_{3,1}$, $x_{3,2}$, and $x_{3,3}$:

$$y_1 = \begin{cases} 1 \text{ if less than 200 units of product 3 are made} \\ 0 \text{ otherwise} \end{cases}$$

$$y_2 = \begin{cases} 1 \text{ if between 200 and 500 units of product 3 are made} \\ 0 \text{ otherwise} \end{cases}$$

$$y_3 = \begin{cases} 1 \text{ if greater than 500 units of product 3 are made} \\ 0 \text{ otherwise} \end{cases}$$

If the revenue per unit of product 3 is \$15, in the objective function, x_3 may now be replaced by

$$0x_{3,1} + 7.5x_{3,2} + 12.5x_{3,3} - 0y_1 - 1500y_2 - 4000y_3$$

Whenever x_3 appears in a constraint inequality, it must be replaced by $(x_{3,1} + x_{3,2} + x_{3,3})$, and the following restrictions must be added:

$$x_{3,1} \leqslant 200$$

$$x_{3,2} \leqslant 500$$

In addition, the integer variables must be constrained:

$$y_1 + y_2 + y_3 = 1$$

$$y_1 \geqslant x_{3,1}/200$$

$$y_2 \geqslant x_{3,2}/500$$

$$y_3 \geqslant x_{3,3}/5000$$

where 5000 is, as before, an upper bound on $x_{3,3}$ to guarantee that y_3 will not be forced to be greater than 1. The first constraint specifies that the value of x_3 must be in at most one of the three regions, and the three others are as in the fixed cost example—that is, no nonzero level of production may be assigned to $x_{3,2}$ or $x_{3,3}$ without incurring the relevant fixed cost. After rearrangement, the complete model can be formulated below:

$$
\begin{array}{llllllllll}
\text{Max} & x_1 + 7x_2 + & & + 7.5x_{3,2} + & 12.5x_{3,3} + 4x_4 & - 1500y_2 & - 4000y_3 & & \\
\text{Subject to} & x_1 + 10x_2 + & 3x_{3,1} + & 3x_{3,2} + & 3x_{3,3} & & & \leq & 5000 \\
 & 2x_1 + 6x_2 + & 4x_{3,1} + & 4x_{3,2} + & 4x_{3,3} + x_4 & & & \leq & 3000 \\
 & 2x_1 & 2x_{3,1} + & 2x_{3,2} + & 2x_{3,3} + 5x_4 & & & \leq & 3000 \\
 & & & x_{3,2} & & & & \leq & 200 \\
 & & & & x_{3,3} & & & \leq & 500 \\
 & & & & y_1 + & y_2 + & y_3 & = & 1 \\
 & & - x_{3,1} & & + 200y_1 & & & \geq & 0 \\
 & & & - x_{3,2} & & + 500y_2 & & \geq & 0 \\
 & & & - x_{3,3} & & & + 5000y_3 & \geq & 0 \\
 & & & & y_1, y_2, y_3 = 0 \text{ or } 1 & & & &
\end{array}
$$

The optimal production policy is as before, so that $x^*_{3,1} = x^*_{3,2} = 0$ and $x^*_{3,3} = 666$ with a resultant contribution of \$5657.

ARCHETYPE INTEGER PROGRAMMING FORMULATIONS

This section contains several examples of well-known integer programming models that have received much attention.

Capital Budgeting

Often, multiperiod strategic planning models involve choosing among several investment alternatives. In most cases, these are "go/no go" situations. Either \$10,000 is spent to buy a new piece of equipment or it isn't.[3] There is no spending \$5,000 to buy half of a machine. Suppose a planner has five possible expenditures under consideration as cost-saving projects, and has calculated the Net Present Value of each at the company's hurdle rate. A budget of \$100,000 has been allocated for the cost reduction program. Given below are the necessary figures.

Projects	NPV	Cost	NPV/$ Invested
1	$ 3,000	$10,000	.30
2	$18,000	$50,000	.36
3	$10,000	$40,000	.25
4	$23,000	$60,000	.38
5	$16,000	$50,000	.32

[3]Another common "go/no go" situation is plant location. Either the new plant is built at a location or it isn't. There is no in-between state.

A simple integer program[4] can solve the planner's dilemma: Let

$$y_i = \begin{cases} 1 \text{ if project } i \text{ is undertaken} \\ 0 \text{ if project } i \text{ is not undertaken} \end{cases}$$

$$\text{Max} \quad 3y_1 + 18y_2 + 10y_3 + 23y_4 + 16y_5$$

$$\text{Subject to} \quad 10y_1 + 50y_2 + 40y_3 + 60y_4 + 50y_5 \leq 100$$

This formulation is flexible enough to allow a myriad of other possibilities. Suppose management wants to specify that at least one project be chosen from the first two. We add the constraint:

$$y_1 + y_2 \geq 1$$

Suppose project 2 cannot be undertaken unless project 1 is also, but project 1 can be undertaken alone. We add

$$y_2 \leq y_1$$

This guarantees that the only way y_2 can be set to 1 is that y_1 be 1 also. With sufficient ingenuity, any logical constraint may be formulated with 0−1 variables.

Set-covering Problem

Consider a community that has decided to allocate funds to build long needed new firehouses. There are several possible locations (numbered in Figure 6-3), and, in each, the firehouse will be able to "cover" a different area.[5] The community has been divided into seven tracts and all seven must be serviced by the two new fire stations. Table 6-2 shows, for each location, which tracts would be covered by a firehouse at that location. Each location has associated with it a known construction and maintenance expense. Where should the firehouses be built so as to minimize cost and, at the same time, provide adequate coverage for the community?

[4]An integer program with a single constraint all of whose variables are 0−1 is known as a *Knapsack Problem*. This name refers to the problem of a hiker who must choose among many items to decide which items will go into his knapsack. If each item is subjectively assigned some value, and a weight limit is imposed, the problem of determining the contents of the maximal value pack can be posed as above. An interesting feature of these problems is that a cost-benefit analysis is not effective. In our example, the planner, peeling projects off the top (in order of decreasing NPV/$ invested) will choose project 4 and then project 1 (project 2 and project 5 will have been eliminated due to the budget constraint) for a total NPV of $26,000. The optimal solution chooses projects 2 and 5 for a total NPV of $34,000.

[5]"Cover," here, means "service within some specified time limit."

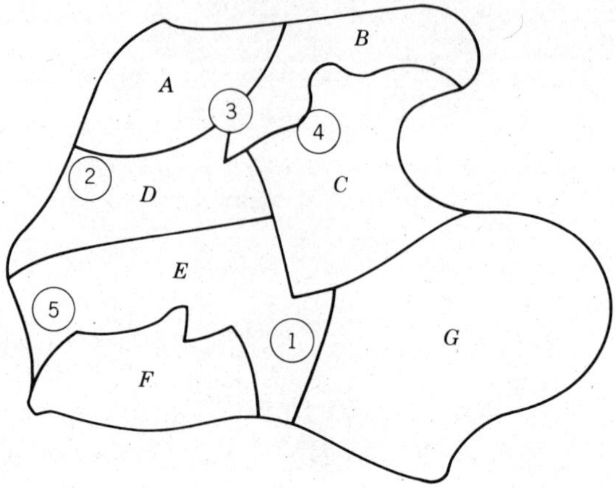

FIGURE 6-3. Possible locations.

TABLE 6-2
Accessible Locations

	Tract							
Location	*A*	*B*	*C*	*D*	*E*	*F*	*G*	Present Cost ($M)
1.			√	√	√	√	√	3.1
2.	√			√	√			2.3
3.	√	√	√	√				3.7
4.		√	√				√	2.6
5.				√	√	√		1.8

Let

$$y_j = \begin{cases} 1 \text{ if a firehouse is to be built at location } j \\ 0 \text{ otherwise} \end{cases}$$

Tract *A* will be covered, then, as long as

$$y_2 + y_3 \geqslant 1$$

Tract *D* will be covered as long as

$$y_1 + y_2 + y_3 + y_5 \geqslant 1$$

and a similar inequality may be written for each of the other tracts. Subject to a constraint of this form for each of the seven tracts, the city planner will minimize the total (present) cost,

$$3.1y_1 + 2.3y_2 + 3.7y_3 + 2.6y_4 + 1.8y_5$$

and the community will discover what funds need be allocated.

The general model is

$$\text{Min} \quad \sum_{j=1}^{n} c_j y_j$$

$$\text{Subject to} \quad \sum_{j=1}^{n} a_{ij} y_j \geq 1, \quad i = 1, 2, \ldots, m$$

$$y_j = 0 \text{ or } 1, \quad j = 1, 2, \ldots, n$$

where the a_{ij} are either 0 or 1. Aptly for the above example called the *set covering problem*, this model is applicable to a wide variety of situations: airline crew scheduling, choosing delivery routes, political redistricting, assembly line balancing, and resource-task matching, to name a few.

Facility Location Models

Perhaps the most successful application of integer programming is facility (e.g., plant, warehouse) location. We have previously seen how the classical transportation problem,

$$\text{Min} \quad \sum_{i,j} C_{ij} X_{ij}$$

$$\text{Such that} \quad \sum_{j} X_{ij} = S_i \quad \text{for } i = 1, 2, \ldots, m$$

$$\sum_{i} X_{ij} = D_j \quad \text{for } j = 1, 2, \ldots, n$$

may be used to allocate product[6] shipped from m supply points to n demand points. X_{ij} represents the amount of product "flowing" from supply point i to demand point j. This formulation is appropriate when all supply and demand locations have been previously established, but often our concern is not only to select an optimal routing plan but also to determine the optimal location of

[6]Assume, for simplicity, a single product. The formulations to follow may be easily expanded to deal with a multiple product situation.

supply facilities (e.g., plants) or demand points (e.g., warehouses) from among a myriad of possibilities.

For example, if the supply points, i, above represent plants (with the demand points, j, representing regions of customer demand), we might want to choose the best m or fewer locations for those plants out of M previously determined possibilities. This problem could be formulated using M 0–1 variables, y_i, defined as

$$y_i = \begin{cases} 1 \text{ if a plant is located at point } i \\ 0 \text{ otherwise} \end{cases}$$

The resulting integer program:

$$\text{Min} \quad \sum_{i,j} C_{ij} X_{ij}$$

$$\text{Such that} \quad \sum_{j} X_{ij} \leq S_i y_i \quad \text{for } i = 1, 2, \ldots, M$$

$$\sum_{i} X_{ij} = D_j \qquad \text{for } j = 1, 2, \ldots, n$$

$$\sum_{i} y_i \leqq m$$

$$y_1, y_2, y_3, \ldots, y_M = 0 \text{ or } 1$$

The first set of constraints guarantees that if no plant is to be located at a given point I—for example, $y_I = 0$—then all X_{Ij} will be zero also; that is, no product will flow from that location. If, on the other hand, $y_I = 1$, denoting a plant at location I, then

$$\sum_{j} X_{Ij} \leq S_I$$

That is, the sum of all flows emanating from location I will be constrained to be less than or equal to the allowable supply at I. The last constraint guarantees that we select no more than m locations.

Perhaps a more realistic formulation, rather than explicitly restricting the number of plants,[7] would recognize fixed as well as variable costs at each location, fixed costs that would only be incurred if a plant was located at that point. For example, let f_i, $i = 1, 2, \ldots, M$, denote the fixed (capital and/or

[7]In the previous formulation, we had to restrict the number of plants or else the model would have minimized total variable costs by "locating a plant" at *every* location. This "solution" is caused by the fact that that model omitted the fixed costs that would naturally cause us to restrict the number of plants.

annual fixed operating) cost of locating and running a plant at location i. The model becomes

$$\text{Min} \quad \sum_{i,j} C_{ij}X_{ij} + \sum_i f_i y_i$$

$$\text{Such that} \quad \sum_j X_{ij} \leq S_i y_i \qquad \text{for } i = 1, 2, \ldots, M$$

$$\sum_i X_{ij} = D_j \qquad \text{for } j = 1, 2, \ldots, n$$

$$y_1, y_2, y_3, \ldots, y_M = 0 \text{ or } 1$$

We can include with this formulation any number of policy constraints that insure desired relationships between plant locations. For example, if locations 7 and 8 are very close, adding the constraint

$$y_7 + y_8 \leq 1$$

insures that we will not locate plants at both points. If, due to our firm's policy on delivery time, we want to insure that at least one plant is located west of the Rockies (let locations 1, 2, 3, and 4 be such), we might add

$$y_1 + y_2 + y_3 + y_4 \geq 1$$

Numerous other combinations are possible.

The above examples have dealt with integer variations to the classical transportation problem. The same concept may be applied to more complex networks. Consider the following example:

Our firm distributes its product from three plants through regional warehouses to customers. (See the following illustration.) Denote the capacity

(supply) at plant i by S_i and customer k's demand by D_k. After much research, the Site Selection Committee has chosen 30 cities in which it seems reasonable to lease warehouse space. In which should we locate warehouses? The following model will give us insight:

$$\text{Min} \quad \sum_{i,j} C_{ij}X_{ij} + \sum_{j,k} \bar{C}_{jk}X_{jk} + \sum_{j} f_j y_j$$

$$\text{Such that} \quad \sum_{j} X_{ij} \leq S_i \qquad \text{for } i = 1, 2, 3$$

$$\sum_{i} X_{ij} - \sum_{k} X_{jk} = 0 \qquad \text{for } j = 1, 2, \ldots, 30$$

$$\sum_{j} X_{jk} = D_k \qquad \text{for } k = 1, 2, \ldots$$

$$L_j y_j \leq \sum_{k} X_{jk} \leq U_j y_j \qquad \text{for } j = 1, 2, \ldots, 30$$

where: C_{ij} = Variable production cost at plant i plus transportation cost from plant i to warehouse j.
$\bar{C}_{jk}$ = Variable warehousing cost at warehouse j plus transportation cost from warehouse j to customer k.
f_j = Annual fixed cost of operating a warehouse at location j.
L_j = Minimum throughput for a warehouse at location j (can be zero).
U_j = Maximum throughput for a warehouse at location j.

$$y_j = \begin{cases} 1 \text{ if a warehouse is opened at location } j \\ 0 \text{ otherwise.} \end{cases}$$

The first three sets of constraints duplicate the standard network constraints of a transshipment problem; the last set guarantees that:

1. The throughput[8] at warehouse j will be zero whenever $y_j = 0$.
2. The throughput at warehouse j, if nonzero, will be between specified upper (U_j) and lower (L_j) limits.

[8]The above formulation represents the throughput at warehouse j as

$$\sum_{k} X_{jk}$$

Alternatively, this throughput could have been written as

$$\sum_{i} X_{ij}$$

Note that all the location models presented above have the characteristic that, if the y variables are all set to 0 or 1, the resulting subproblem is then a network. Due to the computational efficiency of modern-day network algorithms, techniques have been devised for location models which exploit this special structure. These techniques operate by generating a series of network subproblems, each of which can be solved very quickly.

A Special Case

One problem that occurs frequently is assigning workers to jobs, machines to products, and the like. Succinctly, each of n tasks may be performed by any one of n agents. To each agent-task pair is associated a cost. How can agents be assigned to tasks so as to minimize total cost? This problem, aptly called the *assignment problem*, may be modeled as follows: Let

$$x_{ij} = \begin{cases} 1 \text{ if agent } i \text{ is assigned to task } j \\ 0 \text{ otherwise} \end{cases}$$

$$\text{Min} \quad \sum_{i=1}^{n} \sum_{j=1}^{n} c_{ij} x_{ij}$$

$$\text{Subject to} \quad \sum_{i=1}^{n} x_{ij} = 1 \text{ for each } j = 1, 2, \ldots, n$$

$$\sum_{j=1}^{n} x_{ij} = 1 \text{ for each } i = 1, 2, \ldots, n$$

$$x_{ij} = 0 \text{ or } 1$$

The first set of equalities guarantees that every task receives exactly one agent; the second set, that every agent is assigned to exactly one task.

A closer look at this structure bears out an important aspect: If we make a simple transformation—multiplying the second set of constraints by -1—then every variable appears in exactly two constraints, once with a coefficient of $+1$, once with a coefficient of -1!

This is the characteristic that defines a network. In fact, the assignment problem is merely a special case of the transportation problem: each "factory" has a supply of one unit; each "customer" demands one unit. The important point here is that, as we remarked during our discussion of networks, if the assignment problem is solved as a linear program, with the $0-1$ restriction removed, *the solution will be $0-1$ nevertheless*! Thus, the assignment problem is not an "honest" example of an integer program. Similarly, other "seemingly integer" programs can be recognized as networks, and a far more efficient solution technique applied than would otherwise be possible.

INTEGER PROGRAMMING SOLUTION TECHNIQUES

The growing importance of solving combinatorial problems that can only be formulated as integer programs as well as the severe computational difficulties that have plagued integer solution procedures since their development in the early 1960s, has led to keen interest among researchers focused at devising an integer programming algorithm capable of solving problems with many thousands of variables. Unlike the Simplex method for linear programming, there is no single universally accepted algorithm. There are three general approaches:

1. "Smart" enumeration—for example, branch and bound.
2. Cutting plane techniques.
3. Group theoretic techniques.

The latter category, the most recently devised, is based on aspects of mathematical set theory and is sufficiently technically complex to be beyond the scope of what we intend in this text. The second category was hailed as "the answer" when it was first developed by Ralph Gomory in 1963, but has proven to be quite disappointing in practice. The first category, while not as mathematically elegant, has had the best track record in practice—branch and bound, in some form, is the technique used today by most commercially available mathematical programming packages. We will discuss Gomory's cutting plane algorithm briefly, to give the "flavor" of how it operates. We will then delve, somewhat more in detail, into the workings of branch and bound algorithms.

Gomory's Cutting Plane Algorithm

This technique begins, as does branch and bound, with the optimal solution to the continuous version of the problem. If the optimal solution does not satisfy the integer requirements of the problem, new constraints are formulated and added to the problem. These constraints represent lines (in two-space), planes (in three-space) or hyperplanes (in n-space) that effectively "cut away" portions of the feasible solution space while assuring that no feasible integer solution points are eliminated.

The principle behind the use of these cutting planes can be tied back to our discussions of linear programming. In Chapter 3 we concluded that at least one of the optimal solutions to an LP problem occurs at a corner point of the feasible solution space. The Simplex algorithm provides a systematic search of corner points until it identifies the one which is optimal. Since all corner points are not necessarily integer, the cutting plant method attempts to pare down the feasible solution space such that it contains integer corner points. If this is accomplished, then application of the Simplex method will result in the identification of the optimal integer solution.

This approach can best be illustrated graphically with a two-variable problem. Figure 6-4*a* shows the feasible region and the *continuous* optimum for a maximization problem. Aside from the origin, none of the other corner points is all-integer. The cutting plane method reduces the area of feasible solutions to "force" integer corner points. Figure 6-4*b* illustrates the extreme

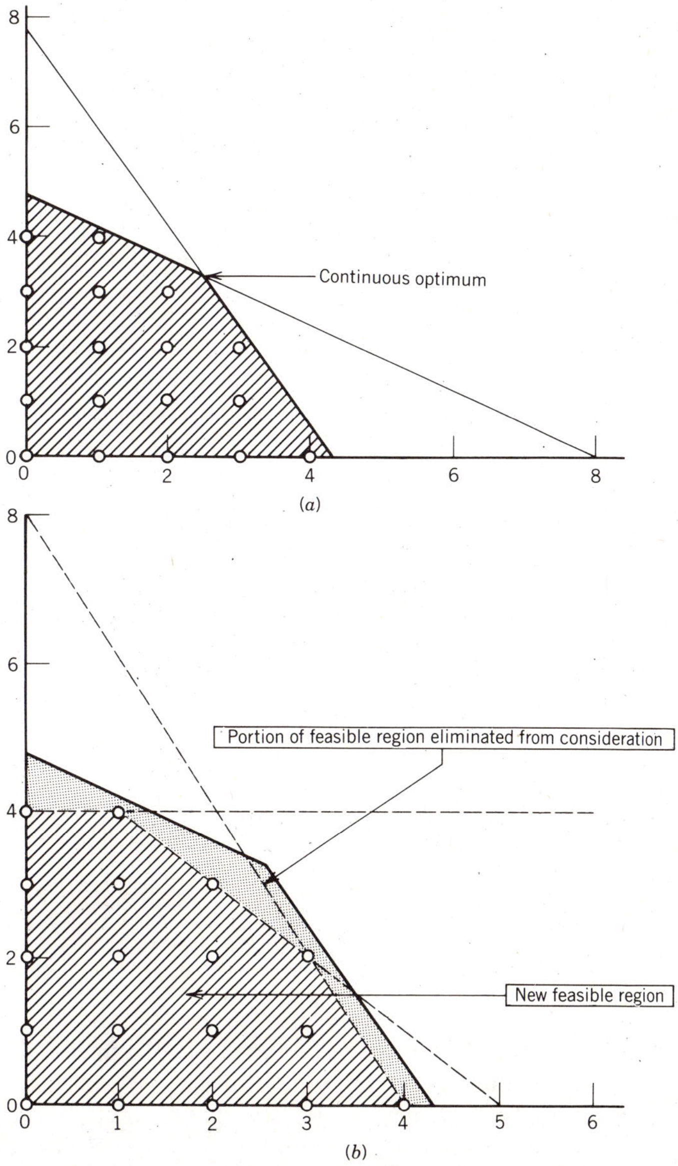

FIGURE 6-4. A cutting plane approach.

case in which all corner points are forced to integer coordinates. Three cutting planes have been added that slice into the solution space, thereby eliminating the shaded portions of the original area of feasible solutions. No feasible integer-valued points have been eliminated. This is an important characteristic of cutting plane algorithms: New constraints are introduced that eliminate continuous feasible points *without* eliminating feasible integral solutions. Application of the Simplex method would now lead directly to the optimal integer solution.

It might not be necessary to add all three cutting planes of Figure 6-4*b*. Depending on the objective function, it might take as little as one cutting plane to reach the optimal integer solution.

We will not present the mechanics of how cutting plane equations are developed; rather, we will offer some feeling for how the procedure works. First, constraints are generated to force the noninteger-valued variables in the current solution to integer values. If a starting solution involves a number of noninteger variables, then generally one variable is selected to provide the basis for developing a new constraint (cutting plane). Once this constraint has been identified, it can be added to the original constraint set and the entire problem can be re-solved by the Simplex method. Or, an advanced simplex procedure, called the Dual Simplex method, can be used to incorporate the constraint in the final tableau of the previous solution and allow the optimal solution of the larger problem to be found without starting from scratch.

Branch and Bound

Like cutting plant techniques, branch and bound adds constraints that eliminate infeasible regions. However, in contrast to cutting plane methods, branch and bound separates a current subproblem into two, adding a different constraint to each.

Branch and bound works especially well with zero-one programs. Let us illustrate.

Consider a simple problem:

$$\begin{array}{ll} \text{Min} & 20y_1 + 64y_2 + 40y_3 + 57y_4 + 10y_5 \\ \text{Subject to} & 2y_1 + 6y_2 + 4y_3 + y_4 + y_5 \geqslant 6 \\ & 2y_1 + 5y_2 + y_3 + \quad + 2y_5 \geqslant 4 \\ & y_1, \ldots, y_5 \ \ 0 \text{ or } 1 \end{array}$$

We might consider solving this problem by means of complete enumeration as pictured in Figure 6-5. While it is theoretically possible to compute the 32 endpoint values (and, simultaneously, check each path for feasibility), this enumeration procedure is clearly unsuitable for larger problems. A decision tree for 20 0−1 variables, for example, would have over a million endpoints.

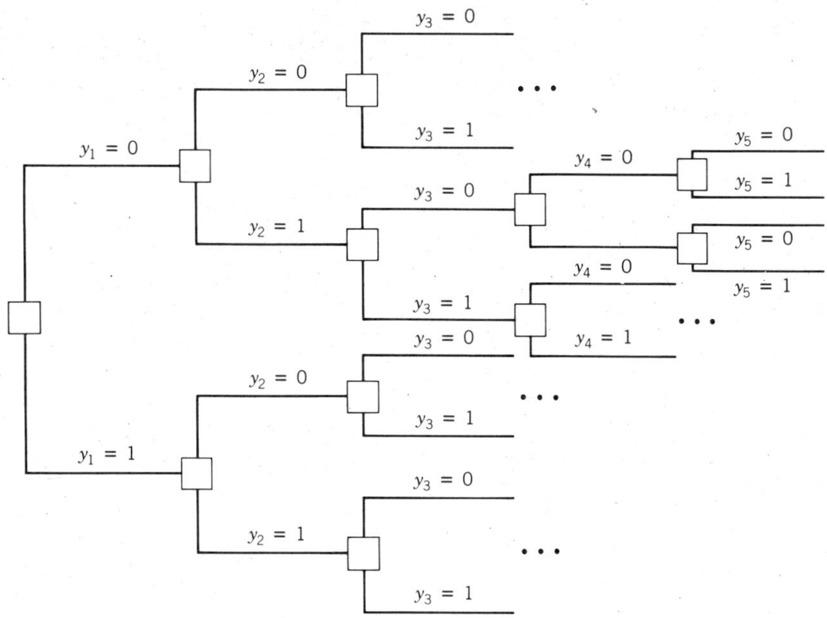

FIGURE 6-5. Complete enumeration.

This methodology does, however, suggest an alternative plan. Suppose for example that we focus on the first fork:

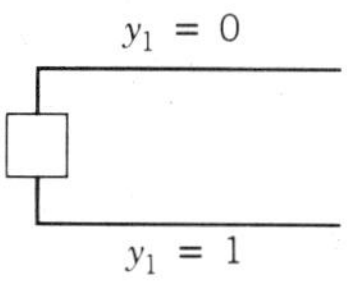

At each branch, rather than enumerate the values of the other four variables, let us split the problem into two problems and solve each linear program obtained after setting y_1 to the specified value (0 or 1) *without requiring that the other variables take on integer values.*[9] If we are very lucky, we might find that the optimal decision variables obtained by solution of the unrestricted problems take on only the values 0 or 1. If both branches yield such solutions, the problem is solved—the optimal solution corresponds to the branch with the lower objective value. In all likelihood, however, we will not be so lucky. We must then continue to branch, solving the appropriate linear program at each step. If at any point, we find a $0-1$ solution, that path may be terminated. Similarly, if an infeasible program is reached, termination is called for: when there exists no feasible solution with some variables unrestricted, further

[9]Solving the linear program without the integer restriction is called a *relaxation*. In this case, the variables must be restricted to be "≤ 1."

restrictions—that is, further branching—will not produce one. In addition, if we reach a point where the objective value for the linear program just solved is larger than the lowest value corresponding to a 0−1 solution, we can terminate that branch; again, further branches impose further restrictions and can, therefore, only make the current solution worse. Continuing with the above example will illustrate these points.

With y_1 set to 0, and the other variables restricted only to be less than or equal to 1, the optimal solution is

$$y_1 = 0, \quad y_2 = .21, \quad y_3 = .93, \quad y_4 = 0, \quad y_5 = 1$$

and the objective function value is equal to 60.86. With y_1 set to 1, the optimal solution is

$$y_1 = 1, \quad y_2 = 0, \quad y_3 = .86, \quad y_4 = 0, \quad y_5 = .57$$

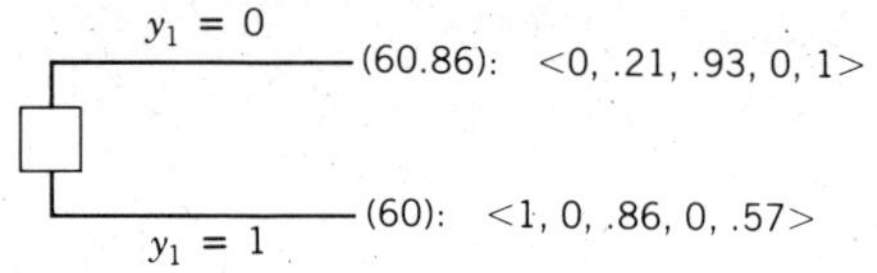

with the objective value equal to 60. Unfortunately, neither branch leads to a 0−1 solution, so both paths must be extended. Let us continue with the top path, branching on[10] y_2:

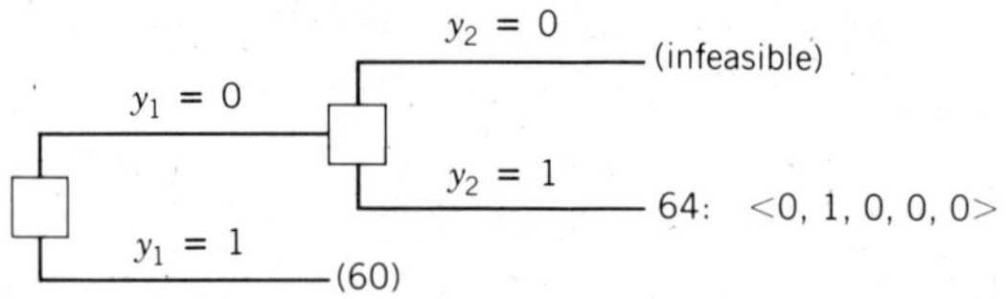

Setting y_2 to 0 as well as y_1 makes it impossible to satisfy the second constraint, so this path need not be pursued further. With y_2 set to 1 (in addition to y_1 set to 0), the linear program yields a solution with *all variables 0 or 1*, and an objective function value of 64. This path may be terminated; *further branching (i.e., restrictions on y_3, y_4, or y_5) cannot lead to a better solution*. We must, however, pursue the lower path. We choose to branch on y_3.

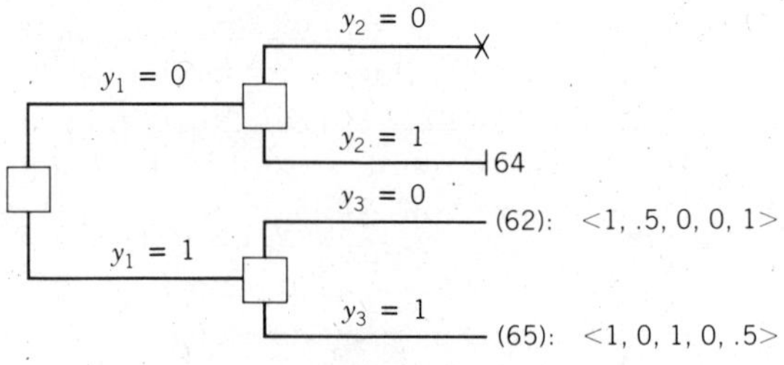

[10]The choice of which variable to branch on next is somewhat arbitrary. In this example, we will choose to branch on the next variable that has taken on a fractional value in the linear program just solved.

Again, neither path leads to a $0-1$ solution; both contain one fractional variable. Note, however, that the objective function value on the lower branch is 65. Further branching will only serve to make this value worse. Currently y_2, y_4, and y_5 are unrestricted; further branching will set restrictions on them and lead to a larger objective. However, as we already have a *bona-fide* integer solution with an objective value of 64, there is no point in pursuing this path: it is *bounded* by the previous integer solution (thus the name *branch and bound*).

The $y_1 = 0$, $y_3 = 0$ path is as yet unbounded, so we must continue, branching on the single fractional variable y_2:

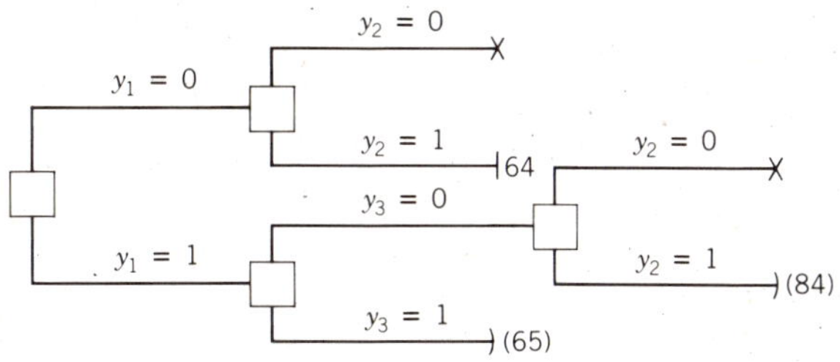

The $y_2 = 0$ branch leads to an infeasible program; the $y_2 = 1$ branch leads to an objective value of 84 and is thus dominated by our previous $0-1$ solution. All paths are now terminated after 7 linear programs have been solved; the optimal $0-1$ solution sets y_2 to 1 and the other variables to 0.

The procedure can easily be extended to variables that take on any nonnegative integer values. If, in any "relaxed" linear program, a variable takes on a fractional value, a two-pronged fork is added to the path at that point. On one prong, that variable is constrained to be less than or equal to the next lower integer; on the other branch, the variable is constrained to be greater than or equal to the next higher integer. For example, let us try to

$$\begin{array}{lllll} \text{Min} & 7y_1 + 3y_2 + 4y_3 & & & \\ \text{Subject to} & y_1 + 2y_2 + 3y_3 - y_4 & & = 8 \\ & 3y_1 + y_2 + y_3 & - y_5 & = 5 \\ & & y_1, \ldots, y_5 \text{ integer} & \end{array}$$

If this problem is solved with the integer restriction removed, the solution is $y_1 = .4$, $y_2 = 3.8$, $y_3 = 0$, $y_4 = 0$, $y_5 = 0$ with an objective value of 14.2. y_1 and y_2 both take on fractional values, so we choose one on which to branch. Say we choose y_2:

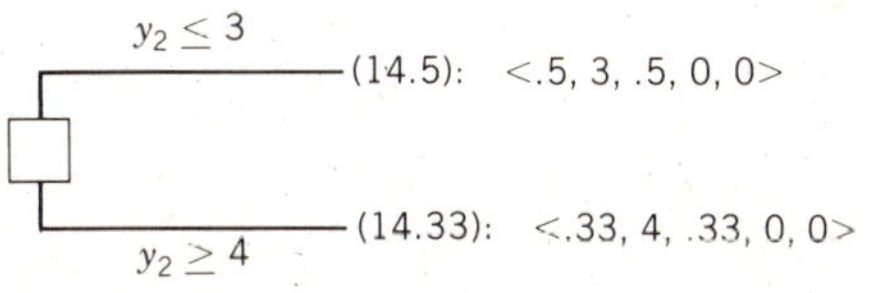

Both solutions involve fractional values, so we must pursue both branches. We will work down from the top:

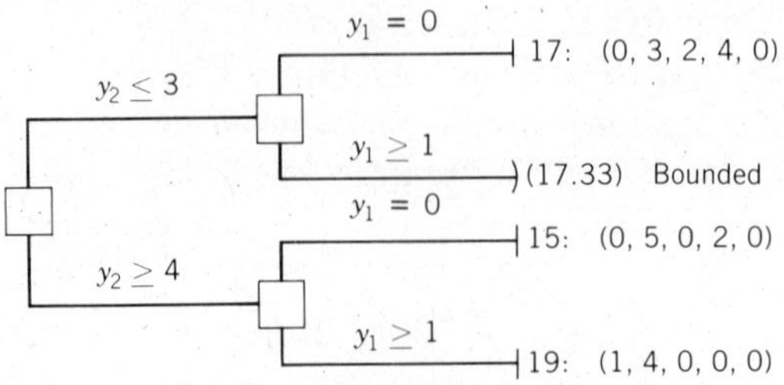

The optimal solution is thus $y_2 = 5$, $y_4 = 2$, all others zero with a minimal objective value of 15.

Either of these methodologies may easily be extended to mixed integer programs. In the case of a mixed integer program, branching is only done on the integer variables.

Computational Difficulties with Integer Programming

Consider again last section's first example. To some extent, we were "lucky" to be able to terminate all paths so quickly. If we had chosen to originate our tree by branching on y_4, and if the succeeding branches had been selected sufficiently badly, we might have concluded with the tree shown in Figure 6-6, involving the solution of 28 linear programs (each node or endpoint corresponds to one). In this tree, the optimal solution is found at the last (feasible) endpoint, and only one path is terminated by bounding. The sequence of bounds—that is, "best 0−1 solutions so far"—is 127,121, 87, 70, and finally 64. What does this example show? Several points emerge:

- The success of the branch and bound procedure is not guaranteed. Theoretically, it could enumerate all possible paths. In practice, however, with real data, this never happens. All successful integer programming solution procedures based on "smart" enumeration are of this type. Be careful, however, not to be led astray by the claim that a given method only enumerates 1/100 of 1% of all possible branches. With only 100 0−1 variables, there are over 10^{30} endpoints, so this technique would have to enumerate 10^{25}. A quick algorithm that could enumerate a billion per second would require over four billion years to solve this.
- The confirmed existence of a "good" bound is crucial. Often the manager will know of some feasible solution (e.g., what the firm is currently doing) that can provide such a bound and serve to greatly shorten computation. Most commercially available codes allow such a bound to be input.

- The order of branching is important. Much research effort has gone into the decision of which variable is to be chosen next. Depth-first? Breadth-first? Most "fractional" variable? Many answers have been proposed and most commercially available systems come equipped with ancillary heuristics that have worked well in practice.

Because solving a single medium-size integer program may require the solution of hundreds of linear problems, it is not surprising that computation for a given integer program will consume far more time than that for linear program of the same size. Nevertheless, programs with up to several hundred integer variables have proved computationally tractable; in fact, oil companies routinely solve tanker scheduling problems involving over a thousand 0−1 variables.

Certain approximation methods are available on some systems that may substantially reduce computation time in the solution of integer programs. For example, several algorithms exploit duality theory to estimate the optimal objective function value as the solution proceeds. Because gains from iteration generally exhibit marginal returns (time spent improving a solution its last 1%

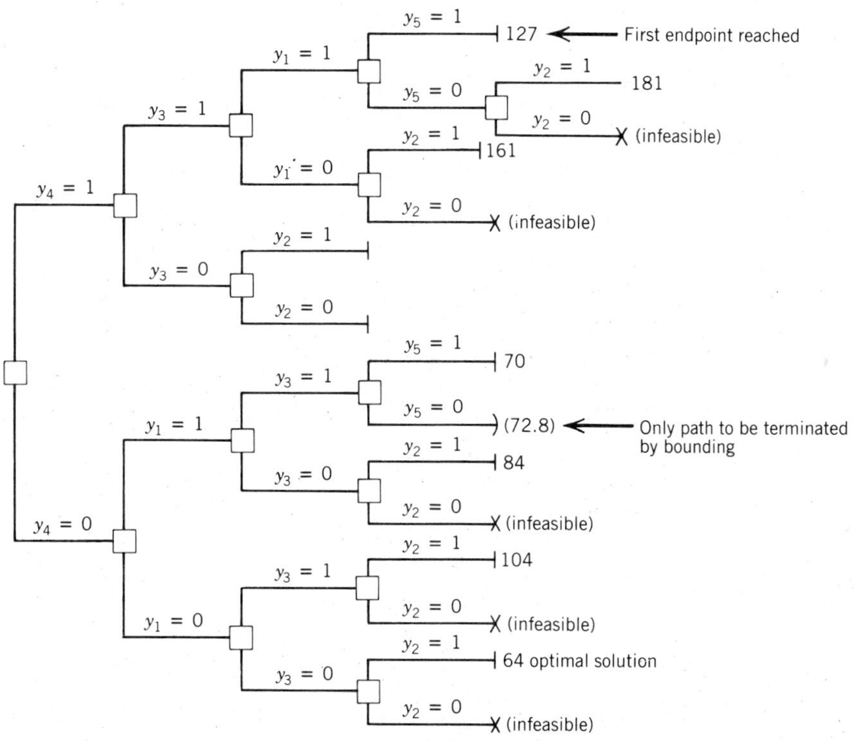

FIGURE 6-6. Complex tree for branch and bound example.

may be greater than the time necessary to get to that point), the user, by requesting an approximate solution (an objective function value within, for example, 1% of the optimum), can often significantly reduce the problem's computation time. Recently, vehicle routing problems involving up to one hundred thousand variables have been solved within a few percent of their optima in a matter of minutes by algorithms that exploit their special structure.

EXERCISES

1. Consider the following integer program.

$$\begin{array}{ll} \text{Max} & x_1 + 5x_2 \\ \text{Subject to} & -4x_1 + 3x_2 \leq 6 \\ & 3x_1 + 2x_2 \leq 18 \\ & x_1, x_2 \geq 0, \text{ integer} \end{array}$$

Graph the feasible region and then apply branch and bound, graphically solving each linear program you encounter. Interpret the branch and bound procedure as well, showing, graphically, how the feasible region changes from step to step.

2. The Board of Directors of a firm that produces industrial power tools has decided that the firm's original East Coast plant, which for some time has been cramped and outdated, must be either modernized or a new plant built. If the new plant is built, a fully automated production line could be installed rather than continue with conventional, less capital intensive methods. The firm has also considered the possibility of investing in a private trucking fleet for delivery of its products. If the private fleet is acquired and the old facility is modernized rather than rebuilt, a new shipping dock would be required.

A somewhat unrelated project, but one that would require a portion of the same capital resources, is the purchase of a local industrial distributor's warehouse to consolidate inventory held in several smaller leased warehouses.

The alternative uses of capital described above are summarized in the following table, along with each project's estimated net present value and the required capital expenditure for each over the next 3 years. The capital budget approved for these projects over the next 3 years is \$1,200,000 in year 1, \$600,000 in year 2, and \$200,000 in year 3.

Formulate a model to help the firm allocate capital to the available alternatives. Solve by branch and bound.

	Expenditure			
Project	Year 1	Year 2	Year 3	Estimated Net Present Value
Modernize Existing Facility	\$400,000	\$150,000	\$ 50,000	\$200,000

(continued)

Project	Expenditure Year 1	Year 2	Year 3	Estimated Net Present Value
Build New Facility	8,000,000	200,000	—	400,000
Automate Line	150,000	50,000	—	110,000
Acquire Fleet	10,000	400,000	100,000	220,000
Build Shipping Dock	90,000	60,000	—	80,000
Buy Warehouse	200,000	100,000	100,000	400,000

3. A manufacturer of baby cereals is test marketing a new product next month and requires 20,000 boxes made to specification. The firm plans to purchase these from a number of small paper processors with whom it has dealt in the past and has requested each of these suppliers to submit a bid. Six suppliers have responded; each has specified the maximum order deliverable by the beginning of next month, and, to guarantee themselves some profit in the event that less than a break-even quantity is ordered, each has specified a setup charge. The bids are shown below:

	Set-up	Variable Cost/100 boxes	Max. Quantity
Supplier 1	\$200	\$2.83	5,000
Supplier 2	\$250	\$2.66	7,000
Supplier 3	\$300	\$1.87	4,000
Supplier 4	\$350	\$1.88	7,000
Supplier 5	\$400	\$1.82	10,000
Supplier 6	\$400	\$1.81	9,000

Formulate a model to determine the order which should be placed with each supplier.

4. The owner of the California Angles, concerned with his team's inability to win the pennant despite his continued efforts to improve the team, has decided to try to sign two more free agents. In preparation, he has started to negotiate with four players. Their career batting averages and his estimate for the bottom line financial remuneration necessary to sign each is given below:

Player	Batting Average	First-year Bonus	Annual Salary	Deferred Compensation
Dave Wonfield	.321	\$600,000	\$600,000	\$1,500,000
Reggie Traction	.285	200,000	800,000	800,000
Pete Nose	.315	100,000	700,000	1,700,000
Carlton Kissed	.275	500,000	500,000	900,000

The Angles owner has determined that, given the state of his current and estimated future financial resources, he can spend no more than \$800,000 in first-year bonuses, no more than \$1.3 million in salaries, and no more than \$3 million in deferred payments.

Formulate a model to determine the players to sign so that he gets the best hitting players subject to his financial constraints.

5. Knighted Parcel Service has deliveries to be made today to five large industrial customers. These deliveries consist of several parcels of various sizes; their total weight is given below:

Customer	Weight of Shipment (lb)
1	1,000
2	2,000
3	3,000
4	5,000
5	8,000

The firm has four delivery vans free to make these deliveries; their capacities and operating costs are given below. In addition, due to distance considerations, a single van cannot deliver to customers 1 and 3 nor to customers 2 and 4. Also, company policy dictates that the shipments to a single customer not be divided among trucks; that is, a customer received its deliveries from one truck. Formulate an integer programming model to determine the least cost allocation of delivery vans that guarantees all deliveries will be made.

Truck	Capacity	Operating Cost/Day
1	2,000	\$48.00
2	6,000	\$52.00
3	8,000	\$55.00
4	11,000	\$67.00

6. Heath Maintenance Organizations (HMOs) to provide prepaid health care are being planning for a large metropolitan area. There are three potential sites and six census tracts to be served by the HMOs. City staff has estimated the number of subscribers from each of the six tracts who would go to a given site if the tract was assigned to the HMO at that site, as indicated by the data below:

	Census Tract					
Site	1	2	3	4	5	6
A	50,000	20,000	30,000	12,000	25,000	35,000
B	14,000	17,000	37,000	45,000	46,000	13,000
C	32,000	45,000	41,000	22,000	31,000	10,000

The cost of building an HMO at site *A* is estimated to be $4.3 million, at site *B*, $4.1 million, and at site *C*, $4.7 million. The budget is $10 million. Also, each new HMO requires a minimum enrollment of 79,000 subscribers.

Formulate a model to help city planners determine where HMOs should be built.

7. An airline is preparing its schedule for counter personnel for the busy Christmas season. The number of people needed during each hour of the day is given below:

Time	Personnel Needed
6 A.M.– 7 A.M.	5
7 A.M.– 8 A.M.	8
8 A.M.– 9 A.M.	8
9 A.M.–10 A.M.	10
10 A.M.–11 A.M.	12
11 A.M.–12 noon	16
12 noon– 1 P.M.	16
1 P.M.– 2 P.M.	14
2 P.M.– 3 P.M.	15
3 P.M.– 4 P.M.	11
4 P.M.– 5 P.M.	13
5 P.M.– 6 P.M.	12
6 P.M.– 7 P.M.	14
7 P.M.– 8 P.M.	13
8 P.M.– 9 P.M.	10
9 P.M.–10 P.M.	8
10 P.M.–11 P.M.	6

Over the years, the company has established 11 shifts, shown on the following page:

Shift	Time Period
1	10 A.M.–6 P.M.
2	6 A.M.–2:30 P.M.
3	10 A.M.–1 P.M. and 4 P.M.– 8 P.M.
4	10 A.M.–1 P.M. and 6 P.M.–10 P.M.
5	9:30 A.M.–5:30 P.M.
6	11 A.M.–7 P.M.
7	12 noon–8 P.M.
8	1 P.M.–9 P.M.
9	4 P.M.–10 P.M.
10	5 P.M.–11 P.M.
11	6 A.M.–9 A.M. and 5 P.M.–9 P.M.

Temporaries can be used for this seasonal schedule, so that split shifts (3, 4, and 11) can be used.

Formulate a model to minimize the number of people needed to meet anticipated demand. Are integer restrictions on the decision variables necessary?

8. Gentle Edison, Inc. plans to build six new nuclear power plants on sites which have been selected. The firm plans to start construction on all plants within a 5-year period, and the director of planning plans to formulate an optimization model to determine how the expansion should proceed subject to a number of company policies and governmental regulations. The model is to be comprised of 0−1 variables y_{ij}, defined as follows:

$$y_{ij} = \begin{cases} 1 \text{ if plant } i \text{ is started in year } j \\ 0 \text{ otherwise} \end{cases}$$

To guarantee that the model requires each plant to be started in one and only one year, the formulation will contain the following constraints:

$$\sum_{j=1}^{5} y_{ij} = 1 \text{ for each } i = 1, 2, \ldots, 6$$

For each of the limitations below, formulate a constraint (or set of constraints) to incorporate that restriction into the integer programming model. Consider each stipulation separately. In other words, each restriction is independent of the previous ones.

a. Plants 1, 2, and 3 must be started no later than year 3.
b. No more than two plants can be started in any year.
c. No more than three plants can be started in the first 2 years, and no more than five plants in the first 4 years.
d. Plant 2 cannot be started before Plant 1, but they can be started in the same year.
e. If Plants 1 and 2 are started within the same year, then no other plants can be started in that year.
f. In the first 2 years, the company starts construction either on Plants 1 and 2 or on Plants 3 and 4.
g. Plant 1 can be started in year 1 if and only if either Plant 2 *or* Plant 3 is started, but not if both 2 *and* 3 are started.

9. An assembly line is comprised of several work stations where one or more jobs may be performed in the process of completing the manufacture of an item. The item typically travels from station to station (on a conveyor belt, for example) and the time it spends in each station is not permitted to exceed the *cycle time* of the assembly line. Consider, as an example, a product whose assembly requires six operations. Certain operations are required before others may be performed, and the table below gives the precedence relations and job durations. For this assembly line, the cycle time is 80 seconds. Formulate an optimization model to schedule the assembly operations such that a minimal number of work stations must be opened.

Job	Immediate Predecessor	Duration (sec)
1	—	40
2	—	49
3	1	16
4	2,3	39
5	1,2	33
6	4	23

***10.** Super Sunday, the once-a-year TV extravaganza focused upon the activities of the NFL Super Bowl, has almost become a tradition. CBS has received the TV contract for this year's family festivities. Producers have identified 12 potential camera locations within the stadium. They have also identified 25 stadium areas which may require camera coverage during the pregame, game, and postgame activities. The accompanying table indicates each camera location and the stadium areas the camera can cover.

Camera Location	Stadium Areas
1	1,3,4,6,7
2	8,4,7,12
3	2,5,9,11,13
4	1,2,18,19,21
5	3,6,10,12,14
6	8,14,15,16,17
7	18,21,24,25
8	2,10,16,23
9	1,6,11
10	20,22,24,25
11	2,4,6,8
12	1,6,12,17

CBS executives are concerned about costs for the production. Consequently, they have set an objective of minimizing the number of camera locations. In seeking this objective they want at least one camera to be available to cover each stadium area. Camera location 9 is the "blimp," and executives have decided that the blimp will be used because of viewer expectation and fascination with the shots from this location. Stadium areas 1 and 2 are locker room locations.

*From F.S. Budnick, R. Mojena, and T.E. Vollman, *Principle of Operations Research for Management*, Richard D. Irwin, Inc., Homewood, Ill., 1977, p. 309.

The viewer interest in football personalities has led the executives to request that at least two camera locations be available to cover each of these areas.

Formulate a model to determine the minimum number of cameras needed for the desired coverage.

***11.** Custom Pilot Chemical Company is a chemical manufacturer that produces batches of specialty chemicals to order. Principal equipment consists of eight interchangeable reactor vessels, five interchangeable distillation columns, four large interchangeable centrifuges, and a network of switchable piping and storage tanks. Customer demand comes in the form of orders for batches of one or more (usually not more than three) specialty chemicals, normally to be delivered simultaneously for further use by the customer. Custom Pilot Chemical fills these orders by means of a sort of pilot plant for each specialty chemical formed by interconnecting the appropriate quantities of equipment. Sometimes a specialty chemical requires processing by more than one of these "pilot" plants, in which case one or more batches of intermediate products may be produced for further processing. There is no shortage of piping and holding tanks, but the expensive equipment (reactors, distillation columns, and centrifuges) is frequently inadequate to meet the demand.

The company's schedules are worked out in calendar weeks, with actual production always taking an integer multiple of weeks. Whenever a new order comes in, the scheduler immediately makes up a precedence tree-chart for it. The precedence tree-chart breaks down the order into "jobs." For example, an order for two specialty chemicals, of which one needs an intermediate stage, would be broken down into three jobs as shown in Figure E-1. Each job requires certain equipment, and each is assigned a preliminary time-in-process, or "duration." The duration can always be predicted with a high degree of accuracy.

Currently, Custom Pilot Chemical has three orders that need to be scheduled as shown in the table below. Orders can be started at the beginning of their arrival week and should be completed by the end of their week due.

Order Number	Job Number	Precedence Relations	Arrival Week	Duration in Weeks	Week Due	Resource Requirements: Reactors	Resource Requirements: Distillation Columns	Resource Requirements: Centrifuges
AK14	1	None	15	4	22	5	3	2
	2	(1)	15	3	22	0	1	1
	3	None	15	3	22	2	0	2
AK15	1	None	16	3	23	1	1	1
	2	None	16	2	23	2	0	0
	3	(1)	16	2	23	2	2	0
AK16	1	None	17	5	23	2	1	1
	2	None	17	1	23	1	3	0

*Based on the Custom Pilot Chemical Co. case, written by Richard F. Meyer and Burton Rothberg, which in turn is based on "Multiproject Scheduling with Limited Resources: A Zero-One Programming Approach," *Management Science* 1969, by A.A. Pritsker, L.J. Watters, and P.M. Wolfe.

For example, AK14 consists of three jobs, where Job 2 cannot be started until Job 1 has been completed. Figure E-1 is an appropriate precedence tree for this order. Generally, the resources required are tied up for the entire duration of a job. Assume that Custom Pilot Chemical is just finishing week 14.

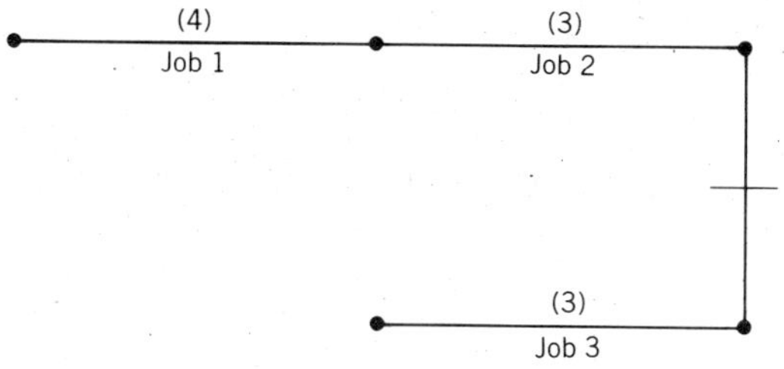

a. Formulate an integer program that minimizes the total completion time for all orders.
b. With a more complicated list of orders, what other objective functions might be appropriate? How would these be formulated as integer programs?

12. Consider the plant location model presented in the text.

a. Consider the situation in which a company is investigating three possible sites for its new New England plant:

Waltham, Mass.	$(i = 1)$
Hartford, Conn.	$(i = 2)$
Concord, N.H.	$(i = 3)$

There are n demand points in New England, each with demand d_j. Write down the resultant model.

b. Rather than assuming a single product, indicate how the model would be modified if there were L different products.
c. Suppose that there exist economies of scale at any of three sites. In particular, let variable production cost be a function of volume as shown in Figure E-2. Formulate the plant location model under these conditions.

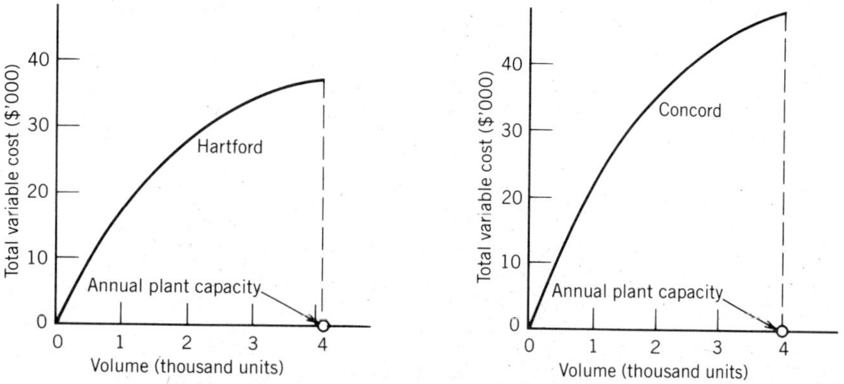

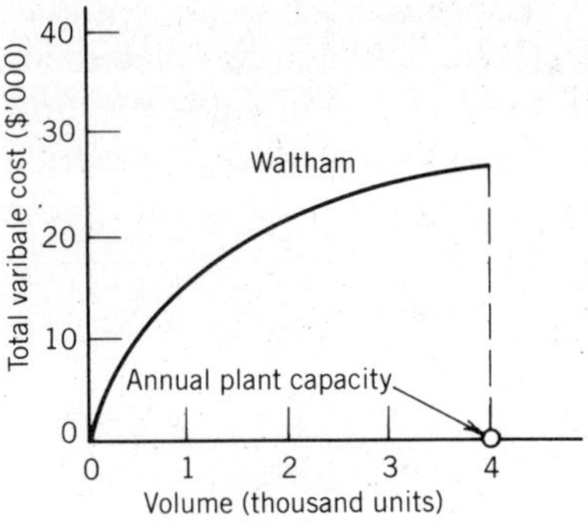

Western Optics, Ltd.

Late in the afternoon of August 17, 1976, Western Optics' production foreman, Bob Race, was wondering how to squeeze in the two special rush orders which had just been received for delivery on or before September 1. Western Optics, located in Calgary, Alberta, is a small producer of custom-ground lenses for use in high quality optical instruments. The firm had been incorporated in 1971, and in its short five-year existence, had earned a reputation for quality and service. Sales had been growing rapidly as Western Optics attracted new customers from the scientific and military communities, and the firm was beginning to strain its capacity limits.

One of the two orders which had been received that afternoon came from the U.S. Department of Defense. That agency had been a steady customer of Western Optics since 1974, and had recently ordered two oval lenses (Specification #1-391-X). Delivery on or before September 1 for that order had been promised earlier in the month. The current order requested up to six additional lenses of the same type, for delivery at the same time.

The second rush order originated from the College Scientific Supply Corporation. They also ordered a lens which had been produced for them in the past, a round lens (Specification #1-901-X). The current order requested as many lenses of that type as could be delivered by September 1. Both orders carried a firm delivery deadline, so Race abandoned any idea he had had of delaying delivery.

During the last two weeks of August, available machine time was even more scarce than usual due to the number of confirmed orders in process. Previously scheduled orders could not be delayed unduly, and try as he might,

Race could free up no more than 56 hours of unscheduled machine time. In addition, the original Department of Defense order for two oval lenses had not, as yet, been scheduled and would cut into this 56 hours. The oval lens would require 8 hours of machine time, while the slightly simpler round lens would require only 7 hours. Each of the oval lenses sold would contribute $110 toward fixed overhead; the round lens' contribution would be $100.

Race decided to formulate a simple linear programming model to determine the production quantities which would maximize contribution.

Altec, Inc.

Altec, Inc. is an Ohio-based producer of precision metal castings for use in aircraft engines and heavy-duty machinery. Formed in 1947 by a group of Air Force Engineers who had left the services after the war, Altec had kept abreast of new technological developments and had been able to sustain continued growth. A cadre of satisfied customers had formed over the years, and since many parts have 25 or more written specifications covering the size of the part, its appearance and metal characteristics, the company prided itself on its reputation for quality.

Since Altec was a capital-intensive firm, its continued expansion required purchases of new machinery almost annually.This year was to be no different, and Ed Hiller, Director of Planning, was looking over the new equipment purchases which his engineers recommended. The Accounting Department had spent some time estimating the value of the equipment over its lifetime in terms of both cost savings and increased revenue, and the figures made all the proposed purchases look quite impressive (See Exhibit 1).

Hiller, however, saw no way to buy all the equipment because of serious space limitations. In recent years, equipment purchases had been made without the concurrent expansion of the physical plant. At the current time, 1200 square feet of space was available in the north wing, and this would allow for the purchase of at most one piece of equipment. To make room for more machinery, Hiller had considered enlarging certain sections of the plant. This would be expensive, but he thought it well worth the money. At his request,

EXHIBIT 1
Net Present Value of Machinery Purchases

Vacuum Melting Furnace, Model 370-2	$22,000
Ingot Press, Model 71	$52,000
Radiac Trimmer	$40,000
5000 Ton Loewy Extrusion Press	$61,000

his engineers and the Accounting Department had provided him with the figures in Exhibit 2.

Hiller wanted to expand the factory for another reason, also. The engineers had balked, originally, at the idea of new equipment purchases because the plant's generators were already overloaded and could not support an additional load. They had agreed that expanding the physical plant would give them the opportunity to increase the power available. The power requirements for the machinery and the possible capacity increases are also shown in Exhibit 2.

Hiller had been convinced, after extensive discussions, that the two expansion projects shown in the exhibits were his only alternatives. It was, of course, possible to undertake both and he thought it probable that the joint expense would be the sum of the expenses for the two individual projects. With this thought in mind, Hiller gathered the figures and sat down to try to make a decision.

EXHIBIT 2

Space Requirements of Machinery Purchases (sq. ft.)

Vacuum Melting Furnace, Model 370-2	600
Ingot Press, Model 71	1400
Radiac Trimmer	800
5000 Ton Loewy Extrusion Press	2000

Power Requirements of Machinery Purchases (KVA)

Vacuum Melting Furnace, Model 370-2	5
Ingot Press, Model 71	10
Radiac Trimmer	10
5000 Ton Loewy Extrusion Press	15

Plant Expansion	Space Available (sq. ft.)	Power Available (KVA)	Present Cost[a]
North Wing Alcove	400	5	$10,000
North Wing Extension	1000	20	$30,000

[a]The Present Cost is the original outlay plus the discounted stream of costs associated with the addition (heat, electricity, etc.).

New York City Environmental Protection Administration (A)

In 1971, New York City disposed of 25,000 tons of solid waste each day, approximately seven million tons annually. Creating a strategy for refuse disposal for New York City through the end of this century has been one of the most difficult tasks facing the New York City Environmental Protection Administration (EPA). As part of their analysis the EPA has examined alternative forms of disposal, such as the use of smaller, efficient incinerators with low pollution levels. In 1971 the Urban and Policy Science group of the State University of New York at Stony Brook was asked to examine the economic feasibility of these "mini-incinerators." Because of recent antipollution legislation, an important question was the tradeoff between new incinerator technology and the long-used landfill disposal methods.

REFUSE DISPOSAL

Each day, New York City's 25,000 tons of refuse is collected by truck, then either taken to incinerators, or to one of six sanitary landfills on the city's perimeter, or to one of the transfer stations located at the water's edge. At the marine transfer stations, the trucks unload the refuse onto waiting barges, and it is then taken to Fresh Kills, a remote landfill on Staten Island.

Apartment incinerators and those in commercial buildings in addition to municipal incinerators accounted for 70% of particulate matter pumped into the air in 1965. In 1966 Local Law 14, an antipollution control bill, was passed. At that time, 30% of all refuse was disposed of in municipal incinerators, and compliance with the new law entailed having the city either shut down or upgrade its low efficiency, high pollution incinerators. Due to high costs and certain technological risks of upgrading, the decision was made to shut down some of these facilities. The removal of some of the City's incinerators, however, would cause an increase in collection cost, since trucks would have to travel larger distances to dump their loads. Increased congestion at the remaining sites could be expected to increase the collection cost further. Since collection accounted for 80% of the overall collection and disposal budget these cost increases were viewed with some apprehension.

Also a problem was the fact that most of the landfills were nearly exhausted, with essentially no new sites available. The most feasible alternative to landfill would be hauling refuse by rail to disposal sites outside the city, at several times the current cost. Hence, the EPA had expressed concern over the early depletion of existing landfill sites. An additional constraint on the system is that the landfills and transfer stations are limited to a maximum daily throughput. For example, a typical marine transfer facility handles only about 1,800 tons per day.

Compliance with Local Law 14 clearly meant that at least part of the approximately 7,000 tons per day processed by New York City's municipal incinerators would have to go to the remaining facilities. Since use of these other locations was physically limited and would have a substantial negative economic impact on both collection and disposal cost, there was a need for analysis of alternative means of disposal.

An important possibility was to build a series of small incinerators which would disperse the air pollution over a larger area, in smaller quantities. The capacity of these "mini-incinerators" would be about 150-500 tons per day, compared to the usual capacity of about 1,000 tons per day. They would burn solid wastes at extremely high temperatures in the absence of air and might have a number of useful by-products such as heavy oil, magnetic metals such as iron filings, and steam to be used for space heating. The residue from these incinerators was virtually negligible and air pollution low.

In late 1971, Edward Beltrami, Brian Hurevitz, and James Moar of the Urban and Policy Sciences Program at Stony Brook began to evaluate the cost-effectiveness of using a large number of these smaller high-efficiency pyrolytic incinerators.

PILOT STUDY DATA

As a first cut to their feasibility analysis, the Stony Brook group designed a pilot study to measure the economic impact of building mini-incinerators in five Northern Queens sanitation districts. As of January 1, 1972, all refuse generated by this area was trucked to a marine transfer station in District 1 (see map—Exhibit 1) and was then barged to the Fresh Kills landfill.

The operating and capital costs of the mini-incinerators was expected to be higher, on a per ton basis, than that for the marine transfer station, but these would be balanced by the reduction in collection time as well as a reduction in the rate of depletion of available landfill if mini-incinerators were built. The economic viability of mini-incinerators for Northern Queens depended on quantifying this tradeoff.

Based on discussions between Mike Hirsch of the New York City EPA and companies who proposed to build low-polluting mini-incinerators, operating costs, including residue disposal, were estimated to be $10/ton. Recycling metal at $2/ton of refuse and selling steam at $2/ton were considered to be promising cost-reducing possibilities, although some doubt had been expressed as to whether the latter would be feasible. Capital costs were made part of operating costs by amortizing repayment over a suitable period of time. Under the range of assumptions suggested by the EPA, mini-incinerator capital costs would be between $3.46 and $5.18 per ton of capacity, for capacities between 150 and 500 tons per day.

The Stony Brook analysis was done with data projected to January 1, 1974, the date when the mini-incinerators could be operational. Exhibit 2 gives the estimated refuse generation in the five districts and the estimated time to travel from each district to a disposal site, dump refuse, and return. From EPA data, the Stony Book group was able to determine that a typical truckload of refuse weighed 4.26 tons. The cost of operating District 1's marine transfer

EXHIBIT 1
Collection Districts in Northern Queens[1]

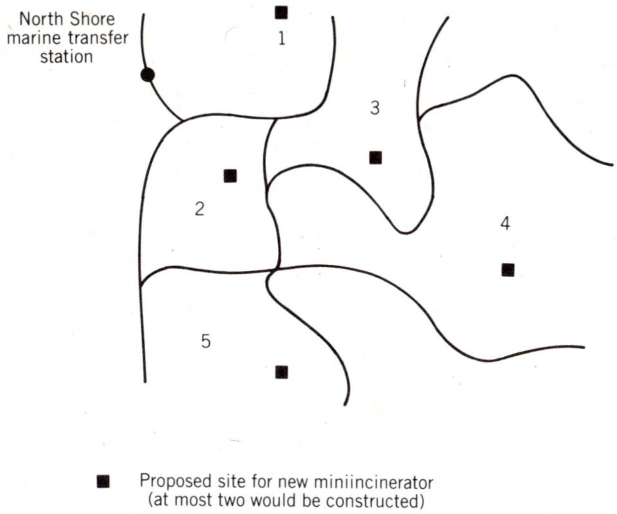

[1] Actual district numbers have been changed.

station was projected to be $4.33/ton, including the actual landfill costs at Fresh Kills. In addition, however, was the less-easily quantified cost of landfill depletion. EPA projections in 1971 indicated that available landfill in New York City would be depleted by 1985. The cost of transporting refuse out of the New York City area once the landfill was exhausted had been estimated to be between $10 and $15 per ton (in constant dollars). If mini-incinerators were to be built,

EXHIBIT 2
Estimated 1974 Refuse Generation and Travel Times

District	Refuse Generated[a] (tons/day)	District					Travel Time (in minutes) to: North Shore Marine Transfer Station
		1	2	3	4	5	
1	216	18	39	39	63	50	31
2	311	39	18	38	39	36	39
3	191	39	38	18	34	34	55
4	160	63	39	34	18	34	75
5	188	50	36	34	34	18	65

[a]Assumes a refuse growth rate of 4% per year.

EXHIBIT 3
Refuse Collection Costs

Three sanitation workers per crew
Average salary of sanitation worker = \$12,000
Fringe benefits @ 23% of salary
214 days worked per year, 6 hours per day
Capital and operating cost of truck = \$5.98/hour

Collection cost/hour

$$= 3 \times \frac{1.23 \times \$12{,}000}{214 \times 6} + \$5.98$$
$$= \$40.47$$

landfill availability (in tons) would be extended by the amount of tonnage handled by the mini-incinerators during 1974–1985. This then allows refuse to be disposed of at \$4.33/ton rather than \$10–\$15/ton from 1985 to whenever the new landfill fill-up date would be.

The hourly labor cost associated with refuse collection is calculated in Exhibit 3.

BIBLIOGRAPHY

Beltrami, E.J., *Models for Public Systems Analysis*, Academic Press, New York (1977).

Beltrami, E.J., Hurevitz, B., and Moar, J., "An Economic Analysis of Using Mini-Incinerators in Northern Queens," USE Technical Report 72-18, Program for Urban Policy Sciences of the State University of New York at Stony Brook.

Russell, J., "Incineration and Local Law 14," *Urban Analysis*, Vol. 1 (1973), pp. 249–264; Vol. 2 (1974), pp. 145–154.

Hunt-Wesson Foods, Inc.

In late 1971, Mr. Steven Niino, Manager of Operations Research at Hunt-Wesson Foods, Inc., met with Professor Arthur Geoffrion, a consultant from

Much of the information contained herein was obtained from the several excellent articles written by Professor Arthur Geoffrion, "Multicommodity System Design by Benders Decomposition" written with G. W. Graves, *Management Science*, January 1974; "A Guide to Computer-Assisted Methods for Distribution Systems Planning," *Sloan Management Review*, Winter 1975; and "Better Distribution Planning with Computer Models," *Harvard Business Review*, July-August 1976. Special thanks are due to Mr. Robert Lowery and other members of Hunt-Wesson management for their cooperation and willingness to provide additional information for this case.

the Graduate School of Management at UCLA, to initiate a comprehensive distribution planning project. In the previous few years, the company had faced pressing distribution-center expansion and relocation problems, and rather than make a series of smaller regional decisions, management had made a commitment to a computer-based model which would rebalance the company's entire distribution system. This involved the potential relocation of distribution centers, reassignment of customers to the centers, and the determination of how products would flow through the system.

The end result of this study was a large-scale mixed integer programming model which has served throughout the 1970s as the major tool for Hunt-Wesson's warehouse location and distribution planning decisions. The cost savings attributable to this model have been estimated to be greater than $1,000,000 annually.

HISTORY OF HUNT-WESSON FOODS, INC.

Hunt-Wesson Foods, Inc. is a subsidiary of Hunt Foods and Industries, Inc., formed by a merger between Hunt Foods and Industries, Inc. and Wesson Oil and Snowdrift Co., Inc. on June 30, 1960. Wesson's nationally branded food products such as Wesson vegetable oil and Snowdrift, a premium quality solid shortening, provided Hunt Foods and Industries, Inc. with additional well-known products to complement the Hunt brand name and add balance and diversification. In addition, Wesson's direct sales organization, with distribution to institutions, industrial customers, and retail outlets, provided economies of scale as well as very promising areas for increasing Hunt's marketing efforts.

Hunt Foods and Industries, Inc. began as Hunt Brothers Fruit Packing Company, a small regional food processor formed in 1890. By 1943, the firm had grown to approximately $10 million in sales, and the company began a series of acquisitions and mergers with the express purpose of promoting additional brand recognition for their products. Between 1943 and 1956, Hunt purchased or merged with several other food producers/processors and in 1955 acquired Glass Containers, Inc. to produce containers for Hunt's products. In 1956 Hunt Foods,Inc. (the name was changed in 1945) acquired the Ohio Match Company, and in 1957 the name was changed to Hunt Foods and Industries, Inc. By 1960, Hunt had sales of $164 million and net income of $3 million and was engaged directly or through its subsidiaries in processing, manufacturing, packaging, and selling tomato products, specialty fruits and vegetables, cans, glasses, and can-making machinery, and in manufacturing and leasing peach-pitters.

Wesson Oil and Snowdrift Co., Inc. was the product of a 1925 merger of three companies, the largest of which was the Southern Cotton Oil Company. Between 1925 and 1960 the company rapidly expanded both through internal growth and by merger and acquisition. By 1959 Wesson had sales of $147 million and a net income of $6.1 million, fully owned seven subsidiaries, and had a controlling interest in three others. Wesson and its subsidiaries manufactured and packaged a wide variety of food products, including vegetable oil, shortening, margarine, mayonnaise, salad dressing, seafood, and vegetables.

After the merger the firm grew rapidly, and in 1968 Hunt Foods and Industries, Inc. merged with Canada Dry Corporation and McCall Corporation to form Norton-Simon, Inc. Since 1968 Hunt-Wesson has continued to grow both internally and through acquisitions (in 1970, Reddi-Wip was acquired), and by 1975 the food and service divisions of Norton-Simon, Inc. (primarily Hunt-Wesson) had sales and profit before taxes of approximately $848 million and $55 million respectively (see Exhibit 1).

THE DISTRIBUTION SYSTEM IN 1970

In 1970, Hunt-Wesson Foods, Inc. manufactured foods at 12 locations (Exhibit 2) consisting of Wesson refineries and Hunt canneries. National distribution from factories to customers was accomplished through 10 distribution centers (Exhibit 2) by rail and by common and contract truck carriers. For large customers, it was possible to eliminate the distribution center and transport directly by making use of a storage-in-transit privilege accorded by the rail carriers. The company's policy dictated that each customer be serviced by a single distribution center. In addition to simplifying accounting and marketing functions and making customer inquiries easier to process, this policy offered several opportunities to take advantage of economies of scale through lower bulk shipping rates.

The firm's continued rapid growth had made it imperative that several distribution centers (DCs) be expanded or relocated and, in addition, the rail carriers' recent pricing policies permitted rate reductions (savings of approximately $.50 per hundred weight) whenever one million pounds or more were shipped via eight or more boxcars on the same train. In view of these developments, Hunt-Wesson's management sought a means of coordinating future expansions of the distribution system and of finding the least cost strategy for expansion. Initially, a "National Distribution Taskforce" was established, consisting of representatives of the Operations Planning, Traffic, Marketing and Distribution Departments. This taskforce recognized that the current

EXHIBIT 1
Food and Food Service Divisions Sales Profit Before Tax For the Years 1971–1975 (in thousands)[a]

	1971	1972	1973	1974	1975
Sales	481,414	510,965	551,909	683,041	847,962
Profit before Tax	32,268	32,772	32,633	28,152	55,362

[a]1975 annual report of Norton Simon, Inc.

EXHIBIT 2
Hunt-Wesson Plants and Distribution Centers in 1970.

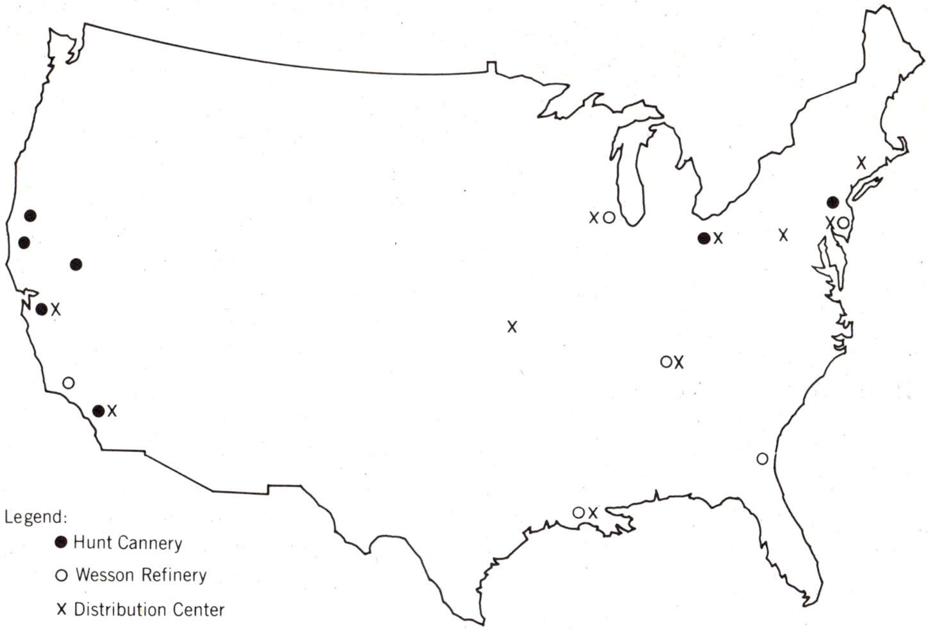

method of doing a number of case studies of individual distribution centers was clearly suboptimal and began searching for techniques to allow them to construct a medium-range (1–3 years) planning model which would interweave the complexities of the distribution process nationally.

Several experts in distribution systems were consulted. An early suggestion by one of these experts was a "planning board." This analogue device was a large wooden board with holes drilled at each customer location. Strings between holes represented the varying distances (approximating transportation charges), and weights hung on pulleys at each "customer" represented demand. The board would then be shaken, the weights would fall, and a central knot would appear—the optimal location for the DC to service that set of customer demands. Unfortunately, however, this method posed a serious drawback—it required that the U.S. be divided into sections; the planning board would then find the optimal DC for each section, one at a time. Since this sectioning operation could be done in any number of ways, there was no assurance that the DC locations proposed by the planning board were optimal when taken in total.

An effort was then made to formulate the problem as a linear program, but this methodology also had serious limitations due to considerations such as mandatory service area constraints (i.e., no customer can deal with more than one DC), and the multicommodity aspect of the problem. At the same time, Hunt-Wesson's Director of Distribution recalled that Professor Arthur Geoffrion of UCLA's Graduate School of Management had been doing research on optimal distribution systems. He contacted Geoffrion and found that Hunt-Wesson's problem meshed quite well with Geoffrion's research, but that the problem was initially too large for Dr. Geoffrion to solve. At this point in time, early 1970, a new group was formed within Hunt-Wesson's Corporate Facilities Planning Department, the Operations Research Department. The OR group's first task was to work with Professor Geoffrion and determine whether a mathematical programming approach might be expected to yield a systematic and cost-effective solution to the problem of DC location and distribution. The verdict was affirmative.

DEVELOPMENT OF THE MODEL

In early 1971, a team was formed to develop a distribution model along the guidelines proposed by Professor Geoffrion. This team consisted of staff members from various corporate groups: Information Systems and Services (ISS); Distribution and Logistics Accounting; Distribution; Traffic; and Corporate Facilities Planning. The specific questions that management wanted the model to address were:

- How many distribution centers should there be?
- In which cities would they be located?
- What size should each distribution center be?
- How should each plant's output of each product be allocated among distribution centers or customers?
- What should the annual transportation flows throughout the system be?
- For a given level of customer service, how do the costs of the distribution system compare with a projection of the current system to the target period?

A summary of the model's constraints, assumptions, and simplifications is given in Exhibit 3.

Because it was impractical to expect the model to distribute each of the many hundred Hunt-Wesson products to the many thousand individual customers, some careful *aggregation* and *preselection* were required. The products were aggregated into17 product groups; individual customers were aggregated into 121 zones based on the total demand associated with each zip code zone, its proximity to DC sites, and its proximity to demand concentrations. Since only a few plants would normally manufacture any one of the product groups,

49 plant-product group combinations were preselected from the 238 (14 plants × 17 product groups) theoretically possible. Possible DCs were limited to 45 preselected locations,including a "dummy" location to permit certain customers to receive commodities directly from the plant, making use of the applicable storage-in-transit rates. Lastly, from the possible 5445 (45 DC sites × 121 customer zones), 682 were preselected to be those in which the customer zone can be served by the DC with a certain maximum prescribed delay by rail or truck. These modeling specifications are described in Exhibit 3. Despite the simplifying effect of this aggregation and preselection, over 24,000 plant-product group-DC customer combinations were permissible, and from this group the model was to choose the optimal set.

THE MATHEMATICAL FORMULATION

A mixed integer program was formulated to minimize total distribution costs subject to the restrictions of Exhibit 3. The program determines which DC sites should be utilized, the optimal size of each DC, which customer zones should be served by each DC, and what the pattern of transportation flows should be for all commodities. The mathematical formulation follows.

Decision Variables

x_{ijkl}—The amount of commodity i shipped from plant j through DC k to customer zone l.

y_{kl} —A 0−1 variable that will be 1 if DC k serves customer zone l, and 0 otherwise.

z_k —A 0−1 variable that will be 1 if a DC is acquired at site k, and 0 otherwise.

The Objective Function

$$\text{Minimize} \sum_{ijkl} c_{ijkl}x_{ijkl} + \sum_{k}\left[f_k z_k + v_k \sum_{il} D_{il} y_{kl} \right]$$

for $x \geqq 0$; $y, z = 0, 1$, where

c_{ijkl} is the average unit cost of producing and shipping commodity i from plant j through DC k to customer zone l.

f_k is the fixed portion of the annual possession and operating costs for a DC at site k.

v_k is the variable unit throughput cost for a DC at site k.

D_{il} is the demand for commodity i in customer zone l.

EXHIBIT 3
Summary of the Model's Facts and Assumptions

SUMMARY OF THE MODEL'S FACTS AND ASSUMPTIONS

1.
The many hundreds of Hunt Wesson's different products are aggregated into 17 product groups. Nearly all products in a given group are similar with respect to the plants which can make them, the production technology by which they are made, and their gross-to-net shipping weight ratios. All quantities are measured in gross shipping hundredweight (cwt.).
Examples of the product groups are: bottled cooking oil, packaged shortenings, ketchup, tomato sauces, puddings, and matches.

2.
The production capacity (cwt./yr.) of each of the company's 14 plants is given for each product group.
Not every plant could make every group—in fact, only about 50 plant-product group combinations were possible.

3.
Distribution center locations to be evaluated by the model are limited to a list of 45 cities composed of current locations, major metropolitan demand concentrations, the locations of major competitors' distribution centers, and other gateway cities.

4.
The size of each distribution center chosen by the model must be between stipulated lower and upper limits expressed in terms of total annual throughput volume.
The lower limit was due to the need to accommodate rail shipments of about eight boxcars, which permits important economies of scale.

5.
The many thousands of individual customers are aggregated into 121 customer zones defined in terms of zip codes.
The zones consist of a key city with a significant demand plus (usually) adjacent areas with relatively small associated demand. Each zone is sufficiently small in geographical area so that the best distribution center for the key city is also the best one for the other customers in the zone.

6.
The annual demand forecasted for each product group in each customer zone must be met.
The demand forecasts were based on historical summaries prepared by computer from magnetic tape archives of shipment invoices. These tapes also included most of the information required to develop a realistic transportation cost structure.

7.
Each customer must continue to be serviced from a single full-line distribution center for all products.

8.
No distribution center location is allowed to service a customer zone unless it is sufficiently close to the zone that normal delivery time would be less than a specified maximum number of days.
This maximum is the model's primary control over the level of customer service.

9.
Differential production costs ($/cwt.) are given when there is a significant difference between unit costs at different plants for a given product group.
Not to be overlooked among the reasons for differential production costs is the influence of plant location when it is significant in determining delivered raw material costs (as it was here in the case of refinery products).

10.
The costs of each distribution center are expressed as a "fixed" charge if a location is selected for use by the model plus a "variable" cost ($/cwt.) applied to the total

(continued)

EXHIBIT 3 (Continued)

annual throughput volume over all product groups.
These costs included all handling, storage, and other operating expenses, inventory carrying charges, and amortized capital costs associated with expanding an existing distribution center or acquiring a new one. A standard throughput cost curve like the hypothetical one shown in the figure was developed based on historical analysis of the current distribution centers' costs. This curve was then modified for each of the 45 locations to reflect geographical differences. A straight-line approximation proved adequate, although a more accurate brokenline approximation would have been used if necessary to portray economies and diseconomies of scale over the range of interest.

11.
For each product group and possible plant-to-distribution center transportation link, a unit freight rate ($/cwt.) is specified.
A judgment was made in each case whether truckload or rail carload lots were the most appropriate mode of supply.

12.
For each product group and possible distribution center-to-customer zone link, a total annual freight cost ($) is specified.
It made sense to specify an annual cost for a link, since each customer had to receive all of a product group from a given distribution center if he received any at all. Most of the delivery costs were generated by computer from specially constructed approximation tables, taking into account the proportions of mode (LTL, TL, CL) and shipment size likely to prevail in each case.

13.
Transit rail rates are used whenever they are advantageous.
The storage-in-transit rate applicable to some product groups is figured as though the shipment moves directly from plant to customer with a small charge for stopping over at the intermediate distribution center (so long as this stop is not too far off the direct line), instead of as the sum of the regular plant-to-distribution center and distribution center-to-customer rates.

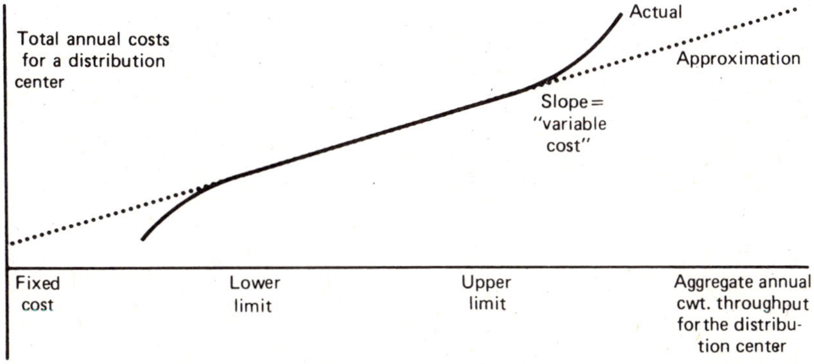

Source: Reproduced with permission from A.M. Geoffrion, "Better Distribution Planning with Computer Models," *HBR*, July-August, 1976, p. 97.

The Constraints

1. $\sum_{kl} x_{ijkl} \leq S_{ij}$ for all i, j
   where S_{ij} is the production capacity for commodity i at j.
2. $\sum_{j} x_{ijkl} = D_{il}\, y_{kl}$ for all i, k, l
3. $\sum_{k} y_{kl} = 1$ for all l
4. $\underline{V_k} z_k \leq \sum_{il} D_{il}\, y_{kl} \leq \overline{V_k} z_k$ for all k, where $\underline{V_k}$ and $\overline{V_k}$ are, respectively, the minimum and maximum total annual throughput for a DC at site k.
5. Also, linear configuration constraints on y and/or z, for example, to ensure that at most *one* DC is opened at each site; upper and/or lower bounds on the total number of DCs allowed, and so on.

DATA COLLECTION

The collection and development of the data required by the model was (and remains) a considerable task, requiring the efforts of much of the OR group over a three- to four-month period.

The 49 capacity figures (S_{ij}) were developed individually based on existing capacities and anticipated changes. To collect the demand data, the firm, fortunately, had the previous year's shipment invoices stored on magnetic tape. The historical demands were calculated using the customer zones' zip codes, the individual commodity codes, and gross shipping weight. The demands (D_{il}) could then be obtained by projecting the historical demands forward in time according to the company's market forecast. The upper and lower bounds ($\overline{V_k}$ and $\underline{V_k}$) on warehouse size were based on management's experience with the present distribution system. The corresponding fixed and variable cost coefficients (f_k and v_k) were determined by developing a function $W_k(w)$ (as pictured in Exhibit 3) defined as the total warehousing costs associated with leasing and operating a DC at size k with an annual throughput of w hundredweight per year. These $W_k(w)$s were estimated by fitting a linear approximation of the form

$$W_k(w) = f_k + v_k w$$

over the range $\underline{V_k}$ to $\overline{V_k}$ using linear regression.

Specifying the transportation costs (c_{ijkl}) was a complex task. As described above, the over 24,000 allowable $ijkl$ combinations had to be identified through a process of aggregation and preselection. Then, for each combination, it had to be determined by hand when the rail transit privilege applied. This privilege applies when the DC is not too far off the most direct rail route between plant and customer and permits substantial savings over simply adding the plant-DC and DC-customer rail rates. When the in-transit rate does *not* apply, c_{ijkl} is the sum of the plant-DC rate and a DC-customer zone rate. For each plant-DC combination, everything moves either by rail or by truck. The DC-customer zone rates, however, involve both rail and truck shipping. The truckload rate and rail rates were generated from tables and a weighted average of the sub-rates was then taken to obtain each DC-customer zone rate with the weights based on past experience of mode and shipment size for each demand center.

When the transit rate did apply, c_{ikjl} was taken to be a weighted average of the appropriate transit rate (computed from tables) and the nontransit rate. There is always some nontransit truck shipping even when the transit privilege is applicable, since not all customers accept direct rail shipments.

Considerable effort was made to ensure that the required data was both correctly developed and correctly transferred to the data sets that would be accessed by the main computer program. The S_{ij}, $\underline{V}_k$, $\overline{V}_k$, f_k, and v_k coefficients were sufficiently few in number that they could be individually audited and entered manually into the program. The demand and shipping cost data, however, were of such a large volume that verification posed a problem. These data were transmitted on magnetic tapes, which then had to undergo considerable preprocessing in order to set up the properly indexed data sets required by the main program. Numerous editing and checking routines were among the programs which produced the demand and rate tapes, and further verification was done during the preprocessing. Selected and random samples were subjected to manual checking as well.

Final validation was done by carrying out a number of runs with many of the variables locked at particular values. The output was then scrutinized for anomalous or seemingly unreasonable results that might be indicative of inaccurate data or program malfunction.

Agua del Pueblo, Inc.

Bruce Clemens, Executive Director of Agua del Pueblo, Inc., sat at his desk in Guatemala City trying to decide how much steel reinforcing to send to the rural village of Panimaquib. The steel reinforcing was to be used in the construction of a holding tank for the water supply system for this rural village. Panimaquib was quite poor and the villagers were paying for a significant portion of the material cost of this system, so Mr. Clemens did not want to send more steel than was absolutely necessary. At the same time, he wanted to be certain that enough steel was sent to complete the job. Panimaquib was not easily accessible, and if sufficient steel was not sent, long construction delays would result.

AGUA DEL PUEBLO, INC.

Upon graduation from Cornell University's College of Civil Engineering, Bruce Clemens toured throughout Central America in 1972. During this trip, Mr. Clemens became astonished at the abysmal poverty and squalid living conditions that existed in the rural areas. Basic sanitation facilities and an adequate water supply were unknown to most rural villages.

Mr. Clemens found that the rural populations were industrious and hard working; however, they lacked the technical expertise and the financing necessary to realize these public works projects. Mr. Clemens believed that his expertise could be used to greatly aid these rural villagers. He settled in San Lucas Taliman in the western highlands of Guatemala and established Agua del Pueblo, Inc. (Water for the Common Man)—a nonprofit organization that provided technical assistance to these rural communities.

By 1977 Agua del Pueblo had aided scores of villages in Central America in the construction of potable water systems and sanitation facilities for their communities. Agua del Pueblo provided three basic inputs—aid in community organization, technical assistance, and access to financing. *Community organization* involved aiding communities in forming the legal committees and the organizational structure necessary to carry out a public works project. *Technical assistance* involved the design and construction supervision of the projects. *Access to financing* involved aiding communities in obtaining grants and/or loans from national governmental agencies, international developmental agencies, or foreign (generally U.S.) foundations. The average family income in many rural communities was only $1 per day and few communities could afford to repay the full cost of a water supply system (average capital cost equaled $120 per family served). Therefore, generally a combination of loans and grants were used to finance most projects.

About half of the grant monies available to the villages were provided from profits from Agua del Pueblo consulting activities. Agua del Pueblo consulted for large U.S. construction firms working in Central America, for local (Central American) construction firms and for numerous international development agencies.

PANIMAQUIB

Panimaquib was a small village of 53 families (330 people) located in an isolated rural area in the Department of Solala, Guatemala. A five-hour drive (part of which was over unpaved roads) and a 45-minute walk over mountain trails were required to reach the village from Guatemala City.

The potable water system to be constructed in Panimaquib would be more costly than usual since the mountain spring that provided the source of water was located far from the village. The estimated cost of the system was as follows:

Purchased Materials (tubing, steel, cement, plumbing accessories)	$3,010
Locally collected raw materials (sand, gravel, building rock, lumber)	600
Labor to construct the water system (to be denoted by villagers, 2,000 man days at $1/day)	2,000
Construction foreman (hire local plumber/ construction supervisor to oversee construction)	700
Technical assistance (survey, design, administrative assistance, construction monitoring)	1,100
Total Cost	$7,410
Cost per family served	$ 140
Cost per person served	$22.45

Both grants and loans were to be used to finance this water project. The technical assistance was donated by Agua del Pueblo. The labor to construct the water system was donated by the villagers themselves (at essentially no opportunity cost since alternative employment opportunities did not exist). The locally available raw materials were collected and donated by the villagers. Of the remaining $3,710 cost, $700 was being provided by the Government of Guatemala and $2,000 was being provided by a United States foundation. The remaining $1,010 was being loaned to the community by a local cooperative group and would be repaid over a period of four years at an interest rate of 10%. The debt obligation of each family was thus significant relative to their resources.

Bruce Clemens had this in mind as he reviewed the tank plans for this village (Exhibit 1). The tank served as a reservoir for collecting and holding the water from the mountain spring. The tank was to be constructed primarily of mortar (rock and a cement mix). The top of the tank was to be constructed of reinforced concrete. Reinforced concrete requires bars of steel to be criss-crossed throughout the structure.

The reinforced steel bar required to construct the tank are listed under the section "Planilla de Materiales" of Exhibit 1. This section lists a reference number ("Tipo"), the diameter of the steel bar required ("Diam"), the length of the piece required in centimeters ("Long"), the number of pieces of that diameter and length required ("Cant"), and a reference number for the diagram depicting how the steel was to be bent ("Detalle").

EXHIBIT 1
Construction Plans

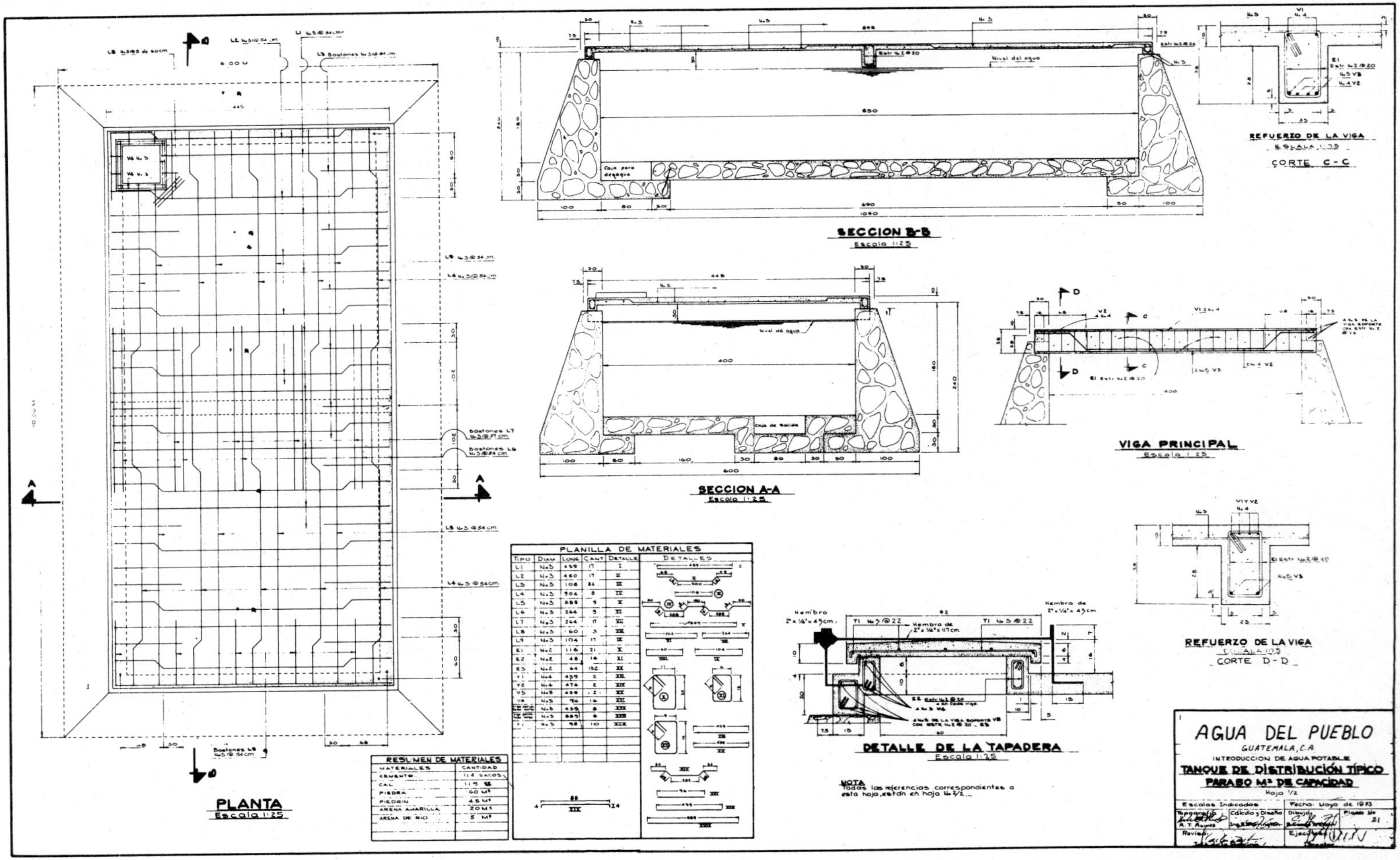

Mr. Clemens needed to determine how many steel bars of each diameter to order. Steel bars could only be bought in six meter (600 cm.) lengths. The bars would be sent directly to the village where they would be cut and bent by the villagers according to a plan submitted by Mr. Clemens. Each of the required pieces would have to be cut from these six meter bars.

For example, if twelve 100-cm pieces were required, they could be cut exactly from two six-meter bars. Similarly, if two 250-cm pieces were required, they could be cut from a six meter bar and a 100-cm piece would be "left over."

The "left-over" pieces could be discarded, cut to fill requirements for shorter pieces, or combined with other "left over" pieces to fill requirements for longer pieces. Left-over pieces were combined by wiring a 30-cm overlap of the pieces together. For example, two 100-cm pieces would be combined to form a 170-cm piece.

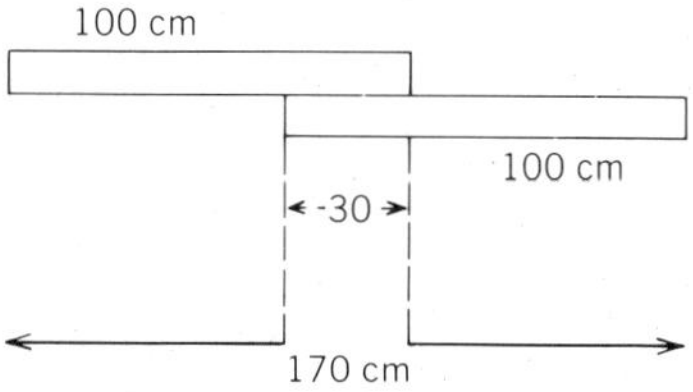

Similarly, a requirement for a 910-cm piece can be met by combining a 600-cm bar with a 340-cm piece, or combining five 206-cm pieces, or by any other combination of pieces that yielded the required total length.

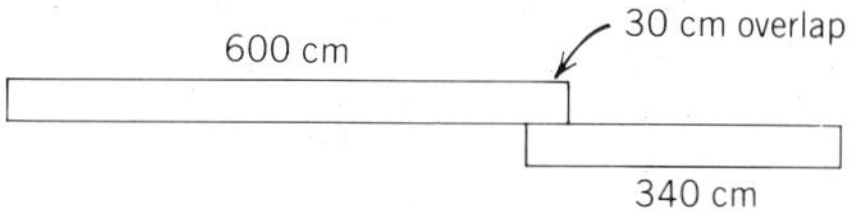

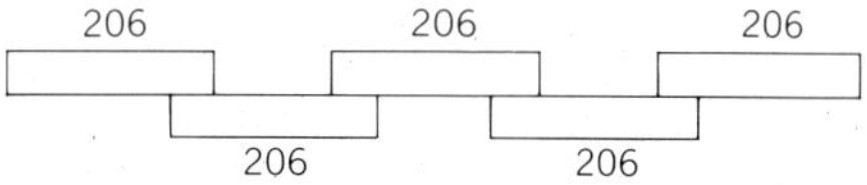

Traditionally, Mr. Clemens had determined the number of bars of steel to order by taking the total length of the steel required (for each diameter), adding 10%, dividing by 600 cm, and rounding upwards. He would then allow the villagers to cut the steel in any manner so long as the final material list requirements were met. Any extra steel was discarded.

However, he had always felt that a more exact method could be found to determine the amount of steel to order. He knew that ordering fewer bars of steel would result in a significant savings to the village. However, he wanted to be sure to send an adequate amount of steel to the village because transportation costs and construction delays caused by a second shipment would negate the savings of minimizing the excess steel sent to the project.

As part of some consulting work with the University of San Carlos, Mr. Clemens had access to computer facilities at essentially no cost. If some computer-aided technique were applicable to this project, Mr. Clemens believed that he could employ the same technique profitably in his consulting business. He had recently encountered a similar problem when finalizing plans and a material list for a large construction project.

The Bubble

In the winter of 1977, Assistant Secretary of the Treasury Anthony Solomon began to study the detrimental economic impact of government regulation on the steel industry. His work led to a recommendation that the Environmental Protection Agency (EPA) study the "bubble" concept which suggests that tradeoffs of control stringency among emission sources in a given plant be made if such trades result in no net increase in emission for the facility as a whole and provide savings in emission control costs. Prior to this recommendation, the EPA and the American Iron and Steel Institute (AISI), the steel industry's lobbying organization, had already discussed the bubble idea. Thus, in 1978, the EPA convened a task force to explore the bubble concept. In January, 1979, a policy proposal was announced.

The steel industry was encouraged by this announcement. However, the publication of a proposed policy did not mean that industry could immediately alter its emission control plans. The bubble policy was open to public comment and dispute—the possibility existed that it might not be adopted; and, in the end, it would be up to individual states to choose whether or not to implement any EPA endorsed policy. Furthermore, the policy as proposed left open the question of which emission types should be considered tradeable. Any limitations with respect to the emittants covered under the bubble policy would diminish the potential savings.

The publication of this proposed policy was, however, of great significance. It represented a conciliation between the steel industry and the EPA, two groups which rarely found themselves on the same side of issues.

BACKGROUND-ENVIRONMENTAL PROTECTION

I am glad to approve this legislation . . . Under this legislation we can halt the trend toward greater contamination of our atmosphere. We can seek to control industrial waste discharged into the air. We can find ways to eliminate dangerous haze and smoke.

With these words, President Lyndon B. Johnson signed the Clean Air Act of 1963, the first major piece of federal air pollution legislation. This act gave the

Department of Health, Education, and Welfare (HEW) the authority to publish nonmandatory air quality criteria and to give states aid for the development of air pollution control legislation. In 1965, the first amendments to this Act were passed. These mandated that HEW promulgate and enforce federal emissions standards. In 1967, the Act was again amended and as a result HEW was assigned the duty as well as the authority to issue air quality criteria, and the responsibility to request the Attorney General's office to bring an abatement action in any air pollution situation which presented danger to health. Finally, the 1970 Clean Air Amendments shifted responsibility for environmental quality issues to the newly formed Environmental Protection Agency.

The principal purpose of the 1970 Amendments was "to protect and enhance the quality of the Nation's air resources so as to promote the public health and welfare and the productive capacity of the population." To this end, the EPA was to establish nationally uniform ambieint air quality standards for major air pollutants emitted from mobile and stationary sources. The Amendments required states to develop and enforce implementation plans for the achievement of these standards. However, the EPA had the authority to accept or reject such plans.

A 1977 report from the U.S. Department of Commerce summed up the implications of the Clean Air Amendments as follows:

> *The Clean Air Act [Amendments] imposes an ambitious and in some measure inflexible scheme for the clean-up of the Nation's ambient atmosphere. Its provisions are designed to "force" technology—a not unworthy legislative objective—to enable this clean-up. The changes in the Nation's energy and economic outlook since its enactment in 1970 continue to compel close scrutiny of the feasibility and consequence of strict adherence to the Clean Air Act's statutory mandate. Creative and balanced administration of the Act by the EPA . . . has avoided gross near-term economic penalties disproportionate to the perceivable near-term benefits of cleaner air. Nevertheless, the statute's literal, rigid requirements—always subject to court mandate that they be fully enforced—can have most serious consequences for the Nation's energy and economic goals.*

In carrying out its mandate, the EPA has been caught between two vocal and powerful interest groups: industry and environmentalists. Major industrial polluters resisted initiating expensive pollution controls because, they claimed, these costs would adversely affect the growth of American industry. Conversely, environmentalists took an active interest in the effectiveness of the enforcement programs of the EPA. The EPA has been the target of legal suits both from those resisting enforcement and those seeking it.

EPA AND THE STEEL INDUSTRY

One week after William Ruckelshaus took office as the first administrator of the EPA, the EPA asked the Justice Department to prosecute Jones and Laughlin Steel for discharging pollutants into the Cuyahoga River. Other companies singled out for early enforcement action were Armco Steel and U.S. Steel. During the subsequent 9 years, the EPA and the steel industry were embroiled in legal and political wrangling as the former insisted on enforcing and the latter insisted on resisting pollution controls.

According to Ed Speer, former Chairman of U.S. Steel,

> *When we point out the technological impossibility of [EPA] standard[s], we are accused of recalcitrance. If we agree to [them] and then fail to achieve 100% compliance, we are prosecuted.*

From the EPA's point of view, the big steel companies had wantonly resisted environmental standards at many plants without just cause.

The 1970s were lean economic years for the steel industry. Major steel producers suffered from increased foreign competition as well as a capital shortage in this country. The imposition of pollution controls required large amounts of this scarce capital. In 1978, capital expenditures for pollution control were 18% of total new capital employed. In some cases, companies chose to close down aging equipment with several years' life remaining because the installation of pollution controls would be too expensive over the life of the machine. This led to decreased productive capacity and the loss of income and jobs at the plants affected.

The economic situation of the steel industry led to a 1977 recommendation by the Steel Communities Coalition, a group of mayors and county officials of steel producing areas, that the EPA be more flexible in their enforcement of pollution control laws with respect to the steel industry. The bubble policy is a partial answer to the Coalition's recommendation.

Under the bubble concept, a single pollution limit would be set over an entire plant; the total level of permissible emissions would remain the same as in the past. (See Exhibit 1.) Each plant would be free to determine the most cost effective way to meet the overall limit.[1] A recent *Fortune* article illustrates:

> *Consider a simplified example, a single plant with two boilers, each of which is limited by current regulations to emitting no more than fifty tons of pollutants a year. One boiler burns high sulfur coal and is producing sixty tons a year; the other is burning low sulfur oil and is putting out forty tons a year. To comply with existing regulations, the plant is forced to burn a mixture of both fuels in each boiler. That means building a pipeline from the oil burner to the coal burner and renovating each boiler to burn the fuel of the other. Under the bubble concept, the limit would be 100 tons for the two boilers together. No investment would be needed, and the overall air quality would be the same.*[2]

[1]This is denoted "plantwide control" in contrast to the EPA's currently required "source specific control."

[2]Peter Nulty, "A Brave Experiment in Pollution Control," *Fortune*, February 12, 1979, pp. 120–123.

EXHIBIT 1
The Bubble Concept: An Alternative Approach to Emission Reduction

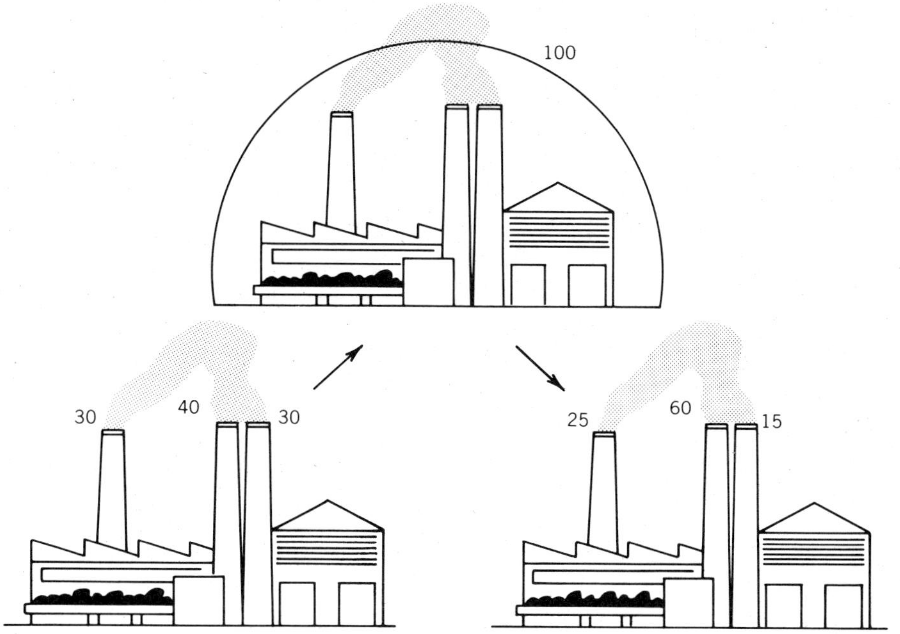

THE BUBBLE MODEL

In 1978, while the EPA was exploring the bubble concept in detail, the management consulting firm of Putnam, Hayes and Bartlett, Inc. (PHB) was asked to analyze the cost impact of plantwide (i.e., bubble) emission control versus source-specific control.

PHB provides management and economic counsel to business and government policymakers. In addition, PHB's highly regarded staff of business and economic analysts has been called upon to provide expert economic analysis and counsel in a large number of legal and regulatory proceedings. PHB has particular expertise in the steel industry, where they have acted as consultants to AISI and several of the large steel companies, and in environmental regulation, where they have done substantial work for the EPA.

PHB analyzed the cost impact of the bubble concept using a plant-specific computer model which determined the pollution control technologies which would minimize pollution control costs for a fixed emission level. Input for this model was provided by a plant-specific model which PHB had previously developed and used successfully in its work with the steel industry,[3] to determine optimal production flows. For this analysis, four domestic steel plants were selected to represent the range of domestic plant types:

- Plant A, which is similar in configuration to Jones and Laughlin Steel Corporation's plant at Indiana Harbor, Indiana.
- Plant B, which is similar in configuration to Bethlehem Steel Corporation's plant at Burns Harbor, Indiana.
- Plant C, which is similar in configuration to Inland Steel Corporation's plant at East Chicago, Indiana.
- Plant D, which is similar in configuration Borg-Warner Corporation's plant at Chicago Heights, Illinois.

Each plant configuration is described in more detail below.

Plant A is an integrated plant, meaning that its coke ovens and blast furnaces allow it to make steel from iron ore products and coal. The plant is configured to have an annual raw steel capacity of 3.8 million net tons, which is considered medium-sized in comparison with other domestic integrated mills. All steelmaking is carried out in Basic Oxygen Furnaces (BOF).

Plant B is also fully integrated. With an annual raw steel capacity of 4.4 million net tons, it also is considered a medium-sized integrated mill. As with Plant A, all steelmaking capacity is BOF. Exhibit 2 depicts the configuration, production capacity, and optimum production flows of this plant.

Plant C is a large fully integrated steel plant, with an annual raw steel capacity of over 10 million tons. Three types of steelmaking capacity exist at Plant C: open heart, BOF, and electric furnace.

Plant D is a nonintegrated plant; that is, it uses electric furnaces to melt scrap or direct-reduced iron. Its annual raw steel capacity is .21 million net tons, which makes it a typical sized "minimill."

TECHNOLOGY OF POLLUTION CONTROL

Measuring changes in air quality due to emissions from an industrial source is a difficult and expensive task. For this reason, the EPA has developed technology-based rather than emissions-based standards for the enforcement of air quality regulations. Thus, a facility is in compliance if it has the appropriate pollution control technology; compliance is not a function of what is *actually* emitted from the facility.

[3]Alternatively, this model might have been used to minimize pollution levels for various predetermined cost levels.

EXHIBIT 2

Plant B Plant Configuration—Production Capacity (million net tons)

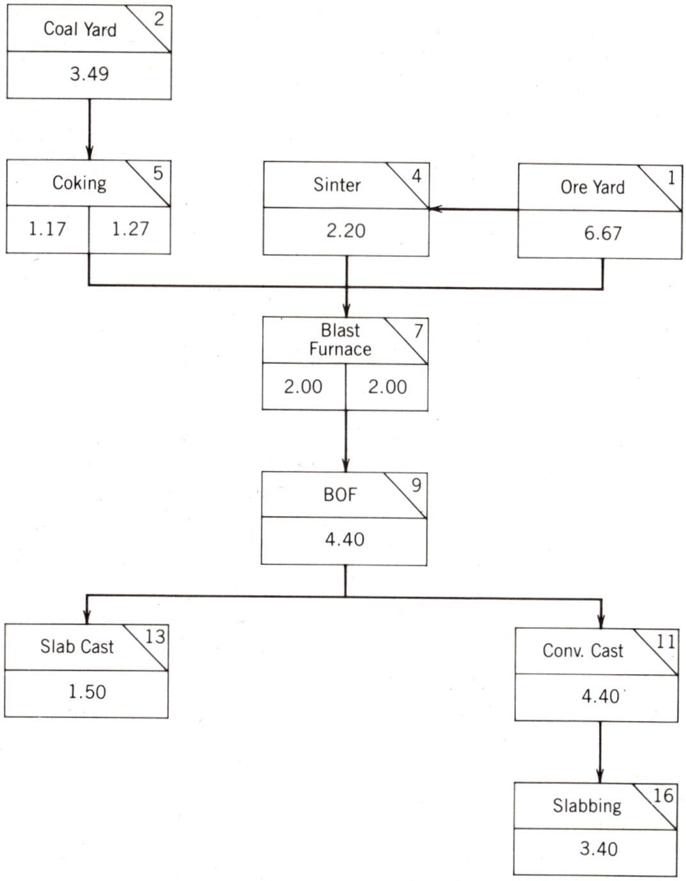

Plant B Plant Configuration—Production Flows (million net tons)

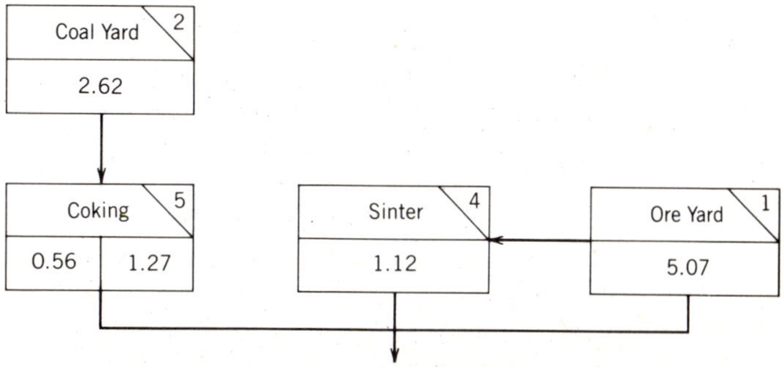

(continued)

EXHIBIT 2 (Continued)

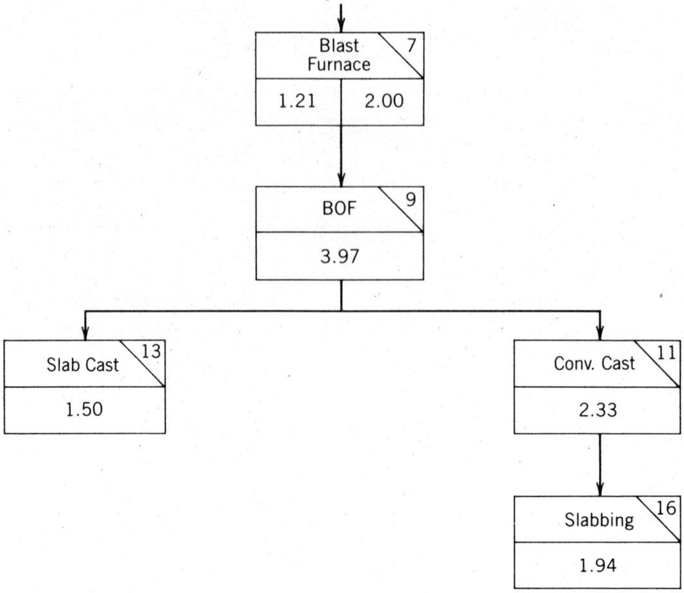

Four levels of pollution control technology have been designated by the EPA:

- No Control (N)
- Reasonably Available Control Technology (RACT)
- Best Available Control Technology (BACT)
- Lowest Achievable Emissions Rate (LAER)

In general, compliance with the Clean Air Amendments as interpreted by the EPA and implemented through the State Implementation Plans (SIP's) required that all sources be controlled at the level of RACT at the time of the study.

For each emissions source, there are a number of possible pollution control devices or processes. In some cases, the control technology is the same for more than one EPA recognized level (i.e., the same device may be designated BACT and LAER); in other cases a different level signifies a totally different type of equipment with different costs and emission rates. (See Exhibits 3 and 4.) As an example, the costs and emission rates associated with each level of control at Plant B are shown in Exhibits 5 and 6.

EXHIBIT 3
The Bubble

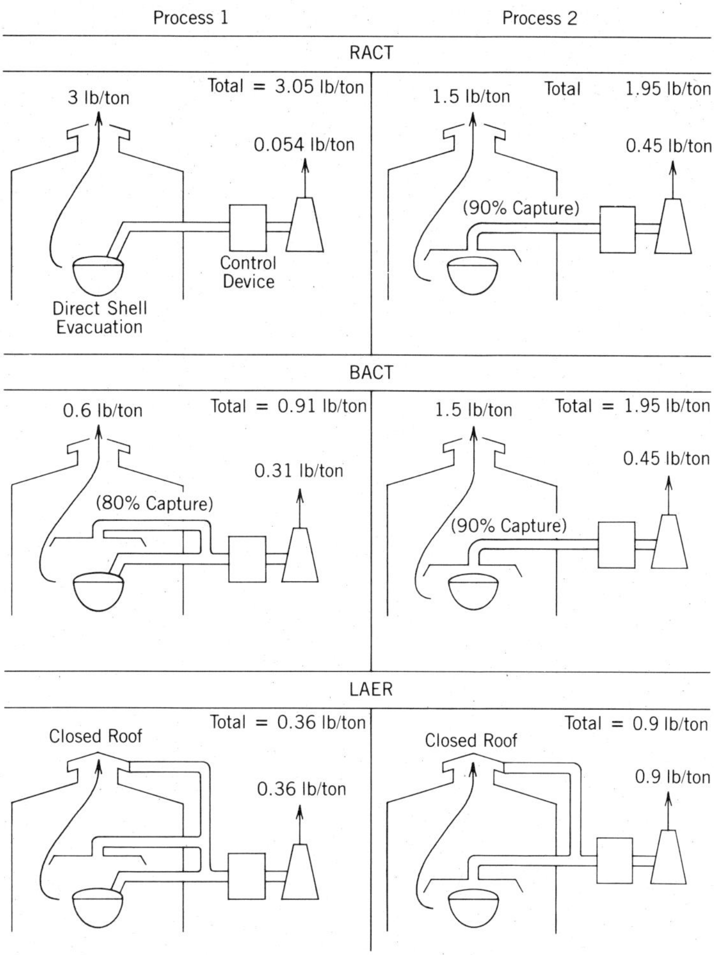

EXHIBIT 4
Example of Differences in Control Technologies

Process or Operation	Basis for Emission Measurement	Uncontrolled Emission Rate, TSP[a] unless Otherwise Noted	RACT Control	RACT Emission Rate	BACT Control	BACT Emission Rate	LAER Control	LAER Emission Rate
Coke quenching	Coke produced		Baffles and clean water	1.0	Baffles and clean water	1.0	Diverted flow baffles and clean water	0.43
Door emissions	Coke produced	0.71	Door maintenance	0.14	Door maintenance	0.14	Door maintenance and auto cleaning	0.07

(continued)

EXHIBIT 4 (Continued)

Process or Operation	Basis for Emission Measurement	Uncontrolled Emission Rate, TSP[a] unless Otherwise Noted	RACT		BACT		LAER	
			Control	Emission Rate	Control	Emission Rate	Control	Emission Rate
Topside leaks	Coke produced	0.49	Good operation and maintenance	0.043	Good operation and maintenance	0.043	Good operation and maintenance	0.043
Underfire stack	Coke produced	1.0	Dry ESP	0.15 or 0.03 gr/scf	Dry ESP	0.15 or 0.03 gr/scf	Dry ESP	0.08 or 0.015 gr/scf
Coke handling	Coke produced	0.03	Baghouse	0.002	Baghouse	0.002	Baghouse	0.002
Coke oven gas	Coke produced	SO 19.0	Desulfurization	2.7	Desulfurization	1.4	Desulfurization	0.43
Coal preheater	Coal used	7.05 HC 1.06	Scrubber	0.35 0.53	Scrubber	0.35 0.53	Scrubber	0.35 0.53
Ironmaking:								
Cast house emissions	Hot metal produced	0.69	Taphole hood and bag house	0.07	RACT and runner covers	0.042	Cast house evacuation	0.042
Slag pouring	Hot metal produced	0.28	None	0.28	Hood and scrubber	0.014	Hood and scrubber	0.014
Slag crushing and screening	Hot metal produced	0.24	Water sprays	0.12	Baghouse	0.025 or 0.005 gr/scf	Baghouse	0.025 or 0.005 gr/scf
Steelmaking:								
Open hearth hot metal transfer	Hot metal used	0.35	Baghouse	0.007 or 0.01 gr/scf	Same as RACT	Same as RACT	Same as RACT	Same as RACT

[a]Total suspended particulate.

EXHIBIT 5
Plant B: Unit Control Costs

Process and Unit	Source	RACT	BACT	LAER
		($/ton production)		
1.1 Ore Yard	1. Fugitive	.0090	.0134	.1012
2.1 Coal Yard	1. Fugitive	.0171	.0262	.2267
	3. Preparation	.0257	.0267	.0267
4.1 Sinter Plant	1. Windbox		.5758[a]	.7589
	2. Discharge	.0850[a]	.0946	.0946
	3. Fugitive	.0000	.0314	.0314
5.1 Coke Battery	1. Charging	.2266[a]	.3238	.3238
	2. Pushing	.6200	.6200	.6200
	3. Quenching[a]		.6042[a]	.6042
	4. Doors	.2690	.4043	.4043
	5. Topside	.1322	.1322	.1322

(continued)

EXHIBIT 5 (Continued)

Process and Unit	Source	RACT	BACT	LAER
		($/ton production)		
	6. Underfire Stack	.3120	.3249	.3455
	7. Handling	.0725	.0725	.0725
	8. Gas		.4295[a]	.5098
5.2 Coke Battery	1. Charging	.2124[a]	.3034	.3034
	2. Pushing	.5892	.5892	.5892
	4. Doors	.2454	.3699	.3699
	5. Topside	.1180	.1180	.1180
	6. Underfire Stack	.2982	.3058	.3316
7.1 Blast Furnace	2. Cast House	.3055	.4912	.4912
	3. Slag Pouring	.0000	.6671	.6671
	5. Slag Processing	.0000	.0181	.0181
7.2 Blast Furnace	2. Cast House	.3055	.4774	.4774
	3. Slag Pouring	.0000	.6671	.6671
9.1 BOF	1. Hot Metal Transfer	.0283	.0283	.0283[a]
	2. Refining	.1982[a]	.7756	.7756
	3. Charging, Tapping	.0370	.2896	.2896
	4. Slag Pouring	.0000	.0645	.0645
	5. Slag Processing	.0000	.0134	.0134
11.1 Ingot Casing	1. Teeming	.0000	.0000	.0000
13.1 Continuous Slabber	1. Emissions	.0000	.1970	.1970
16.1 Slabbing Mill	1. Soaking Pit Stack	.0000	.0000	.0000
	3. Automatic Scarfing	.1889	.1889	.1889[a]

[a]In-place.

EXHIBIT 6
Plant B: TSP Emission Rates

Process	Source	Uncontrolled	RACT	BACT	LAER
		(pounds/ton production)			
1. Ore Yard	1. Fugitive	.185	.110	.045	.019
2. Coal Yard	1. Fugitive	.119	.046	.028	.012
	2. Preparation	.400	.040	.006	.006
4. Sinter Plant	1. Windbox	4.300	.472	.270	.081
	2. Discharge	7.000	1.100	.046	.046
	3. Fugitive	.700	.700	.007	.007

(continued)

EXHIBIT 6 (Continued)

Process	Source	Uncontrolled	RACT	BACT	LAER
		(pounds/ton production)			
5. Coke Battery	1. Charging	1.140	.160	.021	.021
	2. Pushing	5.700	.043	.043	.043
	3. Quenching	8.600	2.100	1.000	1.000
	4. Doors	.710	.140	.070	.070
	5. Topside	.490	.043	.043	.043
	6. Underfire Stack	1.000	.146	.146	.073
	7. Handling	.033	.002	.000	.000
	8. Gas	.000	.000	.000	.000
7. Blast Furnace	2. Cast House	.690	.070	.042	.042
	3. Slag Pouring	.280	.280	.014	.014
	5. Slag Processing	.250	.120	.001	.001
9. BOF	1. Hot Metal Transfer	.280	.006	.006	.006
	2. Refining	51.000	.340	.026	.026
	3. Charging, Tapping	1.000	.400	.080	.080
	4. Slag Pouring	.120	.060	.010	.010
	5. Slag Processing	.170	.080	.010	.010
11. Ingot Casting	1. Teeming	.060	.060	.060	.060
13. Slab Caster	1. Emissions	.120	.120	.010	.010
16. Slabbing Mill	1. Soaking Pit Stack	.030	.030	.030	.030
	3. Automatic Scarfing	.240	.030	.030	.030

MODEL RESULTS

The PHB model showed that considerable cost savings occur when pollution is controlled on a plantwide rather than source-specific basis.

Pollution Control Costs: Source Specific vs. Plantwide (all tradeable)
(millions of 1977 dollars)

Plant	Source Specific	Plantwide (all tradeable)	Savings
RACT			
Plant A	10.419	8.455	19%
Plant B	8.053	6.674	17%
Plant C	17.903	13.516	25%
Plant D	NA	NA	—
BACT			
Plant A	16.491	15.510	6%
Plant B	15.213	14.270	6%
Plant C	30.022	28.079	6%
Plant D	.511	.407	20%

The results summarized above are based on equal total plant emissions for the source specific and plantwide cases. However, although emissions do not increase, there remains a potentially adverse effect on ambient air quality when tradeoffs occur under the bubble. Air quality is affected by emission source characteristics such as stack height, emission temperature, and exit velocity. As these measures increase, particles tend to travel a greater distance. Thus, if tradeoffs take place among sources with different individual impacts on the surrounding air quality, overall air quality may be impacted even though total emissions remain constant.

Two alternative bubble concept policies were also evaluated at Plants A and B, as shown below. First, coke ovens were removed from tradeoffs due to the possibly carcinogenic nature of their emissions. Second, while seemingly innocuous, road dust is characterized by small particulate size and broad dispersion. Thus, several groups have argued that open dust sources (ore and coal yards) be excluded from the bubble.

Pollution Control Costs: Source Specific vs. Plantwide (no coke) and Plantwide (no coke or dust) (millions of 1977 dollars)

Plant	Source Specific	Plantwide (no coke)	Savings	Plantwide (no coke or dust)	Savings
RACT					
Plant A	10.419	9.483	9%	9.853	5%
Plant B	8.053	7.093	12%	7.192	11%
BACT					
Plant A	16.491	15.945	3%	16.186	2%
Plant B	15.213	14.561	4%	14.565	4%

Sectoral Planning for the Indian Concrete Industry

On a warm day in April 1974 in New Delhi, Mr. Bala Chatterjee, Mr. Shankar Mahini and Mr. Raju Parpia were preparing their strategy for an important meeting that was to take place the next day to discuss a mathematical model they had developed for investment planning in the cement industry.

The three young men had obtained MBAs from a leading school of management in western India and had been working as a team for the last few months in the Investment Appraisal Division (IAD) of the National Planning Board, Government of India. They were part of the multi-disciplinary group of engineers, economists and management graduates that made up IAD, a relatively new department of the National Planning Board. Their task, when they took up the current assignment, was vaguely described as: "developing quantitative and other aids to improve investment planning in key industries of the Indian economy."

The National Planning Board provided staff support to the Cabinet for drawing up the country's Five-Year Plans. Within this broad mission, the task of the IAD, in particular, was to prepare detailed plans for individual sectors of the economy and to use these in appraising specific investment proposals made by private and state-owned corporations. After some preliminary search, Bala, Shankar and Raju agreed with Mr. Kapoor, the head of the IAD, that they would first take a shot at the cement industry. Nearly two months had gone by and, by April 1974, the IAD team was quite familiar with the technoeconomic issues surrounding the cement industry. It seemed to them that a mixed integer programming model was the most appropriate tool to use for long-term investment planning for that industry. They had put down their ideas on the proposed model, the kind of data they would use, the likely outputs of the exercise, etc. and circulated it to a team of experts from other government departments and private and state-owned cement companies. It was to this panel of experts that they would be "defending" their model the next day.

THE POLITICAL SYSTEM IN INDIA

India is a Parliamentary democracy in which representatives are elected every five years to two assemblies—one, the local State Legislature, and the other, the national "Lok Sabha" ("House of the People"). The Government, both at the national level (Centre) and at the State level, is formed by the party which has the greatest strength in the respective assembly. State Governments are headed by Chief Ministers and the Central Government by the Prime Minister. The Head of State is the President who is elected by Members of Parliament and State Legislatures.

Under the constitution, there is a complicated delineation of authority between the Centre and the States. There are some subjects, such as National Defense and Foreign Policy, which are exclusively under the control of the Centre. Others, like Education and Agriculture, are wholly within the jurisdiction of the States. Finally, there is a "concurrent list" covering matters over which both the Centre and the States have jurisdiction.

PUBLIC AND PRIVATE SECTORS

The delineation of roles between public and private sectors in the Indian economy is as important, from IAD's point of view, as the division of powers between the Centre and the States. India has a mixed economy, one in which public and private sectors have a complementary role to play in economic development. Public sector control arises from the need to insure the social objectives of national development (e.g., to eradicate poverty, eliminate regional imbalances in development, provide equal opportunities, attain technical and economic self-sufficiency). The founding fathers of the country felt that the public sector could be an important tool in realizing these goals, a view still widely held by politician in power.

The Public Sector spearheads the national effort in health, education and infrastructure development. It has also been actively involved with Industry, as laid out in the "Industrial Policy Resolution (1956)" which divides industries into three categories: the first, those industries whose development is the exclusive responsibility of the public sector (defense, atomic energy, aircraft manufacture, etc.); the second, industries in which the Public Sector takes most of the initiative in establishing new firms, but where the private sector is expected to supplement these efforts (e.g., machine tools, fertilizers, drugs, etc.); and third, the remaining industries (cement, paper, light engineering, etc.), whose future development is generally left to the initiative and enterprise of the private sector. Thus, both sectors are expected to play an important role in industrial development.

The National Planning Board (NPB), located in New Delhi, is at the apex of India's national planning system. It was formed in 1950 with a long list of functions (Exhibit 1). Its primary objective is to assist the Prime Minister and his or her Council of Ministers—formulating, appraising, and reviewing long-term economic plans for the country.

NPB draws up long-term plans with a 20-year horizon. Detailed plans are then mapped out for each five-year period. This broad outline and priorities for each plan are provided by the National Development Council (NDC), a body presided over by the Prime Minister and composed of Union Cabinet Members, Chief Ministers of State, and Members of the Planning Commission.

The NPB acts as a coordinator for input from all sectors of the economy, both government and private. Decisions are complicated because of the trade-offs between different sectors of the economy, between the Centre and the States, and between the public and private sectors. The NPB's recommendations are presented to the NDC for discussion and debate. Final approval authority rests with the Parliament.

Each Five-Year Plan sets targets only for the terminal year of each period. A steady building in production is assumed to take place from current output levels to target levels during the period. With respect to investments by the private sector, the Plan does not specify which firm will increase capacity, at what cost, or where, to achieve targeted production levels. In fact, there is no guarantee whatsoever that private firms will fill the roles expected of them by the government.

EXHIBIT 1
Functions of the National Planning Board of India

The formal functions of the National Planning Board:

1. to make an assessment of the material, capital and human resources of the country, including technical personnel, and investigate the possibilities of augmenting such of these resources as are found to be deficient in relation to the nation's requirements;

2. to formulate a Plan for the most effective and balanced utilization of the country's resources;

3. on a determination of priorities, to define the stages in which the Plan should be carried out and propose the allocation of resources for the due completion of each stage;

4. to indicate the factors which are tending to retard economic development and determine the conditions which, in view of the current social and political situation, should be established for the successful execution of the Plan;

5. to determine the nature of the machinery which will be necessary for securing the successful implementation of each stage of the Plan in all its aspects; and,

6. to appraise from time to time the progress achieved in the execution of each stage of the Plan and recommend the adjustments of policy and measures that such appraisal may show to be necessary.

Since 1950, each Plan has marked an important phase in India's economic history. The beginning and the end of a Five Year Plan are regarded as vital dates in the national life, each Plan being at once an appraisal of the past, a guide map for action in the future and a reformulation of basic national policies in the light of experience and evaluation. Events now move faster than a few years ago, and adjustments over shorter periods have gained in importance.

Source: Government of India, Planning Commission, *The Planning Process* (1975), p. 3.

INDUSTRIAL REGULATION

Indian industry is subject to considerable regulation, the objective of which, broadly speaking, is to ensure that the development and conduct of industries proceeds according to Plan priorities and targets. Both public and private firms are subject to these regulations. The three major regulation categories are industrial licensing, price and distribution controls.

The framework for industrial licensing is provided by the Industries (Development and Regulation) Act of 1951. The main objectives of licensing are: to prevent concentration of ownership of industries; to ensure the balanced economic development of different regions of the country; and to protect "small-scale industries" from unfair competition from medium and large industries. As per this Act, no new industrial undertaking can be begun, nor substantial expansion of any existing plant undertaken, without obtaining a license from the Central Government.

Price and distribution controls have been used in a number of industries in India: for example, iron and steel, nonferrous metals, coal, fertilizers, cement, paper, sugar, drugs and medicines, bicycles, tires and tubes, soap, matches, and so on. Some controls, for example, those on iron and steel or cement, were prompted by a desire to ensure allocation of adequate amounts of these scarce inputs to "priority" sectors of the economy at a "reasonable" price. Other controls, such as those on motor cars, were prompted by "equity" considerations. Finally, price controls have been used to curb inflationary trends. These controls have had an important effect on the willingness of the private sector to invest resources in price-regulated industries.

THE INVESTMENT APPRAISAL DIVISION

The Investment Appraisal Division (IAD) is one of the newest divisions in the National Planning Board. Set up in 1972, its two main functions are: (i) to draw up long-term investment plans (choice of location, size, technology, etc.) for key industries of the economy; and (ii) to appraise specific investment proposals submitted to Government by various public enterprises.

Investment plans prepared by IAD help to translate the broad, terminal-year production targets laid out in the Five-Year Plans into concrete, time-phased goals. Thus, they serve as a framework for IAD to perform its second function, namely Project Appraisal. Sectoral studies are carried out by IAD with the active involvement and participation of representatives of the concerned industry and the corresponding administrative Ministry, since the latter have better access to data and ultimately have to take on responsibility for implementing the studies' recommendations.

For project appraisal, IAD employs the technique of social cost-benefit analysis. In this approach, projects are evaluated in terms of their net benefit to society as a whole, rather than in terms of their net benefit to the sponsoring organization alone. IAD evaluation is believed to have an important influence on the final decision taken by Government as to whether to undertake any particular public sector project.

SELECTION OF AN INDUSTRY FOR STUDY

The first job of the Task Force on sectoral planning was to select the industry which should be studied. Several candidates offered themselves: oil refineries, petrochemicals, aluminum, zinc, copper, engineering, iron and steel, coal, power, cement, paper and others. Of these, the cement industry appeared to be the most attractive to the IAD team for a number of reasons. First, there were a large number of potential locations for new cement plants.

This was so because limestone, the main raw material for cement manufacture, is quite widely distributed over India. Also, the economic size of a cement plant is quite small compared to the total additional capacity proposed to be installed in the Fifth Plan. Second, transport costs are significant for cement, given its bulk. Therefore, the trade-off between economies of scale and transportation costs had to be carefully considered. Finally, relatively more data was available on the cement industry, particularly with regard to past patterns of cement movement and consumption, than for any of the other industries. Thus, the IAD team decided to try to use mathematical modelling techniques for preparing a long-term investment plan for the cement industry. Bala, Shankar, and Raju began systematically collecting background information on the cement industry.

CEMENT INDUSTRY BACKGROUND

Capacity and its Utilization

Cement is an important intermediate good for any developing economy. With the advent of rapid industrialization after India became independent in 1947, the production of cement increased several-fold. At the end of the First Five-Year Plans (1951–56), installed capacity for cement was 5.4 million tons. This increased to 9.2 million tons by 1961, 12.0 million tons by 1966 and 19.7 million tons by 1973. The distribution of installed capacity in 1973 (plant location, size, and age) is set out in Exhibits 2, 3, and 4 respectively.

In 1973, cement was produced in 52 plants located in 14 states. Associated Cement Companies (ACC), a private sector firm, owned and operated 17 of these plants and was the largest single supplier. The public sector had entered this industry in the late fifties and by 1976 had nine plants in operation. Of these, the Cement Corporation of India, a public enterprise owned by the Central Government, operated two plants; the other public sector firms were owned by the State Governments.

Exhibit 5 shows the geographical distribution of cement capacity and consumption in India in 1972. Exhibit 6 provides figures on the interregional movement of cement in the same year. Officials in government were quite concerned about the regional imbalance in supply and demand for cement and the sharp rise in the average distance of cement haulage in recent years.

Most of the cement factories produced what is known as ordinary Portland Cement (OPC), the most commonly used variety of cement. Some other plants produced special cements such as white cement, rapid hardening cement, hydrophobic cement or oil-well cement, in addition to OPC. Two plants produced only Portland Blast Furnace Slag (PBFS). These plants mixed slag, a throw-away by product of steel production, with regular clinker to produce PBFS, which is as effective as OPC for many applications.

EXHIBIT 2
State-wide Distribution of Cement Manufacturing Capacity (as of September 1973)

	Name of State[a]	Annual Capacity (thousand tons)	No. of Cement Plants
i.	Jammu and Kashmir	20	1
ii.	Kerala	51	1
iii.	Meghalaya	84	1
iv.	Maharashtra	600	2
v.	Haryana	545	2
vi.	Orissa	797	2
vii.	Uttar Pradesh (U.P.)	907	2
viii.	Andhra Pradesh (A.P.)	1926	6
ix.	Rajasthan	1966	4
x.	Karnataka	2034	6
xi.	Gujarat	2281	6
xii.	Bihar	2408	7
xiii.	Madhya Pradesh (M.P.)	2729	5
xiv.	Tamil Hadu	3412	7
	Total	19760	52

[a]See also map of India in Exhibit 5. Note that some large states, such as Himachal Pradesh (H.P.) and West Bengal, have no cement plants.

Source: Government of India, Tariff Commission, *Report on Comprehensive Review of the Cement Industry and Revision of Fair Ex-Works Prices Payable to the Producers* (1974), Bombay, India, p. 31.

EXHIBIT 3
Distribution of Cement Plant Capacities

Annual Capacity Range (000 tons)	Number of Plants
Below 100	4
100–199	4
200–299	12
300–399	11
400–499	10
500–599	4
600–699	1
700–799	1
800–899	2
900–999	1
1000 and Above	2
Total	52

Source: Government of India, Tariff Commission, *Report on Comprehensive Review of the Cement Industry and Revision of Fair Ex-Work Prices Payable to the Producers* (1974), Bombay, India, p. 29.

EXHIBIT 4
Age Distribution of Cement Plants

Age range (years)	Number of Plants
Below 10	14
10–19	14
20–29	8
30–39	9
40–49	1
50 and above	6
Total	52

Source: Government of India, Tariff Commission, *Report on Comprehensive Review of the Cement Industry and Revision of Fair Ex-Works Prices Payable to the Producers* (1974), Bombay, India, p. 28.

EXHIBIT 5
Production and Consumption of Cement in India (1972)

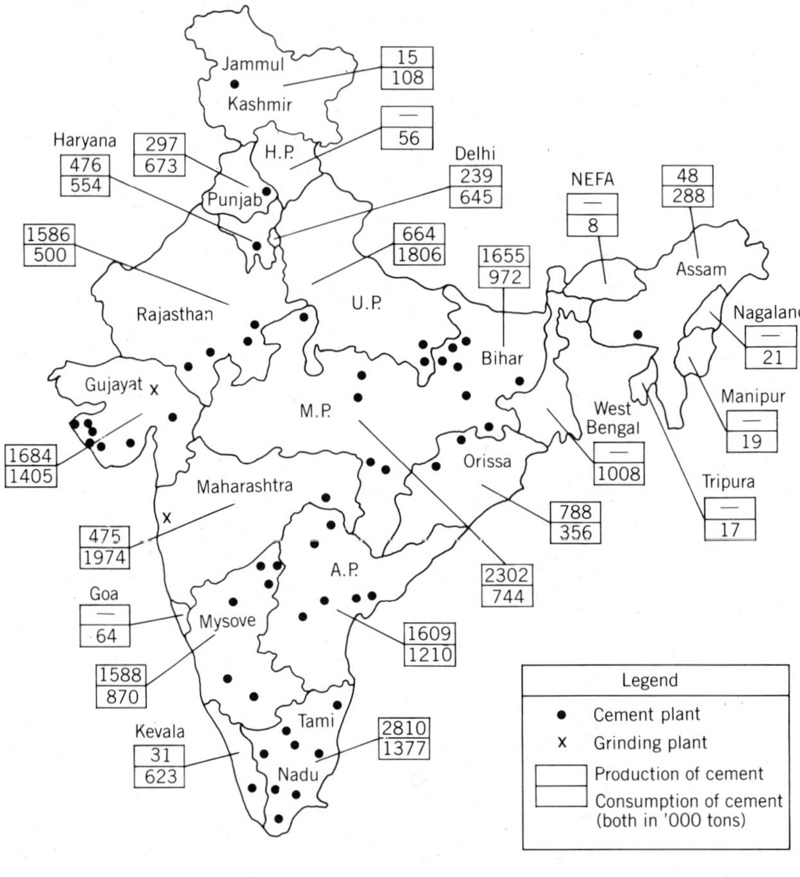

Source: Government of India, Tariff Commission Report on the Cement Industry (1974), Appendix 12.

EXHIBIT 6

Consolidated Inter-Regional Movement of Cement (1972)[a] (thousand tons)

From	To North	To East	To West	To South	Total
North	2,737	23	101	—	2,861
East	234	2,290	26	—	2,550
West	1,231	343	2,785	48	4,407
South	300	215	1,380	4,072	5,967
Total	4,502	2,871	4,292	4,120	15,785

[a]Regions are defined as follows: Northern region includes all states north of U.P. and Rajasthan; Western region includes Gujarat, Maharashtra and M.P.; Eastern region includes Bihar, Orissa, West Bengal and the other eastern states; Southern region includes Mysore, A.P., Tamil Nadu and Kerala.

Average Movement of Cement

Year	Approximate average distance moved (kms)
1969	480
1970	495
1971	525
1972	535

Source: Government of India, Tariff Commission, *Report on the Cement Industry*, Table 30.6.

Sixteen cement firms in the private sector manufactured refractories, machinery, jute goods, sugar, foundry, iron, and steel, in addition to cement itself. ACC had its own cement machinery manufacturing plant in the south which also sold its products to other cement firms.

Capacity utilization in the cement industry has been poor (see Exhibit 7). The highest utilization was in 1969 when it reached 94%. However, in some years, as in 1971 or 1973, it fell to 77% and 74% respectively. Individual plants have been able to operate at or above 100% rated capacity, but several plants operating at 60% or below have brought down the overall average. Supply shortages are believed to be a major cause of the underutilization of capacity: these include power shortages, short supply of wagons to move cement, short supply of coal, water, and so on. In 1973, for instance, over 50% of the total loss in production was due to power breakdowns/restrictions. Short supply of wagons accounted for another 12% of the loss.

Manufacturing Processes

The main raw materials for cement production are limestone and coal. Limestone containing calcium carbonate is reduced in a kiln in the presence of

EXHIBIT 7
Trend in Capacity Utilization of Cement Plants

Year	Annual Capacity (thousand tons)	Annual Production (thousand tons)	Percentage
1968	14,314	11,892	82
1969	14,514	13,576	94
1970	16,916	13,912	82
1971	19,349	14,904	77
1972	19,705	15,713	80
1973 (Jan.– Sept.)	14,783[a]	10,893	74

[a]Pro rata basis.

Source: Government of India, Tariff Commission, *Report on Comprehensive Review of the Cement Industry and Revision of Fair Ex-Workers Payable to the Producers* (1974), Bombay, India, p. 33.

coal to produce calcium oxide. This combines with soil and laterite to produce what is known as "clinker." Clinker is lumpy and has to be ground to produce cement. A small quantity of gypsum is added during clinker grinding to prevent the cement from setting too easily. The ground cement is then packed in sacks and dispatched.

There are two major processes for cement manufacture—the wet and the dry processes. In both cases, limestone is first quarried from a nearby deposit and sent to a crusher where it is reduced in size. It is then blended with inferior/superior grades of limestone and additives such as clay so that the mixture has the desired composition. The mixture is then sent to the Raw mill for grinding. In the wet process, water is added to insure a homogenous character for the slurry that emerges from the raw mill. In the dry process, however, this is achieved by using compressed air. The wet process requires more coal per ton of cement produced to cause the excess water to evaporate. It also requires a longer kiln than the dry process.

The dry process is the newer technology. It involves lower capital and operating costs and makes better use of the waste heat carried by flue gases leaving the kiln. However, it requires relatively more sophisticated equipment to pre-heat the "raw meal"[1] before it enters the kiln. Ten of the plants in operation in 1974 employed the dry process and the rest employed the wet process.[2] Most of the new plants under construction, however, were based on the dry process.

There was also the possibility of wet process plants converting to the dry process, especially at a time when they were carrying out plant rehabilitation

[1]The term "raw meal" is used to describe the raw mill output to be fed into the kiln. In the wet process, this is a slurry since water is added in the raw mill. In the dry process, it is a dry mixture.
[2]Three plants used a combination of both wet and dry processes in different sections of the factory.

programs. Conversion appeared economically very attractive since it could almost double the capacity of a wet process plant at a capital cost per annual ton of capacity that was only 50% of that required for setting up the same capacity in a new dry process plant. Moreover, it would reduce operating costs by lowering coal consumption per ton of cement. Conversion did pose a number of technical problems and, for one reason or another, few plants had actually converted from wet to dry process.

Two other manufacturing possibilities are of interest. First, under "split location production," clinker is not ground into cement at the cement plant, but at a location closer to the point of actual consumption. For example, the factory at Sawaimadhopur (Rajasthan State) might produce clinker and send it to a grinding unit in the outskirts of New Delhi for converting into cement. The advantage of this arrangement is that clinker can then be conveniently shipped in open railway wagons which are more readily available than the closed wagons required to transport ground cement.

The other important manufacturing possibility is to use slag produced in steel plants to make Portland Blast Furnace Slag (PBFS) cement. Millions of tons of slag are dumped every year by Indian steel plants. If this slag is granulated in a Slag Granulation Plant, it can be inter-mixed with clinker up to a ratio of 1:1 and ground in the normal cement mill to produce P.B.F.S. cement. Because the cost of granulating slag is only 15%–20% of the cost of producing clinker, production costs are significantly reduced. Despite these advantages, less than 15% of the slag available in 1974 was being granulated and mixed with clinker to produce PBFS cement. The production of slag cement can be more efficiently achieved by using split location production. Slag is best granulated next to the steel plant itself and clinker is best produced close to a limestone deposit. Therefore, instead of moving granulated slag to the cement factory for inter-grinding with clinker, it is more cost effective to transport granulated slag and clinker to a demand center for inter-grinding there.

Table 1 summarizes the various production possibilities available in the cement industry.

Raw Materials for Cement Production

Exhibit 8 shows the amount of raw material required per ton of cement. Since every ton of cement requires about 1.4 tons of limestone, cement plants have invariably been located close to limestone deposits.

The percentage of calcium carbonate Caloz in limestone varies from deposit to deposit. The higher the Caloz content, the richer the deposit. In regions where the quality of limestone is poor, a process known as beneficiation can be carried out to improve its quality. This involves crushing the inferior ore and forming a slurry in water which is then slowly stirred so that richer slurry is separated from contaminations. The enriched slurry is then sent to the kiln and the rest of the manufacturing process is the same as in the Wet process. (Note that when beneficiation is necessary, the Dry process cannot be used.)

TABLE 1
Alternative Production Possibilities in the Cement Industry

Legend:	□	Signifies a clinker production unit
	○	Signifies a clinker/slag grinding unit
	△	Signifies a granulated slag production unit
		Signifies distribution
	===⇒	Signifies transportation to another zone

1. The Integrated OPC Plant

Clinker is manufactured and ground into OPC at the same location.

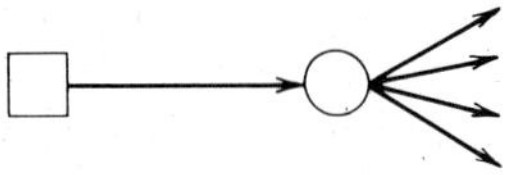

2. Split Location OPC Plant

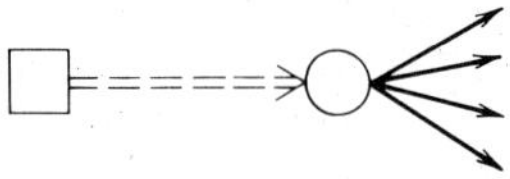

3. Integrated Slag Cement Plant

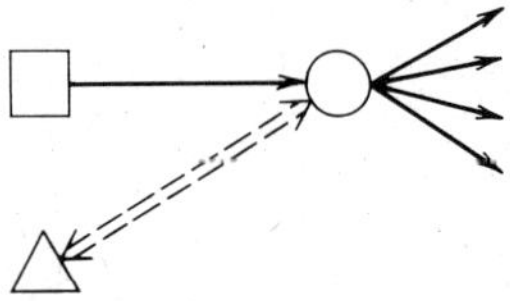

4. Split Location Slag Cement Plant

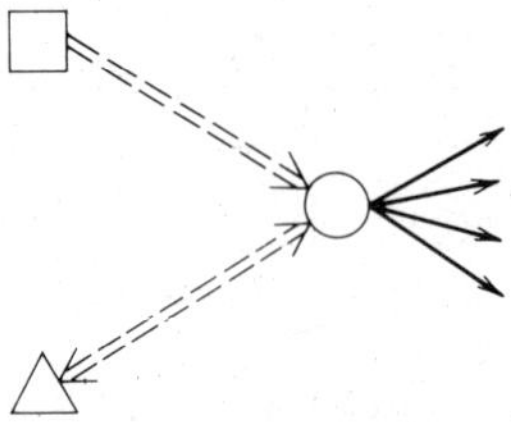

As a rough estimate, beneficiation may increase operating costs by $2 per ton of cement and capital costs by $0.25 million.

Information on limestone resources in the country is periodically released by the Geological Survey of India (GSI). Reserves are indicated under three broad headings: Measured, Indicated and Inferred. Measured reserves are very reliable reserves since they have been proven by detailed studies on surface samples, drilling of bore holes, and so on. Indicated reserves are estimated on the basis of how much proven reserves exist in the neighborhood. Only a few drill holes are made to actually test the quality and quantity of deposits. Inferred reserves are estimated on the basis of surface samples only and are quite uncertain. GSI data also indicates which reserves are considered to be of "cement-grade."

Limestone deposits also vary in terms of how mineable they are. For instance, an excellent deposit may not be economically mineable if it is at a depth of more than 40 meters. There is also the danger of hitting underground water reserves if the mine has to be deep. Generally, overburden ratios (i.e., tons of overburden per ton of limestone) of 1:4 to 1:3.5 are preferred although a ratio of 1:2 may be acceptable if the limestone is of superior quality and the quarry is near the factory. In hilly regions, limestone may have to be transported from the quarry to the factory by aerial ropeways. This could increase the cost of the project by as much as $2.5 million. Open cast mines are usually easier to exploit. If the quality of a deposit is uniform, it can be mechanically mined. Manual mining requires four to five times the manpower needed for mechanical mining.

Limestone, like all other minerals, is a national asset and private parties can mine deposits only with the approval of the government and by paying appropriate royalties.

The other important raw material for cement manufacture is coal. Coal is mined chiefly in eastern and central India. Cement plants located in the south often run out of coal as there are few coalfields nearby and railway wagon

EXHIBIT 8
Materials Input for Manufacture (dry process)

Material	Tons Used (per ton of output)
Limestone	1.44
Laterite	0.04
Soil	0.06
Gypsum	0.03
Coal	0.19[a]
Power	128 kw
Water	300 gallons

[a]In the wet process, this would be approximately 0.33 tons.

shortages delay dispatch of coal to their factories. The cost of transporting coal, especially for plants in the south which are distant from coalfields, can be as high as the pithead cost of coal. Thus pithead coal costing $5 per ton could cost $10 by the time it is transported to a cement plant.

Other raw materials are required in very small quantities and are easily available.

Finally, a critical input is Power. Cement plants are power-intensive and a power-cut often chokes production.

Capital and Operating Costs

Capital costs for a typical 1200 tons per day Dry process plant is set out in Exhibit 9. Corresponding operating costs are set out in Exhibit 10. Economies of scale in capital and operating costs can be significant, as shown by the figures in Exhibit 11. The exact costs will vary from site to site depending on the manufacturing process employed, quality of limestone, the cost of mining and moving it to the plant, distance from coalfield, local electricity tariffs and wage rates, length of railway siding to be built, degree of mechanization, and so on.

A cement plant can normally be constructed within three years. In practice, it often takes up to 4 years due to construction delays. After the plant has been commissioned, it usually takes a while to achieve its rated capacity.

Foreign collaboration is not allowed in the cement industry, nor is the industry permitted to import plant or any machinery. However, a small percentage of the total cost, approximately 5%–8%, may be released to manufacturers of cement machinery for importing components which are not available indigenously.

According to a study of the cement machinery industry, the seven firms in the industry claim they can supply machinery for 17 "standard" size cement plants (600 TPD or 1200 TPD) every year. The estimate is almost certainly optimistic. The largest machinery manufacturer, ACC, claims to have the capacity to supply seven plants a year.

At one time, the government recommended a standard plant size of 600 TPD; more recently, 1200 TPD plants have been considered most desirable. There have been problems in setting up plants bigger than 1200 TPD since some of the components required were larger than what could be transported by rail. Some machinery manufacturers, however, claimed they could manufacture equipment for even 2000 TPD plants. In developed countries, plants of 3000 TPD are quite common.

Transportation of Cement

Cement is transported from factories by road, sea or rail, or a combination of these. Rail movement is the most common since cement is a very bulky commodity and long-distance movement by other modes is very expensive.

Cement is one of the most important commodities carried by rail, second only to coal. Cement is usually moved in closed wagons, but clinker can be

EXHIBIT 9
Capital Costs for Typical 1200 TPD Dry Process Cement Plant (thousand Dollars)

Plant and Machinery	12,250[a]
Civil Engineering Works	6,100
Other Expenses	300
Interest during Construction period	600
	19,250

[a]Includes $1.6 million sales taxes.

EXHIBIT 10
Operating Costs for Typical 1200 TPD Cement Plant[a]

	Quantity per Annum	Cost per Annum (thousand dollars)
Raw materials & additives	See Exh. 8	454
Coal (Pit-head cost $8/ton plus average transport cost $5/ton)	74,000 tons	1,036
Power	51.2 GWh	640
Water	120 M gallons	22
Total (raw materials, fuel, power, etc.)		2,152
Salaries and Wages		259
Repair and Stores		275
Overhead (including interest)		100
Cement Packing charge		100
Selling expenses		67
Depreciation		1,000
Total cost of Production		3,953
Cost of Production per ton		$9.88

[a]Equivalent to 400,000 tons annual production capacity.

EXHIBIT 11
Estimates of the Economies of Scale in Capital and Operating Costs

According to a study cited in the Tariff Commission Report, the following economies of scale in capital and operating costs are estimated for cement plants in India:

Annual cement capacity in tons per year (tons/day in brackets)	Capital Cost per ton of annual capacity, as a % of capital cost per ton of a 200,000 TPA plant	Operating cost per ton of annual capacity, as a % of operating cost per ton of a 200,000 TPA plant
60,000 TPA (200 TPD)	135	120
200,000 TPA (600 TPD)	100	100
400,000 TPA (1200 TPD)	85	72

moved in open wagons which are more abundant. Wagons are loaded at the railway siding next to the factory and then attached to regular goods trains at the nearest railway station. If a factory does not have a railway siding, the cement would have to be moved by road to the nearest railway station.

From the factory, cement might be shipped to a government depot, to a selling agent's warehouse, or directly to a private consumer. In the process, a wagon might have to be sequentially connected to different trains.

The Indian railway specifies tariffs for different classes of goods for different distances of movement. The rates are telescopic—that is, unit transportation cost decreases with increasing haulages and they are the same for all routes in the country. Exhibit 12 sets out the railway rates for cement transportation. The railway rates are subsidized for essential commodities like sale and foodgrains, but this is compensated for by charging higher rates on other commodities.

Movement by road is becoming more popular, especially over long distances, since railway service has been unreliable.

Regulation of the Cement Industry

Under the Industrial Policy Resolution, cement is a sector open to both the public and private sectors. Like all other major industries, it is subject to the Licensing system.

Since 1942, the cement industry has been faced with elaborate controls. Originally, price and distribution controls were imposed during the second World War, largely to ensure the supply of cement for defense users at controlled prices. After the war, prices continued to be fixed through executive orders until 1956 when prices and distribution of cement came under the provisions of the Cement Control Order promulgated in exercise of the Powers conferred under the Industries (Development Regulation) Act, 1951. Since 1953, four enquiries have been conducted by the Tariff Commission, Govern-

EXHIBIT 12
Railway Freight Rates for Cement Movement

Distance (km)	Closed Wagons ($/ton)	Open Wagons ($/ton)
100	1.68	1.54
200	2.90	2.26
300	3.24	2.94
500	6.61	4.16
750	6.23	5.60
1000	7.75	6.98
1500	10.55	9.48
2000	12.96	11.64
2500	14.85	13.33

ment of India, into the question of prices payable to cement producers. The ex-factory price has been the same for all producers, with the exception of a very limited number of producers who have been allowed higher prices. The last enquiry was completed in 1974, when ex-works prices of $17 per ton were recommended for existing units, $17.60 per ton for units then under expansion or construction and $18.25 per ton for units to be built in the future.

The price charged for cement from final consumers was much greater than that allowed for producers because the former included the cost of transportation of cement, stocking and distribution expenses, and excise duties. In 1974, for instance, the consumer price of cement was close to $30 per ton.

In 1974, distribution controls on cement were being managed entirely by the Cement Controller at New Delhi, subordinate to the Ministry of Industrial Development. The Cement Controller has four regional offices in the north, south, east and west. He has powers to prohibit removal, to direct sale or transport or disposal of stocks of cement and to issue necessary directives to the producers in this regard. The control on distribution of cement is designed to be a simple mechanism for equitable distribution, particularly to make cement from surplus regions to deficit regions.

THE MATHEMATICAL MODEL

Based on their extensive reading on the cement industry, Bala, Shankar and Raju developed a tentative mathematical model for long-term planning for this industry. They planned to discuss the outline of their model with a cross-section of experts from industry and government before proceeding further.

Summary of the Model's Goals, Facts, and Assumptions

The main features of the model are set out below. A detailed mathematical formulation is given in the Appendix.

1. The planning horizon of the model would be the ten-year period from 1974 to 1984. Two varieties of cement would be considered: Ordinary Portland Cement and Slag Cement. The model was expected to recommend *when* and *where* new cement/clinker/grinding plants should be constructed, *what their capacities* should be, and *what distribution pattern* should be followed for cement, clinker and slag each year.

2. The objective to be minimized was to be the present value of capital, operating and distribution costs over the ten-year period, subject to the constraints of the model. The main constraints would be: for each period and each zone, demand constraints, supply constraints, and slag availability constraints; and slag/clinker ratio constraint in the production of slag cement.

3. All new cement plants would be assumed to employ the more efficient Dry process. Three standard sizes of cement/clinker plants would be considered: 200 tons/day, 600 tons/day and 1200 tons/day; typical capital and operating costs for each would be used.

4. It would be assumed that cement plants could be constructed in three years; that is, an investment decision in year 1 would result in cement production in year 4. All new plants would be assumed to operate at 85% of capacity.

5. It would be assumed that all slag produced in steel plants could be granulated in Slag Granulation Plants located next to the steel plants. This slag could then move either to integrated cement plants for inter-grinding with clinker or to demand zones where separate grinding units would convert it to slag using clinker received from some other zone. Slag would be priced at roughly 20% of the cost of clinker production in a 1200 tpd plant. Split location possibility would be confined to a limited number of zones in the east and north. Capital and operating cost for three standard size grinding units (200, 600, and 1200 tpd) would be considered in the model.

6. The proportion of slag in slag cement would be fixed at 50%.

7. All existing cement/clinker/grinding plants would be assumed to operate at a capacity utilization level consistent with the past several years. The option to close down some existing plants would not be considered, since this would be very difficult to implement. Cement plants under construction would be assumed to go on stream as per their latest completion schedules. All other projects at very preliminary stages of implementation, including licensed projects on which progress had been negligible, would be ignored for the purposes of the exercise.

8. The over 300 districts in the country would be grouped into 100 demand zones of approximately equal size. Demand projections for cement contained in the Five-Year Plan for 1978–79 and 1983–84 would be used to develop estimates of district demand. The process of developing year-by-year district demand forecasts would involve use of regression models, analysis of past trends in consumption, judgments about likely change in trends in the future and so on. In fact, developing these demand forecasts would be a major sub-task in applying the mathematical model.

9. Supply zones would be formed around limestone deposits. Given the geographical distribution of limestone in India, the number of supply zones would be around 40. But this number could be decreased to around 25 if deposits which did not have surplus limestone (after fully providing for the future needs of existing cement plants) were excluded.

10. For each supply and demand zone, a representative city/town would be selected. For the purposes of the model, it would be assumed that the entire supply and demand of cement were concentrated at that city/town. Distance between supply and demand zones would be measured by the distances between their respective representative points. In this way, a "distance matrix" for pairs of zones would be prepared. Then, using the distance matrix and the railway tariff for haulage of cement/slag/clinker over different distances, transportation cost matrices for cement/clinker/slag would be developed and used in the model.

11. The model would be constrained to allow a maximum of 8 new cement plants (of all sizes) in any year since it was unlikely that cement manufacturers could supply more than that many plants in any year.

12. To maintain regional balance in the construction of new cement plants, no single state would be allowed to have more than six new cement plants in the ten-year period.

Meeting with the Experts

The IAD team felt it worthwhile to discuss their formulation of the mixed integer programming model, its goals and restrictions, with a panel of experts from Government and outside before proceeding further. They expected this to serve two important purposes: it would provide useful feedback on the limitations of the model as presently formulated and possibly generate ideas for improving the model; secondly, it would help to create greater commitment for the IAD exercise from key individuals in Government, the private sector and the public sector.

A note was therefore prepared by the IAD team outlining the goals, facts and assumptions of the model, its mathematical formulation and the salient features of the cement industry. The note was circulated to the following persons by the head of the IAD:[3]

1. Mr. P. N. Sethi, a senior civil servant in the Ministry of Industry (MOI) in charge of the cement industry.

2. Mr. R. Kumar, Director (Planning) in Bharat Cements Ltd. (BCL), a major private-sector cement manufacturing company.

3. Mr. D. S. Gupta, a senior executive in the Geological Survey of India (GSI).

4. Mr. R. Badvi, head of the Department of Transportation in the Ministry of Shipping and Transportation (MOST).

[3]All names and designations are disguised.

5. Mr. S. V. Sami, Deputy General Manager, Cement Corporation of India (CCI) (a state-owned enterprise).
6. Mr. V. K. Rao, Senior Economist, National Planning Board (NPB).
7. Mr. S. Singh, Head, Delhi Computing Centre (DCC).

With the consent of the concerned persons, a mutually convenient time for a meeting was fixed. The IAD team was concerned about how their presentation to this distinguished panel of experts would go. They waited for D-day—Monday, September 15, 1974.

The Meeting

The meeting began punctually at 10 A.M. in a compact conference room. Fortunately, all invited participants were present. Mr. Kapoor, Head of IAD, welcomed the experts and following a few words of introduction, gave the floor to the IAD team. The team made a 30-minute presentation reviewing the basic material in the note they had circulated. The meeting was then thrown open to the experts for comments, criticisms, questions, and suggestions for improving the mixed integer programming model.

Mr. P. N. Sethi from The Ministry of Industry, began with a fundamental question.

Mr. P. N. Sethi: (MOI) *I must say at the outset that I am no mathematical wizard and frankly, most of the fancy mathematics and symbols in your note went right over my head. But I think I understood what you were driving at. And that left me somewhat uneasy.*

I have been dealing with the cement industry for the last three years. It is crystal clear to me that we have a very lop-sided situation today in the supply and demand for cement. There's surplus cement in the south and a deficit in the north and to the west. As a result, millions of tons of cement have to be transported every year from surplus to deficit regions. We all know why: it's because there's plenty of good limestone in the south and a lot less in the north, so manufacturers have set up several new plants in the south and few in the north. Since the transport charges are pooled under the Government's freight equalization scheme and cement sells at the same price all over the country, manufacturers have been indifferent to the cost of transport. They have tried to minimize their own costs of production (by exploiting the best limestone deposits) rather than minimize the final cost to customers (i.e., including both production and transportation costs).

The way out of this unbalanced supply-demand situation is clear. In the next ten years, Government should encourage firms to set up new plants in the north in states like Madhya Pradesh,

Rajasthan and Uttar Pradesh, by providing financial incentives, if necessary, and no more cement plants should be allowed in the south. What is the need to use complicated, sophisticated models to reach such an obvious conclusion?

I am sorry if I sound discouraging. We quite appreciate the enthusiasm and efforts of IAD's young officers, but frankly, it seems like a very devious and messy way to reach a straightforward conclusion. In fact, my Ministry is already working on ways to promote cement plants in northern India.

Bala, who was sitting restlessly throughout Mr. Sethi's monologue, could hardly wait to answer the question.

Bala: *The point you have raised, sir, is one that initially bothered us, too. But we now believe that it is worthwhile developing this model.*

Today, the supply and demand for cement are spatially so much out of balance that a sophisticated model might well throw up the same kind of results that intuition leads us to expect. However, in the longer term, it may be very difficult to make optimal investment decisions simply by inspection or by straightforward manual methods. Even in the present situation, it is not easy to decide where exactly in MP, Rajasthan or UP, new cement plants should be located. As you know, sir, MP itself is several hundred miles across in length and width.

Besides, you mentioned, sir, that Government should give financial incentives, if necessary, to encourage investors to set up new plants in the north, but how are we to decide the size of these incentives ought to be? Here the model can come in handy. From what are known as the "shadow prices" in the model's output, one can estimate that it would be worth the country if additional cement capacity were available in the north rather than in the south, and therefore, what additional prices should be allowed to cement plants located in say Rajasthan, where beneficiation of limestone may be absolutely unavoidable.

Mr. P. N. Sethi: *Well, it is still not clear to me exactly what the benefits of using the MIP model are. As for the quantum of incentives that ought to be given, we have already made some calculations in my Ministry. Our approach is essentially as follows: about one million tons of cement are shipped annually from deep in the south to the north, at an approximate average cost of $6 per ton; if the same cement had been manufactured in the north, the transport cost would not have exceeded $2–$3 per ton. So, as a rough cut, we are willing to add half the difference, that is $2 to $3 per ton to the price approved for plants in say Rajasthan. We know that this would more than cover* *the extra cost of beneficiation in most cases. So, to repeat what I said earlier, I don't see why one needs to invest so much time and energy in developing complicated models when equally good rough-and-ready methods are available.*

Mr. Kapoor: (Head, IAD) *Sethi, I think your point is well taken. Although I don't agree with you entirely, I think we should constantly ask ourselves the question you are raising: do the benefits of this model exceed its costs? If we eventually find that it does not—and I personally believe that this cannot be decided in just one or two years—we should be willing to scrap and adopt less sophisticated methods.*

Mr. S. V. Sami: (CCI) *I had a somewhat different problem. As it raised basic questions in my mind about the utility of the MIP model, it may be appropriate for me to express my apprehensions at this stage itself.*

I have serious doubts about the reliability of the forecasts for the spatial distribution of cement demand that you hope to make. In a moment, I will explain why. But assuming this is true, I find that it completely alters the relevance of the model's output. An important feature of the elaborate model you propose to use is that it will give the optimal spatial distribution of new capacity addition between 1979 and 1984. This will be critically dependent on the assumed spatial distribution of cement demand between now and 1979. Therefore, with unreliable demand forecasts, the model will only throw up meaningless conclusions.

Now let me explain why your forecasts are likely to be unreliable. I do not know if you have already looked into this matter. It may seem a peripheral issue to you, but it seems to me to be at the heart of your exercise. Just last week, we at the Cement Corporation had to make detailed demand forecasts while preparing a project report on a new cement plant to be submitted to the Government and the problems we faced are still fresh in my mind.

You see, there is no agreement among different experts even on the total demand for cement in 1979, not to mention 1984. Few people, therefore, have ventured to forecast how this demand will be spatially distributed. Crude methods are available, but they are inadequate: for instance, some people assume that current shares of different States in the total cement consumption will hold in the future. In your case, the problem is compounded by the fact that you want to forecast demand at the level of smaller zones which, on the average, are one-sixth or one-seventh the size of a typical Indian State. Moreover, you want to do this on a year-to-year basis up to 1984. I am very pessimistic about the quality of the forecasts you will be able to make and therefore, of the real utility of the MIP model you propose to run.

Mr. Raju: (IAD) *Sir, I agree with you that demand forecasting will be a tough job. We are going to be forced to make several arbitrary assumptions, especially because of the inadequacy of data on the spatial distribution of past consumption of cement. But that may not vitiate the entire exercise because we hope, in any case, to run the model with different sets of demand forecasts. We will then see how sensitive the conclusions are to those forecasts; where they are sensitive, a "best guess" will have to be made. The point is that we can't be better off by ignoring the uncertainties surrounding future demand than we would be by dealing with it explicitly.* A priori, *we doubt if our errors in forecasting the spatial distribution will be so large as to make the entire exercise meaningless.*

Moreover, we see the model as a tool to be repeatedly used and perfected over the years. It can take a few years to get a large mathematical model such as this one working usefully—both in terms of its design and the quality of the data fed into it. But we have to make a beginning somewhere and hope that with time, many of the problems will get sorted out.

Mr. D. S. Gupta: (GSI) *Talking of uncertainties, I would like to bring up the question of uncertainties in the supply of limestone. I take it that for your exercise, you will be using data on limestone deposits published by the Geological Survey of India (GSI).*

Bala: *Yes, that's right. We've assumed that all three categories of reserves—measured, inferred, and indicated—are available for exploitation.*

Mr. D. S. Gupta: (GSI) *Well, I hope you realize that GSI's estimates of reserves, especially in the last two categories, are very rough. Even if the quantitative estimates turn out to be correct, there is no telling whether the reserves will be of cement-grade and whether they will be economically mineable. I think you would be overestimating reserves by assuming that the total reserves under all three categories will be available for cement production.*

Shankar: *We might indeed be overestimating the supply of limestone, but whether or not that matters for a particular location will depend on how many plants the model recommends should be set up at that location. If it does seem to matter, then detailed investigations could be carried out by the GSI to assess more reliably the reserves at that location. Thus, instead of wasting time now on improving the reliability of data on reserves, it would be better to make a few runs of the model and then, depending on the results, make detailed investigation selectively. In fact, we hope that eventually the model will be useful to GSI in selecting a limited number of regions for detailed investigations, instead of conducting less detailed investigations all over the country, as is presently the practice.*

Mr. D. S. Gupta: (GSI) *I see what you mean. Yes, I guess it might be a good idea to use the model to provide some guidance in the future to GSI on our prospecting program. Why, it would be good if other users of GSI's data could provide similar guidance. That way GSI could provide them better service.*

Mr. R. Kumar: (BCL) *With Mr. Gupta's permission I would like to steer the discussion to another issue. The model formulated by the PAD team seems to overlook what I consider a very important aspect. In PAD's model, capital and operating costs are assumed to be the same in all zones, for a given technology or scale of plant. As a man from the field I can tell you this is very unrealistic and unreasonable. Whenever we evaluate a site for a new plant, one of the important factors we consider is the likely capital and operating costs at that particular location. The capital cost can vary from site to site due to soil conditions, distance between limestone quarry and factory, mode of transportation of limestone from quarry to the factory, length of railway lines to be constructed, whether or not a beneficiation plant is required, and so on. The operating costs can vary also, due to the landed cost of coal, the quality of limestone in the quarry, local power and water, tariffs, labor rates, and so on. To ignore all these important cost differences between different supply zones is, in my personal view, a serious mistake.*

Bala: (IAD) *Actually, sir, we entirely agree with you that it is a serious defect. It would not be difficult to make the necessary changes in the formulation, but the real problem is in obtaining location-specific cost data. We have tried and given up. For example, we compared the capital and operating costs of existing cement plants located in the same supply zone and found them to be widely different. Thus, the location of a plant is pinpointed; it will not do merely to say that it is located in a particular supply zone since the latter is too large to be characterized by a typical capital or operating costs.*

Mr. R. Kumar (BCL) *Well, I can see you have some problems there. But I still feel uncomfortable about the compromise you are making. I leave it to you to decide whether the model is meaningful despite the unreasonable assumption you are forced to make. I would submit, however, that you should at least use differential landed cost of coal for different locations. This is one location-specific cost which does not depend on the pinpointed factory site, but rather on the zone in which the plant is located. It is not difficult to calculate what the lowest landed cost of coal would be for any zone. I think this is important because the cement industry consumes one-third ton or so of coal for every ton of cement output and it is important from an overall point of view, in choosing plant locations, to minimize the transportation costs not only of cement and clinker but also of coal. This is all the more important because of the perpetual shortage of wagons faced by the Indian Railways.*

Bala: (IAD) — *That's a very useful suggestion. We will think further about your comments on the unreasonableness of ignoring location-specific cost variations.*

Mr. R. Kumar (BSL) — *I would also like to point out that your formulation ignores two important alternatives for producing cement. Firstly, it ignores the possibility of expanding cement output in an existing plant at a lower capital and operating cost than that involved in creating the same additional capacity in a new location. Secondly, it ignores the possibility of converting existing Wet Process cement plants to the dry process. As you may know, conversion can double the output of a Wet Process plant at a very low investment per ton of additional capacity. Also, it takes a lot less time than the 3-year gestation period involved in setting up new cement plants, and it also lowers the operating costs.*

Bala: (IAD) — *We must admit we overlooked these possibilities while designing the model. My immediate reaction to your points, however, is that they would unnecessarily increase the complexity of the model. It would also be difficult to obtain estimates of the costs of expansion or conversion for each existing plant hence the costs would vary considerably from case to case. Therefore, it might be better to ignore these possibilities in the model, but to encourage manufacturers to adopt them in actual practice, that is, to encourage construction of plants in any zone only if no possibility is in sight for plants in that zone to expand capacity or convert from Wet to Dry process.*

Mr. V. K. Rao: (NPB) — *I have a basic question about the kind of cost data you plan to use in the model. Since you propose to take the rational point of view in your analysis, I think you should value costs and benefits in social terms rather than in private terms. For instance, there may be taxes or excise duties included in the capital cost of a cement plant; you would want to correct for these before plugging in the capital cost coefficients in the objective function. Similarly, you may want to value limestone at a higher price than that actually paid for it by manufacturers if you think that the country's reserves of limestone are fast running out. You might also want to plant a premium on foreign exchange inflows and outflows because the country is facing a severe shortage of hard currency. And so on. How do you propose to deal with these issues?*

Mr. R. Badri: (MOST)

Before you answer that, let me add another question to Mr. Rao's list: how do you propose to value the cost of rail transportation? Of course there is the official rail tariff rate, but as you might know, they do not reflect the Railway's actual costs. For instance, government policy requires certain commodities such as salt, firewood and charcoal to be carried at rates much below cost. Cement, too, is carried at rates somewhat below cost. Moreover, rail tariffs are known to be more telescopic than haulage costs actually are. The actual costs of transportation vary quite a bit depending on the gauge of railroad (broad gauge or meter gauge); the kind of terrain involved, the kind of traction employed (diesel, steam or electric) and the empty-return ratio for wagons on individual routes. Further, if transhipments from broad gauge to meter gauge or vice versa is necessary, as often happens in the South, the costs would be higher. Finally, certain routes are more congested than others, but the railways charge the same rates on all routes; since cement is one of the bulkiest commodities carried by the railways (next only to coal and foodgrains), it might well turn out that in certain heavily congested regions, a new track may have to be laid just to handle the anticipated volume of cement movement. In view of these complications, how do you propose to cost rail transportation in your model?

And before I forget, let me add two more questions I intended to ask. First, your model completely ignores road transportation although this accounts today for some 20% of the total cement movement. Over short distances, it is in fact cheaper than rail. Secondly, how do you propose to determine the shortest rail distance between pairs of cities? In a number of cases, this cannot be done satisfactorily by visual inspection.

Mr. Kapoor: (Head, IAD)

I think Rao has made an important point. We will have to see how close we can get to using social rather than nominal prices. However, since we are not really concerned with absolute costs as much as we are with relative costs in choosing plant sizes, locations, movement patterns, etc, perhaps it will not make too big a difference to our decisions if we used nominal prices. But I really don't know yet.

Badri has also raised some interesting and difficult conceptual questions about rail tariffs which I think we need to think about further. Regarding road transportation, Badri, it is desirable only over short distances, possibly only for intra-zone movements. Any movement over say 100 km or so should be by rail, cement being a bulky, low-value per unit weight commodity. Our main concern in using the model is to minimize long-distance movements of 500 to 800 km which are currently taking place. Since road transportation is clearly uneconomical over such long distances, we can afford to ignore it in the model.

Mr. Rao: (NPB) *I also noticed that the production costs are linear and do not distinguish fixed operating costs from variable ones. Shouldn't you be using cost functions of this sort? (Walks up to the board and draws the graph below.)*

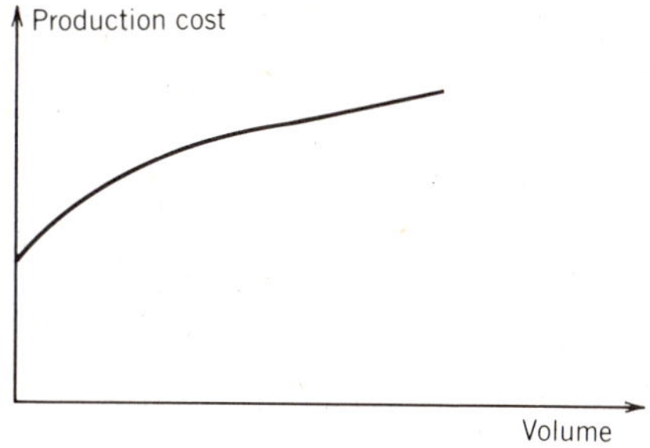

Shankar: (IAD) *I don't think that is really necessary. One can simply compute the present value of fixed cost of operation of new cement plants from the year production begins to the terminal year of the model and include it along with the capital cost of the project. The unit production cost could then be comprised of only the variable elements. That seems like a reasonable way to go about it, or doesn't it? I guess we will have to think it over.*

Mr. S. V. Sami: (CCI) *If I may raise a couple of minor points now, I think the model makes an unreasonable assumption about capacity utilization. You assume that right from the first year of production, plants operate at 85% of capacity. In actual practice, a number of teething problems are faced in the first one or two years. It would be more reasonable to assume that it gradually increases from 60% in the first year to 75% in the second and 85% in the third. Of course, I do not know how you would incorporate this into the model.*

Secondly, your model assumes that none of the existing plants will be shut down. Several of them are already more than 30 years old. In the last decade, a couple of plants did shut down. You might want to consider this matter.

Finally, your model ignores export opportunities altogether. It also does not consider the possibility of building or depleting cement inventories to meet demand over the years.

Mr. P. N. Singh: (DCI) *If I may interrupt Mr. Sami, the model is already very large and complicated. I have been hearing a number of suggestions to complicate it further. While these may be theoretically desirable, there is another side to the problem, and that is the question of what computing capabilities are available to the IAD team in New Delhi. A rough calculation I made indicates that the model already has more than 2000 integer variables. Given the size of the largest computer in Delhi, this is awfully big.*

I can think of a way to eliminate 30% of the integer variables. Since the model minimizes cost and the construction period for new plants is three years, the model will not recommend any new plants during the last three years of the planning horizon. Therefore, we need not incorporate variables for these years. However, that reduction by itself may not be enough.

In the case of continuous variables, a number of cells in the transportation cost matrix can be left empty since the model is unlikely to throw up movements in excess of 600 kms. or so. Perhaps this could reduce the number of continuous variables to 40,000.

In any case, I think you need to pay greater attention to the computing capabilities available to you before designing the model and certainly before collecting data to run it. Otherwise, after lots of effort you may find that you can't compute the optimal solutions to your problem. That would be very unfortunate, wouldn't it?

Mr. P. N. Sethi: (MOI) *While Singh cautions you about problems in running the model, let me add my own words of caution about how feasible it might be to implement the conclusions and recommendations of the model. These are my personal views, and presently I am only thinking aloud.*

Most of the cement industry is in the hands of the private sector. Sure enough, we regulate their activitives, but most of our power is negative in nature; we can say "no" to them, but we cannot easily get them to do something positive. Our major means for intervention is through licensing, but there is always a great deal of uncertainty about which of the licensed parties will actually set up a cement plant and which will not. We will have to think more carefully about the role that public instruments such as the cement corporation of India and the Geological Survey of India can ply in the implementation process.

Again, in the case of cement distribution I don't know how far the Cement Controller will be able to enforce the movement pattern recommended by the model. And if he cannot, the entire basis for determining the optimal distribution of new cement plants may be nullified. Then again, there is the problem with slag cement. Steel plants fall under the Ministry of Steel and Mines and implementation of the MIP model's recommendations will hinge on the assumption that the steel plants will granulate all the slag that is available in future years. An important controversy in this regard is the price that steel plants should receive for granulated slag. The key question here is how the benefits of using slag (over clinker) should be split between the steel plants and the slag cement manufacturers. Inability to resolve this conflict has resulted in considerable delay so far in setting up slag granulation plants. Will this old controversy get resolved in the near future? I really don't know.

I also don't know whether the cement freight equalization system will continue forever? If it does not, how will it alter the model's conclusions? And incidentally, can the model be used in any way to evaluate the merits and demerits of current policies of the Government on freight equalization or price and distribution controls?

I think I should stop rambling now. But, I did want to share my apprehensions with all of you. I am still very skeptical about the utility of such mathematical models, but I must confess, today's discussion has set my mind thinking on a number of questions.

Mr. Kapor: (Head, IAD) *Well, thank you Sethi for your very thought-provoking observations. I think they are very pertinent questions indeed and should be occupying our minds too.*

As it is getting late now, I must draw the meeting to a close. I am sure my IAD colleagues will agree with me that it has been a very stimulating discussion. We still have a lot to think about and resolve. Better now than later! Thank you very much gentlemen for your comments and for sparing your valuable time. We will probably have to bother you all once more after a month or so. Meanwhile, thank you once again.

Bala, Shankar, and Raju remained in the conference room after everyone had left. Bala glanced through his notes. There were several points they had to think about. Shankar remarked: "If you had talked to me yesterday, I would have told you we were close to completing our exercise. Today, I think we still have a very long way to go. The end of the tunnel is nowhere in sight. Anyhow, let's not waste time worrying. What is it we need to do?"

And so the members of the IAD team put their heads together once more.

APPENDIX **Details of the Mathematical Formulation**

MAIN FEATURES OF THE MODEL

1. Time horizon of model: 10 years
2. Decision Variables

- Number of integrated cement plants/clinker plants/grinding plants/slag cement plants to be set up in each zone each year.
- Quantity of ordinary cement/clinker/slag cement to be produced in each plant each year.
- Quantity of cement/clinker/slag to be moved from various supply zones to various demand zones each year.

3. The Objective Function: To minimize the present value of capital, operating and distribution costs of meeting the projected demand for cement over the 10-year period.

4. Constraints

- Production should be less than capacity in each zone each year.
- Supply should equal demand in each zone each year.
- Clinker flow constraint (inflow into each zone plus local production should equal outflow plus local consumption).
- Clinker/slag ratio in slag cement should be greater than one.
- Maximum number of new cement plants in any year should be 8.
- Maximum number of new cement plants over the 10-year period in any state should be less than 6.

Decision Variables

Investment Decisions

x_{it}^{α}—A 0, 1, 2 variable that represents the number of new integrated ordinary Portland cement plants of capacity α tons per day in zone i in year t.

y_{it}^{α}—A 0, 1, 2 variable that represents the number of new clinker plants of capacity α tons per day in zone i in year t.

z_{it}^{α}—A 0, 1, 2 variable that represents the number of new grinding units of capacity α tons per day in zone i in year t.

w_{it}^{α}—A 0, 1, 2 variable that represents the number of integrated slag cement plants of capacity α tons per day in zone i in year t.

(α will take three values: 200, 600, and 1,200 tons per day.)

Production Decisions

$P^{x\alpha}_{it}$, $P^{y\alpha}_{it}$, $P^{z\alpha}_{it}$, $P^{w\alpha}_{it}$—Total production in tons of ordinary cement/clinker/slag cement in plants of capacity α tons per day in zone i in year t.

P^{s}_{it}—Amount of slag.

Shipment Decisions

q^{x}_{ijt}, q^{y}_{ijt}, q^{w}_{ijt}—Amount in tons of ordinary cement/clinker/slag cement shipped from zone i to zone j in year t.

The Objective Function

Minimize:

$$\sum_{i,t,\alpha} B^{e}_{t}(x^{\alpha}_{it}\,k^{\alpha}_{x} + y^{\alpha}_{it}\,k^{\alpha}_{y} + z^{\alpha}_{it}\,k^{\alpha}_{z} + w^{\alpha}_{it}\,k^{\alpha}_{w})$$

$$+ \sum_{i,t,\alpha} B^{e}_{t}(r^{x\alpha}\,P^{x\alpha}_{it} + r^{y\alpha}\,P^{y\alpha}_{it} + r^{z\alpha}\,P^{z\alpha}_{it} + r^{w\alpha}\,P^{w\alpha}_{it} + r^{s}P^{s}_{it})$$

$$+ \sum_{i,j,t} B^{e}_{t}\{c^{xzw}_{ij}(q^{x}_{ijt} + q^{y}_{ijt} + q^{w}_{ijt}) + c^{ys}_{ij}\,(q^{y}_{ijt} + q^{s}_{ijt})\}$$

for all q's $\geqslant 0$, all P's $\geqslant 0$, and $x, y, z, w = 0, 1,$ or 2.

Where:

B^{e}_{t} is the time discount factor corresponding to a discount rate of e% for year t.

k^{α}_{x}, k^{α}_{y}, k^{α}_{z}, k^{α}_{w} are respectively the capital costs in dollars of new integrated ordinary cement, clinker, grinding unit, and integrated slag cement plants of capacity α tons per day.

$r^{x\alpha}$, $r^{y\alpha}$ are unit production costs in dollars for ordinary cement and clinker; $r^{w\alpha}$ is unit production cost in dollars of slag cement in an integrated slag cement plant and $r^{z\alpha}$ is unit cost of grinding clinker or a mixture of clinker and slag in a grinding unit.

c^{xyw}_{ij} is the unit shipment cost in dollars for ordinary cement or slag cement moving from zone i to zone j.

c^{ys}_{ij} is the unit shipment cost in dollars for clinker or granulated slag moving from zone i to zone j ($c^{ys}_{ij} < c^{xzw}_{ij}$ since the former requires open wagons while the latter requires closed wagons).

The Constraints

Total Installed Capacity: Definitions

$$\sum_{\alpha} Q_{it}^{n\alpha} \begin{cases} = Q_{it}^{n^*} & \text{(for } t = 1, 2, 3) \\ = Q_{it}^{n^*} + \sum\limits_{\alpha} \sum\limits_{t'=1}^{t'=t-3} \alpha n_{it'}^{\alpha} & (t \geq 4) \end{cases} \quad \text{for all } i \text{ and } t$$

$$\sum_{\alpha} Q_{it}^{y\alpha} \begin{cases} = Q_{it}^{y^*} & \text{(for } t = 1, 2, 3) \\ = Q_{it}^{y^*} + \sum\limits_{\alpha,t'=1}^{t'=t-3} \alpha y_{it'}^{\alpha} & (t \geq 4) \end{cases} \quad \text{for all } i \text{ and } t$$

$$\sum_{\alpha} Q_{it}^{z} \begin{cases} = Q_{it}^{z^*} & \text{(for } t = 1, 2, 3) \\ = Q_{it}^{z^*} + \sum\limits_{\alpha,t'=1}^{t'=t-3} \alpha z_{it'}^{\alpha} & (t \geq 4) \end{cases} \quad \text{for all } i \text{ and } t$$

$$\sum_{\alpha} Q_{it}^{w} \begin{cases} = Q_{it}^{w^*} & \text{(for } t = 1, 2, 3) \\ = Q_{it}^{w^*} + \sum\limits_{\alpha,t'=1}^{t'=t-3} \alpha w_{it'}^{\alpha} & (t \geq 4) \end{cases} \quad \text{for all } i \text{ and } t$$

where $Q_{it}^{x\alpha}, Q_{it}^{y\alpha}, Q_{it}^{z\alpha}, Q_{it}^{w\alpha}$ are total installed capacity of integrated ordinary cement plants, clinker plants, grinding plants and integrated slag cement plants of capacity α tons per day in zone i in year t; and $Q_{it}^{n^*}, Q_{it}^{y^*}, Q_{it}^{z^*}, Q_{it}^{w^*}$ are total *existing* capacity of the above in zone i in year t ($t \leq 3$). This includes plants already in operation in $t = 0$ plus new plants under construction in $t = 0$ that will get commissioned in $t = 1$, 2, or 3.

Production Constraints

$P_{it}^{s} \leq Q_{it}^{s}$ for all i and t ($Q_{it}^{s} \geq Q_{\min}^{s}$, where $Q_{\min}^{s}$ is the minimum economic size for a slag granulation plant).

$P_{it}^{y\alpha} \leq 0.85\, Q_{it}^{y\alpha}$ for all i, t, and α.

$P_{it}^{z\alpha} \leq 0.85\, Q_{it}^{z\alpha}$ for all i, t, and α.

$P_{it}^{w\alpha} \leq 0.85\, Q_{it}^{w\alpha}$ for all i, t, and α.

$P_{it}^{x\alpha} \leq 0.85\, Q_{it}^{\alpha}$ for all i, t, and α.

Demand Constraints

$$\sum_{\alpha} P^{\alpha}_{it} + \sum_{j} q_{jit} - \sum_{j} q_{ijt} + \sum_{\alpha} P^{z\alpha}_{it} + \sum_{j} q^{z}_{jit} - \sum_{j} q^{z}_{ijt}$$

$$+ \sum_{\alpha} P^{w\alpha}_{it} + \sum_{j} q^{w}_{jit} - \sum_{j} q^{w}_{ijt} = D_{it} \text{ for all } i \text{ and } t$$

Clinker Balance Constraint

$$\sum_{\alpha} P^{y\alpha}_{it} + \sum_{j} q^{y}_{jit} - \sum_{j} q^{y}_{ijt} = 0 \text{ for all } i \text{ and } t$$

Clinker:Slag Ratio in Slag Cement

$$\sum_{j} q^{y}_{jit} - \sum_{j} q^{y}_{ijt} \geqslant \left\{ \sum_{j} q^{s}_{jit} + P^{s}_{it} - \sum_{j} q^{s}_{ijt} \right\} - \tfrac{1}{2} \sum_{\alpha} P^{w\alpha}_{it} \text{ for all } i \text{ and } t$$

$$\sum_{\alpha} P^{z\alpha}_{it} = \sum_{j} (q^{y}_{jit} - q^{y}_{ijt}) + \left(\sum_{j} q^{s}_{jit} + P^{s}_{it} - \sum q^{s}_{ijt} \right) - \tfrac{1}{2} \sum_{\alpha} P^{w\alpha}_{it}$$

for all i and t

Maximum Number of Cement Plants

$$\sum_{\alpha,i} x^{\alpha}_{il} + w^{\alpha}_{it} + y^{\alpha}_{it} \leqslant 8$$

$$\sum_{i,\alpha} \sum_{t} x^{\alpha}_{it} + w^{\alpha}_{it} + y^{\alpha}_{it} \leqslant 6$$ for sets of i's corresponding to different states of the country

Chapter 7

INCORPORATING UNCERTAINTY INTO A LINEAR PROGRAM

Several years ago, a new federal agency, the Strategic Petroleum Reserve Organization (SPRO), was charged with constructing a plan for obtaining and storing emergency crude oil stocks in case of another embargo of foreign oil. The plan required the specification of storage locations, which crude oils should be purchased, from whom, and in what amounts. It was proposed that a large-scale mathematical program be constructed to help deal with these questions. Unfortunately, there were two major uncertainties: demand for petroleum end-products and crude oil availability. Since the *raison d'etre* for the plan was protection against future political or economic shocks that would cut off the supply of oil to the U.S., these uncertainties could not very well be ignored. Instead, in order to be meaningful, the model had to deal with them explicitly.

As we have seen, one of the major drawbacks of mathematical programming models is the assumption that all the input data (objective function coefficients, right-hand sides, and constraint coefficients) is known with certainty. In some cases, we might be willing to accept this because the uncertainty is small relative to the quantities involved, and sensitivity analysis made possible by the LP output is sufficient to resolve any doubts about the validity of the data. Often, however, as in the situation SPRO faced, there are major uncertainties that warrant explicit consideration. This chapter will describe a few ways in which this can be done.

Stochastic programming is inherently more difficult than deterministic linear programming[1]:

1. There are new conceptual difficulties—the decision maker must often assess probability distributions (or discrete probabilities) for the major uncertain quantities. The assessment process is often difficult in practice—perhaps you have experienced the misgivings of managers who are guided to decisions based on analysis that depends on these assessed probabilities.
2. Stochastic programming models may create a computational nightmare since many different possible scenarios—that is, combinations of events—have to be considered, and since nonlinearities are often introduced.

Before we can proceed, we must resolve the question of what an optimal decision is. As in a deterministic problem, it is essential to select a criterion or objective. When uncertainties are considered, this choice becomes more difficult. Now it is not enough, for example, to minimize cost, since cost is uncertain; possible criteria might be:

- Minimize expected cost.
- Minimize expected cost subject to satisfying demand 90% of the time.
- Minimize the probability that cost will exceed $100,000.
- Minimize the quantity c for which there is a 90% chance that the cost will be below c (i.e., minimize the .90 *fractile*).
- Minimize "risk," (i.e., some function of the cost that measures the decision-maker's preferences).

Any of these are reasonable; the latter three often lead to highly nonlinear objective functions and other complications outside the scope of this chapter. We will focus, therefore, on those situations where the decision maker might reasonably choose the expected value of the criterion as the appropriate objective.

UNCERTAINTY IN THE OBJECTIVE FUNCTION COEFFICIENTS

The simplest case is when the only major uncertainties occur in the objective function coefficients. This is often the case since, for example, raw

[1]A decision problem is *deterministic* if we are willing to assume that the uncertainties are small enough to be ignored. If the uncertainties are to be considered explicitly, we call the problem *stochastic*.

material costs are often unknown and revenues depend on prices that are functions of unknown market conditions. As another example, if resources are to be allocated to five projects x_1, x_2, x_3, x_4, and x_5—each with uncertain returns r_1, r_2, r_3, r_4, and r_5 per dollar invested, then the total return would be

$$r_1x_1 + r_2x_2 + r_3x_3 + r_4x_4 + r_5x_5$$

Since the expected value of a sum is the sum of the expected values, maximizing the expected return—that is[2],

$$\text{Max } E\,[r_1x_1 + r_2x_2 + r_3x_3 + r_4x_4 + r_5x_5]$$

is the same as maximizing the sum of the expected returns—that is,

$$\text{Max } E\,[r_1x_1] + E[r_2x_2] + E[r_3x_3] + E[r_4x_4] + E[r_5x_5]$$

But the coefficients r_i are the uncertain quantities; the x_i are decision variables, so the above objective is the same as

$$\text{Max } (E\,[r_1])x_1 + (E[r_2])x_2 + (E[r_3])x_3 + (E[r_4])x_4 + (E[r_5])x_5$$

Or, in general,

$$\text{Max } E\,[\Sigma r_ix_i] = \text{Max } \Sigma(E[r_i])x_i$$

In other words, if a linear objective function has uncertain coefficients, and we wish to maximize (or minimize) its expected value, *simply substitute for each coefficient its expected value (mean) and run the program as a standard linear program.*

In practice, if, for example, we are concerned with raw material prices, we merely need assess a *point forecast* of each to run the program. There is no need for a full probability distribution. If we are concerned with which projects to invest in, we must make a "best guess" of each project's return. Unfortunately, as we will see, uncertainties in yields, demands, or resources availabilities complicate matters greatly.

CHANCE CONSTRAINED PROGRAMMING

Several years ago, the city of Baltimore was investigating changes in its reservoir system. A mathematical programming approach was used to decide what reservoir levels should be and what volumes of water should flow from one reservoir to another at certain times during the year. Not only is the demand for water in the Baltimore area uncertain, but so is the rainfall. There

[2]The notation $E[\;\;]$ denotes the expected value of the quantity inside the brackets.

is, however, abundant historical data for both of these uncertain quantities so that the probability distributions could be easily assessed. A constraint which would *guarantee* that demand be satisfied would be absurd in a probabilistic sense, since the one-in-a-thousand combination of very hot weather, coupled with drought conditions, might conceivably make it impossible to satisfy the resulting demand. Instead, city planners decided to add a constraint which required that demand be satisfied *95% of the time*.

This approach is called *chance constrained programming*. In the private sector, especially, it is often unrealistic or overly expensive to insist that demand always be satisfied. Instead, a manager (judgmentally or by using underage and overage estimates) might set a *service level* which would require that the chance that customers go unsatisfied not exceed, say, 20%. Instead, a manager might be concerned with overproduction and want the probability that production exceeds demand to be no more than 20%. The uncertain quantity need not be demand—yield, capacity, or labor required are other possibilities. Let us look at a simple example.

XYZ Corporation produces gas turbine blades for several types of aircraft engines. Over the past several years the plant has been running a two-shift operation, and though, on occasion, a third shift has been used, the organizational upset involved (rather than purely the extra expense) is great enough so that it is rarely done. October, however, is particularly busy, and, in scheduling that month's orders, the production manager recognizes that the number of available man-hours is, unfortunately, unknown, due to sick days, vacation days, and so on. Based on previous months' actual man-hours worked versus theoretical capacity, a distribution of man-hours available (shown in Figure 7-1) had been assessed. The labor required for each of the firms products is known exactly and is shown in Table 7-1.

In accordance with plant policy, the production manager wants to add to the production scheduling linear programming model a constraint:

$$\text{Prob}\,\{\text{time used} > \text{2-shift time available}\} \leq .10$$

From Figure 7-1, we see that the .10-fractile[3] is 8660 hours. That is,

$$\text{Prob}\{\text{actual time available} \leq 8660 \text{ hours}\} = .10$$

Thus, if the time schedule to be used is under 8660 hours, there will be less than a 10% chance of it exceeding the time available. Putting this together with the information in Exhibit 1, we may write the constraint:

$$27x_1 + 23x_2 + 28x_3 + 19x_4 + 20x_5 + 22x_6 \leq 8660$$

[3]The *p*-fractile of a distribution is defined as that outcome Z such that Prob {actual outcome $\leq$ Z} = p.

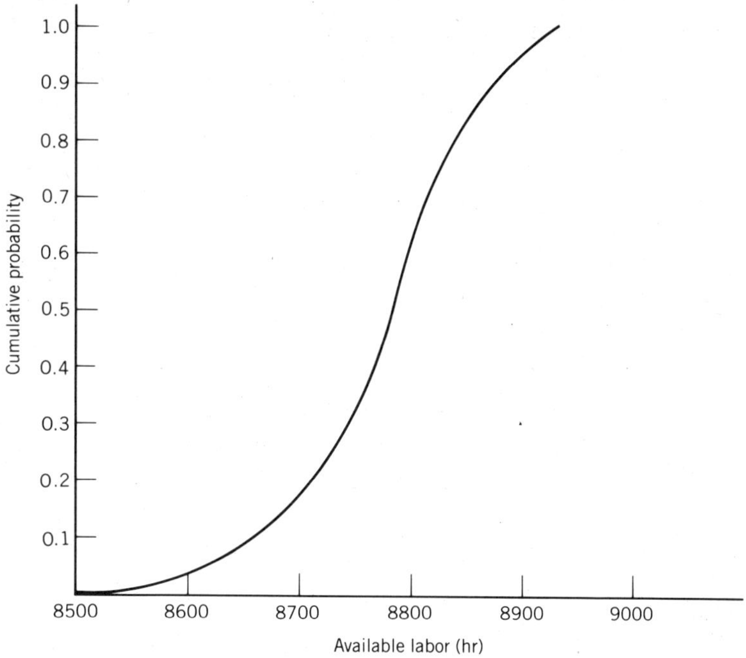

FIGURE 7-1. A cumulative probability distribution.

TABLE 7-1

Product	Labor Required (hr)
1	27
2	23
3	28
4	19
5	20
6	22

where x_i is the amount of product i to be produced this month.

In general, a *chance constraint* has the form

$$\text{Prob}\,\{\Sigma\, a_i x_i \geq b\} \leq p$$

where b is an uncertain quantity for which a distribution has been assessed. This may be replaced by a deterministic constraint

$$a_i x_i \leq B$$

where B is the p-fractile of the distribution.

There are two major drawbacks to chance constrained programming, one technical, one conceptual. First, the uncertain quantity may depend on the level of the decision variables. For example, in a portfolio planning problem where x_i represents the amount to be invested in Project i, we cannot write

$$\text{Prob}\ \{\text{return} \geq \$1000\} \geq 90\%$$

as a linear constraint, since the return depends on the particular allocation of funds. In such cases where a distribution must be derived, we are often faced with replacing the probabilistic constraint above with a nonlinear albeit deterministic one. Similar problems arise when we require, for example, the joint probability that several constraints hold be at least α.

Secondly, and perhaps more important, a chance constraint rarely reflects a manager's true feelings; it is unlikely that satisfying the customer at the 90% level is necessary, whereas an 89% level is intolerable. Chance-constrained programming does not allow for the direct evaluation of the economic consequences of violating one of these constraints. The optimization thus seems somehow incomplete.

MULTISTAGE PLANNING

Perhaps the approach most suited to a manager's way of thinking is *Multistage* (or *Scenario*) *Planning* by which several different scenarios for the future can be made to influence a decision which must be made now. Scenario planning is useful whenever an immediate decision is incumbent on later resource allocation decisions. Figure 7-2 depicts this situation in decision tree terms.

For example, imagine a manager who is contemplating expanding productive capacity (the "initial decision" of Figure 7-2) so as to introduce a new product. It may be that if production resources are allocated efficiently (the "resource allocation decision" of Figure 7-2), less new capacity will be needed. The crucial point is that the decision as to how much capacity to add depends critically on how resources will be allocated thereafter.

In order to develop such a scenario planning model, it is necessary to first identify the key areas of uncertainty, or risk, facing the firm and then indicate how these risks can be combined to form scenarios. There are four areas of risk

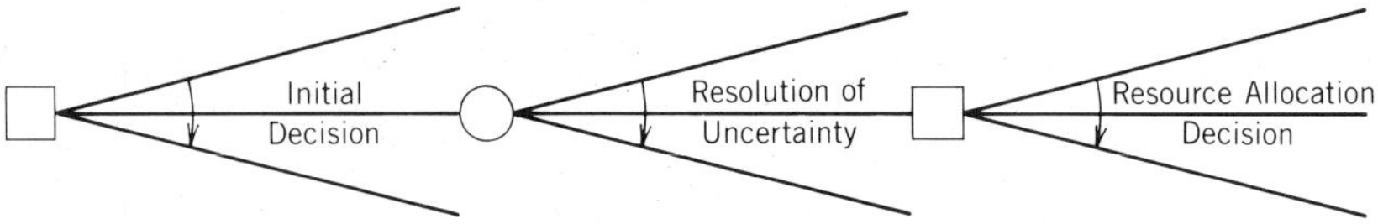

FIGURE 7-2. A two-stage decision.

that could play important roles in any strategic decision: economic, market, technological, and financial risks.

First, at the highest level is the uncertainty in the *general macro-economic* outlook and specifically how this outlook will impact the industry in question. In particular, the timing of investments is crucial. Other uncertainties that are part of the general macro-economic environment include overall interest rates, inflation rates, the possibility of wage and price controls, regulatory requirements, and so forth. Often as important, the availability of raw materials, labor, and recently, energy are uncertainties which influence and are influenced by the economy. In addition, if multinational issues are being considered, *relative* inflation rates, interest rates, currency devaluation rates, elimination or establishment of various tariffs and export/import incentives might have to be considered. In any particular planning environment, not all of these uncertainties will be important, but in most instances some of these should affect the planning in a significant way.

The second group of uncertainties are those involving *market dynamics*. The level and geographical distribution of the demand over time for the products offered are not known with certainty in any event; and if the products are relatively new, these aspects might be particularly uncertain.

The third area of uncertainty involves *technological and manufacturability risks*. There could be uncertainty as to the productivity of various technologies. For example, one technology might be a conventional approach with fairly certain costs and benefits, while another technology might be radically new with quite an uncertain range of both costs and benefits. If the radically new technology is successful, then its costs would be reasonable and its benefits would be substantial; however, higher costs and only marginal benefits could be possible, as well as a very unsuccessful implementation involving substantially higher costs and even negative "benefits."

The fourth area of uncertainty is *financial*, which to a certain extent is derivative of the three proceeding areas. However, the firm's overall financial health usually depends on a number of attributes not related to the particular decision under consideration. Hence, the rate at which a firm can borrow money both now and in the future, the availability of industrial bond financing, the internal funds generated and made available for investment purposes are key uncertainties that may impact the timing and scope of the decisions at hand.

From a strategic standpoint, the critical aspect of scenario planning models is that they can be structured in such a way as to appropriately hedge against those uncertainties that are most important in the specific firm's assessment. It is important to recognize that the appropriate hedging strategy cannot usually be determined from finding the optimal plans for each scenario. If the optimal plans were identical for each scenario, then, of course, no hedging strategy would be necessary. However, in general, these plans will be very different. In a pessimistic scenario involving economic downturns, high interest rates, and little market penetration, perhaps it would be optimal to not expand capacity at all. However, in a more optimistic scenario, several new plants might be the recommendation. What is needed is not the optimal plan for a known scenario, but the optimal hedging plan against the uncertainty as to which scenario will occur.

A Simple Example in Detail

Consider a pharmaceutical firm that must order its raw materials at the beginning of each month for two very perishable products that are produced to order. Any leftover raw material decomposes by the end of the month and must be discarded. The costs, raw material requirements, and contributions are given in Table 7-2. There are two uncertainties that must be considered:

1. Demand is highly uncertain. However, orders for the two products are highly correlated; that is, typically when orders for product 1 are high, so are orders for product 2. Management has postulated four possible demand scenarios for next month.[4]

	Demand (in lb)			
	High	Medium		Low
	Scenario 1	Scenario 2	Scenario 3	Scenario 4
Product 1	4000	2000	2000	800
Product 2	1500	1200	600	200

2. Capacity is uncertain since one of the blending machines is under repair and may not be usable in the coming month. Without this machine, capacity is 2400 lb. The blender, if repaired, would provide an additional 900 lb.

TABLE 7-2

		Raw Material Requirements (pounds needed per lb of product)	
Raw Material	Cost/lb	Product 1	Product 2
A	\$10	.236	.367
B	\$27	.079	.337
C	\$7	.786	.396
Contribution excluding raw material costs		\$11.39	\$23.59

[4]Of course, we recognize that these four scenarios do not describe *all possible* demands. Instead, they are meant to be representative.

After some thought, management agrees that high demand is more likely than low demand, but unless demand for product 1 is high, more than likely demand for product 2 will be low. These feelings have been quantified as:

Scenario	*Probability of Occurrence*
1	.4
2	.1
3	.25
4	.25

In addition, the maintenance engineers say that there is a 70% chance that the broken blender will be repaired.

The decision problem is to decide *now* how much of each raw material is to be purchased, consistent with possible demand levels.[5] In decision tree terms, we might represent this problem as shown in Figure 7-3. One solution procedure would be to *choose* a possible combination of amounts of raw materials to be purchased, r_A, r_B, and r_C; solve eight linear programs to determine the production levels, then fold back to evaluate that choice of raw material levels. This could be repeated for as many choices of r_A, r_B, and r_C as desired; that is, we could conduct a three-dimensional search for the best decision. Clearly, this would require running an enormous number of LPs and not guarantee an optimal solution. Instead, we can devise a single linear program that will.

The important fact to recognize here in this *two-stage* decision problem is that one set of decisions (the raw material purchase) must be made *prior* to the resolution of the uncertainties; the other set of decisions (the production levels) are made *after* the resolution of the uncertainties.[6] Thus we would expect that in addition to the raw materials decision an optimal solution would consist of a decision rule or strategy; for example:

> *Produce* X *product 1s and* Y *product 2s if Scenario 1 or 2 occurs and the blender is repaired; otherwise product* X′ *product 1s and* Y′ *product 2s.*

Let us formulate this problem:

Decision Variables

x_{ij} = amount of product i to be produced *if* Scenario j occurs and the blender is repaired.

[5]It may happen, of course, that too few raw materials are purchased to satisfy the demand that materializes. We will take the "cost" of lost orders to be only the opportunity cost implicit in the lost revenue.

[6]The decision variables for this set of decisions are often called "recourse variables" since they indicate the recourse the decision maker has after observing the outcomes of the uncertainties.

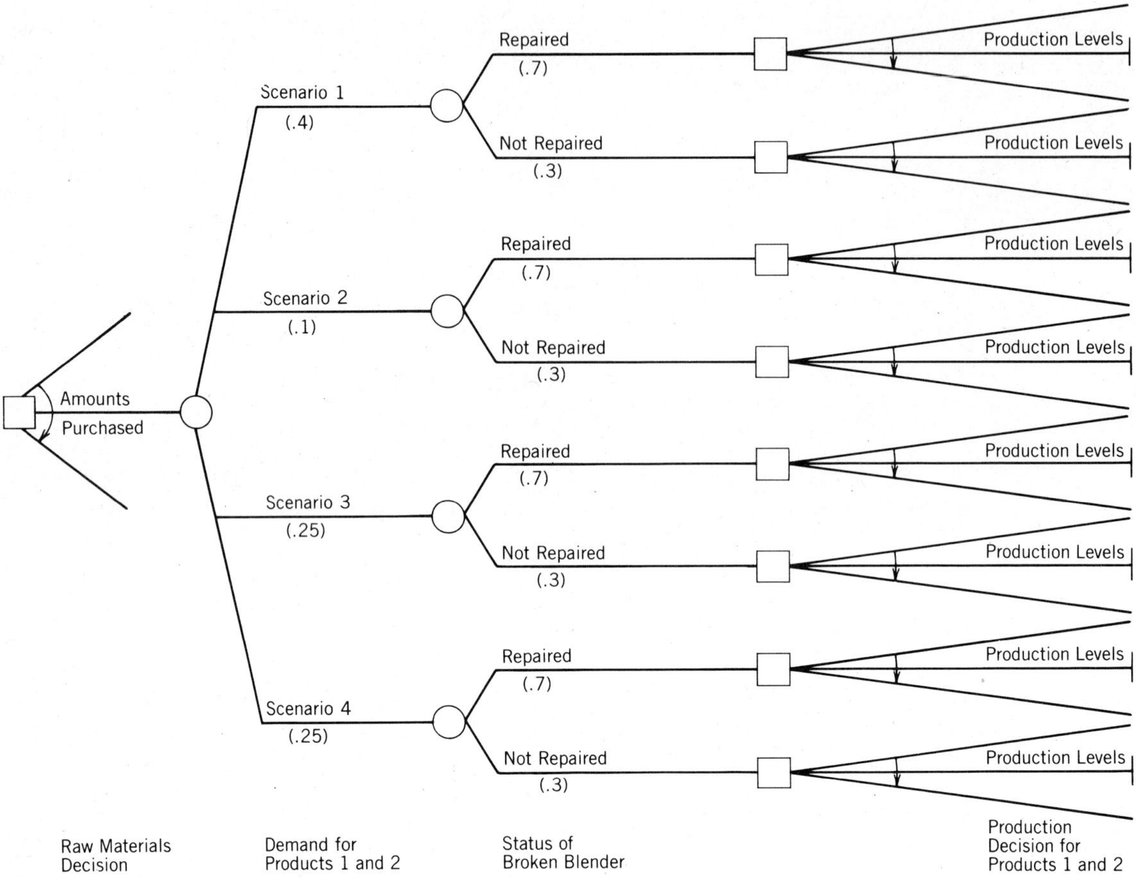

FIGURE 7-3. A scenario-based decision diagram.

y_{ij} = amount of product i to be produced *if* Scenario j occurs and the blender is not repaired.

r_k = amount of raw material k ($k = A, B,$ or C) to be purchased.

Constraints

Raw Material Availability: Amount of Raw Material Used $\leq$ Amount of Raw Material Purchased. This must hold *regardless* of which scenario obtains; that is,

$$A: \quad .236\, x_{1j} + .367\, x_{2j} \leq r_A \quad \text{for each } j$$
$$.236\, y_{1j} + .367\, y_{2j} \leq r_A \quad \text{for each } j$$

$$B: \quad .079\, x_{1j} + .337\, x_{2j} \leq r_B \quad \text{for each } j$$
$$.079\, y_{1j} + .337\, y_{2j} \leq r_B \quad \text{for each } j$$

$$C\colon\quad .786\, x_{1j} + .396\, x_{2j} \leq r_c \quad \text{for each } j$$
$$.786\, y_{1j} + .396\, y_{2j} \leq r_C \quad \text{for each } j$$

Capacity Availability: Amount of Capacity Used ≤ Available Capacity. If the blender is repaired, the available capacity is 3300 lb; if not, 2400 lb; that is,

$$x_{1j} + x_{2j} \leq 2400 \quad \text{for each } j$$

$$y_{1j} + y_{2j} \leq 3300 \quad \text{for each } j$$

Demand:

$$\textit{Scenario 1}\text{—}\begin{cases} x_{11} \leq 4000, & y_{11} \leq 4000 \\ x_{21} \leq 1500, & y_{21} \leq 1500 \end{cases}$$

$$\textit{Scenario 2}\text{—}\begin{cases} x_{12} \leq 2000, & y_{12} \leq 2000 \\ x_{22} \leq 1200, & y_{22} \leq 1200 \end{cases}$$

$$\textit{Scenario 3}\text{—}\begin{cases} x_{13} \leq 2000, & y_{13} \leq 2000 \\ x_{23} \leq 600, & y_{23} \leq 600 \end{cases}$$

$$\textit{Scenario 4}\text{—}\begin{cases} x_{14} \leq 800, & y_{14} \leq 800 \\ x_{24} \leq 200, & y_{24} \leq 200 \end{cases}$$

Objective Function

Recall that we have chosen to maximize the *expected contribution*, so the objective function is where the probabilities are taken into account.

$$\text{Raw material cost: } 10r_A + 27r_B + 7r_C$$

Expected contribution (excluding raw material costs):

$$\text{Prob\{blender is not fixed\}} \times \text{Prob\{scenario 1 occurs\}} \times 11.39 \times x_{11}$$

$$+ \text{Prob\{blender is not fixed\}} \times \text{Prob\{scenario 2 occurs\}} \times 11.39 \times x_{12}$$

$$+ \cdots - - - - -$$

$$+ \text{Prob\{blender is not fixed\}} \times \text{Prob\{scenario 1 occurs\}} \times 23.59 \times x_{21}$$

$$+ \cdots - - - - -$$

$$+ \text{Prob\{blender is fixed\}} \times \text{Prob\{scenario 1 occurs\}} \times 11.39 \times y_{11}$$

$$+ \text{Prob}\{\text{blender is fixed}\} \times \text{Prob}\{\text{scenario 2 occurs}\} \times 11.39 \times y_{12}$$

$$+ \cdots \quad - \; - \quad - \quad - \quad -$$

This may be rewritten as

$$.3(11.39 \sum_{j=1}^{4} p_j x_{1j} + 23.59 \sum_{j=1}^{4} p_j x_{2j})$$

$$+ .7(11.39 \sum_{j=1}^{4} p_j y_{1j} + 23.59 \sum_{j=1}^{4} p_j y_{2j})$$

where p_j = probability of the jth scenario. Thus the entire LP formulation is as shown in Figure 7-4 and the resultant CMP output in Figure 7-5.

The optimal strategy, as given in Figure 7-5, is to purchase 361 lb of raw material *A*, 249 lb of raw material *B*, and 708 lb of raw material *C*, with plans to make 800 lb of product 1 and 200 lb of product 2 if scenario 4 materializes, to produce 598 lb of product 1 and 600 lb of product 2 otherwise. The extra capacity that would become available if the blender is repaired is unnecessary; it will not be used even if the blender becomes operable. Note that even in the presence of essentially unlimited resources (raw materials and capacity) we will produce to demand only 25% of the time—that is, only if scenario 4 occurs. The risk of being left with unused (expensive) raw material if demand is low offsets the chance of receiving more revenue if demand is medium to high.

In practice, of course, these four scenarios would only be taken to be representative of the many possible demand levels which could occur. Essentially, then, this output tells the manager to purchase the raw materials specified above, and produce as many pounds of product 2 as are demanded *up to 600*; then produce as many pounds of product 1 as possible (up to demand).

The major drawback of this technique is that there need be one set of constraints for each scenario (there are really eight scenarios—for each demand scenario, there are two possible blender outcomes) and a set of second-stage decision variables for each scenario. Thus the LP is close to eight times as large as it would be if we chose to ignore the uncertainty. This was, of course, a simple example with only two stages. In general, a multistage planning problem will have the form depicted in the illustration below. Especially in the

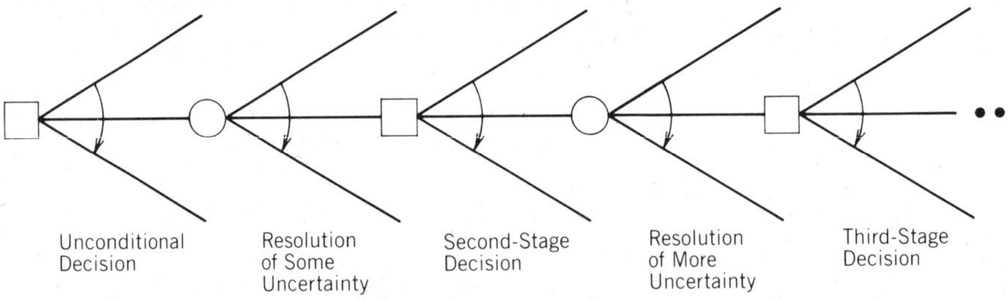

	1	2	3	4	5	6	7	8	9	10	11	12	13	14	15	16	17	18	19			
	r_A	r_B	r_C	x_{11}	x_{21}	x_{12}	x_{22}	x_{13}	x_{23}	x_{14}	x_{24}	y_{11}	y_{21}	y_{12}	y_{22}	y_{13}	y_{23}	y_{14}	y_{24}		RHS	<u>UPPER BOUNDS</u>
Max	-10	-27	-7	1.367	2.831	.342	.708	.854	1.769	.854	1.769	3.189	6.605	.797	1.651	1.993	4.128	1.993	4.128			
A_{x1}	-1			.236	.367															$\le$	0	$x_{11} \le 4000$
A_{x2}	-1					.236	.367													$\le$	0	$x_{21} \le 1500$
A_{x3}	-1							.236	.367											$\le$	0	$x_{12} \le 2000$
A_{x4}	-1									.236	.367									$\le$	0	$x_{22} \le 1200$
A_{y1}	-1											.236	.367							$\le$	0	$x_{13} \le 2000$
A_{y2}	-1													.236	.367					$\le$	0	$x_{23} \le 600$
A_{y3}	-1															.236	.367			$\le$	0	$x_{14} \le 800$
A_{y4}	-1																	.236	.367	$\le$	0	$x_{24} \le 200$
B_{x1}		-1		.079	.337															$\le$	0	$y_{11} \le 4000$
B_{x2}		-1				.079	.337													$\le$	0	$y_{21} \le 1500$
B_{x3}		-1						-.079	.337											$\le$	0	$y_{12} \le 2000$
B_{x4}		-1								.079	.337									$\le$	0	$y_{22} \le 1200$
B_{y1}		-1										.079	.337							$\le$	0	$y_{13} \le 2000$
B_{y2}		-1												.079	.337					$\le$	0	$y_{23} \le 600$
B_{y3}		-1														.079	.337			$\le$	0	$y_{14} \le 800$
B_{y4}		-1																.079	.337	$\le$	0	$y_{24} \le 200$
C_{x1}			-1	.786	.396															$\le$	0	
C_{x2}			-1			.786	.396													$\le$	0	
C_{x3}			-1					.786	.396											$\le$	0	
C_{x4}			-1							.786	.396									$\le$	0	
C_{y1}			-1									.786	.396							$\le$	0	
C_{y2}			-1											.786	.396					$\le$	0	
C_{y3}			-1													.786	.396			$\le$	0	
C_{y4}			-1															.786	.396	$\le$	0	
CAP_{x1}				1	1															$\le$	2400	
CAP_{x2}						1	1													$\le$	2400	
CAP_{x3}								1	1											$\le$	2400	
CAP_{x4}										1	1									$\le$	2400	
CAP_{y1}												1	1							$\le$	3300	
CAP_{y2}														1	1					$\le$	3300	
CAP_{y3}																1	1			$\le$	3300	
CAP_{y4}																		1	1	$\le$	3300	

FIGURE 7-4. Raw material scenario planning formulation.

```
DATA SET NAME? STOC

TITLE: RAW MATERIAL SCENARIO PLANNING
PROCEED, DISPLAY, OR REJECT? P

MAXIMIZE OR MINIMIZE? MAX

OPTIMAL SOLUTION FOUND.
    OBJ            3879.06

OUTPUT OPTION? U

ALL ITEMS NOT LISTED IN SECTIONS 1 - 4 HAVE THE VALUE ZERO.

*1* DECISION VARIABLES
  1. R-A          361.440
  2. R-B          249.479
  3. R-C          708.000
  4. X-11         598.473 \
  5. X-21         600.000 |
  6. X-12         598.473 |  Scenarios 1-3 without additional capacity
  7. X-22         600.000 |
  8. X-13         598.473 |
  9. X-23         600.000 /
 10. X-14         800.000 }  Scenario 4 without additional capacity
 11. X-24         200.000 }
 12. Y-11         598.473 \
 13. Y-21         600.000 |
 14. Y-12         598.473 |  Scenarios 1-3 with additional capacity
 15. Y-22         600.000 |
 16. Y-13         598.473 |
 17. Y-23         600.000 /
 18. Y-14         800.000 }  Scenario 4 with additional capacity
 19. Y-24         200.000 }

*2* SLACK(+) AND SURPLUS(-) IN CONSTRAINTS
  1. +A-X1        .000000     \
  3. +A-X3        .000000     |
  4. +A-X4        99.2397     |
  6. +A-Y2        .000000     |
  7. +A-Y3        .000000     |  slack raw materials
  8. +A-Y4        99.2397     |
 12. +B-X4        118.879     |
 16. +B-Y4        118.879     |
 18. +C-X2        .114441E-04 /
 25. +CAP-X1      1201.53 \
 26. +CAP-X2      1201.53 |
 27. +CAP-X3      1201.53 |
 28. +CAP-X4      1400.00 |  slack capacity
 29. +CAP-Y1      2101.53 |
 30. +CAP-Y2      2101.53 |
 31. +CAP-Y3      2101.53 |
 32. +CAP-Y4      2300.00 /
 33. +UB-X11      3401.53 \
 34. +UB-X21      900.000 |
 35. +UB-X12      1401.53 |
 36. +UB-X22      600.000 |
 37. +UB-X13      1401.53 |
 38. +UB-X23      .000000 |
 39. +UB-X14      .000000 |  slack demand
 41. +UB-Y11      3401.53 |
 42. +UB-Y21      900.000 |
 43. +UB-Y12      1401.53 |
 44. +UB-Y22      600.000 |
 45. +UB-Y13      1401.53 /
```

FIGURE 7-5. Raw material scenario planning solution.

```
*3* SHADOW PRICES FOR CONSTRAINTS
  2. A-X2        1.17226
  5. A-Y1        8.82775
  9. B-X1        7.2Ø792
 1Ø. B-X2        .82339Ø
 11. B-X3        4.5Ø52Ø
 13. B-Y1        9.44914
 14. B-Y2        4.2Ø462
 15. B-Y3        .8Ø9722
 17. C-X1        1.Ø1447
 19. C-X3        .633955
 2Ø. C-X4        1.Ø8677
 21. C-Y1        .457214
 22. C-Y2        .591775
 23. C-Y3        2.45449
 24. C-Y4        .761323
 4Ø. UB-X24      1.33894
 46. UB-Y23      2.88344
 47. UB-Y14      1.3948Ø
 48. UB-Y24      3.82682

*4* REDUCED COSTS FOR DECISION VARIABLES
```

FIGURE 7-5. (continued)

case of multiperiod problems, the number of scenarios quickly multiplies so as to make the LP too large for solution by conventional techniques. Fortunately, there are algorithms (e.g., Decomposition methods) that have been constructed to enable large problems with this structure to be solved more easily. Thus, there have recently been some realistic large problems formulated in this multistage planning way that have been solved effectively. In fact, SPRO's problem, outlined briefly at the beginning of this chapter, was formulated as a scenario planning model. After consultation with a range of experts, a sequence of world-wide political and economic scenarios were generated, each with concommitant implications on the demand for and the supply of crude oil. The probability of occurrence for each scenario was assessed and the scenarios were incorporated into the model as discussed above.

EXERCISES

1. A farmer can choose to plant his 100 acres of farmland with corn, wheat, sorghum and/or soybeans. The yield (in bushels per acre), and therefore the farm's profit, depends on the summer's rainfall, as described by the data below:

	Yield (bu/acre)			
	Heavy Rainfall	Medium Rainfall	Light Rainfall	Contribution ($/bu)
Corn	300	600	400	$1.23
Wheat	400	500	300	$1.38
Sorghum	500	400	200	$1.63
Soybeans	100	300	500	$1.79

Consulting his *Farmer's Almanac*, the farmer has assessed the probability of heavy rainfall in the upcoming summer at 20%, the probability of medium rainfall at 40% and the probability of light rainfall also at 40%. Due to the limited storage facilities on the farm as well as the available harvesting equipment, no more than 20 acres of soybeans can be planted and at least as much wheat as corn must be planted. Formulate a model to help the farmer decide how many acres of each crop to plant.

2. Prior to planting, the farmer described in Problem 1 above, has heard about a new miracle fertilizer spray which will reduce the negative impact of a wet summer. The fertilizer must be applied in equal amounts at seeding time and toward the end of the summer *if* the rainfall has been heavy. If not, the farmer need not apply any additional fertilizer; the fertilizer applied at the beginning of the season will have no effect in that case. The farmer can purchase fertilizer at $2.80 per gallon prior to the season or $4.60 per gallon during the summer when it is in greater demand. The fertilizer required depends on the crops planted; requirements and the impact of the fertilizer are summarized in the data below:

	Total Fertilizer Required[a] (gal/acre planted)	Impact (percentage increase in yield if rainfall is heavy)
Corn	18	20%
Wheat	25	25%
Sorghum	18	20%
Soybeans	15	20%

[a]Applied in equal amounts before and during the season. For example, if corn were planted, 9 gallons per acre would be sprayed prior to planting; 9 gallons per acre during the late summer.

Given this new information, formulate a model to help the farmer decide how much fertilizer to purchase before planting, how many acres of each crop to plant, and how much fertilizer to buy during the summer.

3. Consider the critical path scheduling example on page 242 of Chapter 5. The problem data, as given in Table 5-5 of Chapter 5 is reproduced below:

	Task	Immediate Predecessors	Duration (days)	Earliest Starting Times
1	Framing	0	12	t_1
2	Roofing	1	6	t_2
3	Siding	1	16	t_2
4	Windows	3	13	t_3
5	Plumbing	3	8	t_3
6	Wiring	2,4	11	t_4
7	Inside Finishing	5,6	23	t_5
8	Outside Painting	2,4	15	t_4
9	Finish	7,8	0	t_6

Suppose, as is likely in reality, that the task durations are uncertain. For simplicity assume that each is distributed uniformly between the duration shown above −20%, and that duration +20%. For example, the cumulative probability for the time necessary for framing is as shown in Figure E-1.

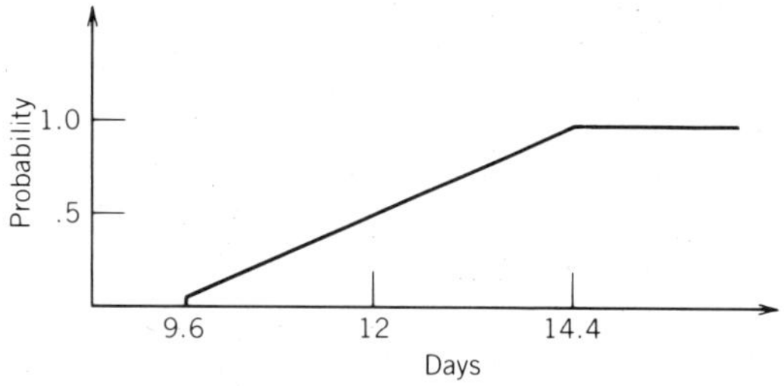

FIGURE E-1. Framing duration.

Assuming that durations are independent random variable, formulate a model to determine the least-time schedule that would guarantee that, for each task, the probability that its predecessor tasks be completed before the scheduled starting time for the task under consideration is at least 75%. Is this (or can this be reformulated as) a network?

4. Instead of the above formulation, suppose that, if the project is delayed, the construction supervisor can hire additional labor to speed up certain of the tasks, as specified below:

	Task	With No Additional Labor	Duration (days) with 1 Additional Person	With 2 Additional People
1	Framing	12, with prob. ½ 13, with prob. ½	12, with certainty	11, with prob. ½ 12, with prob. ½
2	Roofing	6, with certainty	NA	NA
3	Siding	16, with prob. ⅓ 17, with prob. ⅓ 18, with prob. ⅓	16, with prob. ½ 17, with prob. ½	16, with certainty
4	Windows	13, with prob. ½ 14, with prob. ½	13, with certainty	12, with certainty
5	Plumbing	8, with certainty	NA	NA
6	Wiring	11, with prob. ⅓ 12, with prob. ⅔	11, with certainty	NA
7	Inside Finishing	23, with prob. ⅓ 24, with prob. ⅓ 25, with prob. ⅓	23, with prob. ⅔ 24, with prob. ⅓	23, with certainty
8	Outside Finishing	15, with certainty	NA	NA
9	Finish	0	NA	NA

Unfortunately, the construction firm has to contract for the additional labor in advance, at $100 per person per day, whether that labor is needed or not. The firm has agreed to pay $500 per day penalty for each day, beyond 75 days, which the project is delayed. Formulate a model to decide how much extra labor, if any, to contract.

5. A firm ships its product from two plants to four regional warehouses for distribution directly to the customer, typically by one or both of two transportation modes: rail (inexpensive but slow) and air (expensive but fast). Based on preliminary forecast assessments, shown in Figure E-2, the firm is about to schedule its rail shipments. There is no penalty for overshipment, and any excess product will be discarded with no salvage value. If, on the other hand, demand exceeds the amount of product shipped by rail, the firm will air freight the needed amount. In the event that there is no available product at either plant to satisfy unanticipated demand, the East Coast plant will expedite

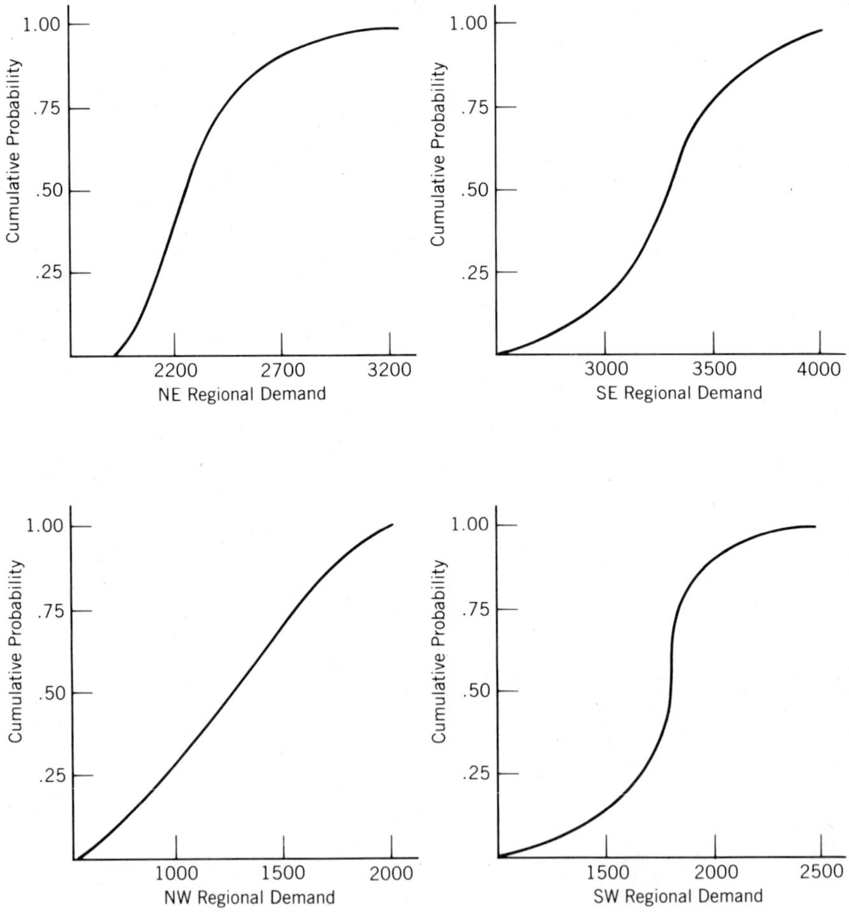

FIGURE E-2. NE, SE, NW, SW regional demand.

production to manufacture what is needed at the premium of $10 per unit. That rush order will then be air freighted to the appropriate regional warehouse.

Transport rates and normal supplies are shown below:

Plants		Transport Rates for Regional Warehouses				Plant Supply (units)
		NE	SE	NW	SW	
East Coast:	Rail	$1.30	$1.60	$4.60	$4.90	5000
	Air	$5.40	$7.20	$10.10	$11.20	
West:	Rail	$5.20	$5.10	$2.20	$1.80	4000
	Air	$10.90	$10.90	$5.90	$5.60	

Formulate a model to help the firm schedule its transport of this product.

6. The planners for the Israeli kibbutz described in exercise 4 of Chapter 2 recognize that their crop planning must incorporate the water consumption of the three possible crops. Unfortunately, this water consumption is uncertain and depends on the weather conditions during the year. The cumulative probability distributions in Figure E-3 represent the planners' assessments of the water needs for each of the three crops. Formulate a model with the constraints as in exercise 4 of Chapter 2 as well as the chance constraint that the probability that the water available for irrigation is not sufficient be no more than five percent. What underlying assumptions are you making about the nature of the uncertainties?

7. A firm's purchasing manager is about to contract for the supply of a needed alloy. The four suppliers have submitted the following bids:

	Price/lb	Maximum
1	$ 9.80	20,000 lb
2	9.90	30,000 lb
3	10.20	30,000 lb
4	10.60	50,000 lb

The purchasing manager recognizes that the firm's needs for this alloy are uncertain and has probabilistically assessed the requirements as shown in Figure E-4. Suppliers 1 and 2 are less expensive but insist on firm contracted amounts. Neither more nor less can be obtained. Supplier 3, on the other hand, has agreed to supply any amount (up to the 30,000 pound maximum) on demand, at a 10% premium, as long as 15,000 pounds are guaranteed in advance. Supplier 4 will supply any amount (up to the 50,000 pound maximum) on demand, at a 20% premium, and does not require a minimum pre-contacted amount. To complicate matters, the purchasing manager has heard rumors of an impending strike at supplier 1's foundry. He has assessed a 10% probability that supplier 1 would be unable to supply any alloy.

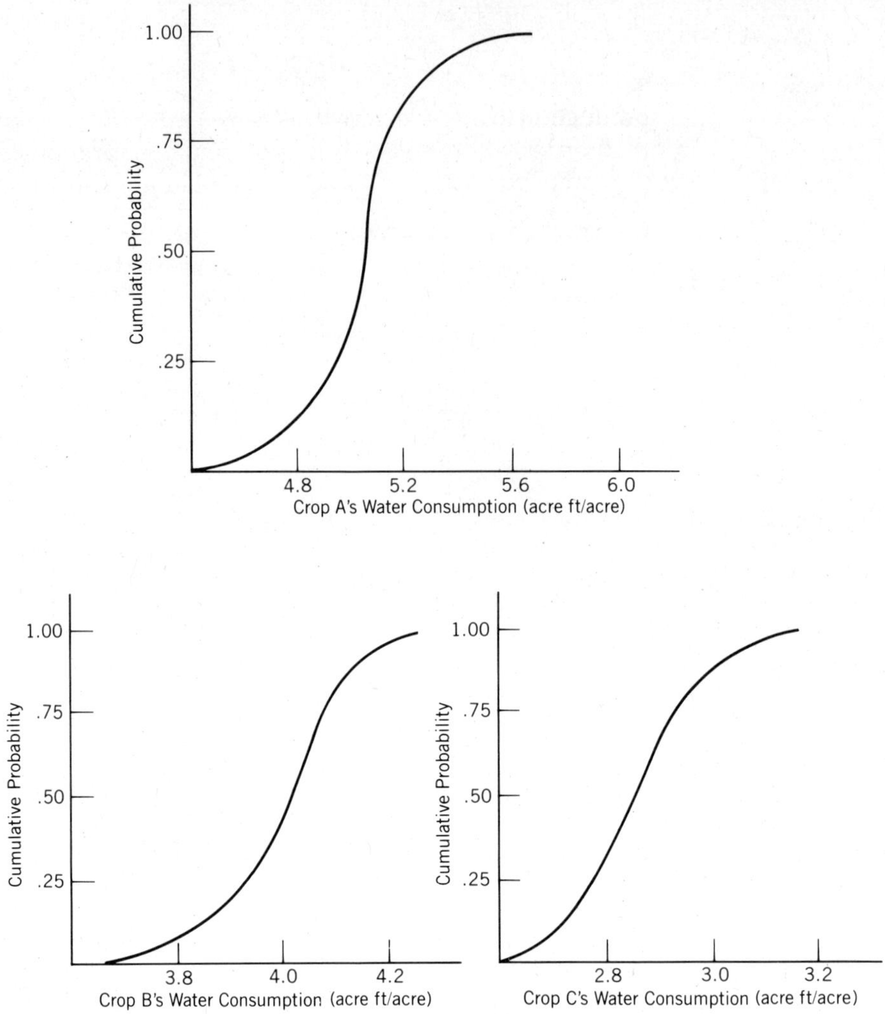

FIGURE E-3. Crop A,B,C's water consumption.

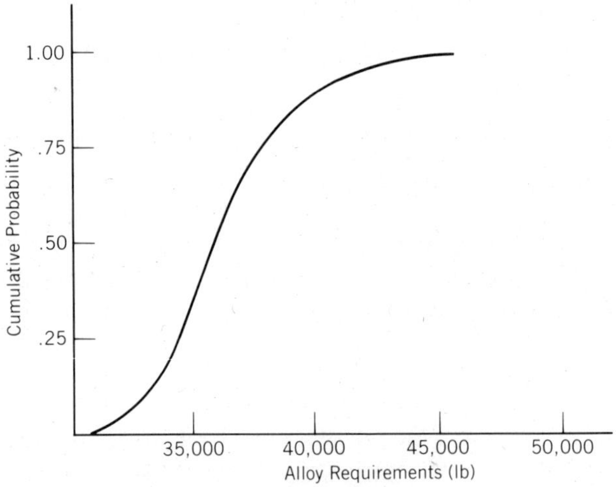

FIGURE E-4. Alloy requirements.

Formulate a model to help the purchasing manager to decide what contracts to let. Is this a linear program?

8. How might the production scheduling example formulated in Chapter 2 be altered to explicitly account for the fact that future demand is uncertain? What problems might complicate this approach?

Red Brand Canners (B)

In April 1966, Dan Tucker, Production Manager of Red Brand Canners, was preparing for his annual spring negotiations to purchase tomatoes from Sullivan Farms. Though the quality of the crop varied from year to year, the price and quantity were negotiated in advance. Tucker was scheduled to meet that afternoon with William Cooper, Controller; Charles Myers, Sales Manager; and Mitchell Gordon, Vice President for Operations. For the past several years Red Brand had contracted to buy 3,000,000 pounds of tomatoes from Sullivan to be delivered in September at a fixed price and Tucker felt that the main point of this year's negotiation would be that price. As he walked down the hall to the meeting he remembered last September's heated discussions about production planning and how Gordon had had the idea of using linear programming. "Well," he thought, "at least I won't have to get involved in all that stuff until the end of the summer."

At the meeting, Tucker presented his proposal to buy 3,000,000 pounds at 6½¢ per pound. He explained that his preliminary conversations with Sullivan as well as other growers and pickers had convinced him that Sullivan would not accept a price of less than 6½¢. He was sure however, that he could successfully negotiate for the 6½¢ price. He was slightly distressed to hear Gordon say:

> *Remember last year when we used the LP we had all those arguments about how to calculate costs and finally decided that we had to take into account the fact that we had already committed ourselves to buying 3 million pounds. Maybe we should really be worrying about how* many *tomatoes to buy, since the price seems to be pretty well settled. Maybe we can use the same method we used to optimize production to try to optimize our purchasing decision. I know we've been regularly contracting for 3 million pounds, but I don't like getting "stuck" with Grade B tomatoes that we don't really want to process. On the other hand, maybe we should consider buying more, although Sullivan can't guarantee more than 3 million. The manager of Myrtle Farms, right down the road from Sullivan, has been trying to get some business from us for years. He'd probably be able to contract for up to*

Source: Copyright © 1978 by the President and Fellows of Harvard College. Harvard Business School Case 9-179-048. See Red Brand Canners (A).

> *another 2 million pounds and might even give us a price break. Let's see if we can use linear programming to figure out how much to buy.*

Charles Myers quickly responded that that would be impossible since at this point in the season, even though the selling price per case was going to remain the same as last year, the demand forecasts were really vague.

> *We won't have firm forecasts until August and if we don't know how much we can sell, we can't decide what to make; and if we can't decide what to make, how the heck can we decide how much to buy?*

Bill Cooper said:

> *There's another problem, too. It bothers me that we sign these purchasing contracts without knowing what we're going to get. There's always that risk of getting a terrible crop, and I for one hate to see the money going out for a bunch of Grade B tomatoes. Can't we be more specific about that risk?*

Tucker replied:

> *Sure we could, if we could forecast the weather for the next five months and believe me, if I could predict the future that well, I wouldn't be in this business.*

Gordon tried to calm everyone down.

> *Look, we've been buying our tomatoes from this supplier for eighteen years, so we have some idea of what the crop will be like. There also isn't that much uncertainty about the demand. We know that we can sell as many canned tomatoes as we can make; and we've just had that big grocery chain say they'd buy up to 200,000 cases of juice. So our upper limit on juice sales will be about 250,000 cases. So the only serious questions are demand for paste and the quality of the crop. Let's see if we can collect some data and then try to figure out a way to incorporate that into a linear programming analysis. Charlie, why don't you try to give us some probabilities on different levels of paste demand. Dan, dig up the data on quality of the tomato crop for the past 18 years.*

While going through the data on past crops Tucker made an interesting observation. There were basically two types of years. In very rainy years like 1965, the crop was bad, and produced about 20% Grade A tomatoes, and in years with good weather the percentage of A tomatoes was always very close to thirty. He realized with relief that this would simplify the analysis greatly. Of the 18 years for which he had data, the crop had been "good" in twelve.

After analyzing the firm's history of past demand, Myers was able to characterize his assessment of this year's demand with the probability distribution shown in Exhibit 1.

EXHIBIT 1
Cumulative Demand Forecast for Paste Market, 1966

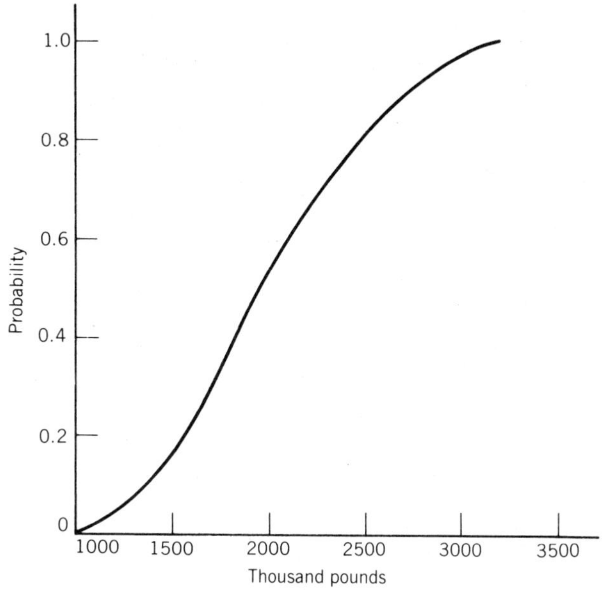

Kemtron, Inc.

Ralph Willis was worried. Yesterday was the day he should have placed the monthly order for the chemicals needed for next month's production of Kemtron's aldehyde and ketone lines. He was tempted to put it off again today. Lately, he had been under pressure from the division manager to keep his orders in line with Kemtron's production needs. Last month he had ordered 1,000 pounds too much of sodium borohydride which was now useless, and the previous month, he had been short by over 200 pounds.

Kemtron is a major producer of pharmaceutical and fine chemicals, textile chemicals, rubber lubricants, and industrial biocides. Their line of textile chemicals include specialized, often short-lived, high performance dyeing auxiliaries and finishing chemicals such as flame retardants and cellulosic biocides. Their rubber chemicals are used as anti-stick agents for uncured rubber, mold release agents, and plasticizing and tactifying agents. Kemtron biocides provide protection in textiles, rubber, wood, food processing equipment, adhesives and papers.

Several of Kemtron's fine chemicals required the use of Sodium Borohydride ($NaBH_4$) and Lithium Aluminum Hydride ($LiAlH_4$) in the reduction and conversion of aldehydes and ketones to the corresponding alcohols. In the past, the firm had, on occasion, experienced difficulty in maintaining these chemicals at a level of purity and quality which would be appropriate for the stringent specifications to be met by these Kemtron products. For several years, Kemtron's policy had been to order these compounds from its supplier in the middle of the month for the next month's production and discard any unused amounts after the next month's production runs. A three to four week lead time between order and delivery necessitated this timing. In addition, since Kemtron' aldehyde- and ketone-derived products were characterized by the same short shelf-life, they were produced to order to satisfy all orders received before the end of the month. A typical schedule is shown in Exhibit 1.

Recently, orders for Kemtron's ketone- and aldehyde-derived products had fluctuated greatly. Several customers had begun ordering sporadically. As a result, orders might be high for several months or might unexpectedly drop off for a month or two. Since the hydrides necessary in production had to be ordered before Kemtron's customers placed their orders, this put Willis under tremendous pressure. This current month had been slow, and the division had been caught with 1000 pounds of unusable $NaBH_4$.

"I can't anticipate a month like this," he had told Bob McAllister, the division manager, "nor could I guess that last month would leave us short so much." "I expect that sometimes you'll order too much, or sometimes too little.

EXHIBIT 1

Typical Order/Production Schedule for Kemtron's Ketone- and Aldehyde-derived Products

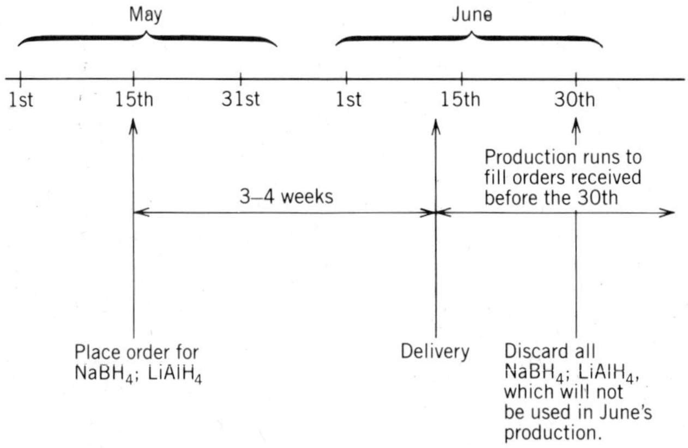

There's a lot of uncertainty in this business," McAllister retorted. "But I don't think you're using the estimates you get from the marketing staff during our weekly meetings. Pete made a pretty good stab last month, and he's laid out the possibilities for next month pretty carefully." Pete Staggs was the division's Director of Marketing. "If you can show me how you're using that information to give us the best shot, I'll be satisfied even if you're way off."

Willis had returned to his office to review the marketing director's comments (Exhibit 2). As Willis had seen in the past, demand for the ketone and aldehyde lines were highly correlated; when demand for one was high, so was demand for the other. Staggs had laid out four possible scenarios: high demand for both; average demand for aldehyde-derived products; or, low demand for both. Willis had called Staggs and asked him how likely these various possibilities were. After some vacillating, Staggs had indicated the probabilities shown in Exhibit 2.

Now, Willis was reviewing Staggs' estimates and other relevant information. Exhibit 3 shows the amounts of $NaBH_4$ (in two concentrations) and

EXHIBIT 2
Likely Order Possibilities (pounds)

	Ketone-derived Products	Aldehyde-derived Products	Probability
Scenario 1 (high demand)	4000	1500	.4
Scenario 2 (average demand)	2000	1200	.1
Scenario 3 (average demand)	2000	600	.25
Scenario 4 (low demand)	800	200	.25

EXHIBIT 3
Hydride Requirements (pounds required per pound of product)

Hydrides	Ketone-derived Products	Aldehyde-derived Products
Sodium Borohydride ($NaBH_4$) (cost: $10.00/lb)	.236	.367
Hydralls®($NaBH_4$-$CoCl_3$ Mixture) (cost: $7.00/lb)	.786	.396
Lithium Aluminum Hydride ($LiAlH_4$) (cost: $27.00/lb)	.079	.337
Contribution per pound of Product (excluding hydride costs)	$11.39	$23.59

$LiAlH_4$ needed for Kemtron's two chemical lines. There was another problem: Willis did not know what the available blending capacity would be for next month; currently, a 900-pound capacity blender was about to go into repair, and the engineering staff had estimated a 30% chance that the equipment would not be ready for next month's production run. The remaining blending equipment had the capacity to process 2400 pounds.

Willis wondered how he could incorporate this information with Staggs' scenarios to come up with a reasonable order and especially, to satisfy McAllister. He thought that since everyone agreed that the most likely outcome was high demand for both chemicals, he would probably place a large order for the hydrides; but, nevertheless, he was worried.

Agricultural Chemical International

Agricultural Chemical International (ACI) is the disguised name of a U.S. based multinational chemical company whose major product lines are manufactured and sold both domestically and overseas. It is early 1973. The enterprise has over 30 manufacturing facilities overseas compared with 10 in the U.S., and overseas sales represent 30% of the total. This case focuses exclusively on the company's Product F business which is part of the Agricultural Chemicals Group.

ORGANIZATION AND MANUFACTURING POLICY AT ACI

ACI's organizational structure consists of Domestic Product Divisions and an International Division broken down into three geographical areas: Latin America, Europe and the Far East. All divisions have profit-center responsibilities for production and marketing. Profit responsibility on exports from the U.S. is given to the domestic product division. World wide coordination for products and functions is assured by strong staffs at the corporate level and with the International Division.

The design and the implementation of manufacturing policies at ACI is coordinated by the staffs who act as mediators between the domestic product divisions and the international areas. The following general rules have developed:

1. *Transfer prices.* They are set centrally at a level which will, in each case, optimize total after-tax profits of the enterprise and take into considera-

Source: Claude L. Pomper, *International Investment Planning: An Integrated Approach,* Chapter 3, North Holland Publishing Co., Amsterdam, 1976.

tion duties, and U.S. and foreign tax laws. No such analysis is made for royalties and fees.

2. *Sourcing pattern*. Customer orders for ACI products are satisfied by a local plant whenever possible. If a product cannot be made available locally, then it is imported from the U.S. No attention is given to the problem of optimizing the flow of products between sources and markets to optimize company profits. An exception exists within the European Area where some transfers between countries are arranged at the discretion of the European regional staff.

3. *Capacity expansions*. Plants have traditionally been built to satisfy demand in the country of construction only, with no provision being made for overseas sales. Expansions have resulted from sequential review of investment proposals prepared locally, which included recommendations on the choice of a capacity and a technology. For one recent investment decision, however, an integrated point of view was taken for a product line in which economies of scale are clearly the paramount factor.

   ACI classifies all countries and therefore overseas investment projects, into one of three risk categories: high, medium, and low. Low-risk projects have to meet the same hurdle rate of return as U.S. projects. Medium- and high-risk projects must meet an incremental hurdle of 5% and 10%, respectively.

THE PRODUCT F BUSINESS

Product F is a mature agricultural chemical whose consumption has experienced a modest but steady growth for many years. The major markets are in countries with a large agricultural base such as the U.S., France or Italy. New markets are also developed in other countries. An increase in the utilization of Product F worldwide is expected, due to the following factors:

1. World production of agricultural products is growing rapidly and particularly grain crops, which use substantial quantities of Product F.

2. New markets are being developed as new application equipment becomes more widely used.

3. Product F is not expected to be affected by more stringent rules and regulations for the use of agricultural chemicals. ACI's product does not contain any of the basic chemicals that are currently either banned or under consideration to be banned in any country.

4. ACI's Product F is slightly superior to competitive products offered by other producers. ACI does not expect any competitor to develop a new product which would jeopardize ACI's market position.

The location and size of markets for ACI's Product F are presented in Exhibit 1. This exhibit also lists the present manufacturing facilities. When world total consumption is compared with aggregate effective capacity; that is, theoretical capacity corrected for operational constraints, it appears that ACI's Product F business has a shortage of capacity. This shortage can temporarily be taken care of by reducing ACI's position in markets where contracts are open for bids—that is, Eastern Europe, the Middle East or Africa.

The problem currently facing ACI is to decide when, where and how to expand capacity for Product F business within the framework of an investment strategy. The timing of new facilities is an important consideration because a worldwide shortage of Product F could result in an erosion of market share in all agricultural product lines. The choices of location, capacity and technology for new plants are felt to be somewhat related but management considers an integrated analysis to be an intractable problem. ACI's worldwide headquarters has just received a detailed capital appropriation request from the French subsidiary for a major capacity expansion. Several analyses for the construction

EXHIBIT 1

Expected Consumption of Product F for 1973 and Structure of the Present Manufacturing System of ACI (in metric tons)

All Markets		Primary Markets[a]		Manufacturing Facilities	
Country	Size	Country	Size	Country	Effective Capacity
U.S.	18,200	U.S.	21,000	U.S.	27,900
Canada	1,500				
Australia	1,300				
Brazil	5,300	Brazil	7,500	Brazil	9,630
Argentina	1,200				
Venezuela	1,000				
Colombia	5,300	Colombia	5,300	Colombia	9,630
France[b]	36,900	France	38,400	France	32,760
UK	1,500				
Italy[c]	14,000	Italy	14,000	Italy	6,965
South Africa	2,900	South Africa	2,900		
	89,100		89,100		86,885

[a]Primary markets result from the aggregation of every small market with a major market of similar characteristics.

[b]Includes Eastern Europe and West Africa.

[c]Includes the Markets around the Mediterranean and East Africa.

of small facilities in Latin America or North Italy have also been prepared. All proposals, however, respect the corporate "rules" of manufacturing policy described earlier. South Africa has been discarded as a possible manufacturing site because the local market is too small and, more importantly, because a key raw material is not available locally and is not transportable. Southern Italy, on the contrary, has been added to the list of potential locations because of the investment incentives offered in this region.

THE MANUFACTURING CHARACTERISTICS OF PRODUCT F

The manufacturing process is based on one well established technology. Management does not expect any new technology to be developed in the next five to ten years. The process is a succession of batch and continuous operations, and a plant manufacturing Product F can be described as the following "train" of units:

1. Reactor 1 where three basic materials are reacted under appropriate conditions to form a solution.
2. Reactor 2 where the output from the first reaction tank is reacted with a metal salt to yield a slurry form.
3. Preliminary drying units where the precipitate is concentrated and reduced to a rough powder.
4. Final drying units where final drying and size reduction take place. The flow from stage 3 to stage 4 is continuous.

All units are available with alternative capacities and several units of the same type can be installed in parallel. The total capacity of a plant can therefore be increased by adding a unit in parallel at the bottleneck stage of the process. Because of consideration of economical size for the various units, however, ACI management has rejected all but two alternative plant configurations (we shall refer to them as Technologies even though they differ only in scale) for present planning purposes.

The first alternative represents the smallest economical capacity for the whole train of units and is a fully balanced plant; that is, all stages of the process have equal capacities. The theoretical output of such a plant is 10.7 kT per annum (one kiloton, or 1 kT, represents one thousand metric tons), and capacity can only be increased by duplicating the plant. This type of facility has been designed for small markets as plants have traditionally been built to satisfy demand in the country of construction only. Such plants are to be considered only for possible capacity expansions in the smaller Latin American markets. They will be referred to as small plants or Technology 1 plants.

There exists a definite and critical economical size for the preliminary drying unit. Considerable experience in the manufacture of Product F has led

ACI to select 31 kT per annum as the optimal capacity for this unit for future plants. As a result, this is the maximum theoretical capacity of a train. Alternative capacities for the other units are derived from this choice of 31 kT. The first reactor is available with capacity 15.5 kT or 31 kT p.a. The final drying unit has an economical size of 15.5 kT, and the second reactor is best built in 7.75 kT units. Total capacity of a plant can therefore be chosen to be either 31 kT per annum, or 15.5 kT per annum. Since 31 kT per annum is such a large capacity (⅓ of world total consumption) only 15.5 kT per annum plants are considered for capacity expansion. They are the alternative to small plants and will be referred to as large plants or Technology 2 plants. A large plant can be expanded into a 31 kT per annum fully balanced plant.

One of the French plants is a Technology 2 plant. Building a second large plant in France will mean expanding the existing facility to a 31 kT per annum facility. All the other plants in ACI's manufacturing system are fully balanced plants of various capacities.

"Shared economies" result from the sharing of common facilities by several plants. Large shared economies in investment and fixed costs exist when expanding a large plant into a 31 kT per annum plant. The capital necessary for the second phase of investment in plant and equipment is only 50% of the initial capital outlay, and the fixed cost of operating a fully balanced system is only 160% of the same cost for a large plant.

Several factors conspire to keep actual production below the theoretical capacity of plants:

1. Plants have to be closed down for maintenance on a regular basis.
2. ACI's Product F business is highly seasonal. Company practice has been to accumulate inventory for the peak season rather than build extra capacity to meet the peak. Because of scheduling inefficiencies and constraints that have been put on the amount of inventory which any given plant can hold, the effect has been to constrain the maximum annual output. The effect of factors 1 and 2 has been to reduce effective capacity to 90% of theoretical capacity.
3. Labor laws in certain countries may reduce the effective capacity of local plants. In Italy for example, plants have to close for weekends. Thus the effective capacity for all facilities built in that country is reduced by 29%.

Investment costs are of two categories: plant and equipment, and working capital. Costs of the first category inflate over time as a function of the rate of increase in the cost of the equipment and of the rate of change in installation costs in the various countries considered. The investment in working capital inflates at some weighted average of the rates of inflation in plant costs and prices of direct production inputs. ACI considers working capital as an initial capital outlay assuming operations at capacity, rather than a direct cost of doing business. Since there is no erosion taking place at any stage of the manufacturing process, and since ACI does not expect any technological change, plants do not age and therefore need no further capital allocation after the initial investment.

Operating costs are classified into fixed and variable costs. Fixed costs are comprised of personnel costs and other costs including process and administrative supplies, property tax and social services, which are very important in certain countries. Variable costs include labor, material, utilities and others. The cost of raw materials is of paramount importance for Product F, and there is no world market for the basic raw material, which is a non-transportable commodity. These different components of manufacturing costs are in principle subject to different rates of inflation at any given location. For a given plant size, ACI expects no improvement in efficiency over time in any mature country since the technology is so well known already.

Exhibit 2 displays the family of curves representing the estimated total

EXHIBIT 2

Estimated Total Production Costs of ACI's Product F as a Function of Volume of Output for All Alternative Capacity Expansions

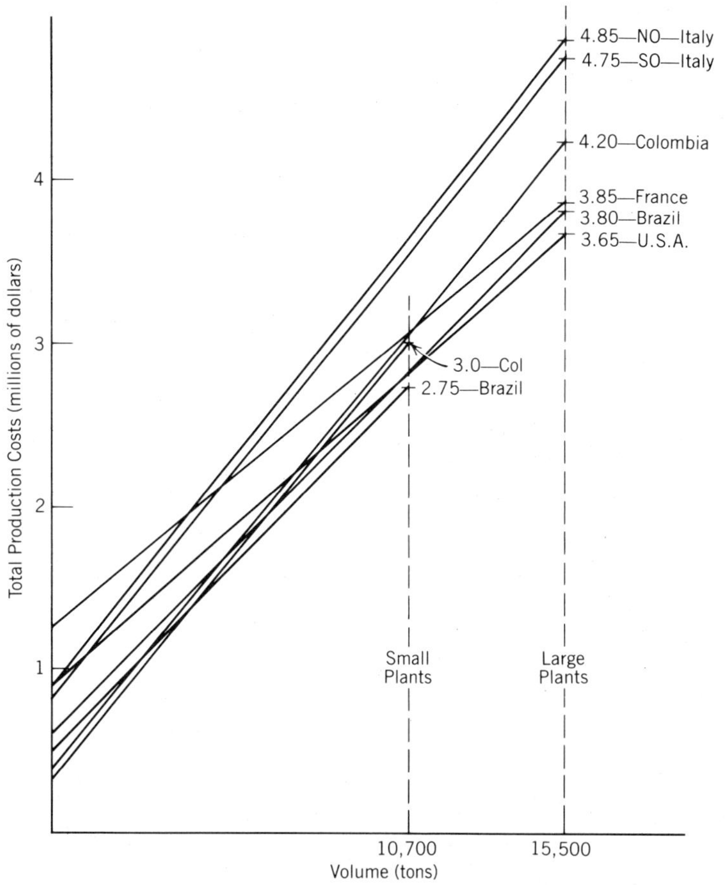

production costs as a function of the volume of output for all alternative capacity expansions. Exhibit 3 shows a breakdown of manufacturing costs for each alternative. It assumes that plants operate at capacity and that the capital charge is 10% of the initial capital outlay. Below are examples of conclusions that can be drawn by reading these charts:

1. A large facility in the U.S. would be the most efficient if a new plant were to produce 15.5 kT of Product F. It is the alternative presenting the lowest production costs at this volume.
2. Italy would be the least efficient producer because of both high operating costs, and very high capital charges due to low effective capacity (5 day week).
3. Both in Brazil and Colombia, large plants are only marginally more efficient than small plants.

EXHIBIT 3

Breakdown of Estimated Manufacturing Costs of ACI's Product F for Alternative Capacity Expansion

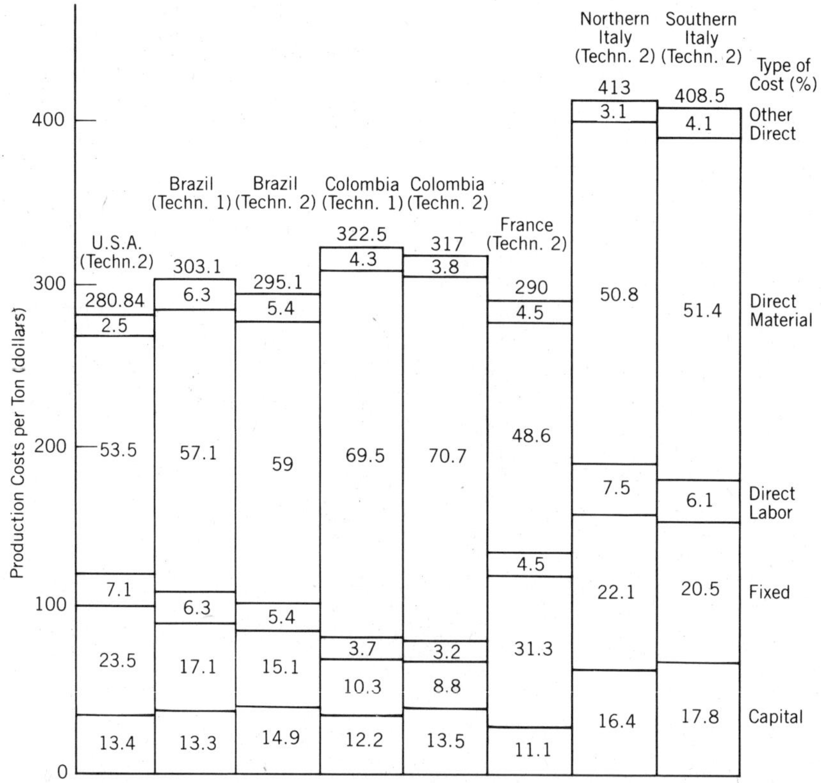

Final packaging is not discussed above because the following packaging decisions result immediately from the choice of a country and a capacity:

1. Whether packaging is subcontracted or not. Any cost of subcontracting is included in the other direct costs category.
2. The choice of a packaging technology if this task is performed internally. Costs incurred at this stage of the process are then distributed to the appropriate categories.

Product F as currently produced by all plants is essentially identical. Different plants may be producing slightly different grades of Product F in order to meet special needs of the local market, but are capable of adapting their operations to the needs of any other market of ACI. Alternate production of several grades of Product F is also possible given the batch nature of the manufacturing process. Problems due to differences in packaging can also be easily resolved.

THE CURRENT BUSINESS ENVIRONMENT AND ITS EXPECTED FUTURE

A country by country summary of expected business conditions and factors influencing profitability is given below as assessed by ACI's management. More detailed information about the environment is presented in the Appendix, Exhibits 6 through 11.

The United States

The U.S. market is expected to grow 1% to 2% annually. The long term inflation rate will average 5% per annum with no further devaluation of the dollar. The U.S. tax legislation on foreign source income will not be altered, nor will domestic investment credits. The DISC (Domestic International Sales Corporation) regulations will remain in effect, resulting in an effective tax rate of 36% on profits from exports. A 5% duty will continue to be applied on imports.

Brazil

Of all markets, Brazil will have the greatest growth because of multiple new applications. Current growth is in the 4% to 5% range but could accelerate to as high as 10% per annum. A continued economic boom is expected to be accompanied by a heavy trade deficit. The government is unlikely to depart from policies of emphasis on export and foreign direct investment that have proved successful in promoting rapid economic growth. Hence, the 30% direct export incentive will remain protected by a 60.5% tariff barrier. There will be continued success in Brazil's struggle against inflation which will average 12%

per annum. The average devaluation rate against the U.S. dollar is expected to be about 7% per annum because of national gains in productivity and in response to the upwards pressure on the European and Japanese currencies where some 45% of Brazil's too large imports come from. The corporate tax rate is expected to be maintained at 30%.

Colombia

The Colombian market is expected to experience a moderate growth of 1% to 2% per annum. An unstable political future is seen for Colombia, but foreign investment will continue to be welcome for what it brings: technology, jobs and an export base other than coffee. Exports are further encouraged by a 16% direct export incentive, and imports are discouraged by a 50% tariff barrier. A high inflation rate, about 13%, is expected to prevail despite stringent monetary policies. Colombia's exchange rate policy has been altered by the government along the lines of Brazil's strategy. The government is expected to devalue the peso at a rate of 10% per annum. The corporate tax rate in Colombia is 50%.

France

The market in France is expected to grow 2% to 4% annually due to the further development of new applications. France is experiencing rapid economic growth resulting in a strong currency. Inflationary pressures are likely to develop, but anti-inflation measures would probably be taken in order to maintain the inflation rate at about 6%. France, however, may experience social unrest and growing labor trouble in response to the government's conservative policies. France imposes a 9.6% duty on imports from sources outside the EEC; the corporate tax rate is 50%.

Italy

An annual growth of about 2% is expected for the Italian market. Italy is attempting economic recovery under continued political instability and a difficult labor situation. Inflation is expected to run at an average of 7% p.a. and the outlook is still uncertain on the Lira which is expected, however, to remain almost at par with the dollar in the long run. Italy offers preferential treatment to imports from developing countries in the form of no tariff. The corporate tax rate is 20%. In order to promote the economic development of the rural South, the government offers a 50% investment grant in Southern Italy, with a ten year tax holiday and free transportation of manufactured output to northern markets.

South Africa

The market in South Africa will probably grow only 1% annually. The local corporate tax rate is 50% and there is no tariff on imports.

STRUCTURING UNCERTAINTY

ACI's management made the following assessments of the structure and timing of actions and events, including assessments of uncertainty through alternate scenarios.

Exhibit 4 presents the time structure of the situation. The present is taken as mid-1973—that is, the time when the analysis is to be completed in time for the first decision. Strategic decisions as well as the observation of environmental changes are assumed to take place every 18 months, which is the average construction lead time for any facility in any country. It is assumed that the environment for Period 0 is already known and that no capacity expansion will take place at Time 3. The duration of Period 3 has also been chosen longer than 18 months since less detailed knowledge is available for more distant periods.

ACI's management decided that the only uncertainty worth considering in the analysis was the uncertainty in the volume of sales in each market. Exhibit 5 presents the uncertainty tree for the Product F business as assessed by management. There is only one possible environment for Period 1, which is expected to be identical to the current environment, since ACI believes that the present developments in agricultural markets will take at least three years to substantially affect the growth in the consumption of Product F. For each

EXHIBIT 4

Timing of Events and Decisions for the Implementation of the Planning Model to ACI's Product F Business

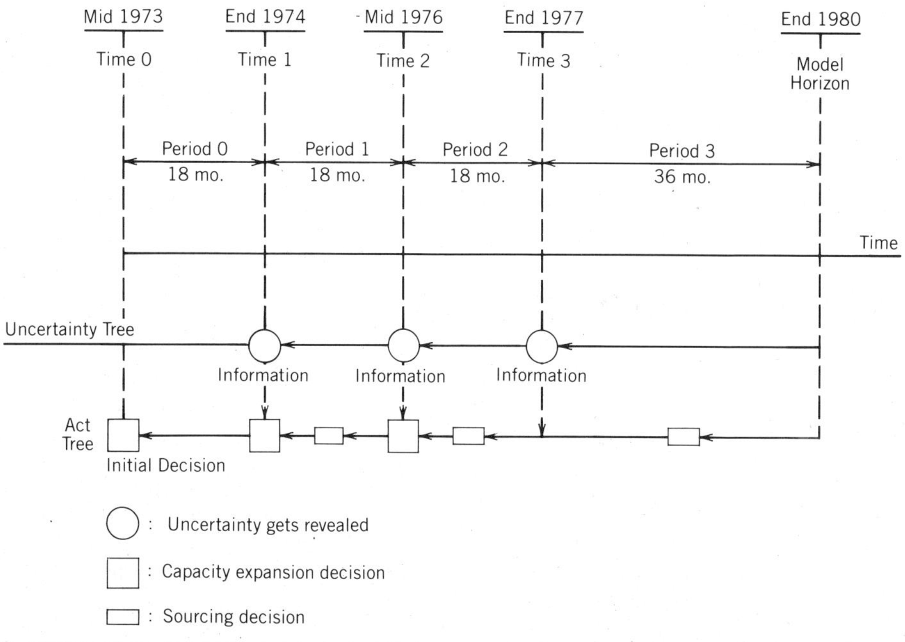

environment, annual market growth is given for every market. The probability of occurrence of each environment is conditional on the previous environment having taken place. The high correlation among markets is due to the existence of world markets for most agricultural commodities and the resulting parallel development of all Product F markets.

EXHIBIT 5
Uncertainty Tree Assessed by ACI's Management: Future Rates of Growth of All Markets for Product F

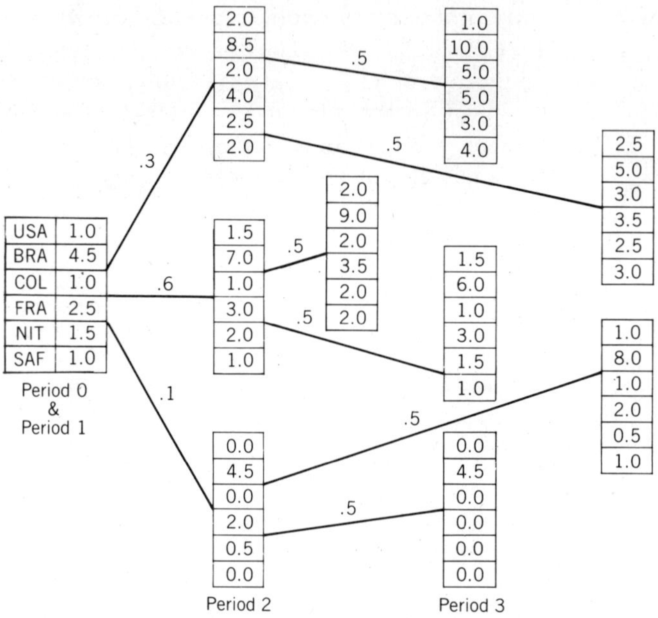

APPENDIX

Exhibits 6–11 list environmental data complementing the brief summaries in the case. They should provide adequate information for a detailed analysis.

Special attention is drawn to the first row of Exhibit 9. Because of differing political risks, capital repatriation restrictions, deferred U.S. taxes, and so on, ACI management, when calculating consolidated worldwide cash flows for decision-making purposes, values one dollar of cash flows differently for different countries.

EXHIBIT 6

Investment and Manufacturing Costs for All Existing Facilities and All Alternative Capacity Expansions for the Production of ACI's Product F (Assumes Early 1973 Prices)

		Investment Cost ($000)		Annual Fixed Costs ($000)		Direct Costs per Unit of Output ($ per ton)			
Country	Standard Capacity (000 tons)	Plant and Equipment	Working Capital	Personnel	Others	Labor	Material	Utilities	Others
U.S.[a]	31.0	—	—	400	1,440	20	150	6	1
U.S.	15.5	3,800	1,430	200	720	20	150	6	1
Brazil[a]	10.7	—	—	130	370	19	173	4	15
Brazil	10.7	2,750	1,125	130	370	19	173	4	15
Brazil	15.5	4,500	1,600	160	460	16	173	4	12
Colombia[a]	10.7	—	—	110	210	12	224	4	10
Colombia	10.7	2,750	1,030	110	210	12	224	4	10
Colombia	15.5	4,500	1,500	130	260	10	224	4	8
France[a]	36.4	—	—	545	2,420	13	141	4	9
France	15.5	2,450	2,050	230	1,035	13	141	4	9
Northern Italy[a]	10.9	—	—	130	625	31	210	6	7
Northern Italy	15.5	3,965	2,730	155	750	31	210	6	7
Southern Italy	15.5	4,450	2,730	230	600	25	210	10	7

[a]Existing facilities.

EXHIBIT 7

Estimated Annual Rates of Inflation of Prices and Rates of Change of Parity Against the U.S. Dollar in Countries of Operations of ACI's Product F Business (percent)

Country	Industrial Prices	Wages	Material	Utilities	Investment Cost	Product Prices	Parity Against $
U.S.	5	6.7	5	4	5	5	0
Brazil	12	15	12	12	12	11	−7
Colombia	13	16	13	13	13	12	−10
France	6	9	4	4	6	5	1
Italy	7	10	3	4	7	7	0
South Africa	—	—	—	—	—	5	.5

EXHIBIT 8

Transportation Costs per Metric Ton of ACI's Product F for Shipments from Plants to Markets (percent)[a]

	From Brazil	From Colombia	From France	From Italy	From U.S.
To Brazil	—	181	48	56	66
To Colombia	181	—	100	101	54
To France	46	36	—	10	47
To Italy	20	36	10	—	40
To U.S.	101	101	56	56	—
To South Africa	35	65	48	48	67

[a]Annual rate of inflation of transportation costs = 10%.

EXHIBIT 9

Tariffs Imposed By Receiving Country On Shipments of ACI's Product F from Plants to Markets (percent of invoiced price)

	From Brazil	From Colombia	From France	From Italy	From U.S.
To Brazil	—	60.5	60.5	60.5	60.5
To Colombia	50	—	50	50	50
To France	9.6	9.6	—	0	9.6
To Italy	0	0	0	—	9.6
To U.S.	5	5	5	5	—
To South Africa	0	0	0	0	0

EXHIBIT 10

Listing of Financial Parameters for the Countries of Operations of ACI's Product F Business

Parameter	U.S.	Brazil	Colombia	France	Italy	South America
Weight of Country Cash Flow in Consolidated Cash Flow Objective	1	.8	.7	.9	.7	.75
Corporate Tax Rate (percent of profits)	48	30	50	50	20	40
Lower Bound on Export Transfer Price (percent of export market price)	80	50	60	85	65	—
Upper Bound on Import Transfer Price (percent of local market price)	90	90	90	90	90	90
Upper Bound on Royalties (percent of local sales)	—	5	5	5	5	5
Upper Bound on Fees (percent of invested capital)	—	—	—	—	—	—
Depreciation (number of years of depreciation using the sum-of-the-years'-digits method)	10	10	10	10	10	10

EXHIBIT 11
Listing of Non-Financial Parameters for the Countries of Operations of ACI's Product F Business

Parameters	U.S.	Brazil	Colombia	France	Italy	South America
Market Price for Product F (dollars per ton)	430	586	532	324	320	463
Selling Cost (percent of market price)	15	15	15	15	15	15
Shortage Penalty (percent of market price)	40	40	20	30	30	20
Export Incentive (percent of invoiced price)	0	30	16	0	0	0
Labor Disruption Index (percent of annual output)	0	0	0	0	29	0
Plant output Index (percent of theoretical capacity)	90	90	90	90	90	90
Rate of Improvement in Efficiency (percent per annum)	0	0	0	0	0	0
Ownership Constraint (percent ownership of ACI)	100	100	100	100	100	100

Carnation International

Carnation International, a division of Carnation Company, is a U.S.-based multinational engaged in the processing and sale of dairy and nondairy food and grocery products, pet foods, containers, and products for scholastic market. As of 1979, Carnation Company had 67 subsidiaries and affiliates located in the U.S.A., Canada, South America, Europe, Asia, Austrialia, and South Africa. Carnation International looks after all the overseas operations of the parent company, except Puerto Rico and Canada.

In 1978, the net sales of Carnation Company were $2,576 million and the net income was $123.7 million. Nearly 32% of its sales were in foreign countries. 85% of the company's business was in food and grocery products, more than half (54%) of which was in dairy-based products.

This case deals with Carnation International's dairy products business, especially with condensed milk. Due to the nature of its operations, many of the products of the company must be processed near the source of raw material or near the market for the products. Also, because of political and organizational reasons, the operations of Carnation International are, for the most part, considered separate from those of the parent company. Carnation International carries out plant location studies separately and submits its proposals to the parent company which makes final decisions based on the recommendation of Carnation International. The early 1970s experienced a considerable shift in the global demand patterns for milk with the demand stable or decreasing in the developed countries but increasing in the developing countries. The envisaged entry of the U.K., Ireland, and Denmark into the European Economic Community (EEC) in early 1973 also was expected to affect the profitability of milk products in Europe. The relative stability of foreign exchange conversion rates generally characteristic of the 1960s was replaced in the early 1970s by a situation of relative instability, certainly for most of the EEC countries. At that time, Carnation International was analyzing its future investment and sourcing decisions, particularly for condensed milk, its principal product. This case deals with the international, economic, and political environment at that time.

BACKGROUND

Carnation Company was founded in 1899, when it began manufacturing condensed milk. After World War I, sales were expanded into other product lines. Its world headquarters was opened at Los Angeles in 1949. Its sales had gradually grown to $1 billion in 1970, after which it experienced rapid growth, crossing the $2 billion mark five years later. This phenomenal growth would not have been possible but for the right investment decisions made by the company in the early 1970s. Exhibit 1 shows the growth of sales of Carnation Company and the general business activity from 1900 through the 1970s. It is one of the 15 manufacturers in the U.S. to have an AAA rating from Moody's and Standard and Poor.

In 1921, Carnation entered into a joint venture with Pet Milk Company. The new company, called American Milk Products, began to acquire overseas

EXHIBIT 1
Carnation Sales and General Business Activity

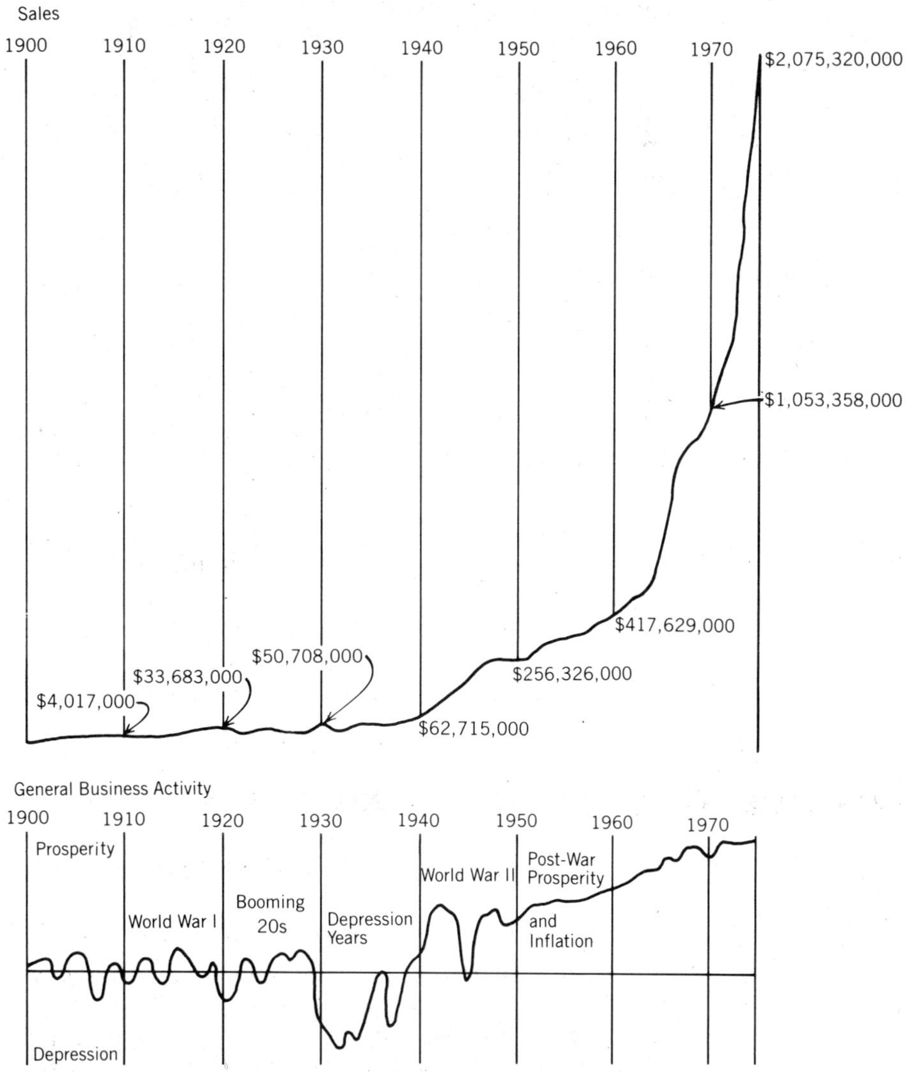

companies for the production and marketing of condensed milk. American Milk Products was renamed General Milk Company in 1930. In 1966, Pet Milk Products sold its interest in General Milk Company to Carnation. This international division was renamed Carnation International in 1972. By 1979, Carnation and its subsidiaries and affiliates had plants in 17 foreign countries and marketed their products in over 130 countries. The net income for the portion of the company's operations outside the U.S. during the year 1978 was

\$26.47 million. Carnation employs about 22,600 persons of whom 8,500 persons are outside the United States. Exhibit 2 gives a list of major facilities of Carnation International.

Carnation operates principally in the food and grocery products industry. Other miscellaneous products include animal feeds, containers and products for the scholastic market, but none of them contributes more than 10% to total sales and revenues or income before taxes. Exhibit 3 gives a list of products marketed by Carnation International in different countries.

ORGANIZATIONAL STRUCTURE

While the parent company in the U.S. is organized on the basis of product divisions, Carnation International is organized geographically. Each international subsidiary is headed by a managing director who reports to an area vice president of Carnation International. The area vice presidents are stationed at the headquarters in Los Angeles and report to the president of Carnation International.

Carnation International headquarters is organized on a matrix structure with the area vice presidents and the functional vice presidents on the two sides of the matrix. While Carnation expects the subsidiaries to operate independently, centralized control is exercised from the headquarters in strategic matters. Thus major investment decisions are studied by Carnation International at Los Angeles to ensure overall profitability.

MANUFACTURING PROCESSES FOR CONCENTRATED MILK

Carnation International manufactures concentrated milk by two processes:

1. Concentrating fresh milk.
2. Recombining skim milk powder with butterfat.

While the first method is universally accepted, the concentrated milk manufactured by the second method—called recombined milk—is not accepted in many of the developed countries, especially the European Economic Community (EEC) countries. In South America, Africa, and Asia, due to nonavailability of sufficient quantities of fresh milk, the recombined milk is accepted, as milk is important for the nutrition of the people. Thus the process used by Carnation International depends on the availability of fresh milk and the host country's policy. Carnation does not manufacture skim milk powder but it purchases its requirements primarily from New Zealand, which is a major producer of skim milk powder. Carnation cooperates with its milk suppliers by

EXHIBIT 2
List of Major Facilities of Carnation International

Baby Food Plants

Germany: Bad Essen

Butter Factory

France: Vire

Can Factories

France: Carentan, Corbie, Guingamp
Germany: Bad Essen, Marktbreit,
Neustadt i/Holst
Holland: Gorinchem
Malaysia: Kuala Lumpur[a]
Mexico: Queretaro
Peru: Arequipa[a]
Philippines: Manila
South Africa: Bergville, Harrismith
United Kingdom: Dumfries (Scotland)

Cheese Factory

South Africa: Afrikaskop

Coffee-mate Plant

United Kingdom: Dumfries (Scotland)

Condenseries

Australia: Merrigum
France: Carentan
Germany: Bad Essen, Marktbreit,
Neustadt i/Holst
Holland: Gorinchem
Malaysia: Kuala Lumpur[a]
Mexico: Durango, Queretaro
Peru: Arequipa[a]
Philippines: Manila
Singapore: Singapore[a]
South Africa: Bergville, Harrismith
United Kingdom: Dumfries (Scotland)

Food Processing Plants

Belgium: Brussels
South Africa: Durban, Johannesburg,
Politsi
Spain: Barcelona, Gijon, Miajadas,
Navelgas

[a]Carnation holds minority interest.

EXHIBIT 3
Products Marketed by Carnation International

Australia
Carnation and Gloria Evaporated Milk
Coffee-Mate Non-Diary Creamer
K-9 Pet Foods

Belgium
Carnation Evaporated Milk, Chocolate Drink, Instant Skim Milk Powder
Friskies Pet Products
Gourmet Canned Cat Food

France
Gloria Evaporated Milk
Gloria and Carnation Instant Skim Milk Powder
Bretel Butter
Gloricao Chocolate Drink
Anhydrous Butterfat
Foranimal Animal Feeds
Friskies Pet Products
Friskies Canned Dog and Cat Foods
Gourmet Canned Cat Food
Cans

Germany
Glucksklee Evaporated Milk, Chocolate Drink and Butter
Gluckslee Strained, Junior Baby and Children's Menu
Grissana Cheeses
Friski Dry Cat and Dog Foods
Cans

Holland
Carnation, Gloria and Red Seal Evaporated Milk
Blue Cross, Favorita and Gloria Condensed Milk
Friskies Pet Products

Italy
Encia Pet Products

Malaysia
Blue Cross, Tower and Peacock Sweetened Condensed Milk
Carnation and Gloria Recombined Evaporated Milk
Coffee-Mate Non-Dairy Creamer

Mexico
Carnation Evaporated Milk, Proteinada and Vitaminada Evaporated Milk
Fido Canned Pet Foods

Peru
Gloria and Carnation Evaporated Milk
Gloria Vitaminada Milk

Philippines
Liberty Filled Evaporated Milk
Liberty Filled Condensed Milk
Carnation Recombined Evaporated Milk
Carnation and Gloria Evaporated Milk
Cans

Scandinavia
Coffee-Mate Non-Dairy Creamer
Friskies Dry Cat Food
Friskies Pet Products
Friskies Dry Dog Food
Friskies Dry Cat and Dog Foods
Powdered and Liquid Non-Dairy Coffee Creamers (institutional)

Singapore
Blue Cross Sweetened Condensed Milk
Carnation and Gloria Recombined Evaporated Milk
Coffee-Mate Non-Dairy Creamer

South Africa
Carnation Evaporated Milk
Carnation and Gold Cross Condensed Milk
Carnation Canned Fruit Juices, Vegetables and Concentrates
Gouda and Cheddar Cheese
Avocado Oil
Coffee-Mate Non-Dairy Creamer
Lopis Pet Products
Lecol Fruit Drinks

Spain
Solis Tomato Paste, Puree, Sauces, Mayonnaise, Jam
Solis and Nuxina Almond, Cream
Litoral Canned Prepared Foods

United Kingdom
Carnation Evaporated Milk
Coffee-Mate Non-Dairy Creamer
Go-Cat Dry Cat Food
Go-Dog Dry Dog Food
Go-Pet Pet Products

providing technical guidance, and by entering into yearly contracts to buy their entire production.

Concentrated milk is produced in two forms—evaporated milk and sweetened condensed milk. In the manufacture of evaporated milk, sterilization is used to prevent multiplication of the bacteria to give a long shelf life. Sweetened condensed milk is not sterile; multiplication of bacteria is prevented by preservative action of sugar. A description of some of the important steps in the manufacturing processes for evaporated and sweetened condensed milk follows.

EVAPORATED MILK

Though a batch or continuous process can be used for the manufacture of evaporated milk, a continuous process is generally used for large-scale production.

Whole milk is first standardized so that the ratio of fat to solids-non-fat is approximately the same as required in the finished product (1 to 2.2). In practice, this is done by estimating the ratio at the preheating stage and separating or adding the necessary amount of cream to give the desired ratio. Clarification or filtering, though not essential, is done before or after standardization. Then the milk is preheated to improve stability against coagulation during sterilization. Preheating also increases the temperature of the milk for entry into the evaporator, where milk concentrate is produced by removing moisture. The product is pumped continuously from evaporator to homogenizer. Evaporated milk is homogenized thoroughly to obtain a uniform fat emulsion and reduce separation of fat during storage to a minimum. The product is then cooled and held in large storage tanks for restandardization of milk fat and total solids by addition of water, skim milk, condensed skim milk or homogenized cream. Sometimes fortification with vitamin D is also done at this stage. Then the evaporated milk is packed in 14½ oz (410 gm) cans called "Tall" or 6 oz (170 gm) cans called "Baby." The filled cans are sealed. Leak detectors reject the occasional improperly sealed cans. These cans are conveyed continuously through a retort for sterilization. In the sterilizer, as the cans move, the evaporated milk is heated rapidly with hot water and steam, held for a few minutes and cooled with cold water. During this stage, both agitation of the product by means of can movement and careful temperature control are important. The cans are then dried and labels are applied.

SWEETENED CONDENSED MILK

The fluid milk is standardized, clarified, preheated and may or may not be homogenized. Pasturization is accomplished during this stage to eliminate enzymes and reduce microbiological organisms. Then the milk is condensed in vacuum pans to slightly higher than a 2:1 ratio. Sugar or another sweetening agent is added at the end of condensing to inhibit bacterial growth. The sweetened condensed milk is cooled and restandardized to the required level. Lactose is very fine powder form, pulverized nonfat dry milk, or sweetened condensed milk from a previous batch is incorporated into the product as nucleus for crystallization. The product is slowly cooled with continuous agitation to complete crystallization. Then it is canned, labeled, and packed in cases.

RECOMBINED MILK

Recombined milk is made from skim milk (reconstituted powder or liquid) with butterfat added in approximately the same proportions as the butterfat removed from the whole milk. The production process consists of mixing, homogenizing and sterilizing skim milk powder, butterfat, water, and vitamins to the consistency of evaporated milk.

MANUFACTURING FACILITIES AND SOURCING

By the end of 1972, Carnation International had manufacturing facilities for condensed milk in U.K., France, Germany, Holland, Mexico, Peru, Australia, Malaysia, the Philippines, Singapore and South Africa. Except for the facilities located in Europe, all others were mainly producing to meet domestic demand.

Of the four locations in Europe, Holland was the only plant manufacturing sweetened condensed milk. Holland's production was exclusively for export, as there is no domestic demand for concentrated milk due to an abundant supply of fresh milk. Apart from meeting domestic demand, Germany exports to other European, African, South American countries and to the Caribbean Islands. Nigeria is a large consumer of condensed milk and the U.K. supplies most of its demand with the balance being met from Germany and Holland.

The sourcing decisions are made so that the cheapest producer, considering manufacturing and transportation costs, gets the orders first. Holland had always been a low-cost producer and, at the same time, a 100% exporter. Hence, first priority was given to Holland for exports, due to management policy which had originated many years ago. Also, certain markets had been generally allocated to specific sources, either due to historic tradition once set due to freight advantages or because of quality requirements. For example, U.K. was the regular supplier to Nigeria, as Nigeria's requirement for fat content (9%) was the same as in the U.K.'s domestic market. Mexico, Peru, and the Philippines were producing to meet domestic demand both from fresh milk and by recombination.

Plants are operated around the clock in Mexico and Peru where the demand for the product is continuously growing. In Europe, the third shift operation is generally restricted due to high labor costs.

The direct costs of production include milk, cans and labor. Freight is another important factor. Quotations are generally submitted for government contracts and to other customers on a c.i.f. (cost, insurance and freight) basis, which means that the freight to the customer's port of entry and insurance have to be borne by Carnation. Thus the sourcing decisions are very important for profitable operation of the company. Once the order is received, duties (tariffs) of the importing countries do not affect the profitability, as the quotations are submitted on c.i.f. basis. However, competitiveness of the product depends on the LDP (landed duty paid) price to the importer. Subsidies offered by exporting countries play an important part. The company estimates the expected demand growth and the realizable c.i.f. prices considering these factors. The EEC offers substantial subsidies for the exporters of dairy products to meet the low world market prices.

In the past investment decisions have been made by Carnation International considering the following factors:

1. The domestic demand should support the plant capacity; that is, exports are not considered in evaluating the profitability of a new plant.
2. Incentives provided by host governments for investments.
3. Availability and cost of labor and land for present capacity and expansion.
4. Availability and cost of raw milk.
5. Political atmosphere in the host company—stability of government, possible policy changes in profit distribution, possibility of nationalization and government control, and so on.

BUSINESS ENVIRONMENT FOR CONDENSED MILK

The Product and Its Life Cycle

E. A. Stuart, the founder of Carnation, first came in contact with evaporated milk in 1889, when he purchased a case of evaporated cream to see if it might help his ailing infant. His favorable experience with the product encouraged him to take up manufacture of concentrated milk. Originally the product was marketed as "evaporated cream" because of its rich cream-like consistency. The name was changed to "evaporated milk" in 1906. In the year 1929, leading pediatricians, after considerable research, confirmed the utility of evaporated milk formula in infant feeding. For two decades from 1930 to 1950, evaporated milk was extensively used for infant feeding.

With the exception of the war years (1944–48) the evaporated milk industry has experienced the classical growth curve:

1890–1900	Years of experimentation
1900–1925	Years of innovation
1925–1940	Years of growth
1940–1955	Maturity
1955 onwards	Decline

While this life cycle has been typical of developed countries, there has been a considerable lag in the developing countries, especially in Africa, Asia, and South America, where the 1970s are a promising growth period. The main factors contributing to this are population growth, shortage of fresh milk and the desire of local governments to improve nutrition.

The Markets

Carnation International sold concentrated milk in nearly 130 countries in 1972. These markets can be grouped into 10 regions as follows:

1. EEC countries[1] (Federal Republic of Germany, France, Italy, Holland, Belgium, and Luxembourg).
2. Other European countries, including U.K., Ireland, Denmark, Sweden, Switzerland, Greece, Norway.
3. Asia and Oceania, including Malaysia, Singapore, Hong Kong, Philippines, Australia.
4. North Africa, including Morocco, Libya, Algeria, Djibouti.
5. Central Africa, including Nigeria, Liberia, Ivory Coast, Upper Volta.
6. South Africa.
7. Gulf countries, including Saudi Arabia, Kuwait, Bahrain, Qatar, U.A.E., Oman, North and South Yemen.
8. South America.
9. Central America, including Mexico.
10. Caribbean Islands.

[1]U.K., Denmark, and Ireland were due to become members of EEC in 1973.

The EEC had always been a surplus producer of fresh milk, necessitating price supports to maintain farm income and payment of subsidies (called restitution payments) to exporters to match the low world prices, primarily set by New Zealand, the cheapest producer. Imports into the EEC from non-EEC countries are subjected to common tariff rates. The EEC dairy policy, which is an important determinant in the successful operation of Carnation International is described briefly in the next section.

The entry of U.K., Ireland, and Denmark into the EEC in 1973 will subject these countries to EEC regulations with a changeover period of five years. The market for condensed milk has been declining in Europe as observed earlier.

EXHIBIT 4

Production Capacity and Demand for[a] Concentrated Milk in 1972

Production Capacity

U.K.	60,000 MT
France	60,000 MT
Germany	60,000 MT
Netherlands	60,000 MT
Mexico	30,000 MT
Peru	30,000 MT
Australia	30,000 MT
Philippines	30,000 MT
Singapore	8,000 MT
South Africa	30,000 MT
Total	398,000 MT

Demand

	MT	Max. Percent of Recombined Milk Acceptable
EEC	125,000	0
Non-EEC Europe	6,000	25
Asia & Oceania	80,000	75
Gulf Countries	7,000	100
North Africa	45,000	100
Central Africa	30,000	100
South Africa	32,000	0
South America	40,000	75
Central America	30,000	50
Caribbean Islands	5,000	100
	400,000	

[a]Certain of these figures have been disguised to preserve confidentiality.

The demand in Asia, Africa and Gulf countries has been growing, providing ample opportunity in these areas. A similar situation exists in Central and South America and the Caribbean Islands. The Caribbean Islands are supplied from Europe as Central and South American production has not been able to keep pace with local demand. Exhibit 4 gives Carnation International's market for concentrated milk in these 10 regions and production capacities in 1972.

EUROPEAN ECONOMIC COMMUNITY AND ITS DAIRY POLICY

The creation of a common economic grouping of the six European States—France, Federal Republic of Germany, Holland, Italy, Belgium and Luxembourg was a major development of this century. The European Economic Community (Common Market) was created by the signing of the Treaty of Rome in 1957. The Treaty includes the following specific objectives:

1. Removal of customs duties and import and export quotas between each of the six.
2. Establishment of a common tariff and commercial policy for states outside the community.
3. Abolition within the community of obstacles to the free movement of persons, services, and capital.
4. Inauguration of common agricultural and transport policies.

To ensure that agriculture is treated as an integrated part of the economy as a whole, the Common Agricultural Policy (CAP) was developed. The CAP required common prices, uniform subsidies, standards, and a common approach to all questions of production and trade. This meant bringing about revolutionary changes, within, as well as between, each of the six states.

In 1962, the first marketing body began to control and supervise agricultural trade in various commodities. The basic management of the dairy market was instituted in 1964. The dairy products are to be controlled by variable levies. This involved the annual fixing of a target price for the milk, and on the basis of this price an intervention price for butter and skim milk powder and threshold price for a list of twelve pilot products, in which condensed milk is included. Differences were allowed in the target and intervention prices among the member states, within specified limits, prior to July 29, 1968, on which day a "single priced" market was established. The principle underlying this policy is that by managing the prices of dairy products, it would be possible to ensure target prices for milk producers.

The EEC price support system can be explained by considering the following hypothetical example on butter prices (Figure 1).

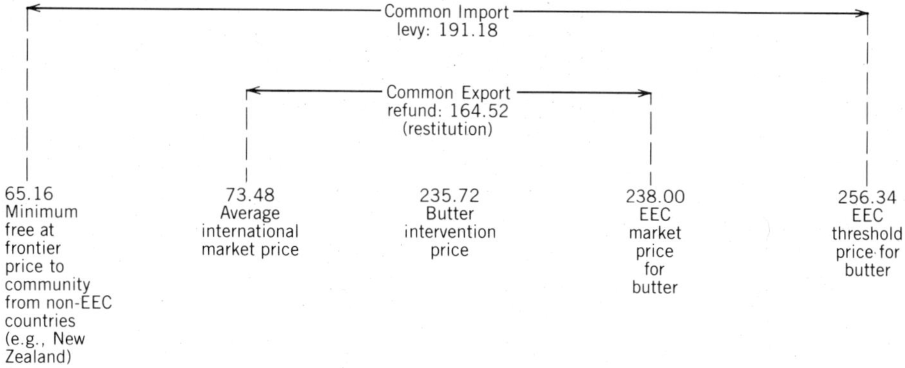

FIGURE 1. EEC price support system for butter (units of account/100 kg).

The threshold and intervention prices for butter are fixed by the EEC. If the EEC market price goes above the threshold price, imports from nonmember countries are allowed. Imports, however, are subjected to a variable levy equal to the difference between the threshold price and the minimum free at frontier price offered by the nonmember country. This will ensure that the imports cannot arrive at a price lower than the threshold price.

During times of surplus, exports are encouraged by payment of subsidies to the exporters. The subsidy is equal to the difference between the EEC internal market price and the average international price. Thus the exporter is assured of receiving a total amount equal to at least the EEC internal price on his exports.

The member states continued with their own currency and exchange policy after the formulation of EEC. This necessitated a mechanism to achieve the uniformity of prices, a basic aim of CAP. This required a set of exchange rates between the currencies of member states that would be used to convert prices, support measures, and so on, from one currency to another, within the framework of the CAP. It also led to the need for a common denominator between these currencies, so that the prices could be fixed on a community basis and subsequently translated into member state currencies. The common denominator employed was the agricultural "unit of account" and the exchange rates fixed for the devaluation of the French franc in 1969. Variation in exchange rates caused losses to some member countries and profits to others. This is compensated through the application of the Monetary Compensation Amounts (MCAs) to or from the affected states.

THE DECISION AT HAND

By the end of 1972 Carnation International management intended to complete their analysis of future investment decisions, including the opening of new plants and any decision to close or to reduce the capacity of existing plants. Crucial to them was the timing of these decisions and the type of manufacturing process to be used at the different locations chosen.

It takes about two years to install a plant and bring it to its full capacity. Two years is also the approximate period in which relevant environmental changes take concrete shape. Carnation wants to consider the six years 1973–1978 in their analysis. Plant expansion and contraction/closure decisions would be made at the end of every second year, so that the last decision made at the end of 1976 would be completely implemented by the end of 1978. The initial investment decision, made at the beginning of 1973, would be implemented by the start of 1975.

The uncertainty that is most important for this analysis is the distribution of demand for condensed milk across each of the 10 marketing areas. During 1973–1975 the growth/decline in demand has been estimated to be the same as that of the previous two years (1971–1972). For 1975–1976 three scenarios—high demand, medium demand, and low demand—are being considered, with probabilities of occurrence of 0.5, 0.4, and 0.1, respectively. For 1977–1978, further estimates of growth/decline are made for high and low demand scenarios. Thus six conceivable "states of the world" are considered: high-high, high-low, medium-high, medium-low, low-high, and low-low. The probabilities of occurrence of the 1977–1978 scenarios have been assessed conditional on the 1975–1976 environment that materializes. These conditional probabilities are given in the following table:

Conditional Probability of Third Period Environments

Second-Period Environment	High	Low
High	0.6	0.4
Medium	0.5	0.5
High	0.4	0.6

Thus, the probability of low demand in the 1977–1978 period given low demand during 1975–1976 is assessed at 60%.

From the above table, we can see that there is a correlation among the successive environments, due to successful acceptance of the product in the developing markets and the pattern of development of markets themselves. Exhibit 5 gives the annual growth rates of the 10 markets under each of the above scenarios.

Presently management is considering three locations for their new facilities:

1. Ireland
2. Venezuela
3. New Zealand

EXHIBIT 5
Annual Growth Rates Under Different Environments

					1977–1978					
		1975–1976			75–76 high		75–76 med.		75–76 low	
Area	1973–1974	High	Med.	Low	High	Low	High	Low	High	Low
EEC	10	7	5	3	5	1	2	−2	1	−5
Non-EEC Europe	−10	−3	−5	−6	0	−2	−3	−5	−3	−7
Asia and Oceania	5	7	5	2	4	3	3	2	1	0
Gulf	5	7	5	2	7	5	5	3	3	2
North Africa	10	12	8	5	5	2	7	5	2	0
Africa	10	12	8	5	5	2	7	5	2	0
South Africa	6	5	4	1	3	2	2	1	0	−1
South America	7	8	5	3	6	4	4	3	2	0
Central America	9	8	6	4	6	4	5	4	3	1
Caribbean Islands	9	8	6	4	6	4	5	4	3	1

Various considerations affecting the choice of each of these locations and the type of process to be chosen at each of these locations are described below:

Ireland

In 1972, Ireland imported 63,000 tons of condensed milk. At present, Ireland is not an EEC member and hence is not subject to the Common Agricultural Policy. Carnation exports condensed milk to Ireland from Germany. As Germany *is* in the EEC, Carnation receives restitution payments (subsidies) on its exports. The c.i.f. price of condensed milk imported by Ireland is controlled by its government at a low level.

When Ireland enters the EEC in 1973, the restitution payments on German exports to Ireland will be withdrawn and at the same time the import price into Ireland will be increased. It is expected that these changes will take place gradually within a period of five years. By having a plant in Ireland, savings in freight can be achieved.

Recombined milk is not permitted to be manufactured in any of the EEC countries at present. Hence it will not be possible to manufacture recombined milk in Ireland once it enters the EEC.

Other important factors to be considered in choosing Ireland are the availability of milk and cartons and tin plate for manufacture of cans.

Ireland is considered politically quite stable.

Venezuela

Venezuela's government has promised new investors incentives such as subsidized land, tax credits and the use of facilities; these will reduce the total investment substantially. Presently, Venezuela is being supplied from Europe so that if a plant is opened in that country, the freight savings will be substantial. There is no ban on selling recombined milk in this country, hence any shortage of milk can be supplemented by this method. In addition to the huge demand in Venezuela, the Caribbean Island markets which are presently supplied from Europe might also be served from here. This will also result in savings in freight. However, acceptance of condensed milk manufactured in Venezuela by the Caribbean consumer habituated to European milk is a matter of concern. Raw materials for manufacture of cans and cartons have to be imported. The uncertainty regarding political stability in Venezuela is high.

New Zealand

New Zealand is the cheapest producer of milk and milk products in the world and hence it is an attractive location for a milk manufacturing plant. As there is no internal demand for condensed milk in this country, due to the abundance of fresh milk, the plant would be exclusively for exports, especially to Southeast Asian countries. Carnation International buys most of its requirements for manufacture of recombined milk from New Zealand, so setting up the recombined milk plant in New Zealand exports seems attractive to many.

In 1973, the U.K. will be entering the EEC and hence the preferential treatment received by New Zealand for exporting its milk products to the U.K. will be gradually eliminated. There is a good chance of availability of skilled labor is also assured because of a well-developed milk industry. The political situation in New Zealand is expected to be quite stable.

Exhibits 4 to 8 give various data collected by the staff at Carnation International to facilitate analysis.

While Carnation International wants to maximize profits, they cannot ignore the fact that if they do not build sufficient capacity to meet future demand, they are likely to lose market to the competitors. At the same time, they recognize that they cannot build large capacity in anticipation of market growth.

EXHIBIT 6

Corporate Tax Rates, Expected Annual Inflation Rates, and Variation in Exchange Rates

	Corp. Tax Rate (%)	Inflation Rates (%)				Variation in Exchange Rates (%)[a]
		Cost			Product Prices	
		Equipment	Labor	Materials		
U.K.	52	7	10	4	5	+2
France	50	6	9	4	5	+10
Germany	41	8	11	4	5	+9
Netherlands	48	6	8	4	5	+8
Mexico	42	10	12	6	7	0

(continued)

EXHIBIT 6 (Continued)

	Corp. Tax Rate (%)	Inflation Rates (%)				Variation in Exchange Rates (%)[a]
		Cost			Product Prices	
		Equipment	Labor	Materials		
Peru	42	10	12	6	7	0
Australia	43	6	5	3	3	+5
Philippines	55	12	6	5	6	−4
Singapore	55	12	6	5	6	+7
South Africa	43	8	9	4	5	−4
Denmark	45	6	8	4	5	+6
Bolivia	33	10	12	6	7	−2
New Zealand	45	6	5	3	3	+5

[a]+ indicates that the currency is stronger.

EXHIBIT 7
Estimated Transport Cost Per MT[a]

From \ To	E.E.C.	Non E.E.C. Europe	Asia and Oceania	North Africa	Central Africa	South Africa	Gulf Countries	South America	Central America	Caribbean Islands
U.K.	0.75	4.50	16.50	5.25	10.05	11.85	8.25	10.05	8.85	8.85
France	0.30	3.75	15.75	4.50	9.30	11.10	7.50	9.30	8.10	8.10
Germany	0.30	4.80	16.80	5.55	10.35	12.15	8.55	10.35	9.15	9.15
Holland	0.30	4.65	16.65	5.40	10.20	12.00	8.40	10.20	9.00	9.00
Mexico	9.45	10.05	12.45	9.60	15.30	15.00	12.45	5.25	0.30	15.00
Peru	12.00	13.05	17.70	15.90	18.90	16.50	18.90	0.30	5.25	4.20
Australia	18.00	13.05	7.20	13.05	10.50	10.50	10.50	14.25	13.65	15.75
Philippines	18.00	12.00	2.40	12.75	10.65	12.35	9.75	21.00	18.00	18.30
Singapore	14.85	9.75	2.70	10.35	8.25	10.05	8.25	19.50	19.50	20.70
South Africa	12.05	10.50	13.20	12.15	3.90	0.75	8.85	25.50	15.30	14.25
Denmark	5.70	0.30	13.05	12.45	12.75	11.25	3.30	11.85	13.95	11.55
Bolivia	9.15	11.55	18.30	12.45	15.00	14.25	14.25	0.75	2.25	0.75
New Zealand	22.05	17.70	7.75	17.70	14.25	11.55	15.75	10.65	11.55	12.75

[a]Some of these figures have been disguised.

EXHIBIT 8
Estimated Sale Price (C.I.F. Value)[a] in Different Markets

Market	C.I.F. Value ($/MT)
1. E.E.C. countries	620
2. Non-E.E.C. Europe	360
3. Asia and Oceania	460
4. Gulf countries	750
5. North Africa	435
6. Central Africa	480
7. South Africa	355
8. South America	405
9. Central America	410
10. Caribbean Islands	425

[a]Some of these figures have been disguised.

Chapter 8

NONLINEAR PROGRAMMING

In practice, optimization problems frequently involve nonlinear behavior that must be taken into account. Unfortunately, optimization with nonlinear constraints and/or a nonlinear objective is computationally far more difficult than the optimization of a problem posed as a linear program. Thus, it is not surprising that many linear programming applications involve linear simplifications of relationships that, in truth, are nonlinear. A quick review of some of the actual cases with which we have dealt will indicate that linear programs are often formulated even though the situation does not fit the mold of strict linearity.

- Demand (and thus revenue) is typically related to price, advertising expense, and promotional activities in a nonlinear way. Yet, in Scott Paper, the LP formulated incorporated the "value" of promotional activities as a linear term in the objective function.
- In blending problems common to chemical and petroleum processing industries, it is typical to find that key characteristics of a product are nonlinearly related to the amounts of inputs. For example, the octane rating of gasoline product is a nonlinear function of the octane of each of the raw gasoline stocks blended together to produce that particular end-product, since octane varies nonlinearly with the amount of tetraethyl lead added. Yet, the model formulated by *Delta Oil* takes the octane

of its end-products to be linear in the octane of each of those products' inputs. The model formulated in *Lowell Steel* employs another simplifying approach. There, the mix of scrap and "hot metal" is constrained to be between given upper and lower bounds. In reality, performance (e.g., yield, throughput time) varies nonlinearly with the proportion of scrap as illustrated schematically in Figure 8-1. The establishment of upper and lower bounds on this proportion restricts operation to the part of this curve that is "not too nonlinear."

- It is common in a wide variety of production and warehouse environments to find diseconomies and/or economies of scale in relating costs to throughput. For example, the figure in *Hunt-Wesson's* Exhibit 3 repeated as Figure 8-2 show both relationships—the model formulated there restricts throughput to the "linear part" of the curve.

This list is not meant to show how badly we have formulated some of our earlier problems or to imply that such simplification is undesirable. On the contrary, if the analyst is knowledgeable about the range in which the solution is likely to fall, restricting the model's activities to be within that range and then linearizing the relationships there is likely to be a quite effective approach, especially given the difficulty, in general, of nonlinear optimization. For certain situations, however, the essence of the optimization problem is captured only by explicitly including the effects of at least some of the nonlinear relationships. This chapter is aimed at providing an overview of how that might be accomplished. We begin, as usual, with formulation, briefly examining a few models that have found their way into practice. We then break with tradition by delving, in more than our usual detail, into the characteristics of solution procedures—the special difficulties which nonlinear optimization pose war-

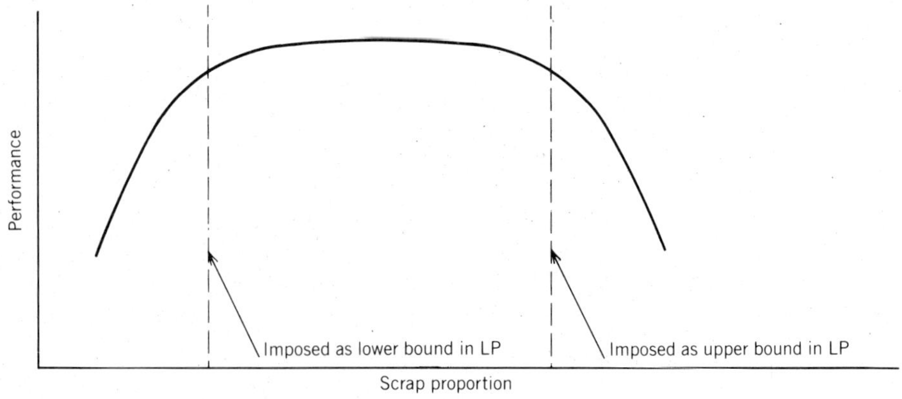

FIGURE 8-1. Actual performance characteristics.

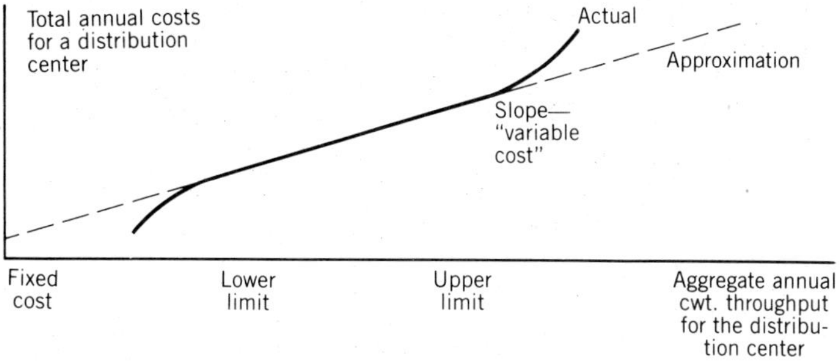

FIGURE 8-2. Warehouse cost for Hunt-Wesson.

rant a modicum of understanding on the part of the manager who will use these computational procedures.

FORMULATION POSSIBILITIES

Product Mix/Pricing Decisions

In all the product mix problems we have considered, the price for each product has been previously established. In practice, the pricing and product mix decisions are often made simultaneously. For example, let decision variables x_i and p_i represent the production quantity and market price respectively, of product i. If c_i is the variable cost (a known constant for each i) associated with the manufacture of product, an appropriate objective would be

$$\text{Max} \quad \sum_i (p_i - c_i)x_i$$

a nonlinear function.

If, in such a situation, demand constraints were present, we would likely want to reflect the fact that demand is a function of price. Thus the formulation for a pricing/product mix problem might contain constraints of the form

$$x_i \leq d(p_i)$$

where d incorporates the (probably nonlinear) interrelationship of demand and price.[1]

[1]If substitution between products is likely, we might want to write

$$x_i \leq d(p_i, \ldots, p_n)$$

Incorporating Uncertainty

Chapter 7 focused on commonly applied ways of incorporating uncertainty without destroying the linear structure. Often this is not possible. For example, consider the general linear program

$$\text{Max} \quad \sum_i c_i x_i$$

$$\text{Subject to} \quad \sum_i a_{ij} x_i \leqslant b_j \quad \text{for } j = 1, 2, \ldots, n$$

Now let us complicate matters by taking the right-hand side coefficients b_i as uncertain and independently distributed according to known cumulative probability distributions F_j; that is,

$$F_j(b) = Pr\{b_j \leqslant b\}$$

The mathematical program above can be formulated so that the joint probability of every constraint being satisfied is at least α:

$$\text{Max} \quad \sum_i c_i x_i$$

$$\text{Subject to}^2 \quad \prod_{j=1}^{n} [1 - F_j(\sum_i a_{ij} x_i)] \geqslant \alpha$$

Portfolio Selection

Portfolio management is perhaps the most common application of nonlinear programming. Substantial attention has been devoted by financial analysts to the development of mathematical models to aid in the selection of stock, bonds, and other financial instruments so as to minimize risk yet earn an adequate return. For example, let x_i represent the dollar investment to be made in financial instrument i, with r_i representing the expected return. A portfolio selection model would then have (1) a budget constraint,

$$\sum_i x_i \leqslant B$$

[2]Analogous to Σ, the summation operator the symbol Π denotes the product operator; that is,

$$\prod_{i=1}^{n} f_i(x) = f_1(x) \cdot f_2(x) \cdot f_3(x) \cdots f_n(x)$$

where B is the maximum amount available for investment and (2) a constraint guaranteeing at least an adequate return,

$$\sum_i r_i x_i \leqslant R$$

where R is the minimum return allowable for the portfolio.

The objective would minimize risk. In particular, risk might be represented as the variance of the *actual* return from the expected return. Since the financial instruments under consideration are undoubtedly such that returns are correlated, we might want to

$$\text{Min} \sum_{ij} \sigma_{ij} x_i x_j$$

where σ_{ij} represents the covariance of returns between financial instruments i and j.[3] The problem formulated above has a nonlinear objective function with linear constraints.

Fractional Goal Programming

Models for personnel planning (see, for example, the Wentworth Pacific Telephone Company case in Chapter 4) often include goals expressed as ratios—for example, that female managers should be 50% of the total. We can write this objective as

$$\text{Min} \left| \frac{\Sigma f_i}{\Sigma(f_i + m_i)} - .5 \right|$$

where f_i and m_i are decision variables representing, respectively, the number of female and male managers in department i. The linearization trick introduced in Chapter 4 will not work when ratios are considered. That is, as before, we denoted by d^+ and d^- the positive and negative deviations from 50%,

$$d^+ - d^- = \frac{\Sigma f_i}{\Sigma(f_i + m_i)} - .5$$

or

$$(d^+ - d^- - .5)\Sigma f_i + (d^+ - d^- + .5)\Sigma m_i = 0$$

[3]A number of other risk measures, usually more complex functionally, have also been used in practice.

This is a nonlinear constraint.

We might go on indefinitely: problems involving nonadditive volume relationships in processing industries; safety stock inventory levels; shared costs; costs that decrease along a learning curve as output increases; media planning models that maximize gross rating points, the product of "reach" and frequency; pricing formulas, such as those used by electric utilities. All these involve nonlinear relationships.

SOLUTION PROCEDURES

Solution of nonlinear programming problems[4] is characterized by an enormous variety of algorithms. Practice has shown that it is not possible to select one or two algorithms that will work best for all nonlinear programming problems. In fact, almost every sensible algorithm is best for some problem. This differs markedly from linear programming methodology, for which all practitioners agree that the Simplex method is *the* algorithm. This difficulty places a tremendous burden on the user, complicated by the fact that a given algorithm can be quite successful for a series of problems, yet fail miserably on the next. Rather than try to inculcate the manager in the difficult and very technical science of nonlinear optimization, this chapter will delineate a few (not necessarily the best, but simply the most commonly used) general solution approaches, and go no further. More detail on algorithm suitability, convergence rates, or computational complexity is best left to the technical expert.

Structural Characteristics

We have made a point of attesting to the difficulty in solving nonlinear programming problems. The problems are twofold:

1. Geometrically, a nonlinear objective function gives rise to behavior very much different from that characteristic of linear objectives, even when the constraints are linear. For example, in solving a linear program, we saw in Chapter 3 that we could greatly simplify the search for the optimal solution among the infinite number of choices defined by the feasible region by recognizing that the optimal objective function value had to occur at one of the finite number of vertices of the feasible region. As illustrated by Figure 8-3, this need not be true if the objective function is nonlinear, even if the constraints are linear. In that case, the optimal solution can be *any* feasible point—the vertex, a point on one of the edges, or any interior point.

2. In the graphical introduction to the Simplex method, we were assured that if we reached a vertex such that traveling in any direction would

[4]Much of the discussion in this section requires knowledge of calculus.

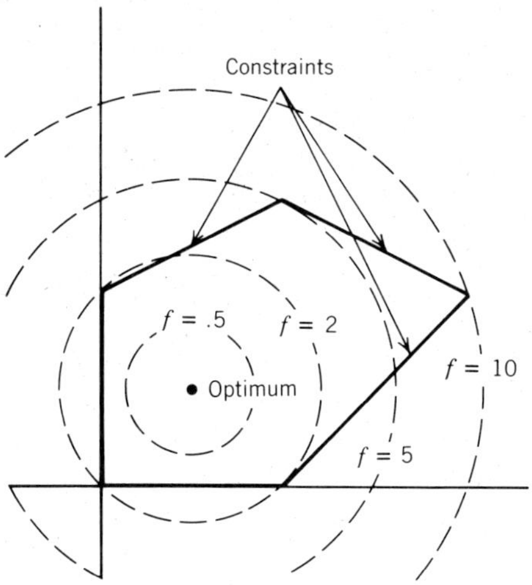

(a)

Min $(x_1 - 1)^2 + (x_2 - 1)^2$

Subject to

$-x_1 + 2x_2 \leqslant 4$

$x_1 + 2x_2 \leqslant 8$

$x_1 - x_2 \leqslant 2$

$x_1, x_2 \geqslant 0$

Solution: $x_1^* = 1$

$x_2^* = 1$

$f(x_1^*, x_2^*) = 0$

FIGURE 8-3(a). An interior point optimum.

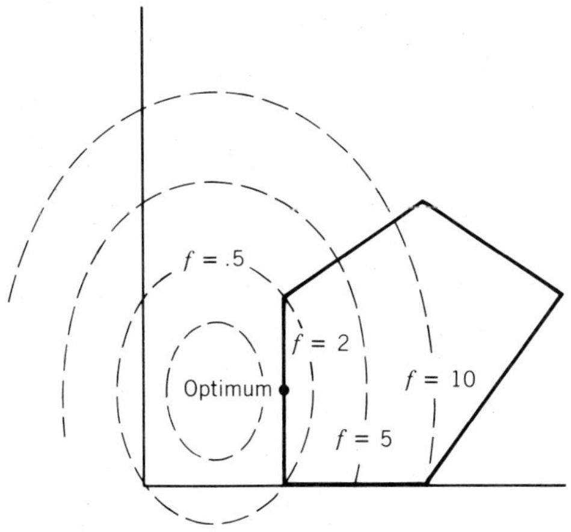

(b)

Min $(x_1 - 1)^2 + (x_2 - 1)^2$

Subject to

$-x_1 + 2x_2 \leqslant 2$

$x_1 + 2x_2 \leqslant 10$

$x_1 - x_2 \leqslant 4$

$x_1 \geqslant 2$

$x_2 \geqslant 0$

Solution: $x_1^* = 2$

$x_2^* = 1$

$f(x_1^*, x_2^*) = 1$

FIGURE 8-3(b). An optimum in the middle of an edge.

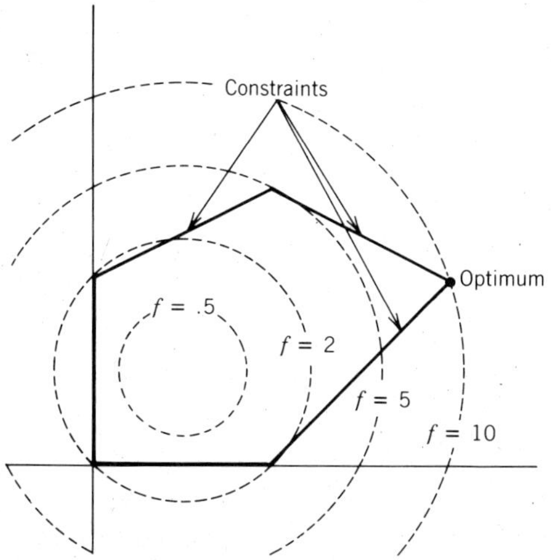

(c)

Max $(x_1 - 1)^2 + (x_2 - 1)^2$

Subject to

$-x_1 + 2x_2 \leqslant 4$

$x_1 + 2x_2 \leqslant 8$

$x_1 - x_2 \leqslant 2$

$x_1, x_2 \geqslant 0$

Solution: $x_1^* = 4$

$x_2^* = 2$

$f(x_1^*, x_2^*) = 10$

FIGURE 8-3(c). A corner point optimum.

only worsen the objective function, then we had found *the* optimal solution. This is an example where any *local optimum* is guaranteed to be the *global optimum*. Figure 8-4*a* illustrates such a situation for a one-dimensional nonlinear objective: in this case, once a search procedure verifies that a given point is optimal locally; that is, within a small range around that point, the search may be concluded—the point is the optimum. Even unconstrained, a general nonlinear objective need not have that characteristic. For example, as Figure 8-4*b* illustrates, there is no guarantee, in general, that the local optimum we have discovered is the global optimum. Figures 8-4*c* and 8-4*d* show how this problem can occur in a different context. Here the objective function *is* linear, but, in both figures, at least one of the constraints is not. Point *A* is a local optimum—traveling in any feasible direction locally (i.e., without going "too far") will only decrease the objective function value; yet point *B* is the global optimum.

Thus inability to guarantee that a just-discovered local optimum is *the* optimum is a serious difficulty. The *only* way to make such a guarantee is to find *all* local optima and compare them, a task that, in practice with multidimensional problems, is computationally difficult. Yet, all general purpose nonlinear programming procedures are "nearsighted" and can only find local optima. Hence, in practice, a particularly important special case is the *convex programming* problem.

Convexity, Concavity, and Convex Programming

A set of points is called *convex* whenever the straight line segment joining *any* two points within that set also falls completely within the set. Mathematically this can be written:

A set X is convex when, for any x_1 and x_2 in X, the point $\lambda x_1 + (1 - \lambda)x_2$ is also in X, for all values of λ between 0 and 1.

Figure 8-5 shows examples of convex and nonconvex sets. Note that the feasible regions in Figures in 8-4*c* and 8-4*d* were not convex. If they had been, any local optimum would have been global. Remember this fact.

A function $f(x)$ is called *convex* if (and only if) for any pair of points x_1 and x_2,

$$f[\lambda x_1 + (1 - \lambda)x_2] \leq \lambda f(x_1) + (1 - \lambda)f(x_2) \quad \text{for any } 0 \leq \lambda \leq 1$$

In words, this says that the line segment joining x_1 and x_2 must lie entirely *above* the function evaluated at any point on that line segment. Figure 8-6 illustrates this. Observe that the set of points *above* a convex function is a convex set.

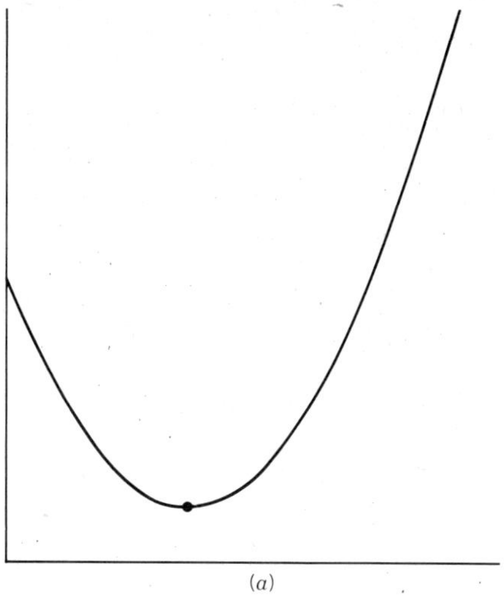

Min $x^2 - 4x + 5$

Unconstrained

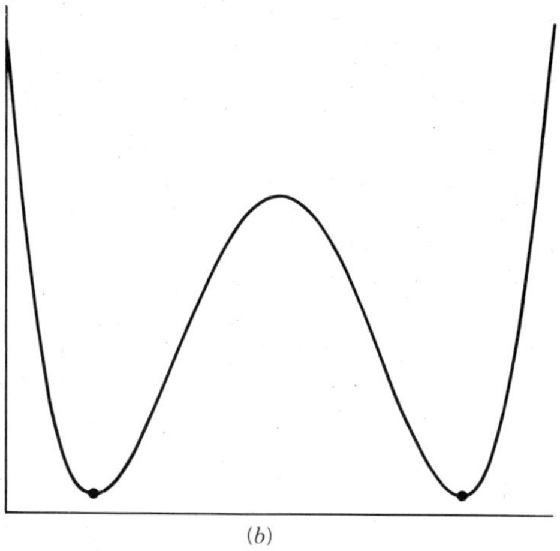

Min $x^4 - 12x^3 + 46x^2 - 60x + 26$

Unconstrained

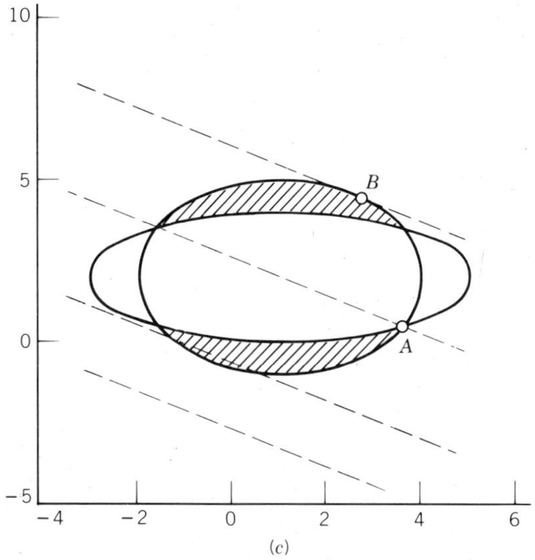

Max $x_1 + 2x_2$

Subject to

$$(x_1 - 1)^2 + 4(x_2 - 2)^2 \geqslant 16$$

$$(x_1 - 1)^2 + (x_2 - 2)^2 \leqslant 9$$

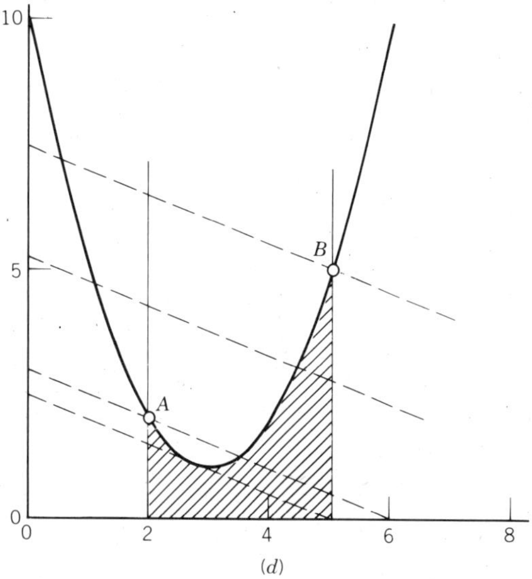

Max $x_1 + 2x_2$

Subject to

$$x_1 - (x_2 - 3)^2 \leqslant 1$$

$$x_1 \geqslant 2$$

$$x_1 \leqslant 5$$

FIGURE 8-4. Local and global optima.

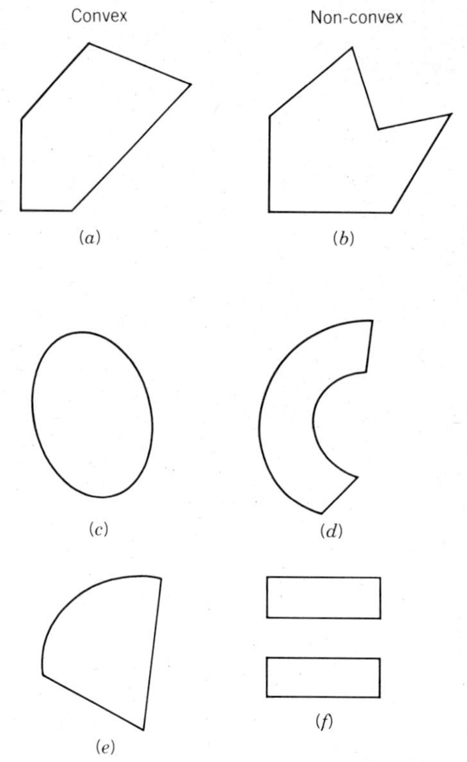

FIGURE 8-5. Convex and nonconvex sets.

A function is *concave* whenever the line segment joining any pair of points lies *below* the function evaluated at any point on the line. Mathematically, f is concave if (and only if)

$$f[\lambda x_1 + (1 - \lambda)x_2] \geq \lambda f(x_1) + (1 - \lambda)f(x_2) \quad \text{for any } 0 \leq \lambda \leq 1$$

The set of points *below* a concave function is a convex set. See Figure 8-6 for several examples of one-dimensional functions that are concave, convex, or neither. Observe that the objective in Figure 8-4*b* is *not* concave. If it were, a local optimum would be the global optimum.

Now that we have defined these fundamental concepts, consider the general nonlinear programming problem:

$$\text{Max} \quad f(x_1, x_2, \ldots, x_n)$$

$$\text{Subject to} \quad g_i(x) \leq b \quad \text{for } i = 1, 2, \ldots, m$$

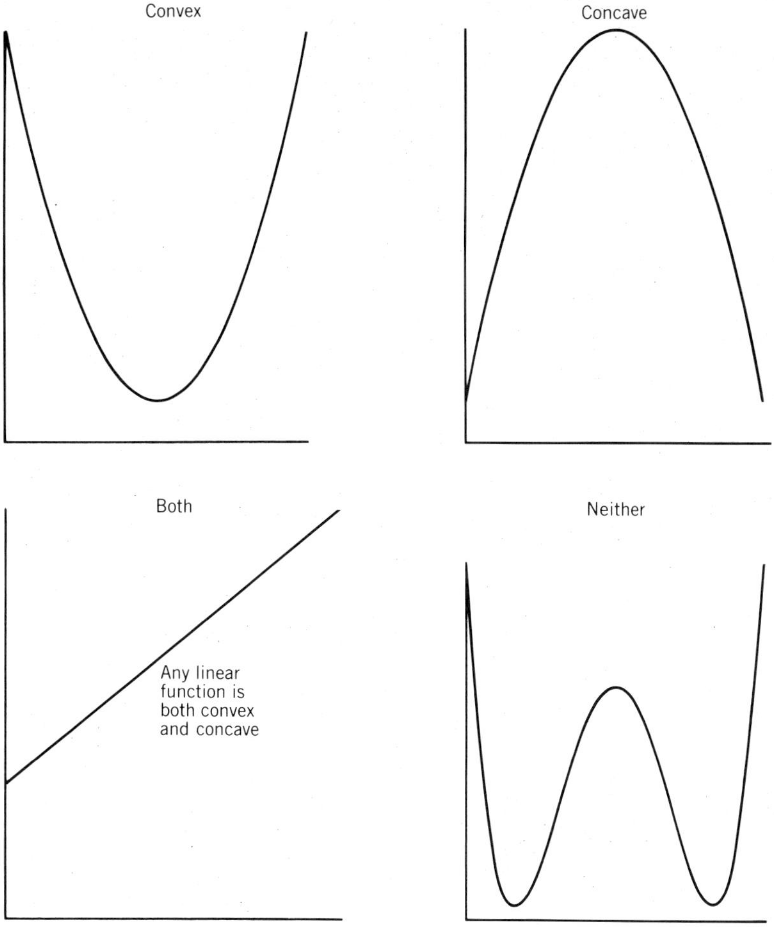

FIGURE 8-6. Four univariate functions.

This is a *convex programming* problem whenever f is a concave function and each of the g_i is convex.[5] The concavity of f guarantees the situation in Figure 8-4*b* cannot occur; the convexity of the g_i guarantees that the feasible region will be a convex set, thus preventing the situations depicted in Figures 8-4*c* and 8-4*d*. In fact, the following statement is true in general:

Any local optimum to a convex programming problem is a global optimum.

The importance of this special case is clear—rather than having to find and compare a large number of local optima, it suffices to find one.

[5]For a *minimization* problem, we require f to be a *convex* function in order for the above properties to hold.

Categorization of Nonlinear Optimization Procedures

We have made the point that there exist a considerable number of algorithms for nonlinear programming. One possible classification follows.

- *Ad hoc algebraic methods.* Techniques tailored to the specifics of the problem under consideration. Examples include reduction and simplification by algebraic elimination, substitution and/or transformation of variables.

- *Approximation.* Substitution of approximating functions that can more easily be optimized (e.g., linearization). This class of methods is often used in conjunction with other approaches on this list. Separable linear programming, introduced in Chapter 2, is a special case.

- *Enumeration.* Branch and boundlike techniques that work particularly well only when there are a finite number of feasible solutions as is the case in most combinatorial optimization problems.

- *Reduction.* The iterative addition of new constraints or conditions that progressively eliminate subsets of the feasible region which do not contain the optimal solution. The cutting-plane method for integer programming, discussed briefly in Chapter 6, is a special case.

- *Relaxation.* As opposed to reduction methods that successively add constraints, these techniques delete constraints or restrictions, adding them back in when it is determined that they are binding.

- *Descent/Ascent.* Often called "hill-climbing" techniques, these are iterative search procedures that operate by determining a "good" direction in which to move next. The Simplex method is a special case. Other special cases include complementarity methods, useful for quadratic programming and fixed point methods that do not require differentiability (or even continuity).

- *Auxiliary function methods.* Techniques that include some or all of the constraints in the objective function in some way, and iterate through a series of unconstrained optimization problems. Commonly used examples include Lagrangian methods and barrier or penalty function techniques.

The latter two categories form the basis for the most commonly applied general nonlinear programming techniques, in particular, those employed by commercially available nonlinear programming packages. The following discussion attempts to give a flavor for the way they work, while purposely sweeping as many of the technical details as possible under the rug.

Ascent/Descent: Unconstrained Problems

We begin by considering the most simple of all nonlinear optimization problems: the unconstrained case. This case has received much study—

thousands of pages have been devoted to it. In addition to setting the stage for more complex constrained problems, unconstrained optimization is worthy of study since many algorithms work by combining constraints into the objective function, thus converting the original constrained problem to a sequence of unconstrained ones. Imagine a hiker climbing a mountain in the dark, his goal to reach the summit as quickly as possible. His basic approach will be to climb in the direction of steepest ascent. That direction can only be determined by "feeling around" at the place the hiker currently is. Thus, the direction that seems best at any moment may not be the best overall. The hiker's success (as measured by how quickly he gets to the top), will depend on the shape of the mountain, but, if the mountain has a single peak[6] and no completely flat places, he will eventually reach the summit.[7]

In essence, this is what ascent/descent methods for unconstrained optimization try to emulate, formally as follows:

1. Select an initial solution $x_1^{(0)}, x_2^{(0)}, x_3^{(0)}, \ldots, x_n^{(0)}$.
2. If the current solution $x_1^{(k)}, x_2^{(k)}, \ldots, x_n^{(k)}$ (where k denotes the iteration number) is optimal, stop. If not, then:
3. Determine a direction $d_1^{(k)}, d_2^{(k)}, \ldots, d_n^{(k)}$ such that, at least locally, the objective function will improve if we move in that direction.[8]
4. Determine a step size $s^{(k)}$, which will control how far we move in the chosen direction for the kth iteration.

[6]That is, any local optimum is the global optimum.
[7]For a minimization problem, the appropriate analogue is the hiker who searches the floor of a valley for its lowest point.
[8]A direction in n-dimensional space can be represented by a vector $(d_1, \ldots, d_n)$. In two-space for example, the directions (1,1) and (1,2) are shown in Figure 8-7*a* as vectors connecting the origin to the points (1,1) and (1,2). If we start at (2,5) and move in the direction (1,2) we pass through the point (3,7) (Figure 8-7*b*), since (3,7) = (2,5) + (1,2).

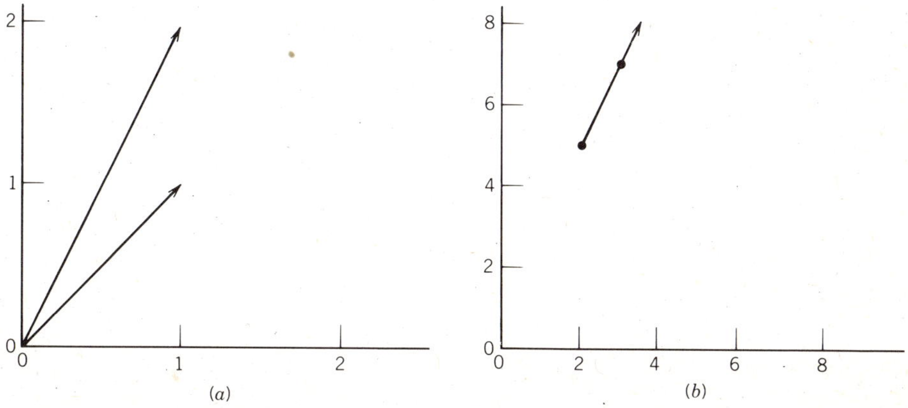

FIGURE 8-7. Directions.

5. Calculate the new solution $x_1^{(k+1)}, \ldots, x_n^{(k+1)}$ by

$$x_j^{(k+1)} = x_j^{(k)} + s^{(k)} \cdot d_j^{(k)}$$

for each j from 1 to n. Then go to step 2.

Various algorithms differ in how they choose the direction in step 3 and how the step size in step 4 is determined. Typically, both are chosen based on behavior of the objective function around the current trial solution. This is analogous to the hiker's inspecting a small area around his current location in order to determine the next direction in which to move.

Most commercially available algorithms make use of the *gradient* to determine the next direction and to decide when to stop. Assuming the objective function $f(x_1, \ldots, x_n)$ is differentiable, the gradient, denoted ∇f, is the vector whose components are the partial derivatives of f—namely,

$$\nabla f = \frac{\partial f}{\partial x_1}, \frac{\partial f}{\partial x_2}, \ldots, \frac{\partial f}{\partial x_n}$$

The gradient can be evaluated at any particular point by evaluating each partial at that point.

The significance of the gradient is that (locally) the rate at which f increases is maximized if we move in the direction of the gradient evaluated at our current location. For example, Figure 8-8 depicts the contours of the function

$$f(x_1,x_2) = \tfrac{1}{2}x_1x_2 - x_1^2 + x_1 - x_2^2 + 3x_2$$

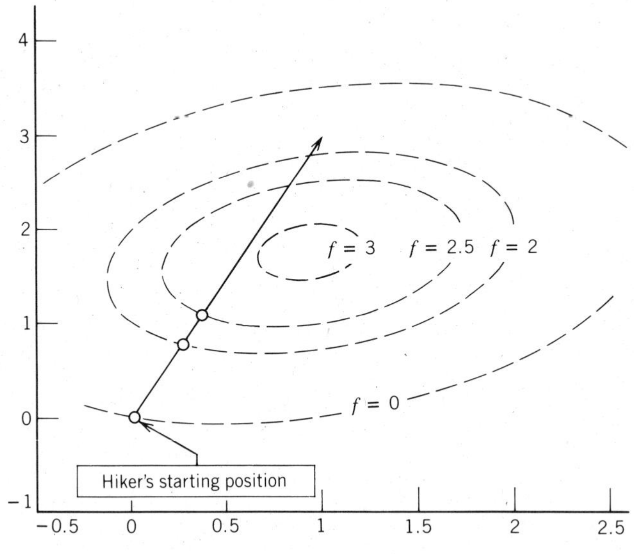

FIGURE 8-8. Hiker's ascent.

Suppose our hiker is standing at (0,0) and wants to start climbing. At that point, the gradient can be evaluated:

$$\left.\frac{\partial f}{\partial x_1}\right|_{(0,0)} = 1$$

$$\left.\frac{\partial f}{\partial x_2}\right|_{(0,0)} = 3$$

so

$$\nabla f|_{(0,0)} = (1,3)$$

Hence, the direction defined by the arrow going from (0,0) to (1,3) is the direction of steepest ascent and our hiker sets off in that direction. His question now is "How far do I go?"

The gradient is also used in the test for optimality. Almost universally, ascent/descent techniques test each trial point to see whether all partial derivatives are zero. In practice, this condition will never be met exactly, so a stopping rule might require that all partials be sufficiently small. If the objective to be maximized (minimized) is not concave (convex), the "solution" at termination need not be the global optimum or even an optimum at all (e.g., the algorithm might terminate at a saddle point[9]). Thus, other tests are often performed.

[9]A saddle point is a point that maximizes the function in one direction but minimizes it in some other direction, as illustrated by Figure 8-9.

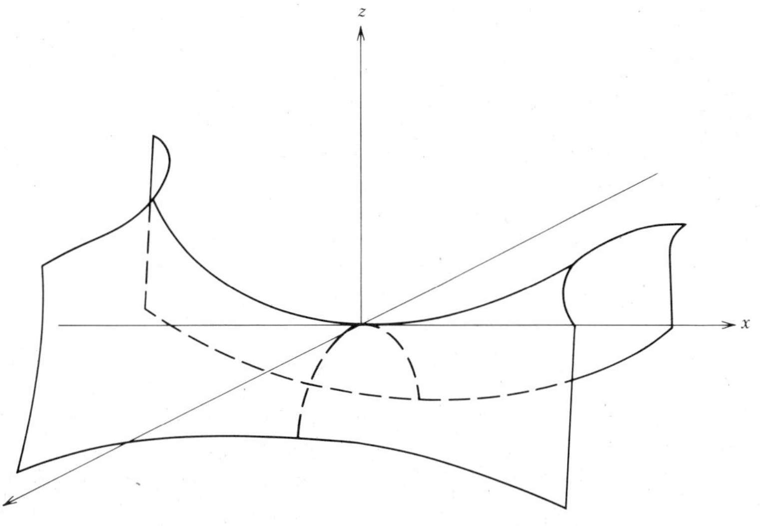

FIGURE 8-9. A saddle point.

We illustrate a gradient ascent technique by optimizing the function of Figure 8-8; that is,

$$\text{Max} \quad f(x_1, x_2) = \tfrac{1}{2}x_1x_2 - x_1^2 + x_1 - x_2^2 + 3x_2$$

Since f is quadratic we could solve it directly by setting $\partial f/\partial x_1 = \partial f/\partial x_2 = 0$. However, to illustrate the operation of a typical ascent algorithm, we solve it iteratively, taking the direction of movement as determined by f's partial derivatives evaluated at the current trial point—that is,

$$d_1^{(k)} = \left.\frac{\partial f}{\partial x_1}\right|_{(x_1^{(k)},\, x_2^{(k)})} = -2x_1^{(k)} + \tfrac{1}{2}x_2^{(k)} + 1$$

$$d_2^{(k)} = \left.\frac{\partial f}{\partial x_2}\right|_{(x_1^{(k)},\, x_2^{(k)})} = \tfrac{1}{2}x_1^{(k)} - 2x_2^{(k)} + 3$$

For simplicity, we take the step size $s^{(k)}$ to be ½ for all k. Starting with an initial trial point of $x_1 = 0$, $x_2 = 0$ at which f is 0, we proceed, as shown in Table 8-1. At iteration 6, with $x_1^{(6)} = .933$ and $x_2^{(6)} = 1.733$, the two partial derivatives are below .001, so we stop. (We might have chosen a stopping rule that terminated the process whenever the further improvement of f was undetected, at three decimal places, and thus terminated at iteration 5.) Figure 8-10 shows the ascent process graphically.

Perhaps the most common ascent technique—the method of *steepest* ascent—is similar to the above except the step size is chosen so that $f(x_1,x_2)$ is maximized along the line given by the current direction. Using our mountain climbing analogy, the hiker chooses a direction and continues along that same direction as long as he continues to ascend. When he starts descending, he immediately chooses a new direction, and so forth. For the above example, this variation leads to the iterative ascent shown by Table 8-2 and Figure 8-11, starting again at (0, 0). Note that, for this example, convergence is somewhat faster than that provided by the first ascent method, and that, characteristic of this variation, there is some "swinging pendulum" behavior, where we overshoot (in the x_2 direction) and then swing back.

TABLE 8-1

Iteration	Direction	New Trial Point	Function Value
1	(1, 3)	(.5, 1.5)	2.875
2	(.75, .25)	(.875, 1.625)	3.055
3	(.0625, .1875)	(.906, 1.719)	3.066
4	(.048, .015)	(.93, 1.727)	3.067
5	(.004, .011)	(.932, 1.733)	3.067
6	(.000, .000)	(.933, 1.733)	3.067

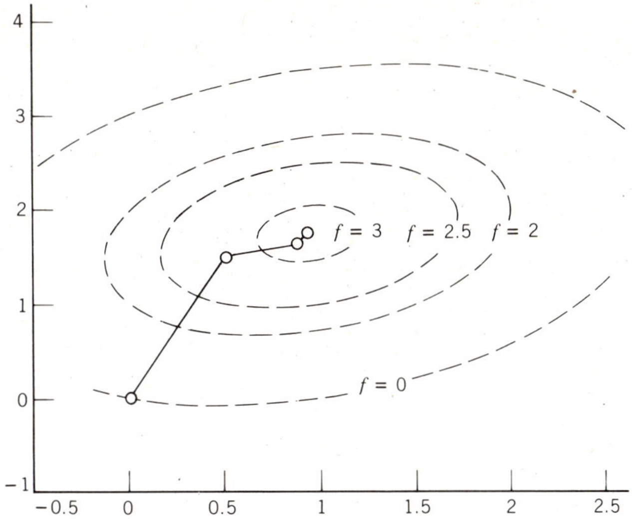

FIGURE 8-10. Ascent with constant step size.

TABLE 8-2

Iteration	Direction	New Trial Point	Function Value
1	(1, 3)	(.588, 1.765)	2.941
2	(.706, −.235)	(.895, 1.662)	3.062
3	(.041, .123)	(.919, 1.735)	3.067
4	(.029, −.010)	(.932, 1.730)	3.067
5	(.002, .005)	(.933, 1.733)	3.067

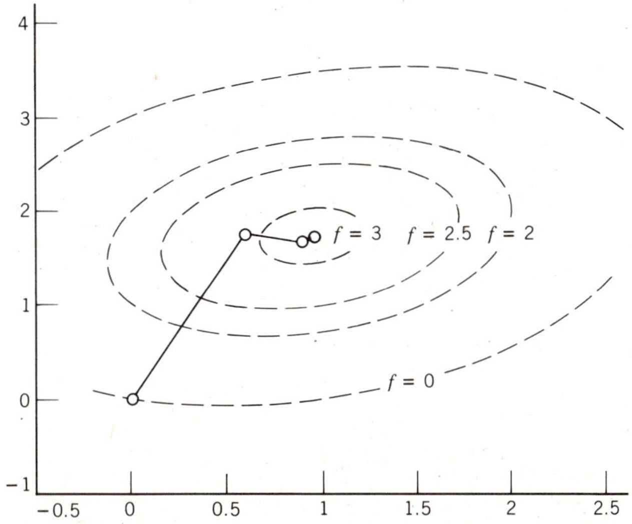

FIGURE 8-11. Steepest ascent.

There are enormous number of variation of the steepest ascent technique along with numerical techniques that reduce the method's inclination to swing back and forth. More detail, however, is beyond the scope of this book.

Ascent/Descent: The Constrained Case

The hiking analogue in the presence of constraints is that there is a fence on the mountain enclosing a large area in which the hiker currently is. His objective is to reach the highest point within the enclosed area. Again, as it is pitch dark, he ascertains a locally desirable direction in which to move and follows an ascent algorithm until he reaches the top *or until he bumps into the fence*. If the latter occurs, he looks for a *feasible* direction in which to move, most likely feeling his way along the fence until no feasible direction leads upward. If the "mountain" is concave and if the "fence" encloses a convex set, then he will have reached the global feasible optimum.

As one might imagine, this is a far more difficult problem than the preceding unconstrained case, and it is beyond the scope of what we intend here to go into much detail relative to the many algorithms which exist. Instead, let us illustrate an ascent by adding constraints to the unconstrained problem we optimized in the preceding section. Our new problem:

$$\text{Max} \qquad f(x_1, x_2) = \tfrac{1}{2}x_1x_2 - x_1^2 + x_1 - x_2^2 + 3x_2$$

$$\text{Subject to} \quad x_1 \leq 1$$

$$x_2 \leq 1$$

We take the starting point, again, to be (0,0). Observe that this initial solution is feasible.

Iteration 1. As before, the optimal direction is $\nabla f(0,0) = (1,3)$. Moving in that direction (see Figure 8-12), we continue to ascend, until we bump into the $x_2 = 1$ barrier and can go no further. The new trial point is $(\frac{1}{3}, 1)$ with $f(\frac{1}{3}, 1) = 2.389$.

Iteration 2. The optimal direction is $\nabla f(\frac{1}{3}, 1) = (\frac{5}{6}, \frac{7}{6})$. Note from Figure 8-12 that *this direction is not feasible*. The best we can do is to move rightward along the boundary; that is, the best feasible direction is (1,0). Moving along the boundary we ascend until we peak at $(\frac{3}{4}, 1)$, and then start to descend.[10] Hence, the new trial point is $(\frac{3}{4}, 1)$ with $f(\frac{3}{4}, 1) = 2.563$.

Iteration 3. The optimal direction would be $\nabla f(\frac{3}{4}, 1) = (0, 1\frac{3}{8})$. Again this direction is infeasible. Observe from Figure 8-12 that this direction is perpendicular to the boundary $x_2 = 1$. Indeed, there exist no feasible

[10]This is easily verified by setting $\partial f/\partial x_1|_{x_2=1} = 0$ and solving for the optimal x_1.

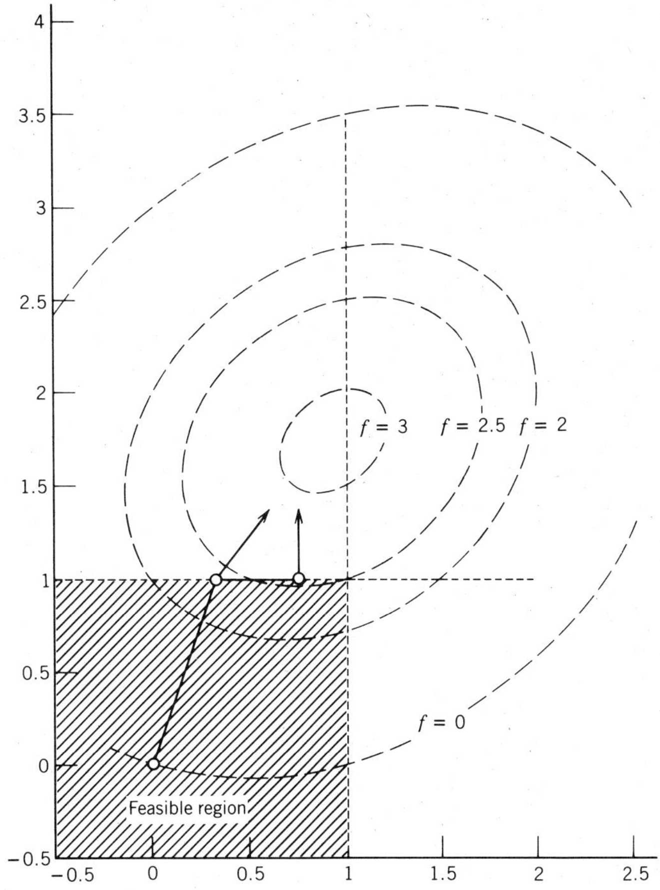

FIGURE 8-12. Ascent with barriers.

direction which will improve f. We are done. (Since f is concave and the constraints are convex, we are guaranteed to have found the global optimum.) $x_1^* = .75$, $x_2^* = 1$, and $f(x_1^*, x_2^*) = 2.5625$.

Auxiliary Function Methods

These methods operate by incorporating some or all of the constraints into the objective function in such a way that the constraints will be satisfied automatically when optimizing the (a series of) new unconstrained problem(s). There are several variations of this method that have been successfully applied. To illustrate the class of techniques, we introduce one of the more simple approaches—a penalty function method.[11]

[11]Penalty function methods are an example of what are often referred to as SUMT algorithms (Sequential Unconstrained Maximization Techniques).

The basis of this method is the introduction of a penalty function into the objective function whose purpose is to penalize the unconstrained optimization algorithm for choosing a solution close to the boundary of the feasible region. Thus, an interior point solution will be chosen. Through a sequence of unconstrained optimization problems, generated in such a way that the penalty becomes less and less, we converge to the true optimum.

One commonly used penalty function is the natural logarithm function, denoted ln and pictured in Figure 8-13. Note that as x decreases toward zero, ln x decreases toward $-\infty$. Thus, if we wanted to assure $x > 0$ in solving a given maximization problem, introducing ln x into the objective function would do the trick. Unfortunately, appending ln x might have too great an effect, so instead we append $r \cdot \ln x$ onto the objective function, where r is a positive parameter. Now, by successively decreasing r toward zero, a sequence of unconstrained optimization problems is generated. For each $x > 0$, ln x is finite yet as r approaches zero, the artificial objective function term $r \cdot \ln x$ will have less and less effect on where the maximum point falls.

We can generalize this concept as follows: Any nonlinear programming problem can be written as

$$\text{Max} \quad f(X_1, X_2, \ldots, X_n)$$

$$\text{Subject to} \quad g_i(X_1, X_2, \ldots, X_n) \geqslant 0 \qquad \text{for } i = 1, 2, \ldots m$$

Consider the new objective function

$$f(X_1, X_2, \ldots, X_n) + r \sum_i \ln[g_i(X_1, X_2, \ldots, X_n)]$$

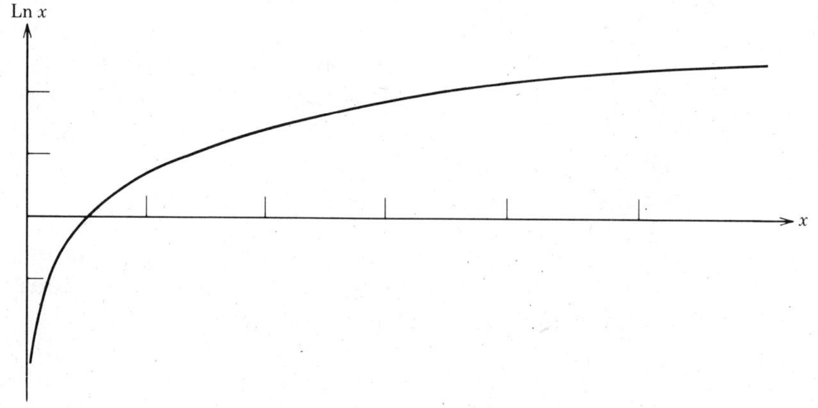

FIGURE 8-13. Ln x.

Maximizing this for any positive r will assure that

$$g_i(X_1, X_2, \ldots, X_n) > 0$$

for each i. If not, there would be a term in the new objective function equal to $-\infty$. Yet, as r decreases toward zero, the penalty function terms will have decreasing impact.

Let us consider a simple example:

$$\text{Max} \quad f(X_1, X_2) = X_1 - 2X_2$$

$$\text{Subject to} \quad X_1^2 \leq X_2$$

$$X_1 \geq 0$$

To be consistent with the notation above, we write

$$g_1 (X_1, X_2) = -X_1^2 + X_2$$

$$g_2 (X_1, X_2) = X_1$$

Then we seek to maximize a sequence of unconstrained problems,

$$\text{Max} \quad X_1 - 2X_2 + r[\ln(-X_1^2 + X_2) + \ln X_1]$$

with r decreasing toward zero. Table 8-3 gives the sequence of solutions starting with $r = 1$ and halving it for each successive iteration. Figure 8-14 shows the feasible region and the trajectory of solutions as r decreases. Note that the entire trajectory is interior to the feasible region even though the optimum is on the boundary. One motivation for the transformation of the

TABLE 8-3

r	X_1^*	X_2^*	$f(X_1^*, X_2^*)$
1	.6404	.9101	−1.1798
1/2	.5000	.5000	− .5000
1/4	.4045	.2886	− .1727
1/8	.3415	.1791	− .0167
1/16	.3018	.1223	.0571
1/32	.2781	.0930	.0922
1/64	.2648	.0779	.1089
1/128	.2576	.0703	.1171
1/256	.2539	.0664	.1211
True Optimum	.25	.0625	.125

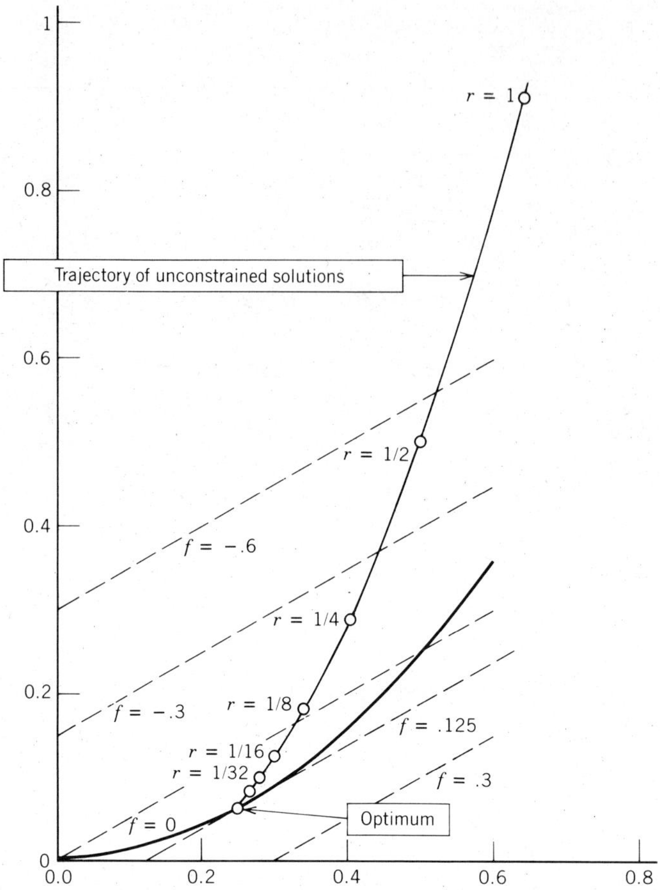

FIGURE 8-14. Sequential unconstrained optimization.

original constrained problem into a sequence of unconstrained problems is that a number of efficient, easy to apply methods for minimizing an unconstrained function are known. Thus, it becomes possible to solve the more formidable constrained problem by utilizing these procedures without recourse to new techniques.

Another highly desirable feature of this approach is that it avoids the necessity of coping separately with the boundary of the feasible region, for instance, attempting to move along the boundary once it is encountered. Such motion is exceedingly cumbersome when the constraining surface is nonlinear. The penalty function couples the objective function and the constraints in a manner that makes it possible to eliminate motion along the boundary completely.

QUADRATIC PROGRAMMING

The quadratic program is an especially simple nonlinear optimization problem. It has been the basis of enough real applications, many of which were outlined at the beginning of this chapter, to warrant special attention. This special case involves the optimization of a *quadratic objective function*[12] subject to *linear constraints*. This implies that the only way in which a quadratic program is different from a linear program is that its objective function is allowed to contain squared terms (e.g., $3x_3^2$) and cross-products (e.g., $2x_1x_4$).

The general quadratic program can be written

$$\text{Max} \quad [\sum_j C_jX_j - \tfrac{1}{2} \sum_j \sum_k q_{jk}X_jX_k]$$

$$\text{Subject to} \quad \sum_j a_{ij}X_j \leq b_i \quad \text{for } i = 1, 2, \ldots, m$$

$$X_j \geq 0 \quad \text{for } j = 1, 2, \ldots, n$$

The q coefficients are such that $q_{jk} = q_{kj}$. Note that for every term

$$q_{jk}X_jX_k$$

there is another *equal* term

$$q_{kj}X_kX_j$$

so that the "$\frac{1}{2}$" in front of the objective function's double summation prevents double counting; that is,

$$\tfrac{1}{2}(q_{jk}X_jX_k + q_{kj}X_kX_j) = q_{jk}X_jX_k$$

The quadratic program has two very desirable properties that are responsible for its success:

1. The objective function is concave if

$$\sum_j \sum_k q_{jk}X_jX_k \geq 0 \quad \text{whenever all } X\text{'s are nonnegative}$$

Hence, since linear constraints are convex, it is very easy to verify when a quadratic program is a convex program.

[12] A quadratic function is one whose terms are second-order or below.

2. The gradient of the objective function is *linear*.

Due to this, procedures have been developed that allow quadratic programs to be solved by a special variant of the Simplex method. In contrast to the other techniques we have encountered in this chapter, this variant, like the Simplex method itself, is guaranteed to converge in a finite number of iterations. In practice, that number is usually quite small. Due to the computational tractability of the procedure, most commercially available mathematical programming packages will solve quadratic programs.

For example, consider the constrained steepest ascent problem we optimize in an earlier section:

$$\text{Max} \qquad f(X_1, X_2) = \tfrac{1}{2}X_1X_2 - X_1^2 + X_1 - X_2^2 + 3X_2$$

$$\text{Subject to} \quad X_1 \leqslant 1$$

$$X_2 \leqslant 1$$

Using the above notation,

$$C_1 = 1 \qquad q_{11} = 2 \qquad q_{12} = -\tfrac{1}{2}$$

$$C_2 = 3 \qquad q_{21} = -\tfrac{1}{2} \qquad q_{22} = 2$$

Since

$$\sum_j \sum_k q_{jk}X_jX_k = X_1^2 - \tfrac{1}{2}X_1X_2 + X_2^2 \geqslant 0 \quad \text{for all } X_1 \geqslant 0, X_2 \geqslant 0$$

$f(X_1, X_2)$ is concave. Using the quadratic programming feature of CMP, we produce the output of Figure 8-15.

Concluding Remarks

Nonlinear programming, in general, is quite difficult. In contrast to the universal success of the Simplex method for linear programming problems, there is no accepted, effective algorithm for the general nonlinear programming problem. In fact, there are many problems that cannot be solved well by *any* known procedure. Considerable effort has been devoted to convex programming, with surprisingly good results. Quadratic programs are another special case for which good procedures exist.

Research continues actively to untie the Gordian knot of nonlinear optimization. Perhaps these research efforts, combined with the ever-increasing speed and ever-decreasing cost of computation will, in the not-too-distant future, allow nonlinear optimization to be feasible in a larger variety of practical applications.

```
                 QUADRATIC PROGRAM


OBJECTIVES           X1            X2

  FUNCTION            1             3
  X1                 -2
  X2                 .5            -2

BOUNDS              NONE          NONE


CONSTRAINTS          X1            X2         RELATION       RHS

  CONST-1             1             0            LE            1
  CONST-2             0             1            LE            1

SOLVE OR REVISE? S

MAXIMIZE OR MINIMIZE? MAX

OPTIMAL SOLUTION FOUND.
     FUNCTION                     2.56250

ALL ITEMS NOT LISTED IN SECTIONS 1 - 4 HAVE THE VALUE ZERO.

 *1*  DECISION VARIABLES
   1. X1                  0.7500
   2. X2                   1.000

 *2*  SLACK(+) AND SURPLUS(-) IN CONSTRAINTS
   1. +CONST-1            0.2500

 *3*  SHADOW PRICES FOR CONSTRAINTS
   2. CONST-2              1.375

 *4*  REDUCED COSTS FOR DECISION VARIABLES

 *5*  GRADIENT OF OBJECTIVE FUNCTION
   2. X2                   1.375
```

As with LPs, the solution algorithm provides this information. Note that the shadow price holds only *__at the solution point__.*

We also get $\partial f/\partial x_i$:

Here, $\partial f/\partial x_1 = 0$

$\partial f/\partial x_2 = 1.375$

at the optimal solution.

FIGURE 8-15. A quadratic program.

EXERCISES

1. Formulate a nonlinear program to find the dimensions of a rectangular tank open at the top that provides the maximum volume and can be constructed from 100 square feet of sheet metal.

2. Plot $f(x) = -12x + x^3$ for $-3 \leq x \leq 3$ and label the global maximum and minimum and all local maxima and minima.

3. Which of the following functions are convex; which are concave; which are neither; which are both?

a. $f(x) = |x|$
b. $f(x) = e^{-x^2}$
c. $f(x_1, x_2) = x_1^2 + x_2^2$
d. $f(x_1, x_2) = 10 + 10x_1 + x_2 - 6x_1^2 - 3x_2^2$
e. $f(x_1, x_2) = 3x_1^2 - 4x_1 x_2 + x_2^2$
f. $f(x_1, x_2) = e^{x_1} + x_2^2 + 1$
g. $f(x_1, x_2, x_3) = 5x_1 + 2x_2 - 12x_3 + 6$

4. Consider the nonlinear programs formulated below. For each: (i) Graph the feasible region. (ii) Consider the point (1,1). Is it feasible? If so, is it an interior point? a boundary (edge) point? a corner point? (iii) Find the optimal solution.

a. Min $2x_1^2 - 2x_1x_2 + 2x_2^2 - 6x_1 + 6$
   Subject to $x_1 + x_2 \leqslant 2$
b. Min $x_1^3 - 3x_1x_2 + 4$
   Subject to $5x_1 + 2x_2 \geqslant 18$
   $-2x_1 + x_2^2 = 5$
c. Min $-5x_1^2 + x_2^2$
   Subject to $\left(\dfrac{x_1}{x_2}\right)^2 - \dfrac{1}{x_2} \leqslant -1$
   $x_1, x_2 \geqslant 0$

5. Consider

$$\begin{aligned} \text{Min} \quad & \tfrac{1}{4}x_1^4 + \tfrac{1}{2}x_1^2 - x_2 \\ \text{Subject to} \quad & x_1^2 + x_2^2 \leqslant 4 \\ & x_1 - x_2 \leqslant 2 \end{aligned}$$

Is this a convex programming problem?

6. Consider

$$\begin{aligned} \text{Min} \quad & 100x_1 + \frac{200}{x_1 x_2} \\ \text{Subject to} \quad & 2x_2 + \frac{300}{x_1 x_2} \leqslant 1 \\ & x_1, x_2 \geqslant 0 \end{aligned}$$

Is this a convex programming problem?

7. Use steepest descent/ascent to solve each of the unconstrained nonlinear optimization problems below:

a. Min $100\,[x_2 - x_1^2]^2 + (1 - x_1^2)$
b. Min $x_1 + \sin x_2$
c. Min $-(x_2 - x_1)^4 + (1 - x_1)^2$
d. Min $-x_1^2 + x_1 - x_2^2 + x_2 + 4$
e. Min $e^{-(x_1-1)^2} - \frac{1}{6}(x_2^2 - \frac{1}{2})^2$
f. Max $x_1^3\, e^{[x_2 - x_1^2 - 10(x_1 - x_2)^2]}$

8. A manufacturer of consumer goods has had its market research staff analyze the relationship between price and demand for the two models in one of its product lines. Experience has shown that the consumer will substitute one for the other; that is, the demand for each is related to the price of the other as well as its own price. After much data gathering, the market research staff has estimated the following relationships:

$$d_1 = \frac{100}{p_1^2\, p_2} \qquad d_2 = \frac{200}{p_1\, p_2^2}$$

where d_i = demand for model i, per week,
p_i = price for model i, in dollars.
The variable production cost per unit for model 1 is \$2; for model 2, \$1.

a. If $p_1 = p_2$, what is the maximum contribution? What is the price which achieves that contribution?
b. If the prices need not be equal, what prices maximize contribution?

9. The R & D team at NO Plastics has synthesized a new plastic with remarkable properties, made from two interchangeable materials. The company has \$400,000 to buy raw materials for next month—raw material 1 costs \$8 per pound; raw material 2 costs \$5 per pound. The amount of the new plastic which can be produced from r_1 pounds of raw material 1 and r_2 pounds of raw material 2 is

$$8r_1 + 2r_2 - r_1^2 - \tfrac{1}{2}r_2^2 \text{ pounds}$$

a. Formulate a nonlinear program that will help NOP maximize the amount of plastic produced subject to the raw material budget.
b. Graph the feasible region and solve the problem using an ascent algorithm. Draw the corresponding trajectory of solution points.

10. The D & E Frig. Co. produces two models in its new Energy-Saver refrigerator line. Unfortunately, the supplier providing much of the new, automated equipment for this line has been stalled by a strike, and DEF has been forced to use badly deteriorated equipment from its older lines. High maintenance

costs and unexpected downtime cause the contribution per unit to decrease as this old equipment is used more intensively. DEF's engineers have estimated contribution (in dollars) to be

$$500—20\,p_1 \quad \text{for model 1}$$
$$700—30\,p_2 \quad \text{for model 2}$$

where p_1 and p_2 are the production levels for models 1 and 2, respectively, during the coming month. DEF's marketing department claims to be able to estimate, quite closely, next month's demand for its two models, as a function of the firm's advertising expenditure:

$$\text{Demand for model 1} = 150 - \frac{10000}{100 + a}$$

$$\text{Demand for model 2} = 250 - \frac{20000}{100 + a}$$

where a is the amount (in \$) spent on advertising the Energy Saver line. DEF's advertising budget for the month is \$30,000 and its machine capacity is 300 units of the two models combined.

Formulate DEF's production problem as a nonlinear program.

11. In Chapter 2, we formulated a model to solve a prototypical multiperiod aggregate production planning problem, allocating workforce, production and inventories to minimize cost and meet seasonal demand. Backorders were not allowed and all costs were assumed to be strictly linear. During the 1950s, Holt, Modigliani, Muth and Simon published their now-famous research on production planning[13] in which they identified the following cost elements for each time period:

1. Regular payroll cost: $c_1 W_t + c_{13}$
2. Hiring and layoff cost: $c_2 (W_t - W_{t-1} - c_{11})^2$
3. Overtime and idle cost: $c_3(P_t - c_4 W_t)^2 + c_5 P_t - c_6 W_t + c_{12} P_t W_t$
4. Inventory and backorder cost: $c_7 [I_t - (c_8 + c_9 d_t)]^2$

where: P_t = production for period t
W_t = workforce for period t
I_t = inventory level at the end of period t
are the decision variables of the problem, c_i, $i = 1, 2, \ldots, 13$ are constants, and d_t denotes the forecast demand during period t. Implicit in cost element 4 above, backorders are allowed (at a cost).

a. Interpret the costs above.
b. Formulate a model to determine production, workforce and inventory levels in each period so as to minimize total costs.
c. How might you solve the problem?

[13] C. Holt, F. Modigliani, J. Muth and H. Simon, *Planning Production, Inventories and Work-Force*, Prentice-Hall, Inc., Englewood Cliffs, N.J. 1960.

12. The DE Foundry engineers have estimated that the probability that a part made of a new alloy will be defective is a function of the proportion of high-grade ore in the charge to the furnace. In particular, DEF's engineers have found that

$$\text{prob(good part)} = 1 - e^{-h/l+h}$$

where h is the amount of high-grade ore in the charge and l is the amount of low-grade ore in the charge. In planning for next month, DEF's purchasing manager can obtain up to 17000 lb of high grade ore at \$1.10 per pound and an unlimited amount of low-grade ore at \$.70 per pound. DEF's furnace takes 1000 lb of ore per charge and, if all parts are good, will produce 600 parts. The presence of defectives will lower that number. For example, if the charge is 50% high-grade ore and 50% low-grade, the expected number of goods parts is

$$600(1 - e^{-1/2}) = 236.08$$

The furnace can be charged 36 times per month at a variable operating cost (not including raw materials) of \$1600; each good part can be sold for \$5; each bad part can be scrapped at 40% of the cost of the input raw materials for that batch.

a. Formulate a nonlinear program to help DEF plan its raw material purchase for next month.
b. Solve the problem using an ascent method.

The Netherlands Dairy Fund

Early in the spring of 1961, Mr. Hans Van Egeraat, a director of the Netherlands Dairy Fund, was reviewing his summary reports on the excessively large milk subsidies being paid to the Dutch farmers. These payments in 1960 had amounted to approximately half a billion guilders or more than 1% of the national income. Modifications to the subsidy program were absolutely necessary.

Milk production in the Netherlands was a major economic activity. Productivity was high due to the advanced dairy techniques employed on the farms; in fact, it was claimed that one Dutch cow produced the same amount of milk as two French cows. Unfortunately, both domestic and export dairy prices were

Source: This case was adapted from an actual situation reported in S. L. Louwes, J.C.G. Boot, and S. Wage, "A Quadratic Programming Approach to the Problem of the Optimal Use of Milk in the Netherlands," *Journal of Farm Economics*, Vol. 45 (1963), pp. 309–317. Harvard Business School Case 9-177-019.

low. Despite this economically discouraging condition, the milk surpluses persisted since Dutch dairy farmers calculated their costs on a marginal cost way, marginal revenue exceeded marginal costs so production continued. Mr. Van Egeraat felt certain that the subsidies could be reduced by changing the price structure for milk products and the resultant product mix. The question that he faced was how to determine these new prices and their related mix.

THE DAIRY INDUSTRY IN THE NETHERLANDS

The Netherlands is a small country, approximately 12,868 square miles, but with its population of 11.5 million people it is one of the most densely populated countries in the world. Due to the country's size, there is no need for regional segmentation of either the productive capacity or the consumptive demand. The country was homogeneous in terms of supply and demand, and the export market was viewed as a second market place.

The Netherlands Dairy Fund is part of the Landbouw Egalisatie Fond,[1] a general fund for agricultural price support, a line item in the national budget. As in other European countries, the Landbouw Egalisatie Fund found its origins in the postwar price support measures of the government. Over the years its regulatory powers had become increasingly more specific and stronger. By 1960, the dairy processing and distribution functions were completely nationalized.

The total raw milk production, excluding that for the dairy farmers' own consumption, is delivered to the Dairy Fund which attempts to maximize its value by processing it into the appropriate products: milk for consumption, butter, cheese, condensed milk, etc. These fund products are sold both on the domestic market and the export market. In addition to its centralized processing and distribution authority, the Dairy Fund has monopolisitic control over domestic consumer prices.

Ultimately, the Dairy Fund's pricing and processing powers determine the revenues received by the producers or, as it has been recently, the size of the government subsidy. The total value of dairy sales divided by the total quantity of raw milk delivered by the farms gives an average price received by the dairy farmers. When the average price is below the guaranteed price the difference is made up by subsidy. The floor prices were determined on commercial grounds taking into account inventory needs, expected future market developments, and the need to keep the dairy farms renumerative.

THE PRODUCT MILK

Raw milk is a mixture of water (87.6% by weight), fat (3.8%), and dry matter[2] (8.6%). For 1961, the quantity of raw milk available can be quite

[1]Agricultural Stabilization Fund.

accurately forecast because the herd size is carefully monitored by the Fund and does not change substantially from year to year. In addition, the output of an "average" cow has been stable over the past two or three years. After making deductions for farm use, exports, and special products, it was felt that the quantities of fat and dry matter available in 1961 would be essentially the same as in 1960. These are shown in Exhibit 1.

In the Netherlands the most important products of raw milk are milk for consumption, butter, fat cheese, and 40^+ cheese. 40^+ cheese is so named because over 40% of the solid components of the cheese is fat. The contents of the final products are given in Exhibit 2. The processing facilities for these products have ample capacity and are sufficiently flexible that change in the products mix is relatively easy. In addition, fixed costs make up nearly all of the processing costs, and variable costs can be taken to be essentially the same for all products.

Price is by far the most important determinant of demand for the final dairy products. As the Fund had expanded its powers over the past decade, it had eventually received authority for the establishment of prices. This was politically a very sensitive power and one that was used with caution. The Dutch government had a long standing economic policy of stable prices, but, in 1960, the consumer price index rose by 2.5%—the largest change since the war years. The current situation was particularly sensitive since the Parliament was controlled by parties whose basic philosophies were in opposition to any action which might lead to a deterioration of the economic position of the working class. Furthermore, two influential advisory organizations to Parliament, The Social and Economic Council and The Foundation of Labor, were on record as being opposed to any policy which would increase the cost of living index.

EXHIBIT 1

The Available Quantities of Fat and Dry Matter in 1960

		Fat ('000 tons)		Dry Matter ('000 tons)
Milk produced (6838 × 1000 tons)		260		588
Deductions:				
1. Farm Use: milk for cattle	11		53	
milk for consumption	16		35	
2. Export: butter	33		1	
fat cheese	23		22	
40^+ cheese	10		15	
milk	3		10	
3. Special products: condensed milk	30		99	
milk powder	15+		102+	
Total Deductions		141		337
Available for Domestic Consumption		119		251

[2] Dry matter is composed of albumen, lactose, and various salts.

EXHIBIT 2
Contents of Fat, Dry Matter, and Water in Milk, Butter, and Cheese (%)

	Fat	Dry Matter	Water	Total
Milk for consumption	2.6	8.6	88.8	100
Butter	80.0	2.0	18.0	100
Fat Cheese	30.6	29.7	39.7	100
40^+ Cheese[a]	24.5	37.1	38.4	100

[a]Note that the solid components of 40^+ cheese are less than 40% fat. This is due to the fact that a small amount of 20^+ cheese has been aggregated into this category.

Despite these opposing forces, the Fund was nevertheless interested in pursuing the use of prices as a means to reduce the subsidy payments. Dairy products were only a small part of the cost of living index and the subsidies were becoming a significant economic factor. Recently the Fund had sponsored an extensive price study whose results were going to form the basis for any new price policies. Some of the outputs of the study effort were the quantities and average consumer prices in1960 and estimates of the price elasticities for the principal dairy products (Exhibit 3). Price elasticities indicate the impact of change in price on quantity where the changes are measured in terms of percentages:

$$\text{Price elasticity} = -\frac{\text{Percentage change in quantity}}{\text{Percentage change in price}} = -\frac{\Delta Q/Q}{\Delta P/P}$$

Since there is an inverse relationship between price and quantity, the minus sign is included in the formula to make the elasticity a positive number. As an example, the price elasticity of fat cheese is .7; a 20% increase in price of fat cheese will result in a $-.14 = -.7 \times .20$ change in quantity—a 14% decrease in demand for fat cheese. The cross elasticities are defined similarly and give the impact on the quantity of one product of changes in the price of another. For example, the cross elasticity between fat cheese and 40^+ cheese is $-.06$, and thus a 50% increase in the price of 40^+ cheese will cause a $.03 = -(-.06) \times .50$ change in the quantity of fat cheese—a 3% increase in demand for fat cheese. The net effect on the demand for fat cheese of the two price changes, the 20% increase in price of fat cheese and the 50% increase in the price of 40^+ cheese, is an 11% decrease (-14% from its own price change and $+3\%$ from the 40^+ cheese price increase).

NEW PRICE STRUCTURE

The strategy for the relief of the subsidy burden that Mr. Van Egeraat wanted to examine was the manipulation of consumer prices and production quantities in such a fashion as to maximize the revenue to the Fund. This was

EXHIBIT 3
Selected Results of Price Study

Elasticities and Cross-Elasticities of Milk Products	
Milk for Consumption	0.3
Butter	1.5
Fat Cheese	0.7
40^+ Cheese	0.4
Fat Cheese and 40^+ Cheese	−0.06
40^+ Cheese and Fat Cheese	−0.3

1960 Average Prices and Quantities

	Price in Guilders (per ton)	Quantity (units of 1000 tons)
Milk for Consumption	400	1645
Butter	4000	54
Fat Cheese	3250	63
40^+ Cheese	2500	17

really the only option available to him since the guaranteed minimum price to dairy farmers was set by the Dutch government with other considerations in mind and was virtually impossible to alter. By maximizing revenues for the raw milk, the subsidy would be minimized. The problem was how to determine the prices and quantities which would achieve that goal.

New York City Environmental Protection Administration (B)

"They do not pick up the garbage on Sunday in New York City, so there is even more on Monday."[1]

In 1968, the sanitation manpower for the City of New York was spread evenly over the days from Monday to Saturday. The "chart-day" system by which this was done had been used for over 40 years. The workload, however was not spread evenly over the days of the week. Primarily due to the lack of Sunday collections, collection demands on Monday and Tuesday were greater than for the rest of the week (Exhibit 1). Normally, there was enough man-

[1]Edward J. Beltrami, *Models for Public Systems Analysis*, Academic Press, 1977.

EXHIBIT 1
Collection Demand, Productivity, and Cost

Average Daily Tonnage
(thousands tons)

Monday	Tuesday	Wednesday	Thursday	Friday	Saturday
17.2	16.1	14.9	14.6	14.0	15.5

Average Productivity
(tons/day per crew[a])

Day Shift	Night Shift
9.35	8.93

Cost
($/crew—1971 rates)

Weekdays		Saturdays	
Day Shift	Night Shift	Day Shift	Night Shift
$127	$134	$143	$150

[a]A crew consists of three sanitation workers: a driver and two loaders.

power to meet collection schedules by diverting effort from other activities such as street cleaning and bulk collection.

The chart day system worked as follows: Each of the approximately 11,000 sanitation men was put in one of six chart groups. In a given week, group one would have Monday off, group two Tuesday off, and so forth. The next week group one would have Tuesday off, group two Wednesday, and group six would have Monday off.

During periods when the total force level was low due to vacations or illness there was not enough manpower to meet Monday and Tuesday collection demands. This was the case in the fall of 1968 when about 10% of the refuse that should have been collected on Monday and 3% of the refuse that should have been collected on Tuesday was collected instead later in the week. (See Exhibit 2.) In fact, in 1969, leaving Monday's and Tuesday's garbage on the streets, with its attendant health hazards and smells, led to riots in the Brownsville section of Brooklyn.

BACKGROUND

John Lindsay, who took office as the Mayor of New York in January 1966, signed a law in 1968 creating the Environmental Protection Administration. This included the Departments of Air Resources, Water Resources, and Sanitation, and had 20,000 employees. In June 1968, Mayor Lindsay announced the

EXHIBIT 2
Sanitation Workers Available for Household Collection

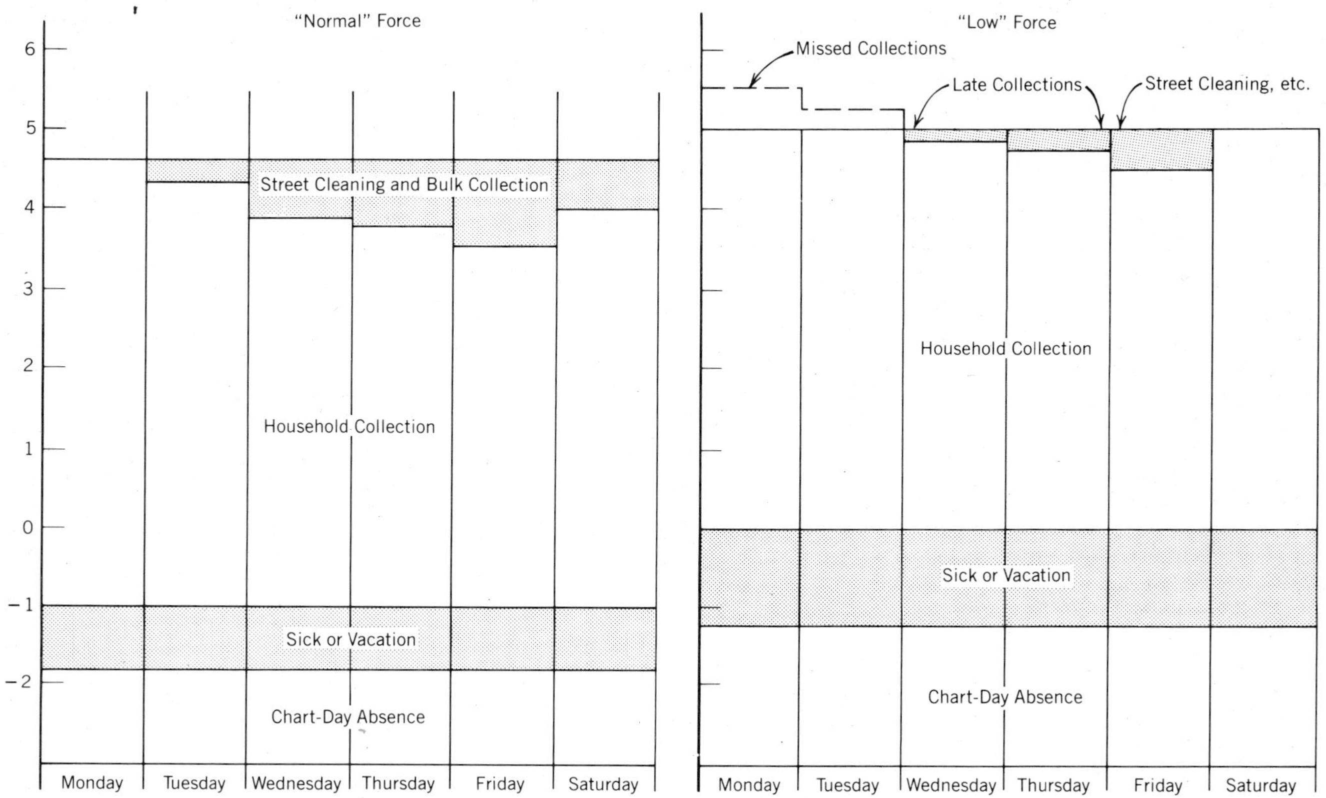

formation of a Planning-Programming Budgeting System (PPB) to increase the analytic and policy content of the budgeting process. Jerry Mechling, formerly an aide to the mayor, was hired to head the PPB office and process for the EPA. In July, 1968, Steve Disman, a Budget Bureau PPB analyst, learned about the 6 group chart day system in discussions with the Sanitation Commissioner. Aware of the extremely tight fiscal situation for the coming year, Disman argued for rescheduling as a means of avoiding expansion of the force. The Department of Sanitation had asked for this expansion in order to keep up with growing refuse loads and to "eliminate missed collections." While the Sanitation Department did not recommend a restructuring of the chart day system in its first PPB submission in December 1968, they did show that the extra men which were requested to "eliminate missed collections" would not increase the total collections, but would allow the collections to be made on schedule. The Department recommended that expanded manpower also be used to increase street cleaning and bulk collections. Exhibit 3 is the diagram used by the Department to illustrate this part of their proposal.

In February 1969, at meetings of the Clean City Task Force, a joint Budget Bureau and EPA Sanitation Group, Disman and the Budget Bureau again argued for restructuring. Disman presented three options to the mayor: (1) Expected collection demands could be met by an expansion of the force at a cost of approximately $5.5 million. (2) Demand could be met by having existing

EXHIBIT 3
Illustration from 1968 PPB Proposal to Expand Sanitation Workforce

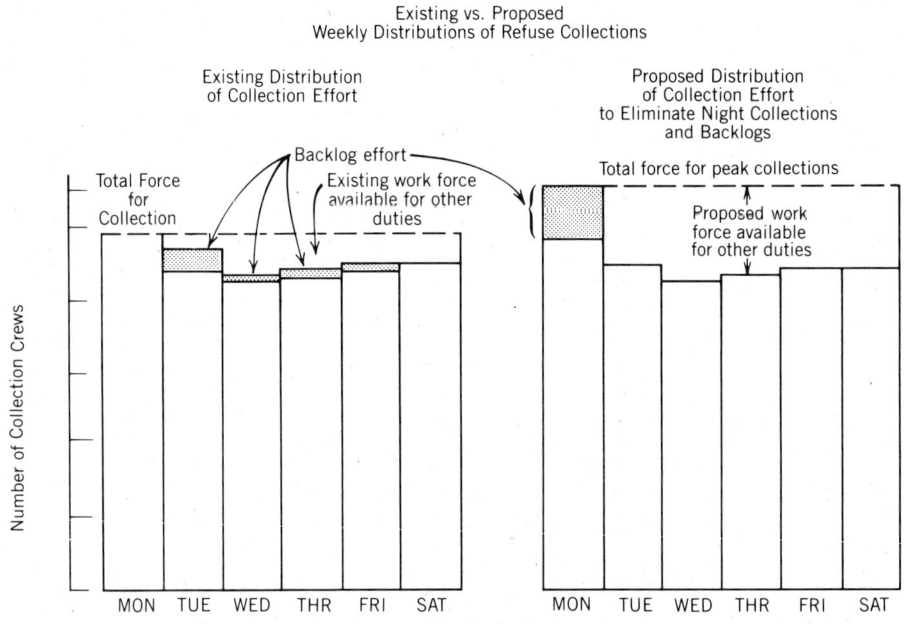

members of the force work on their days off. At standard double-time rates, this would cost approximately $2 million. (3) The chart day schedules could be restructured. This solution was estimated to cost nothing and perhaps to save as much as a million dollars.

The Budget Bureau argued for restructuring, but admitted that there were associated problems. As a followup memo by Disman put it:

> *The main problem with the proposed redistribution of the collection force is a reduction of the number of three-day weekends a sanitationman currently receives. Under the present structure a sanitationman gets an average of eight three-day weekends per year. Under the proposed structure he would get an average of five three-day weekends plus three two day weekends. Thus the 7,730 men assigned to refuse collection would each be required to give up three three-day weekends for three two-day weekends.*

The memo did not explain why the system worked that way, and the details were not well understood by either the EPA or Department of Sanitation policy-makers.

Disman was faced with difficult problems of bureaucratic politics in trying to get his proposal accepted. Strategically, the Department of Sanitation did not want to support manpower restructuring until a decision had been made on whether to increase the total force level. Also, the Department wanted to avoid union problems, and it was felt that the decrease in three-day weekends might be a big problem. The Mayor's history of difficulty with union leaders and his upcoming bid for re-election in November led the Department to feel that he would support them in deferring any consideration of chart-days. In fact, at a later meeting with city department managers, the Mayor decided that the program was an important possibility, but that the time was not right for rescheduling. He decided to defer the issue until negotiations on the union contract which was to expire in October 1970.

In the spring of 1970, however, something unexpected happened. Faculty from the new Urban Science and Engineering Program at the State University of New York at Stony Brook offered to make themselves available to work on EPA problems. In June, Mechling met with Dr. Robert Nathan, Director of the Stony Brook Program, and Dr. Edward Blum of the New York City Rand Institute who was associated with the Program. Blum had made the original contact with Mechling, since Mechling was familiar with work Blum had done with the New York City Fire Department.

Mechling approached the meeting with some pessimism. He felt that outside consultants often created more problems than they solved. It took time to educate them, to get to know their competence, and to trust them. They often promised more than they could deliver, and while they talked about implementation, they rarely stayed around long enough to deliver.

Mechling learned enough about the Stony Brook Group at the meeting to overcome some of his pessimism about outside academic consultants. Stony Brook had recently received a substantial grant from the National Science

Foundation. Up to four professors and six graduate students could be assigned to work on EPA problems that summer if mutually satisfactory problems could be found. The Stony Brook Group was particularly interested in solid waste problems since their expertise was primarily in operations research and solid waste problems seemed particularly tractable to such analysis. Dr. Nathan seemed to Mechling to be sensitive to the difficulties of managing outside studies in a way that would meet the needs of the EPA. He emphasized that he had strong incentives to be responsive to the EPA's needs since his refunding prospects with the NSF would depend on success with the EPA and other local agencies. At the meeting the Group generated a list of projects to be studied further by both sides. First on the list was the chart day problem.

At a follow up meeting later in the month Mechling agreed to accept help from the Stony Brook Group and the chart-day problem was formally structured as follows:

> *The objective of this short-term (summer) project would be to develop manpower allocations such that missed collections would be reduced to zero. The cost of such allocations should be minimized subject to the constraint that the solution(s) must be acceptable to the union. In fact, Stony Brook will look for solutions that may improve the chart-day and vacation package for the men—when and if we learn of effective operations research techniques for generating "solutions," we plan to approach the union to see if we can generate a solution that is particularly attractive from their point of view.*

DATA COLLECTION AND FORMULATING THE MODEL

The initial task was to collect data to establish how much manpower was required for collection and what the associated costs were. With the goal of eventual demonstration districts, the problem would first be solved for District 21 in the Bronx. Stony Brook began by collecting eight weeks of data for the District. The collection manpower needs by day of the week and shift were to be estimated from the following:

- Tons to be collected by day of week: tons actually collected per day plus "missed collections" from the previous day as recorded in the Department's operations book. (Exhibit 1)
- Productivity of the men by day of the week and by shift. (Exhibit 1)
- Truck availability by day of the week. This determines how many men can be fitted on the day shift, and how many must wait for the second shift to get a truck.

Given the fixed manpower available for the district, union acceptability was initially formulated as a series of constraints. These included the number of shifts worked per week per man, and the number of men per crew.

Manpower requirements could be estimated from the above, and costs calculated from daily wage rates and might differentials. The entire problem was formulated as a quadratic programming model. Given total manpower and trucks available, the model would allocate the men over the week so as to minimize missed collections at minimum cost.

On July 17, the Stony Brook Group held a meeting to review their formulation with analysts from Rand. Discussions centered around the necessity for the complex quadratic formulation of the problem, with Rand suggesting a linear formulation and later developing their own model. Stony Brook kept their basic nonlinear model and had it operating in another month. After extensive reviews on August 20, attended by 17 people (8 from Stony Brook and the rest from various city and consulting agencies) estimates of the required manpower profiles were completed and attention turned to the generation of appropriate rotating schedules.

The first schedules called for sixteen rotating "chart groups" instead of the six groups under the old system. The major change is that the number of groups having the day off varied by day of the week. These were presented to the EPA in September 1970, and revisions and early documentation were rewritten during the next two and a half months. Many possible systems were examined before the EPA made the decision to show two new systems to the union in December. The alternatives shown to the union were two thirty-group chart systems which differed in the number of two- and three-day weekends the men would receive. (See Appendix A for a more complete description of the proposed schedules.)

Five criteria were most important in generating new systems:

- Fitting the desired manpower profile as closely as possible.
- Equity.
- An acceptable mix of three- and two-day weekends.
- Conservation of costs and administrative feasibility.
- Keeping control of the negotiation process by presenting a limited number of options. (This final criterion was often discussed at the EPA, but was rarely written down and explicitly analyzed.)

While these criteria could not be included in a single well defined objective function they were nonetheless significant in determining which systems were considered attractive for analysis.

The manpower match with the thirty-group systems was almost perfect. At force levels only ten per cent smaller than "normal" (e.g., at the previous spring's level) the thirty-group system would be essential to permit collections to be made on time. Under the thirty-group system even in a tight resource period street cleaning and bulk collections could be made early in the week when they were most needed.

The need for a close fit ruled out options that workers may have liked better. In comparison with the old six-group system, the thirty-group system would bring the men in on three more Mondays (allowing only two Mondays off rather than five). In return for the three Mondays off that were lost, the new system gave back three days later in the week, two of which were on Saturday (giving seven rather than five Saturdays off every thirty weeks). Exhibit 4 depicts these comparisons. Giving back the third Monday as a Saturday-Sunday weekend was eliminated from consideration because of the need to match the manpower to the workload.

The equity criterion demanded that each sanitationman be treated alike. This was considered important to insure union acceptability. Equity considerations led to a focus on a rotating system where each group would receive a schedule similar to all others, differing only as to the week designated as the beginning of the cycle. Rotating systems excluded some options: the analysts had considered giving a portion of the force, say the half with the most seniority, the option of selecting whatever days off they wanted on a fixed basis. This might have permitted them to work out more attractive moonlighting jobs (a widespread practice in the Department). The remainder of the force could then have been allocated on a rotating basis to fill in the gaps in the desired profile. This possibility was rejected because of equity.

Finding an acceptable, even optimal, pattern of three- and two-day weekends was more complex since, unfortunately, no one knew what either acceptability or optimality consisted of. The Stony Brook and EPA analysts made some guesses to guide their presentation to the union. It became clear during negotiations that the union leaders did not know how the men would react. In the end, it was felt that even the individual sanitationmen would have to live with the new pattern of days off for some time before they could correctly judge their preferences. The Stony Brook data collection team tried to understand this issue better by talking casually with officers and members of Districts 21 and 55. The EPA analysts also informally asked Department members what they liked and did not like about the chart-day system. It was discovered that many sanitationmen aspired to the "normal" Saturday-Sunday weekend where they could be home with their wives and family. An increase in Saturday-Sunday two-day weekends might be an acceptable trade for some decrease in three-day weekends. The men also liked at least some days off in the middle of the week. On these days they could count on not having the kids underfoot, and not finding crowded conditions at shopping and recreational facilities. This knowledge was used in describing the two thirty-group systems presented to the union (whose main difference was in the pattern of three- and two-day weekends).

Other criteria which shaped the development of a new system were cost and administrative feasibility. In general, cost was to be minimized. The impact was to eliminate options that would generate three-day weekends at the additional expense of six-day workweeks. One such plan would have made a

EXHIBIT 4
Effects of the Six- and Thirty-Chart Systems on the Workforce

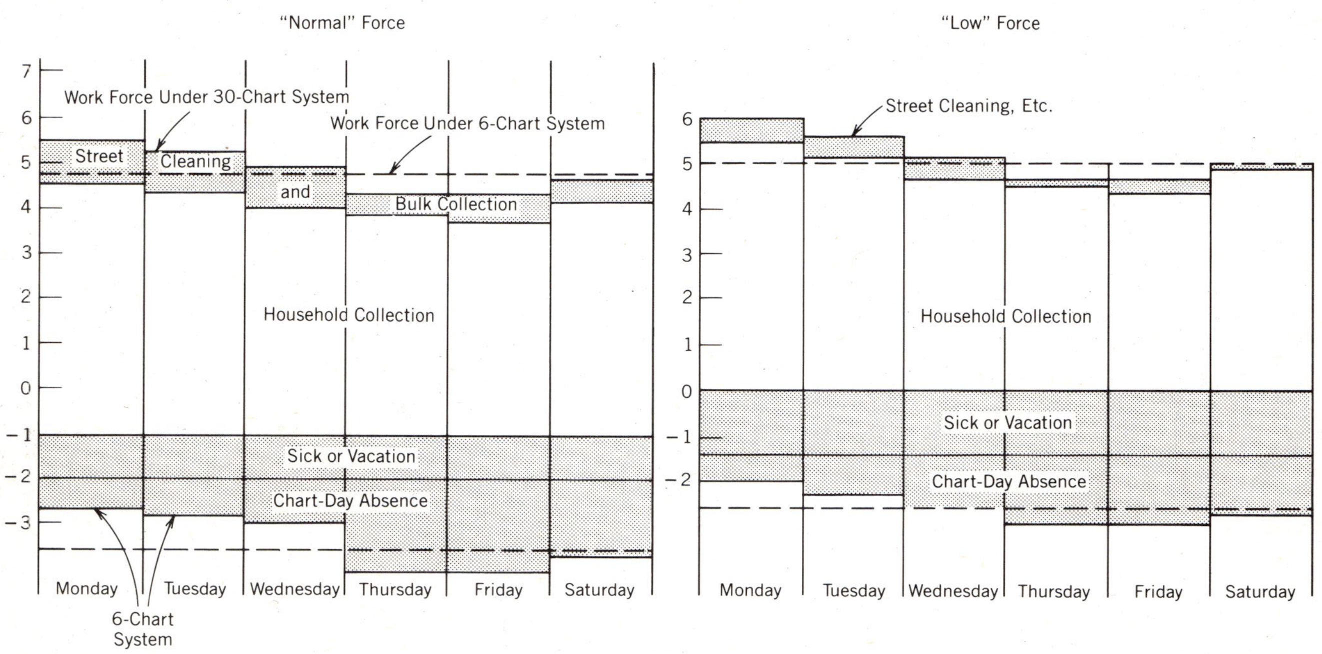

three-day weekend every four weeks, but would have cost over $3 million extra per year.

Administrative feasibility was another elusive criterion. One issue however, concerned "temptation" weekends. A "temptation" was said to occur if a day off fell on a Friday or Tuesday. In such a case the temptation was to call in "sick" on the isolated workday, thus creating a three-day break. During the analysis it was found that more men than average used sick leave on Mondays, but that only an average number called in sick on Saturdays. At least some of the heavy Monday sick leave might thus have been a response to heavier refuse loads on Mondays. The heavy Monday loads would tend to be reduced with the thirty-group system since they would be spread among a larger Monday force. Though the magnitude of the temptation weekend problem remained uncertain, Stony Brook's alternative put all Tuesdays off and as many Fridays off as possible into consecutive weeks. This was done in order to make the misuse of sick leave more easily noticed thereby hopefully creating incentives to avoid temptation.

A final factor shaping the analysis was the upcoming period of union negotiations. One pressure was to get the most information possible on the effects of alternative scheduling systems on the men and on district management. On the other hand there was a need to keep the negotiations under control; not to tip EPA's hand too early or give the union unknown but exploitable opportunities. The EPA decided to limit the number of options prepared for initial discussions, with more expensive and more complex options sacrificed to the simplicity of two variants of a city-wide, thirty-group proposal.

UNION NEGOTIATIONS

Manpower rescheduling claimed a high priority on technical cost-effectiveness grounds. At force levels and productivities of 1969 it was estimated to save about $5 million per year over the alternative of expanding the force. The early problem with the proposal, however, was feasibility of implementation. After the mayor's re-election, the feasibility of rescheduling greatly increased. An important part of the implementation strategy was to make the chart-day system part of the union negotiations. To Mechling, in the summer of 1970, the prospects for eventual acceptance of the proposal looked good.

The Mayor needed to improve his public image vis-a-vis the unions. Early in his first administration, the Mayor had appeared too self-righteous and inflexible and had been caught in a wave of strikes that almost washed him out of office. By the re-election he had made his peace with the important union leaders, and they had supported him, although the rank and file support had never been warm. The price for this support, or so it seemed to much of the public, had been exorbitant (perhaps even extortionate) wage settlements.

It was likely, however, that the union would eventually support manpower rescheduling. Some union members might actually prefer more two-day weekends and more men on Monday to spread the workload. Opposition to the change would certainly develop, but it was not expected to be a gut confrontation issue. If the union were cool or uncertain about the plan, the Department would be willing to try it out first in demonstration districts. If the union were more stubborn, the Department could "buy" the program in the wage negotiations (although the Department's influence on the wage side of the negotiations was limited, with the major responsibilities falling to the Mayor's Office of Collective Bargaining).

In meetings held with Sanitation Commissioner Herbert Elish and EPA Administrator Jerome Kretchmer during the fall of 1970, Professor Stanley Altman's role became increasingly important as the primary spokesman for the Stony Brook work. Altman had the natural self-confidence and ability to talk effectively with Elish as well as Sanitation Department members. During this period, with the analysis essentially completed, Mechling's increasing concern was selling the program to Elish and Kretchmer and helping them sell it to John DeLury (president of the Uniformed Sanitationmen's Association), the union, and the Mayor. Altman was the person most skillful in assisting in this, and he became the prime contact at Stony Brook.

For Elish and Kretchmer, sanitation manpower rescheduling was only one of several priorities in the union negotiations. Kretchmer and Elish had decided that DeLury and the union would be most important in whatever success the EPA might have. After the May garbage riots in the Brownsville section of Brooklyn, Kretchmer began more frequent and direct communications with DeLury as well as regular visits to the sanitation districts to talk directly with the men. By fall, he and Elish had decided that the union negotiations would be too important to be left to the Office of Collective Bargaining, the Mayor's staff, and the Budget Bureau. They put part of a McKinsey group and part of the Program Analysis Office on the job of preparing staff papers for the negotiations.

When the negotiations began formally on the first of November, Herbert Haber, the Director of the Office of Collective Bargaining, had very little time for the Sanitation Department. At this point Kretchmer and Elish made a significant tactical decision. They decided to meet regularly with DeLury and Bigel (the chief union negotiator) and with the outside mediator. Although Elish and Kretchmer were not authorized to make commitments on behalf of the City, they were able to spend a great deal of time discussing issues with the union leaders.

In December, Elish and Kretchmer decided to let Stony Brook present their proposal directly to DeLury and Bigel. This meeting was held at union headquarters without any of the EPA officials in attendance. This made it possible to get a truly tentative proposal to the union for reaction; the union was not put in the position of having to react to a formal proposal. If problems arose that

Kretchmer or Elish wanted to back away from, the EPA could always undercut the legitimacy of the Stony Brook proposal (at almost all other meetings, only Kretchmer and/or Elish represented the EPA). At the same time, having Stony Brook make the initial presentation might lend a note of detached objectivity and professionalism, especially with all those impressive computer print-outs.

Though the Stony Brook presentation seemed to be rather well received by DeLury and Bigel, the almost immediate response of the union was to ask for a five-day Monday to Friday workweek. When Kretchmer later appeared before the entire union negotiating committee of over thirty members to discuss a series of issues, he felt that he had been "beaten up" by the persistent demands for a five-day week.

Discussions with the union continued on a low-key basis throughout December and January. By February 1971, the Sanitationmen were impatient. Several districts walked out over unsanitary garbage conditions but were soon coaxed back to work. DeLury and Bigel said they could not see a way to avoid a total walkout much longer, and they explained to Elish the great flexibility the men had in creating mischief to put pressure on the City for serious negotiations.

Shortly afterward serious negotiations began and Elish brought Mechling to an all-afternoon session with Jack Bigel in order to examine the union's proposal for a five-day work week. At that meeting, Mechling summarized and reiterated arguments that had been written up earlier by Stony Brook. Several days of homework had been done by the Stony Brook group in analyzing the union's five-day proposal. The EPA position on the issue was based on two points. First, not all the Saturday work could be scheduled earlier in the week, and much of what could be done earlier would require undesirable double shifting of trucks and night collections (even requiring some triple-time midnight shift collections). Second, the "essential" Saturday work would take about a third to a half of the force, but the need would be concentrated in a few areas (mostly ghetto) and tasks (mostly collection). The Saturday work would thus not only be expensive for the City (about a 25% cost penalty), but would also create equity problems for the union, since men in some areas, but not all areas, would have to work a six-day week as often as one week in two.

For several hours, Elish, Bigel, Mechling and the mediator played around with numbers seeing why these points were true. On this back-of-the-envelope basis, Bigel was able to see how the thirty-chart system would give District 21 twelve extra men on Mondays. Although the earlier discussions with Stony Brook and Kretchmer might have helped, it was clear to Elish and Mechling that Bigel had not truly focused or fully understood the thirty-chart system until that afternoon. Even Elish was more confident of the proposed thirty-chart system now that he had "been through the numbers" himself.

The following week, DeLury informally promised Elish that the union would go along with some version of the proposed thirty-chart system. He did not, however, want the Department to begin implementation of the new system until the contract was actually signed. This occurred in the summer of 1971.

IMPLEMENTATION

As soon as word was received that DeLury was willing to go ahead Altman and Mechling began to work on implementation. Altman went into the field to become familiar with the scheduling situation at the district level and to preview further work on the problem. This not only provided needed experience with operations but also helped convince the sanitation officers that he had done his "homework" on the issue. Four sets of implementation problems were identified:

- Getting feedback on street cleanliness.
- Insuring that the extra manpower would not be wasted on low-priority tasks.
- Providing for smooth implementation that would give the Commissioner the credit for success.
- Maintaining continuity and pressure from the top.

Altman and Mechling were worried that when they departed Elish might not be able to implement the project alone. The major political benefits had already been achieved and pressure for results in the streets might not arise again from City Hall or the public until there were more missed collections. Meanwhile, Elish would be facing daily operational problems and would be unable to give the manpower rescheduling project much of his attention.

Since chart day rescheduling called for substantial changes in long established operating procedures as well as presenting the potential for conflicts between the needs of the workers, the officers, and the Commissioner, Mechling and Altman classed it as a "difficult" implementation problem. Their strategy called for the creation of an Implementation Task Force, which could make the effort visible and important within the bureaucracy. It could also be used as a means of giving an analyst direct access to potential problems in the field. An analyst could use the Task Force to "gang up" on Elish and demand his continuing attention to the project.

In July, at the urging of Mechling and Altman, Elish got DeLury to agree to the establishment of a Task Force. The first meeting of the Commissioners' Chart-Day Implementation Task Force was held July 28, 1971. At this meeting Elish announced that the job of the Task Force would be to work toward some system similar to the Stony Book thirty-chart system, and then turned the meeting over to Stony Brook Faculty who had come to explore the system.

The Task Force analyst selected for the project was Jim Moar, a Stony Brook Research Associate who was working full-time on EPA projects. Mechling and Altman had selected Moar in order to plan for continuity on the project. They were determined to carefully oversee Moar's development on the project into a role of increasing responsibility.

Throughout this meeting Moar was not convinced that any of the Sanitation personnel, Elish included, understood the Stony Brook presentation. For non-academic people examining the program the first time, the presentation had been too much maximization, minimization, computerization—and much too fast. At the end of the meeting Elish assigned a working group of three, including Moar, to work on an implementation plan.

When the Task Force met for the third time on August 22, Moar had identified numerous operation procedures that would need modification under the thirty-chart system. Unfortunately, at about this time, President Nixon placed the nation under wage and price controls, and the Sanitation contract had to go before the Pay Board for review. At that point DeLury would not permit the chart-day program to proceed until a ruling was made on the contract. The meeting of the Pay Board, which was expected to take a few weeks, did not come until July 1972.

In the year preceding the Pay Boards' decision, changes were made in the chart system. District officers feared that Saturday manpower would not be adequate to handle heavy weekend cleaning duties and litter basket collections. In addition, in order to be able to handle weekend demands to coordinate with community groups on clean-up drives, more men were needed on Saturdays. The union representative indicated that the men did not want to lose too much of the bonus pay they received for Saturday work. Through the Task Force forum Moar took these Saturday problems to Elish, who approved a new thirty-chart system which brought in one additional group on Saturdays. The revised chart system, by day of the week allowed groups off on a schedule of 2, 3, 5, 7, 7 and 6 (Monday, Tuesday, etc.) instead of the original Stony Brook proposal of 2, 3, 4, 7, 7, and 7. This revision also had the effect of cutting potential two-day weekends from nine to eight in a thirty-week period.

A second revision was made in the original thirty-chart system. The original system had given all days off except Saturday in successive weeks. Thus the two Mondays off came in successive weeks, the three Tuesdays in successive weeks, and so on. This had been done to space the "normal" two-day weekends while clustering the "temptation" Fridays and Tuesdays. It was thought that it would be easier to avoid too many absences on any one day if all the days-off were "spaced" throughout the thirty-week period. Perhaps partly because Moar had not known in detail about the thinking behind the original proposal, the revised new system eliminated the incentive to avoid "temptation weekends." The "spaced" thirty-chart system, the system actually implemented, is shown in Exhibit 5.

Moar's field work led him to believe that operational problems with the new system were also likely. He consequently developed a more detailed and more conservative implementation schedule. Under the revised schedule, the new system would be phased in gradually with opportunities for feedback and modification at each phase. The plan covered two major activities: training the clerks, and beginning with demonstration districts before city wide implementation. Moar was convinced that the key task was training the clerks, since they had to run the system and explain its working to the sanitationmen. The

EXHIBIT 5
"Spaced" Thirty-Chart System as Implemented

Weeks or Chart Groups	M	T	W	Th	F	Sa	Su
1	XX						XX
2			XX				XX
3				XX			XX
4					XX		XX
5						XX	XX
6		XX					XX
7				XX			XX
8					XX		XX
9						XX	XX
10			XX				XX
11				XX			XX
12					XX		XX
13						XX	XX
14		XX					XX
15					XX		XX
16	XX						XX
17			XX				XX
18				XX			XX
19					XX		XX
20						XX	XX
21		XX					XX
22				XX			XX
23					XX		XX
24						XX	XX
25			XX				XX
26				XX			XX
27					XX		XX
28						XX	XX
29			XX				XX
30				XX			XX
Total	2	3	5	7	7	6	30

Note: (1)XX indicates the man or the group has the day off. (2) The diagram can be interpreted in two ways: as the basic 30-week schedule or one chart group, and as the schedule for all 30-chart groups for one week.

thirty-chart system would clearly be more difficult than the old six-chart system.

Moar decided to use the Training Division of the Bureau of Administration for the task. The Division trained new sanitationmen and officers for the Department in such areas as equipment operations, safety regulations, etc. As Moar monitored the program his fears about training the clerks at first increased. He held several sessions with Al Hein, the Director of Training, to explain the new system and what he felt should be taught. Hein then made up his own lecture notes. The lecture in front of the first set of clerks went so poorly that Moar had to answer from the back of the room most of the questions the clerks ask Hein. Though Hein had learned his direct presentation rather well, he had not learned enough to answer questions. Once he heard Moar answer the questions, however, Hein quickly learned the answers. By the end of the fifth training session, Moar felt that his own presence was no longer needed. At the conclusion of the twenty training sessions, Moar was pleased with his decision to use the Training Division. Hein was quite effective once he had learned his system and since he was a strong authority figure in the Department, so there was less tendency for the clerks to be rowdy or disruptive than there might have been.

Meanwhile, Moar had also been finding ways to help the clerks learn from each other. He went out to the field and talked personally with those clerks they had observed as "trouble makers" during the early training sessions, since he felt that these were the clerks who were likely to be leaders in their districts. Even if they could not be co-opted into supporting the new system, Moar wanted to be sure that the most talkative clerks at least knew what they were talking about. Another step to encourage the clerks to learn from each other was to make at least one clerk in each borough responsible for answering all questions within his own borough. This tended to create eleven rather knowledgeable clerks and active supporters. It also allowed the other district clerks to bring up questions about "non-official practices" with one of their peers.

The thirty-chart system was put into effect in District 8 on July 17, 1972. The officers and the union shop stewards had been briefed on the system. Then the sanitationmen had been briefed a week in advance of the change-over and allowed to ask questions. Each man had been given a wallet-sized calendar card with his days-off circled in red. Despite all the preparation, there were still some complaints, but no major problems. On the change-over day some of the men had been scheduled for a Monday chart and resented losing a three-day weekend in the summer. Others wanted their paychecks delivered on Thursday instead of Friday, since more men would now be off on Friday. In general, however, District 8 was successful and created an example for the other districts. Instead of being one truck short for collections on Monday, the District was able to put out all its collection trucks plus a hoist-fitted-chassis truck (for servicing detachable containers), a mud truck, and a street flusher, and it doubled the street sweeping effort on both Monday and Tuesday.

After two weeks in District 8, the implementation program added Districts 28 and 51. After two weeks in each District, they added a single district from each of the remaining eight borough commands. Commissioner Elish ordered the city-wide changeover for September 25. By September 11, the district officers were to brief the men on their new chart assignments; they were also to have distributed the new pocket calendar cards. During the first two weeks of implementation, the officers were ordered to announce each evening the names of the men who would be off the next day.

The change-over went smoothly. Moar was pleased but not surprised:

Success came because we spent a lot of time generating a whole set of new rules to insure smooth implementation. These rules provided new security for the men who were perhaps bewildered by the new system. We did not eliminate one set of operational procedures without replacing it with another. And the new procedures were as exact and as detailed as the previous ones.

The results of the thirty-chart system were pretty much on target with no major surprises. A meeting held six weeks later only lasted for five minutes; everyone was satisfied.

The system implemented in September of 1972 was the alternative giving eight two-day breaks but no three-day breaks. After several months the union requested a change to the alternative giving four two-day weekends and two three-day weekends. The men had voiced objections to losing all their three-day weekends. The change was quickly agreed to, with new cards made and distributed in January 1973. Moar and Elish were particularly happy with this change since it indicated their responsiveness to the needs of the men.

Over the next few months the system seemed to Elish, to Moar, and even to the Mayor and some community groups to bring noticeable improvements in street cleanliness. By monitoring the manpower assignments Moar could see that the increased peak-day efforts went to street activities and not into in-house services. Examples of the activities to increase on Mondays and Tuesdays were street flushing and bulk collections (which permitted the Department to encourage more community clean-up drives on Sundays).

The thirty-chart system resulted in 900 more men on the streets on Mondays, and 600 more men on Tuesdays. Of course these men were "stolen" from Thursday, Friday and Saturday, but the shift in time was an improvement as it permitted collections to be met on schedule that otherwise would have been missed. Had the extra peak-day manpower been purchased in overtime, it would have cost $2 million per year. Had it been purchased through expansion of the force, it would have cost $13.5 million per year but of course would have expanded the total quantity of sanitation efforts as well.

Commissioner Elish could also be proud of the smoothness of the implementation. Where the Police Department had had severe problems, the Sanitation Department had had very few problems. Along with his other successes—improvements in vehicle maintenance, his eventual progress with containerization, a twenty-five per cent improvement in overall collection productivity, and continuing good relations with the union—chart-days helped give Elish a reputation as one of the most effective Sanitation Commissioners ever, and one of the best administrators in the Lindsay Administration.

BIBLIOGRAPHY OF SOURCES

S. Altman, E. Beltrami, S. Rappaport, and G. Schoefle, "A Nonlinear Programming Model of Crew Assignments for Household Refuse Collection," *IEEE Trans. Systems, Man, Cybernetics* SMC-1 (1971), 289–293.

E. Beltrami, *Models for Public Systems Analysis*, Academic Press, 1977.

E. Ignall, P. Kolesar, and W. Walker, "Linear Programming Models of Crew Assignments for Refuse Collection," Rand Rep. D-20497 (1970).

J. Mechling, "A Successful Innovation: Manpower Scheduling," *Urban Analysis*, 1974, Vol. 3.

APPENDIX The Stony Brook Proposal

The Stony Brook analysis compared alternative new chart systems to the old six-group system which they displayed over a thirty-week period for comparison purposes. In such a thirty-week period, each sanitationman under the six-group system received the following days off: five Mondays, five Tuesdays, five Wednesdays, five Thursdays, five Fridays, and five Saturdays. Every Monday off followed a Saturday off resulting in five three-day weekends. The other twenty-five Sundays stood alone as isolated days off. This system can be examined in detail in Figure 1.

Weeks	M	T	W	Th	F	Sa	Su
1	XX						XX
2		XX					XX
3			XX				XX
4				XX			XX
5					XX		XX
6						XX	XX
7	XX						XX
8		XX					XX
9			XX				XX
10				XX			XX
11					XX		XX
12						XX	XX
13	XX						XX
14		XX					XX
15			XX				XX
16				XX			XX
17					XX		XX
18						XX	XX
19	XX						XX
20		XX					XX
21			XX				XX
22				XX			XX
23					XX		XX

(continued)

Weeks	M	T	W	Th	F	Sa	Su
24						XX	XX
25	XX						XX
26		XX					XX
27			XX				XX
28				XX			XX
29					XX		XX
30						XX	XX
Total	5	5	5	5	5	5	30

XX indicates the man or the group has the day off.

FIGURE 1. Six-chart system as diagrammed by Stony Brook.

In a thirty-week period under each of the proposed alternative systems, each sanitationman would receive a new pattern of days off: two Mondays, three Tuesdays, four Wednesdays, seven Thursdays, seven Fridays, and seven Saturdays. Under one of the alternatives, none of the Mondays off would follow a Saturday off; this would result in nine two-day weekends (seven Saturday-Sunday, two Sunday-Monday), but no three-day weekends (this is shown in detail in Figure 2). Under the other alternative both Mondays off would follow a Saturday off. This would result in two three-day weekends, and five two-day weekends (Figure 3).

Weeks	M	T	W	Th	F	Sa	Su
1						XX	XX
2		XX					XX
3		XX					XX
4		XX					XX
5	XX						XX
6			XX				XX
7			XX				XX
8	XX						XX
9			XX				XX
10			XX				XX
11						XX	XX
12				XX			XX
13				XX			XX
14						XX	XX
15				XX			XX
16				XX			XX
17				XX			XX
18						XX	XX
19				XX			XX
20				XX			XX

(continued)

Weeks	M	T	W	Th	F	Sa	Su
21						XX	XX
22					XX		XX
23					XX		XX
24						XX	XX
25					XX		XX
26					XX		XX
27					XX		XX
28						XX	XX
29					XX		XX
30					XX		XX
Total	2	3	4	7	7	7	30

Note: (1) XX indicates the man or the group has the day off. (2) The diagram can be interpreted in two ways: as the basic 30-week schedule for one chart group, and as the schedule for all 30-chart groups for one week.

FIGURE 2. Thirty-chart system—alternative 1.

Weeks	M	T	W	Th	F	Sa	Su
1						XX	XX
2	XX						XX
3		XX					XX
4		XX					XX
5		XX					XX
6						XX	XX
7			XX				XX
8			XX				XX
9			XX				XX
10						XX	XX
11			XX				XX
12				XX			XX
13				XX			XX
14						XX	XX
15				XX			XX
16				XX			XX
17				XX			XX
18						XX	XX
19	XX						XX
20				XX			XX
21				XX			XX
22					XX		XX
23						XX	XX
24					XX		XX
25					XX		XX
26					XX		XX
27						XX	XX
28					XX		XX
29					XX		XX
30					XX		XX
Total	2	3	4	7	7	7	30

FIGURE 3. Thirty-chart system—alternative 2.

City of Centerville (A)

Early in May Bruce Forsyth of Ripley Bros., Inc., noticed an announcement in the *Daily Bond Buyer* that Centerville was once again going to the public market to raise $20,000,000 through an issue of serial public improvement bonds. As was the case with past Centerville issues, the choice of underwriter was to be made on the basis of sealed competitive bids for the entire issue, the bids to be opened on Thursday, June 11, 1978. The bonds would go to the bidder with the coupon schedule representing the lowest net interest (NIC) to Centerville. (The appendix describes NIC.) Ripley Bros., as a matter of course, considered all municipal issues of one million dollars or more in order to decide to either form a bidding syndicate or to join a syndicate headed by another firm. Consequently, Bruce wrote immediately to Centerville requesting a copy of the invitation to bid.

Bruce spent most of that morning reacquainting himself with the various municipal issues on which his firm had bid in recent years. He found bids on several Centerville issues over the years. On the majority of these occasions, Ripley Bros. had been a major bracket participant in a syndicate headed by the Southern Oregon Trust Company. Bruce was not in the least surprised, therefore, when later that week he received a letter from Southern Oregon, again inviting Ripley Bros. to be a major bracket participant in a syndicate formed to bid on the latest Centerville issue.

A preliminary pricing meeting was scheduled for Tuesday, June 2 at Southern Oregon's headquarters. Each member of the syndicate, in formulating pricing ideas to be discussed at this meeting, was expected to have analyzed specific characteristics of the Centerville issue and more general information related to current status of the bond market.

After he received Southern Oregon's invitation, Bruce began to consider current trends in bond prices, yields, and market receptivity to new issues. Two of the most important barometers of the status of the bond market were the size of the current floating supply of municipal bonds available in the secondary market and the visible supply of new municipal issues coming to the market within the next thirty days. Bruce found, however, that neither factor (alone or when considered together) indicated an unusual situation which required any special pricing considerations.

In addition to the aggregate supply of bonds, Bruce also reviewed the current reoffering yields on bonds of various maturities already on the market. In conjunction with a bond buyer index of the average 20-year maturity yields for a selected group of securities, he developed what he thought the yield curve for the Centerville issue should be.

On the afternoon of June 2, Bruce arrived at Southern Oregon's headquarters for the general syndicate pricing meeting. After some discussion, it became apparent that there was some divergence of opinion between various under-

Table 1

Year	Reoffering Yield
1	3.10%
2	3.35
3	3.55
4	3.75
5	3.95
6	4.10
7	4.20
8	4.30
9	4.35
10	4.45
11	4.55
12	4.65
13	4.75
14	4.85
15	4.90
16	4.95
17	5.00
18	5.00
19	5.05
20	5.05

writers as to both what the proposed yield and the appropriate spread[1] should be.

Early in the meeting it was announced by Jerry Lerman of Southern Oregon that another major syndicate had been formed by the Valley Bank and Trust Company to bid on the Centerville issue. The underwriters agreed that this information necessitated contemplation of a more aggressive bid (i.e., lower NIC) than many of them had considered in their initial pricing analysis.

Discussions continued for the rest of the afternoon and finally the group agreed on the reoffering schedule outlined above in Table 1 and a gross underwriters' spread of \$15 per \$1000 bond.

With the reoffering schedule agreed upon, all that remained to be done at the pricing meeting was to establish a coupon schedule to yield the lowest possible net interest cost to Centerville. Appropriate to this discussion, Jerry Lerman announced the restrictions that Centerville had communicated to him with regard to coupons, bid premiums and discounts. First it had been decided that due to the difficulty of placing them in the market, they would accept no discounted bonds.[2] Second, there was to be an upper limit on the premium for which any bond could sell. This limit was set at 8.5% above par.[3] Furthermore,

[1]Underwriters' profit.
[2]Bonds that had coupon rates lower than the reoffering yields at which they were planned to sell.
[3]For example, no \$100 face value bond could have a coupon such that the bond would sell for more than \$108.50.

the gross receipts to Centerville after underwriters' profit was to be at least the par value of the issue. These constraints together with the proposed spread of $15 per $1000 bond provided the context in which the syndicate could develop a bid with the lowest net interest cost to Centerville.

APPENDIX

Municipal bonds are generally sold on a serial basis. That is, a $20 million, 20 year issue normally consists of 20 separate groups of $1 million in bonds. Each group has its own coupon (interest) rate and its own maturity. The municipality pays interest semi-annually at half the coupon rate until the bond matures, at which time the par value of the bond is repaid to the investor.

Under the net interest cost (NIC) method a municipality minimizes the net interest it will pay. Net interest cost is defined as the total of interest payments less an adjustment for proceeds at the time of issue. This adjustment is equal to the gross proceeds from the issue less par value and the underwriters' spread (profit).

Exhibits 1 and 2 show net interest cost calculations for a previous Centerville issue of $38 million over a 20 year period, on which the underwriters took $13 per $1000 bond. As Exhibit 2 suggests, there are many possible coupon schedules given a reoffering yield schedule and an underwriters' spread. Centerville's policy has been to select the coupon schedule with the lowest NIC (Column D in Exhibit 2). As is evident from these examples, the time value of money is not considered in the NIC calculations.

EXHIBIT 1
Production and Net Interest Cost

(1) Years To Maturity	(2) Number of Bonds	(3) Reoffering Yield	(4) Coupon	(5) Price of Bond[a]	(6) Production (5) × (2)	(7) Bond Years	(8) Interest Payments
1 (1971)	1900	5.40%	6¾%	101.30%	$ 1,924,700	1900	$128,250
2	"	5.50	"	102.33	1,944,270	3800	256,500
3	"	5.60	"	103.14	1,959,660	5700	384,750
4	"	5.65	"	103.99	1,973,910	7600	513,000
5 (1975)	"	5.70	"	104.51	1,985,690	9500	641,250
6	"	5.80	"	104.75	1,990,250	11400	769,500
7	"	5.85	"	105.11	1,997,090	13300	897,750
8	"	5.90	"	105.36	2,001,840	15200	1026,000
9	"	5.95	"	105.51	2,004,690	17100	1154,250
10 (1980)	"	6.00	"	105.58	2,006,020	19000	1282,500

(continued)

EXHIBIT 1 (Continued)

(1) Years To Maturity	(2) Number of Bonds	(3) Reoffering Yield	(4) Coupon	(5) Price of Bond[a]	(6) Production (5) × (2)	(7) Bond Years	(8) Interest Payments
11	"	6.00	"	105.98	2,013,620	20900	1410,750
12	"	6.05	"	105.91	2,012,290	22800	1539,000
13	"	6.10	"	105.77	2,009,630	24700	1667,250
14	"	6.15	"	105.58	2,006,020	26600	1795,500
15 (1985)	"	6.20	"	105.32	2,001,080	28500	1923,750
16	"	6.25	"	105.01	1,995,190	30400	2052,000
17	"	6.30	"	104.65	1,988,350	32300	2180,250
18	"	6.30	6¾%	104.81	1,991,399	34200	2308,500
19	"	6.70	4	71.22	1,353,180	36100	1444,000
20 (1990)	"	6.70	4	70.49	1,339,310	38000	1520,000
	38,000				$38,498,790	399000	$24,894,750

Gross Production	$38,498,790	Gross Interest Cost	$24,894,750
Less Spread ($13 × 38,000)	$ 494,000	Less Premium	$ 4,790
Net Proceeds to Issuer	$38,004,790	Net Interest	$24,889,960
Premium	$ 4,790	Net Interest Cost/Bond Year	$62.38
Average Reoffering Price[a]	101.3126%	Net Interest Cost Rate	6.238%
Average Price to Issuer[a]	100.0126%		

[a]Expressed as percent of face value of bond.

EXHIBIT 2

(1) Year to Maturity	(2) Number of Bonds	(3) Reoffering Yields	(4) Bond Years	Alternative Coupon Selections: A Coupon	B Coupon	C Coupon	D Coupon
1 (1971)	1900	5.40%	1900	6¾%	6¾%	6½%	7%
2	"	5.50	3800	"	"	"	"
3	"	5.60	5700	"	"	"	"
4	"	5.65	7600	"	"	"	"
5 (1975)	"	5.70	9500	"	"	"	"
6	"	5.80	11400	"	"	"	"
7	"	5.85	13300	"	"	"	"
8	"	5.90	15200	"	"	"	"
9	"	5.95	17100	"	"	"	"
10 (1980)	"	6.00	19000	"	"	"	"
11	"	6.00	20900	"	"	"	"
12	"	6.05	22800	"	"	"	"
13	"	6.10	24700	"	"	"	"

(continued)

EXHIBIT 2 (Continued)

(1) Year to Maturity	(2) Number of Bonds	(3) Reoffering Yields	(4) Bond Years	Alternative Coupon Selections A Coupon	B Coupon	C Coupon	D Coupon
14	"	6.15	26600	"	"	"	"
15 (1985)	"	6.20	28500	"	"	"	"
16	"	6.25	30400	"	6¾	"	"
17	"	6.30	32300	"	6¾	"	"
18	"	6.30	34200	6¾	5	6½	7
19	"	6.70	36100	4	5	5½	2¾
20 (1990)	"	6.70	38000	4	5	5½	2¾
Gross Production[a] (000)				38,499	38,555	38,544	38,556
Net Interest Cost (000)				24,890	24,976	25,144	24,719
Net Interest Cost Rate				6.238%	6.275%	6.314%	6.210%
Bond with Maximum Premium (a) Maturity (year)				1981	1981	1981	1982
(b) Value (% face value of bond)				105.98%	105.98%	103.98%	108.02%
Bond with Maximum Discount (a) Maturity (year)				1990	1990	1990	1990
(b) Value (% face value of bond)				70.49%	81.42%	86.88%	56.83%

[a]Assuming a spread = $13/bond or $494,000 in aggregate.

City of Centerville (B)

On June 11 at 12:01 P.M. Jay Hellman, Centerville's City Manager, opened the two bids which had been submitted for the $20 million serial improvement bond issue. It was readily apparent that the Southern Oregon bid represented the lowest net interest cost to Centerville. Jerry Lerman was informed and placement of the bonds was commenced that afternoon. By Monday afternoon all of the bonds had been sold and all concerned agreed that the bond issue had been handled in a particularly satisfactory manner.

Later in the week of June 15, Hellman sat back to analyze further the two bids that had been submitted. He was curious in learning more about the unusual coupon schedule which had been employed by the Valley Bank and Trust Syndicate. This group had insisted on submitting a bid representing the lowest true interest cost (TIC) to Centerville.

True interest cost denotes the internal rate of return characterizing the cash flows associated with the bond issue. In the case of the Centerville issue, there is an inflow at "time 0" of $20 million Par Value plus any premiums paid for the bonds, less the underwriter's spread. In each of the following 20 years there is an outflow of interest payments and principal repayments. The "TIC" method seeks to minimize the internal rate of return of these payments. Valley Bank and Trust felt that this method would be much more appropriate since it took into account the time value of money. In the cover letter, they said:

> *Using the appropriate constraints on coupons, premiums and underwriter's profit, a computer program has been written to minimize TIC by the University of Oregon's Center for Capital Market Research. The developers contend very strongly that since TIC does consider the Time Value of Money, it is a much better criterion for Municipal Bond Bid Awards than NIC.*

Hellman was concerned with explaining a higher gross interest cost to the public even if in present value terms the interest payments under the TIC schedule would be less than those with the NIC schedule. Exhibit 1 gives the NIC and TIC rates for the two bids submitted for the Centerville issue.

EXHIBIT 1

	Valley's Bid			Southern Oregon's Bid		
Year	Reoffering Yield	Coupon	Bond Price	Reoffering Yield	Coupon	Bond Price
1	3.1	7.0	103.81	3.1	7.0	103.81
2	3.35	7.0	107.00	3.35	7.0	107.00
3	3.55	6.5	108.33	3.55	6.5	108.33
4	3.75	6.0	108.29	3.75	6.0	108.29
5	3.95	5.75	108.09	3.95	4.59	102.6
6	4.1	5.6	107.91	4.1	4.1	100.0
7	4.2	5.5	107.81	4.2	4.2	"
8	4.3	5.4	107.38	4.3	4.3	"
9	4.35	5.4	107.75	4.35	4.35	"
10	4.45	5.3	106.80	4.45	4.45	"
11	4.55	5.25	106.01	4.55	4.55	"
12	4.65	5.25	105.47	4.65	4.65	"
13	4.75	5.25	104.81	4.75	4.75	"
14	4.85	4.85	100.00	4.85	4.85	"
15	4.9	4.9	100.00	4.90	4.9	"
16	4.95	4.95	100.00	4.95	4.95	"
17	5.0	5.0	100.00	5.0	5.0	"
18	5.0	5.0	100.00	5.0	5.0	"
19	5.05	5.05	100.00	5.05	5.05	"
20	5.05	5.05	100.00	5.05	5.05	"
Gross Production			20,894,600			20,300,300
Gross Spread			300,000			300,000

(continued)

EXHIBIT 1 (Continued)

Gross Production	20,894,600	20,300,300
Gross Spread	300,000	300,000
Net Proceed to Issuer	20,594,600	20,000,300
Premium	594,600	300
Gross Interest Cost	10,917,000	10,189,500
Less Premium	594,600	300
Net Interest	10,322,400	10,189,200
No. Bond Years	210 000	210 000
Net Interest Cost/Bond	49.15	48.52
NIC	4.915%	4.852%
TIC	4.78020%	4.84896%

Electricité de France

Electricité de France (EDF) is the French public company that operates the nationalized electric power industry in France. Top management at EDF is responsible to the government for adequate development and efficient operation of the electric power production system to serve a rapidly growing demand. This demand for electricity is highly variable during the year and as electricity is nonstockable, only considerable overcapacity would enable the system to meet demand during the few hundred most heavily loaded hours of the year. EDF competes with other nationalized sectors of the economy for allocation of funds from the government, and a problem of its top management has been to limit investment programs in the best possible way. Investment programs have to be submitted for approval to the government and the parliament. The managers in charge of these programs not only have to determine the most economical solutions to their problems but also to justify these solutions to the authorities in charge of approving the investments.

Toward the end of 1965, the results from the "Investment '85 Model," a mathematical programming planning model, were available to the top management of EDF. This model was the result of a major effort to devise a long-range planning tool that would shed some light on alternative investment programs from 1970 to 1985 and on the complex tradeoffs that had to be made while facing an uncertain future. The main results of the model had essentially confirmed the decisions of the Fifth National Plan (which covered equipment

Source: This case was prepared by Stephen P. Bradley. Harvard Business School Case 4-173-080.

to be operational before 1975) and gave valuable insight into the needs of subsequent years.

The management at EDF felt that the model gave a reasonable geographical distribution of the aggregate amount of investment to be made but that regional results could not be used for any particular purpose. The assumptions made were considered too crude, the modeling of inter-regional transfers of electricity much too simplistic and the data too vague. Yet management at this time attached a particularly high value to more detailed and precise results on the regional distribution of equipment since controlling political bodies were beginning to emphasize planning on a regional basis. The problem now was whether to refine this model or build a new one that would better serve top management at EDF in the future. With the recent introduction of the third generation of computers, larger and more complex models would undoubtedly become practical computationally.

COMPANY BACKGROUND

Electricité de France was founded in 1946, the result of a series of nationalizations which involved basic industries and services in such sectors as transportation, energy, insurance, and deposit banking. After World War II the national emphasis was on reconstruction and nationalizations were accepted by a national coalition government as a means to economic recovery.

The merger of approximately 1400 companies was the first difficult task facing the Board of Directors of EDF when appointed by the government. Other problematic issues concerned planning, organizing, and financing the development of a modern electric power production and distribution system. The objectives given to EDF were not to maximize financial return to the government, but rather to meet the rapidly growing demand for electric power (Exhibit 1) and to render the best possible service to its customers given budgetary constraints.

By 1970, expansion had led EDF to an enviable position among French companies. Its annual value-added was around 10 b.F (billion Francs), highest in the country. As the electric power industry requires expensive equipment, EDF has been investing every year around 50% of its earnings. Further, the productivity of invested capital in the company has been much higher than the French average: EDF has satisfied a demand that has grown at a rate of 8.5% for the last ten years, with investment in constant Francs growing only at 3.6% a year. In France for the same period, net investments have grown to 9.2% to yield a 5.9% rate of growth in GNP. In spite of the growing importance of these investments in absolute terms, their importance as national investments has been steadily decreasing.

EDF supports a particularly heavy financial burden, which includes:

1. The obligation to buy large quantities of French coal at a price much higher than any competitive combustible.

EXHIBIT 1
Electricity Output

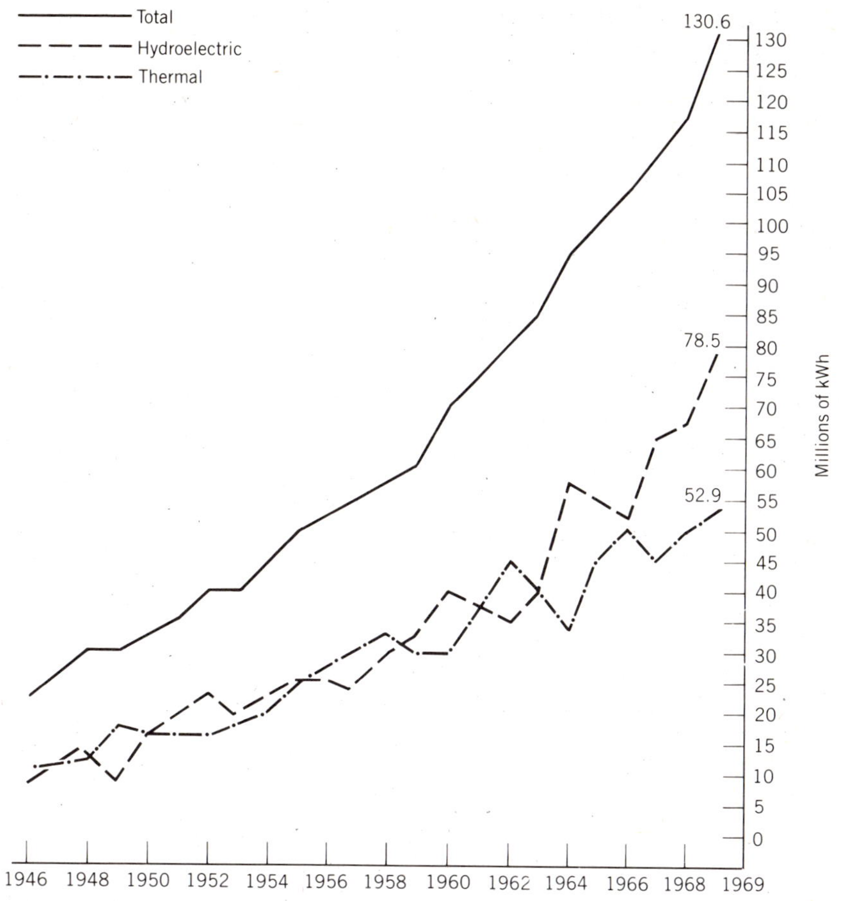

2. The development of nuclear power stations which still have a marginal cost which is much higher than conventional thermal plants. The new technology, different from the widely spread U.S. system, is being developed as a joint effort with the French Atomic Energy Commission under government pressure. The cost imposed to EDF is estimated at 300 M.F. a year.
3. An excise tax burden in excess of the national level.
4. Constant pressure on prices from the public through the government.

In spite of these impediments, EDF has been able to generate a cash flow covering 54% of its funds requirements of 7 b.F., which include 5.3 b.F. of investments in new assets. Financial statements and funds flows for 1969 are given in Exhibit 2.

EXHIBIT 2

Financial Statements for 1969: Balance Sheet as of December 31, 1969 (millions of Francs)

Assets		
Current Assets		
Cash		417
Treasury bills and marketable securities		310
Accounts receivable		2,690
Inventories		740
Regularizing accounts (pre-paid expenses, unbilled accts., etc.)		1,350
Total Current Assets		5,507
Fixed Assets		
Paid-in expenses	4,166	
less depreciation	−1,076	3,090
Plants and equipment service	49,984	
	−12,667	37,317
Plants and equipment under construction		7,141
EDF share in public distribution equipment financing		
(a) on service	5,403	
less depreciation	−1,824	
	3,580	
(b) under construction	378	3,959
Total Fixed Assets		51,507
Long-term loan	1,024	
Joint affiliates	687	
Less allowance for depreciation	5	1,705
Total Assets		58,719

Liabilities and Equity	
Current Liabilities	
Accounts payable	1,659
Notes payable-banks	459
Accruals	1,065
Short-term debt	698
Total Current Liabilities	3,881
Long-term liabilities	
Long-term loans	15,100
Debentures	13,760
Total Long-term Liabilities	28,860
Equity	
Capital stock	2,206
Capital allocations from government	13,944
Capitalized equipment subsidies	989
Retained earnings	7,884
Legal reserves	955
Total Net Worth	25,978
Total Liabilities and Equity	58,719

Sales		
Energy and services	13,695	
less taxes	(2,881)	
		10,814
Cost of Sales		
Direct material	(2,545)	
Personnel expenses	(3,255)	
Depreciation	(2,533)	
Less capitalized expenses	1,581	
		(6,752)
Gross Margin		4,062
G.A.S. Expenses		(2,268)
Income from Operations		1,794
Other Income		
Financial earnings	85	
Profit from equipment subsidies	30	
Net past deferred profits	130	
		245
		2,039

EXHIBIT 2 (Continued)

Other Costs

Interest expenses	(1,893)	
Net extraordinary loss	(29)	
Compensation for nationalization	(15)	
Reserves for depreciation	(79)	
		(1,996)
Profit before tax		43
Income tax		(22)
Net Profit		21

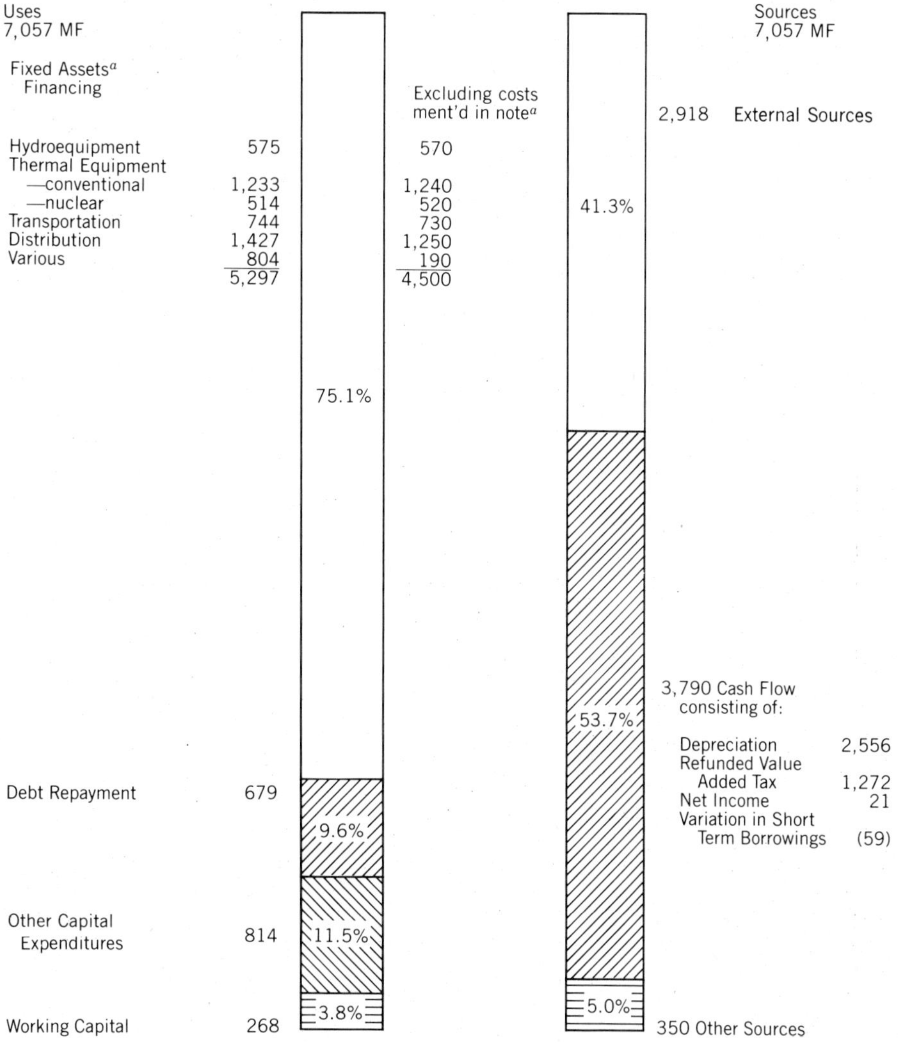

(a) These figures include: interest costs; participations by other companies (mainly for distribution); management costs; various expenses; taxes which are excluded from the figures of Exhibit 3.

Planning for such large investments and choosing among alternative investment programs has been difficult. Given the evolution of the complex demand pattern for electricity and the long construction lead-time, the bulk of investments in each category must be decided upon well in advance. The evolution of the various categories of investments is shown in Exhibit 3. Individual projects (such as those shown in Exhibit 3) are then selected with reference to the planned system evolution. Sophisticated applications of operations research have aided in the investment planning process.

EXHIBIT 3
Capital Expenditures

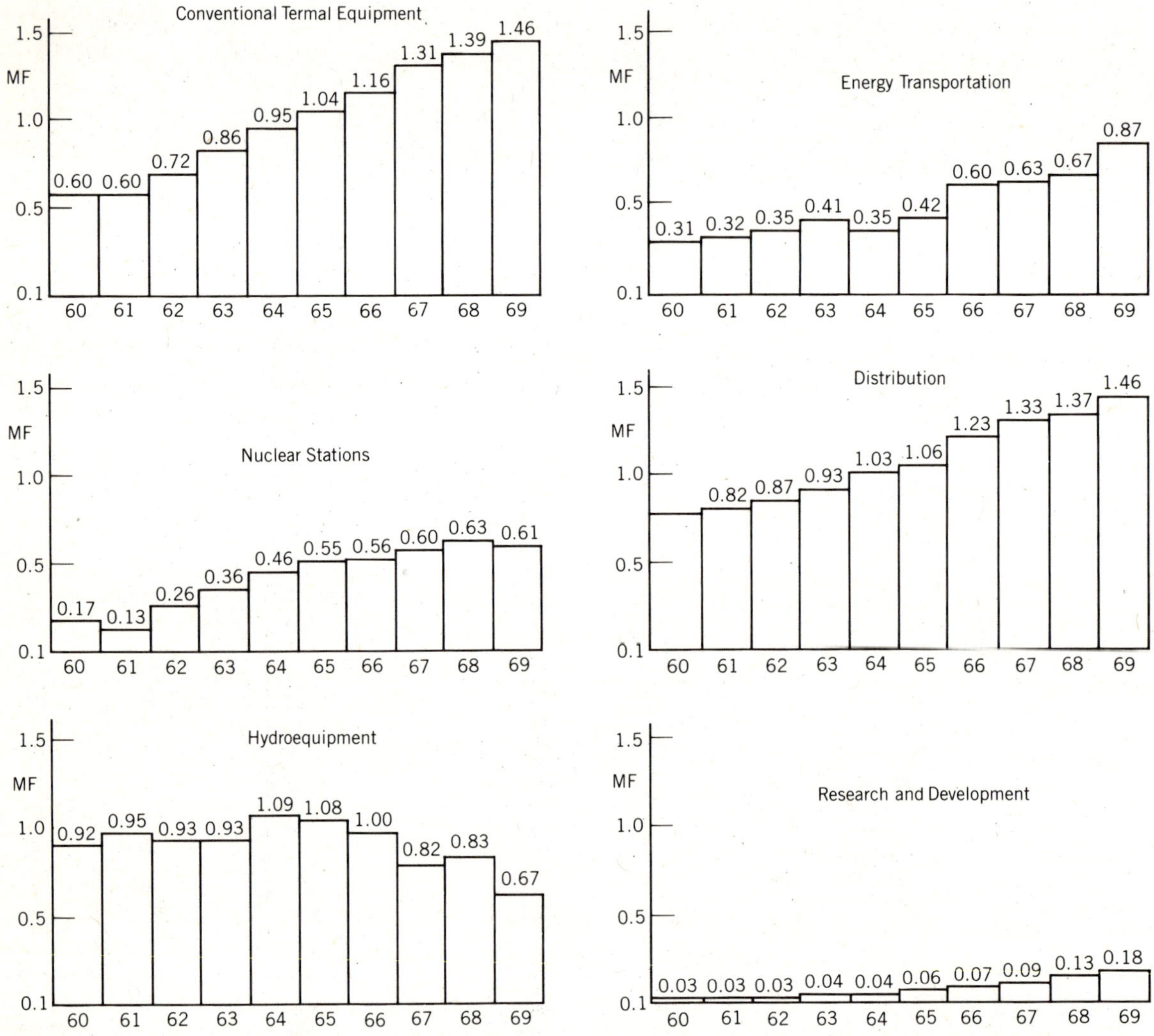

ORGANIZATION AND THE INVESTMENT DECISION PROCESS

EDF is administered by a tripartite Board of Directors including five government administrators, five union representatives and five consumers—three of whom have official duties in city governments. They are all appointed by the government. The responsibilities of the Board are becoming greater because of government has adopted a new policy, lessening tight state control and encouraging more autonomy for nationalized companies. For EDF the government will only outline broad objectives coherent with National Plans and approve the pricing structure.

As shown in Exhibit 4, the organizational structure adopted for EDF separates equipment development from production and distribution management. Top management concentrates on corporate planning, finance, and personnel management in collaboration with other government agencies, such as the Ministère des Finances and the Commissariat au Plan (national planning agency).

EXHIBIT 4
Organization Chart for 1969

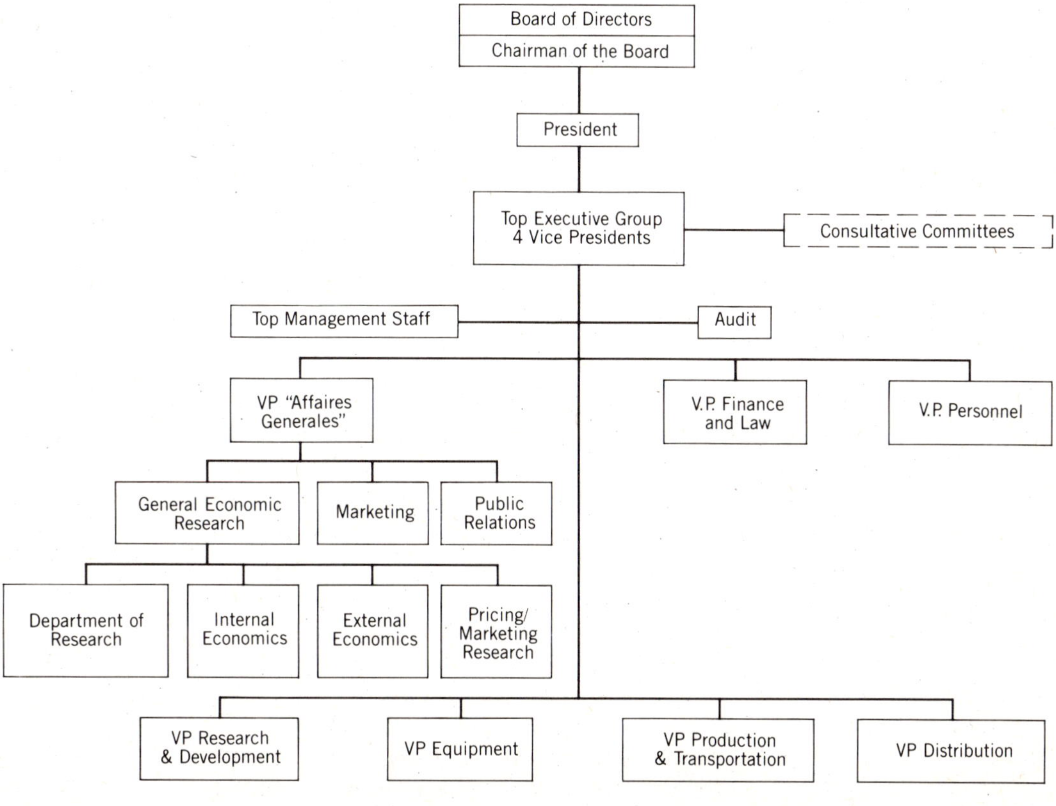

As the present Chairman of the Board puts it, "Electric energy cannot be stocked. Any incident can have very severe consequences and therefore management control at every level is strong; the balancing factor has always been a powerful union organization." In spite of abundant regulation and very centralized unions, decentralized decision making and modern control systems are being implemented. For example, capital budgeting is performed on a decentralized basis within the Equipment Division. There are eight conventional thermal equipment geographical regions, three for nuclear equipment and a single one for hydroequipment. The regional managers decide each year on the capital projects to be proposed for their regions. Yardsticks insuring congruence with a national investment strategy are given by an EDF reference manual called the "Blue Note." The shadow prices given in this Note permit regional managers to calculate for any prospective investment the value of its contribution in power and energy to the system over a year. The Note was developed in the fifties, and turned out to need only slight revision when modified to be consistent with the long-range plans developed at EDF headquarters with mathematical programming models.

Regionally prepared proposals are reviewed by top management at the Paris headquarters, and then discussed with the Commissariat au Plan and with the Ministère des Finances and compared with available resources. After necessary adjustments, the proposals are approved by the Board which is responsible for meeting budget constraints set by the government and parliament.

Most of the corporate planning work at EDF is done by the General Economic Research Group with extensive use of operations research methods. It should be stressed that the very important role of operations research at EDF, since its introduction by Pierre Massé, a former president of the company, receives strong support from top executives, most of whom are former engineers.

Within the General Economic Research Group, the Department of Research (see Exhibit 4), with its many influential mathematicians and economists, plays a rather methodological role. They initiate new approaches and develop models using more appropriate new methods to analyze problems facing top management. The "Investment '85 Model," which is of special interest, was designed in this department under Francis Bessière, a well-known French quantitative economist. The purposes of this Model (already mentioned above) are: (1) to shed light on the evolution, up to 1985, of the optimal structure of electricity production equipment in France; and (2) to outline its geographical distribution by main regions.

Although such models are initiated in the Department of Research, they require many more inputs than the department can provide. The Department of Internal Economics collects technological data while the Department of External Economics deals with the problems of forecasting demand and procurement costs. The Department of Internal Economics has several other tasks including: (1) the elaboration of models, such as production and distribution management models, when their structure is well known and does not pose any difficult methodological problem; (2) responsibility for yearly "Program Adjustments"; (3) liaison with the Commissariat au Plan and the Ministère des Finances; (4) responsibility for the preparation and revision of the famous "Blue Note," on which decentralized decisions are based, with the help of the Department of Research and the Equipment Division. Finally, the Equipment Division, which has always been very receptive to the development of such models, requests its regional managers to participate regularly in seminars on managerial economics where the relevance of the global planning approach is establishing guidelines for decentralized decision making is discussed.

Among the most important consultative committees advising the president is the powerful Committee on Corporate Planning which includes four commissions:

1. The commission on personnel policy.
2. The commission on commercial policy.
3. The commission on industrial policy.
4. The commission on investments, whose secretary is a member of the Department of Research.

THE EDF APPROACH TO DECISION MAKING

Francis Bessière has emphasized that managerial problems and hence supporting operations research studies at EDF reflect the twofold responsibility that the company has. The company's first responsibility is internal: provide the best management and development of the power generation system given environmental constraints. EDF's second responsibility concerns its external impact: the company's activities bear heavily on the economic development of France and affect the entire economic system. Management at EDF might have concentrated on internal and administrative problems, leaving to national politics the strategic decisions involving other subsystems of the national economy. However the past top executives of EDF have encouraged the company to assume both responsibilities.

Internal Management

The solution of decision problems in this case consists essentially of selecting means which allow EDF to reach a particular objective. Problems are characterized by an unidimensional preference structure; that is, find that solution to the problem that minimizes total costs.

In the electric power industry the choice of investment programs for production equipment is particularly important as it accounts for the bulk of total investments. The operations research studies which support decision making in this area assist with the selection of equipment to be developed and provide models of production and distribution management. Francis Bessière defines the investment problem in these terms:

> *Theoretically, if it were possible to draw up a complete list of all feasible projects, the problem would be to select the least expensive combination for given services rendered. But the possibilities multiply as soon as several types of equipment each with special characteristics, are considered. A yet more difficult problem arises if we attempt to find simultaneously for each plant its optimal size, location, date of starting service and even date of retirement.*

The selection of production equipment involves four basic approaches.

1. The *marginal approach* began in 1955 through the operational use of the so-called "Blue Note." This analytic approach "consists of comparing individual projects to a 'yardstick' defined in terms of standard equipment. The method attributes to each particular project a profitability coefficient within a certain context. Although this method helps in making a good choice between projects on a decentralized basis, it is not adequate to insure the best solution. It rests on a number of assumptions about the nature of bottlenecks and the over-all characteristic of the future production system." Indeed to determine whether two projects are equivalent requires reference to a global program which shows how the two projects contribute to the easing of bottlenecks in the production system (i.e., the energy and/or power capacity limit during a certain number of the most heavily loaded hours of the year).

2. The *structural or global approach* complements this approach since marginal analysis cannot define the optimal structure of the production system. Because of the large number of possible equipment configurations, the problem necessitates mathematical programming. Linear programming was first introduced in 1955 by Pierre Massé before he became President of EDF. The first model was very simple but because of the growing complexity of the subject, the successive models grew in size. The "Dantzig Model" appeared in 1957 and considered the timing of investments. The "Three Plan Model," as of 1961, already had 224 constraints and 253 variables including tactical and operating unknowns. This model also introduced elements of uncertainty (meeting a certain fractile of demand) and established the relationship between investment problems and pricing (dynamic pricing at the marginal cost of developing the production system); nonlinear programming appeared in 1963 in the "Investment '85 Model" to account for the nonlinearities in the cost function in a more precise way. Geographical distribution of demand and equipment was the next step to make the model more realistic and hence more useful to planners. This series of programming models has been the cornerstone of the planning efforts at EDF.

3. The *coherence* of the marginal and global approaches is of considerable

importance because consistency of decentralized decision making with a central plan must be insured. This problem has motivated extensive theoretical research which has led to the Principle of Separability. Using this principle it was provided à posteriori that the basic assumptions of the "Blue Note" had been very well chosen and the purpose of its subsequent revision was only to match numerical yardsticks with optimal values of the shadow prices of the "Investment '85 Model."

4. The *"Studies for Program Adjustment"* that are initiated on an annual basis by the Department of Internal Economics, complement the global and marginal approaches. The purpose of these studies is to perform a sequential adjustment of plans for new production equipment to a demand which is increasingly better known. There is a lag of about four years between the time construction of traditional equipment is started and its coming on stream. One might suppose that EDF would launch construction of all equipment which according to the global plan is needed four years later. In the event that construction of a plant were postponed for one year, the present savings would have to exceed the increase in risk of failure to meet demand four years later. But there are means with shorter lead-time, such as certain thermal plants (3 years), gas turbines (2 years), or plant retirements (1 year). It is therefore incorrect to assume that decisions have to be final four years before capacity must be on stream. New estimates of demand can be used to update existing plans and to realize substantial savings on the order of 5% of annual investments. In summary, plans should be adjusted progressively in order to cope with the French demand for electricity in a particular year through a sequence of decisions made in prior years. This problem has been handled at EDF since 1964 using dynamic programming.

The second major category of problems facing management at EDF is of a tactical rather than strategic nature. The issue here is to operate a complex system of production and distribution of electric power as efficiently as possible. Here also operations research models have been developed and implemented at EDF in order to aid in operating decisions. Simulation models have been designed with which to explore alternative policies for operating the transportation system and for scheduling plants across the country when facing changing patterns of demand for electricity. Such problems as the seasonal regulation of hydroelectric reserves and the monthly scheduling of maintenance shutdowns for thermal plants have been tackled using dynamic programming.

External Management

External management concerns the relationship of the company to its economic environment: that is, the succession of alliances and conflicts EDF has with other agencies or groups (government, competition from companies selling different forms of energy, suppliers, and customers). Given the nature of EDF, certain key points emerge as considerations in its strategy decisions:

1. Strategies are limited by the overriding role of the government.
2. Customers may not be excluded from any coalition entered into by EDF.
3. EDF has a monopoly situation in electric power but faces competition from other forms of energy.

Thus far, operations research has been used (1) to aid in defining pricing strategies, and (2) to guide procurement negotiations.

1. As far as EDF is concerned prices are not subject to market regulation, but public services in monopoly situations may not take advantage of this situation. According to classical economics, EDF should react to demand by pricing at marginal cost; however, in order to capture the dynamics of the situation, it is the marginal cost of development of the productive system which is used in setting pricing levels. In fact, the models used for investment planning yield the required marginal values. However, price levels have to be anticipated because customer response to them is not immediate.Here also the accuracy of methods for forecasting demand is critical. Furthermore, possible effects of marketing and sales policies have to be taken into account in pricing decisions.

2. For most of its oil supply the French economy depends on foreign sources. This means that EDF is subject to the effect of price instability on international crude oil markets. This problem caused top management to define a long-term procurement strategy and to negotiate contracts with French-based oil refiners. A key point in the EDF bargaining position was the possibility of putting forward a threat of massive development of nuclear stations. From the results of long-range planning models it was clear to EDF management that nuclear plants would enter the optimal equipment mix before 1958 at the present high level of conventional thermal equipment variable costs. The introduction of nuclear plants would mean a sharp decline in the basic level of nuclear plants would mean a sharp decline in the basic level of demand for fuel in France. It turned out that the analysis of the sensitivity of the optimal program of investment given by the global models was a critical input into EDF negotiation strategy with oil companies. Results were long-term procurement contracts, some 20% below prevailing prices in an attempt by oil producers to delay the massive introduction of nuclear power.

THE "INVESTMENT '85 MODEL"

Introduction

The purpose of the Model can be defined in these terms: to find the optimum balance between expected cost of production and equipment development over a period of 15 years on one side, and the expected cost of failure to meet a particular demand pattern on the other. In fact, there are two main types of decision variables:

1. Strategic variables that represent quantities of different categories of equipment to be constructed.
2. Tactical variables that represent quantities of electrical energy to be produced and transported.

The objective function to be minimized comprises three types of terms investment costs, operating costs, and cost of failure to meet demand. They appear through the discounted sum of their expected values over the 1971–1985 period. The presence of terms representing saturation phenomena (e.g., gradual exhaustion of sites for hydroelectric plants) or random phenomena (e.g., effects of rainfalls of hydroelectric production; magnitude and pattern of demand for electricity) give the objective function a strongly nonlinear character.

Unknowns must obey linear constraints insuring the nonnegativity of equipment and production variables, and expressing either physical restrictions (e.g., capacity limits on peak production or seasonal compensation) or structural assumptions (e.g., nuclear power stations will be scheduled to have a constant output).

The Characteristics of Demand and the Resulting Modeling Effort

The demand for electricity—or the load on the production system—is a very complicated stochastic process. It varies according to the hour of the day, to the day of the week, and to the season of the year. Furthermore, the system's load in a given hour of the year is an uncertain quantity; that is, it is subject to random fluctuation around a mean that can be estimated.

Let us consider the 8,760 hours of the year. With each hour h let us associate the mean electric power demand during that hour which we denote $W(h)$. Now let us plot the 8,760 hours on the horizontal axis in decreasing order of their $W(h)$. One obtains a line called the "load curve" which has the shape of the smooth curve shown in Exhibit 5. The next step is to approximate the load curve by a simpler function in order to be able to model the problem. The EDF specialists have grouped the hours by categories having the same demand characteristics, and $W(h)$ has been averaged out over each one of these categories. The load curve $W(h)$ is therefore transformed into a stepwise decreasing function $W'(h)$ also shown in Exhibit 5.

The categories are the following:

1. The "maximal peak" that includes only the few moments of the year during which the electric power demand on the system is out of range.
2. The peak hours (400 hours).
3. The winter capacity hours (2,000 hours).

EXHIBIT 5
Load Curve

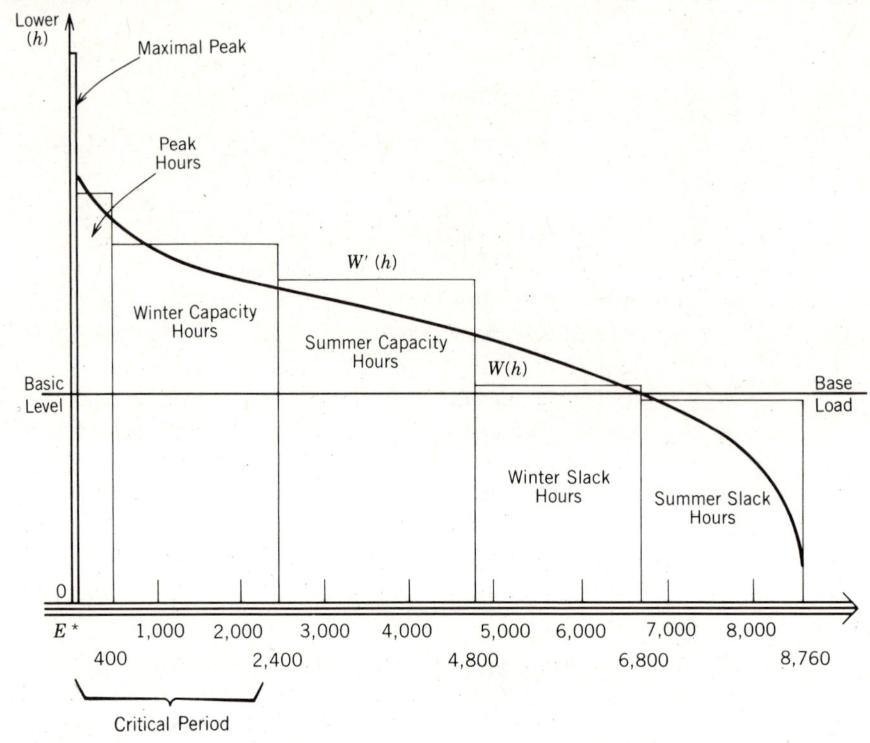

Both of these categories of hours [(2) and (3)] are characterized by a demand substantially higher than the average and because of their 2,400 hours' length constitute an energy demand which is critical for the production system. Hence the name of "critical period."

4. The summer capacity hours.
5. The winter slack hours.
6. The summer slack hours.

The average power demand over the last two categories is referred to as the "base load."

Given the uncertain character of demand combined with random fluctuations in hydroelectric production, it is not realistic to expect the production system always to meet demand, which would require considerable over-capacity at an unacceptable cost. Therefore, EDF has to accept a non-negligible risk of a shortage. A balance must be found between the cost of over-capacity and the cost of a shortage.

Clearly, the risk of a shortage is not constant throughout the year and it is possible to define those periods when the risk is maximal. If a positive balance is achieved for those periods with high probability, then the risk will be negligible for the slack periods. As a consequence, two bottlenecks in the production system have been defined at EDF, which correspond to periods when there is a binding capacity constraint:

1. The "maximal peak," as defined above, where *power* capacity is binding.
2. The "critical period" where the capacity of the system in terms of energy is binding.

In fact, the separability theorem referred to earlier shows that the whole production system can be decomposed for the purpose of analysis, and hence that an investment planning model can be reduced to one that would only take into account the bottlenecks in the system. This is basically what has been done at EDF in the "Investment '85 Model."

The Model

As we have seen the basic problem to be solved in performing the planning activity is to find the optimal tradeoff level between the expected cost of not meeting a stochastic demand either in power or in energy during a certain number of hours in the year, and the expected cost of reducing this risk of a shortage through expensive over-capacity. The evolution of the optimal structure of the electricity production equipment depends critically on this tradeoff. The purpose of the "Investment '85 Model" is to shed light on this issue.

Time Dimension and Regional Decomposition. The life span of productive equipment at EDF is at least 30 years. Technological innovation bears heavily on the development of the entire production system. Keeping these considerations in mind, it is difficult to develop a five-year investment program that does not distort what should be the optimal long-term development strategy. A compromise between the two issues of extending the study over a theoretically infinite period of time and of solving the difficulties in demand and technology forecasting suggests the mid '80s as the time horizon. By that time, substantial nuclear power generation capacity is expected to have been introduced and sites for hydroelectric plants are expected to be exhausted, leaving peak load as the main bottleneck. Furthermore, electricity demand is expected to grow from the 1964 100 TWh to 500 TWh around 1985. In order to match national plans, the three time periods of the model have been chosen as 1971–1975, 1976–1980, and 1981–1985.

France has been divided into six wide regions of reasonable homogeneity (Exhibit 6), grouped into two zones. There is no geographical distribution for 1981–1985, which is considered as a buffer period.

EXHIBIT 6
Regionalization of the "Investment '85 Model"

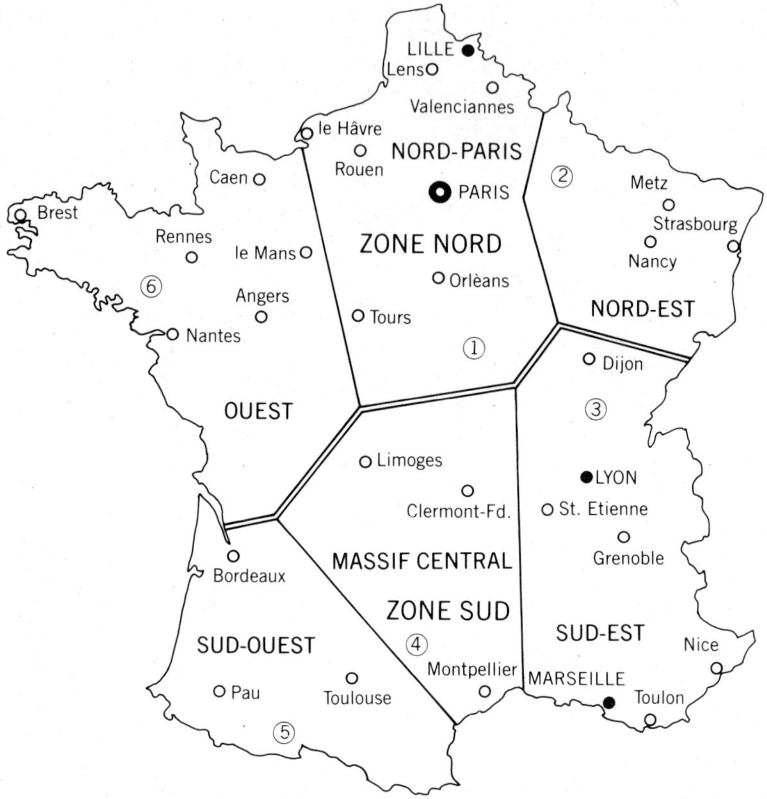

Categories of Equipment. Sources of electric power have been grouped into seven categories of equipment, according to technical and operating characteristics.

Source A: Thermal Equipment

1. Category C: Conventional Thermal Equipment where the only critical parameter is the evolution of the price of fossil fuel, given that this technology is already very well known. It is the only category of equipment to give rise to retirements which are, in fact, considered as data by the model.
2. Category G: Gas Turbines characterized by low investment costs and very high variable costs which are not very well known at this time.

3. Category N: Nuclear stations, expensive to build but much less expensive to operate. The evolution of the cost for the natural uranium-graphite system under consideration is the most significant uncertainty in the model.

Source B: Hydroequipment

4. Category F: Non-modulated production is distributed annually in a uniform manner and comes mainly from "run of stream" installation.
5. Category M: Modulated production can be transferred by reservoirs from slack hours to peak hours.
6. Category R: Capacity of seasonal reservoirs is used for transfer of energy from summer slack period to winter critical periods.
7. Category S: Hydroelectric "over-capacity" on existing facilities helps to meet peak demand requirements.

Hydroelectric production possibilities have a seasonal and random character due to the critical effect of rainfalls. The large number of available observations, however, enables future rainfalls to be probabilistically predicted.

Variables. *Strategic variables* are associated with investments to be made in each period. Thermal equipment is expressed in GW of additional capacity to be constructed; for example, $s(C, r, i)$ = additional capacity of category C to be launched during period i in region r.

Hydroelectric equipment is included in the model in terms of cumulative capacity in TWh to facilitate the formulation; for example, $S(M, r, i)$ = total capacity of category M equipment in region r at the end of period i.

It is necessary at this point to introduce the concept of "standard year" developed at EDF. The operating strategy for the production system and hence the operating cost depend very much on the level of demand. Four "standard years" of scenarios have been defined for each time period of the model (only two for the last time period) representing each a well-defined potential pattern of rainfalls and demand for electricity. To each of these standard years is attached a different load curve—but the same categories of hours—and a probability of occurrence. Production and transportation activities will naturally have a standard year index.

Tactical variables represent quantities of energy to be produced and transmitted. There are two sets of those variables in the "Investment '85 Model."

1. Production variables

   a. $t(R, i, j)$ = energy transfers through reservoirs from low demand to high demand periods of the year for each time period and each standard year.

b. $t(H, i, r, j)$ = hydroequipment output available during peak hours of standard year j in region r for period i.

2. Transportation variables

a. $t(i, r, j)$ = energy imported or exported by region r given normal operating strategy during peak and winter capacity hours. All transfers are made through the Paris region with given energy loss rates.

b. $t(i, r, 0)$ = energy (or power) transported during bottleneck periods given that a particular operating strategy is used when critical period (or maximal peak) shortage is expected. These variables appear only within the shortage terms of the objective function, and describe the emergency transfers made to reduce shortages.

The model includes four categories of constraints.

1. As hydroequipment strategic variables represent cumulative capacity, constraints must prevent these variables from decreasing over time periods.
2. It has been assumed in the model that nuclear plants will have a constant output. A constraint (one for each time period) limits nuclear power plus constant hydroelectric production from run of stream installations to the "base load" defined above from the right tail of the load curve. This assumption represents what is viewed as a reasonable strategic decision at EDF.
3. Seasonal energy transfers through reservoirs [variables $t(R, i, j)$] are of course limited by the available reservoir capacity, or rather 70% of this capacity because of other problems caused by the management of reservoirs.
4. Hydroproduction during peak hours is limited by total hydroequipment capacity. The reservoir capacity is not included because reservoirs are managed and the corresponding productive hydroequipment always operates at capacity during those hours.

The Objective Function. The objective function to be minimized includes four categories of terms.

1. Investment Terms: These terms include construction costs, start up costs, and discounted operating fixed costs up to 1985 less a terminal value of these investments at the end of the third time period when depreciated at the rate of 2% a year.

   As the total installed hydroelectric capacity grows, the adequate sites left are exhausted. The cost of building additional hydroelectric plants in the remaining sites increases very rapidly as more sites are developed. As shown in Exhibit 7(a), these construction costs are strongly nonlinear as a function of total capacity installed.

2. Operating Terms: These represent the expected (over standard years) operating costs for normal operation of the production-transportation system when no shortage is expected.

   For a given year the production system is facing a particular pattern of demand. Nuclear plants and run of stream hydroelectric installations are operated on a year-round constant output basis. Above this basic level other hydroequipment is operated first, up to their capacity because there is no variable cost involved. The complement is then produced by thermal equipment. In fact, thermal plants have different operating costs depending on their age. Because of technological improvement, these costs decrease with the total thermal capacity which has already been installed. For obvious economic reasons the most recent thermal plants will be operated in priority when demnd is larger than the combined output of nuclear and hydroelectric equipment. As a result the marginal cost of producing electricity from thermal plants decreases as a function of the thermal production capacity. Exhibit 7(b) shows this relationship.

3. Cost of Shortage during Critical Period: The consequences of this type of shortage in France are seen as mainly psychological by EDF management. Thus it is possible to assume that the cost of shortage does not depend on the extent of the insufficiency. Hence the expected cost of energy shortage is assumed to be the product of a constant term and the probability of a shortage which is a strongly nonlinear function of the strategic variables as shown in Exhibit 7(c).

4. Cost of Shortage at Maximal Peak: It has been assumed at EDF that the cost to the economy of short-duration power shortage is proportional to the insufficiency.

The model has 153 variables, 93 of which are strategic and 66 tactical, and 53 constraints, 17 of which concern only strategic variables. The cost function has a complicated structure consisting of the sum of one linear investment term corresponding to thermal equipment and of 140 nonlinear terms. Of the latter, 24 are for hydroelectric investment costs, 18 for shortage costs, and 98 for operating costs.

The algebraic formulation of the "Investment '85 Model" was finished in the spring of 1964 and complete data were available a few months later. Operations Research specialists at EDF has been working on the choice and coding of an algorithm for solving the model since 1962. Comparative tests of various computational methods were pursued until the end of 1963 and the "Reduced Gradient Method" due to Philip Wolfe was selected because of its relative efficiency and ability to solve larger problems.

EXHIBIT 7
Nonlinearities in the Costs

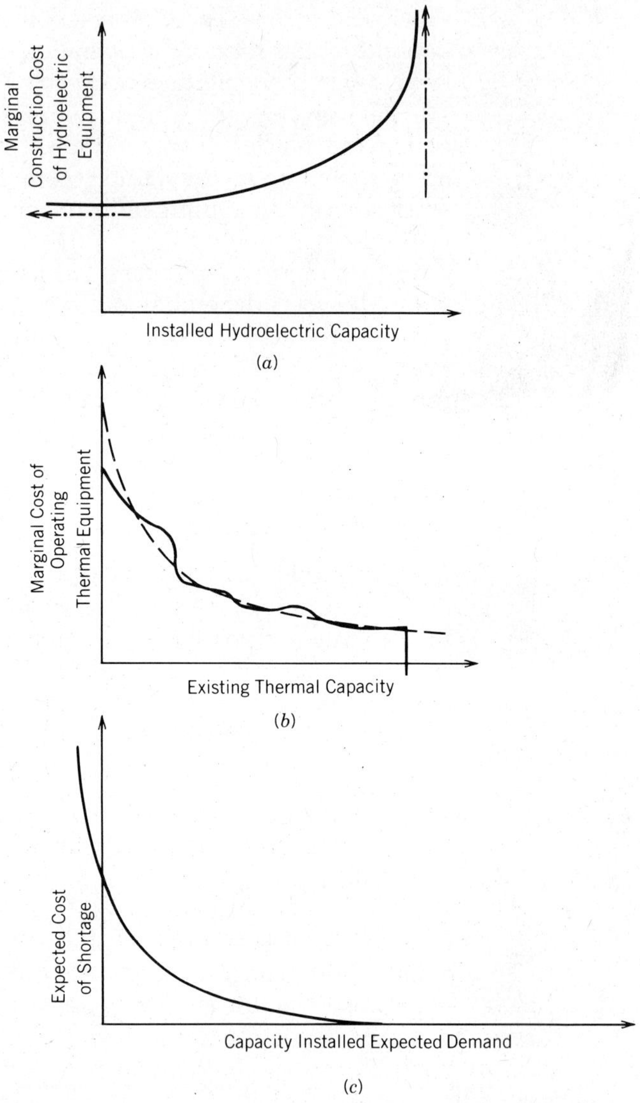

The first runs of the model started in the end of August 1964 on EDF's IBM 7094 computer. By April 1965, complete numerical results with sensitivity and parametric analysis were available to management and a report written by the end of the year.

From the beginning of 1966, EDF management and staff had been concerned about potential improvement of the existing "Investment '85 Model." The results had been very encouraging and given some valuable insight into the period to be covered by the Fifth and Sixth National Plans (1971–1980). However, pressure came from the political level for more regional planning and decentralized decision making. At issue was a possible extension of the model to yield more precise results on the regional distribution of equipment. Further, a larger model was now conceivable as the forthcoming third generation of computers would permit the more rapid solution of such a model.

EXHIBIT 8
Categories of Equipment

(continued)

EXHIBIT 8 (Continued)

Nuclear Station

(continued)

EXHIBIT 8 (Continued)

Conventional Thermal Equipment

Chapter 9

DYNAMIC PROGRAMMING

Often managers face situations that involve a repeating pattern of interrelated decisions and events. For example, the plant manager forecasts demand for the coming month, looks at current inventory levels, and then schedules the month's production. At the end of the month, he will *repeat* the process, but the starting position will be different; that is, the new inventory level will have been determined by his previous decision as well as the actual demand that materialized. Again, taking into account the demand forecast for the next month as well as the new inventory position, he will schedule the next month's production. And so forth. We might use the decision diagram of Figure 9-1 to schematically represent this *sequential* decision problem. At a given point in time, the decision maker must choose among a set of available alternative actions, his choice being dependent on the decisions and events that have gone before; events then occur, partly as a consequence of that decision; another choice is then made; and so forth. As usual, he would like his sequence of choices to be optimal with respect to some chosen criterion, (e.g., least cost, maximal capacity utilization) over the *planning horizon*.

Sequential decision problems that have this structure of repeated interrelated decisions and events can be effectively optimized using *dynamic programming*. Dynamic programming is a recursive optimization approach that operates by breaking a problem down into smaller subproblems called *stages* (e.g., Figure 9-1), then optimizing on a stage-by-stage basis utilizing information derived from the optimization of the previous stages. If the structure of each

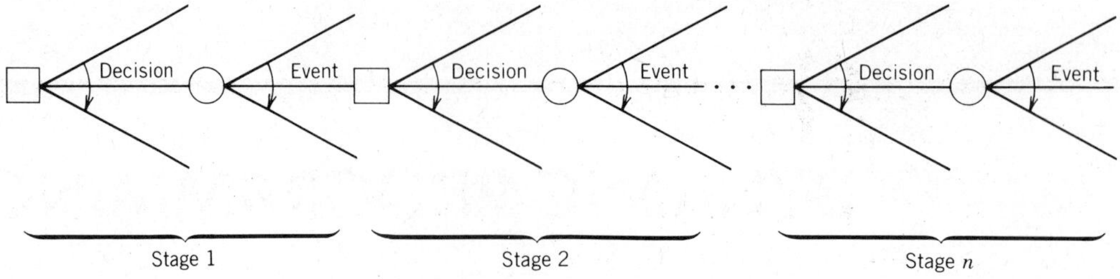

FIGURE 9-1. Sequential decision diagram.

stage is similar—that is, the pattern of decisions and events is repeated in each time period—then dynamic programming is an especially effective tool for determining the optimal action to be taken at each stage. Unlike analyzing a decision diagram, dynamic programming does not require the evaluation of all strategies and can thus allow a tremendous computational advantage for problems that would be too large for a decision tree analysis. Sequential decision problems where dynamic programming is particularly useful include inventory management, production scheduling, manpower planning, long-range planning for replacing depreciating assets, and others.

Dynamic programming is a rather unique approach, quite different from the other techniques presented in this book. The first difference, as is evident from the preceding discussion, is the optimization in stages in contrast to the simultaneous optimization characteristic of all other algorithms we have discussed. The key implication of sequential optimization is that, whereas *each* iteration of the Simplex method, for example, produces a complete solution (not necessarily optimal), dynamic programming produces no solution until it has terminated. The second, more important, difference is that dynamic programming is *not* an algorithm like the Simplex method, branch and bound, or steepest ascent, but rather a solution *strategy*—each problem requires a slightly different solution procedure customized for it. This fact implies strengths and weaknesses.

The key weakness is that the decision maker or analyst cannot simply plug his model into a commercially available dynamic programming package. Nor can he expect to solve any sequential decision problem by mechanically following a sequence of previously determined steps. Although we will, in this chapter, identify a basic set of steps that help in applying dynamic programming, the *details* of these steps depend on the nature of the problem being solved and may require some "art" and intuition to develop.

The major strength of dynamic programming as a solution strategy is its tremendous flexibility and versatility. Dynamic programming will solve problems involving integer restrictions, nonlinearities in both constraints and objective function, or stochastic elements almost as easily as strictly linear, deterministic problems. A second strength is that dynamic programming's solutions are often policy statements rather than specific answers. These statements are of the form: "If beginning inventory is x and forecasted demand is d, do such-and-such." "Such-and-such" will be a function of x and d. Thus, rather than resolve a linear (or other) program *each* time a decision has to be made, the manager can look to the policy statement.

An example will best serve to illustrate the dynamic programming approach.

A SHORTEST ROUTE EXAMPLE

A European trucking company is planning to open a new route from Paris to Istanbul. There are many possible routes, and management has decided the truck should stop only in two cities. The costs associated with the different routes are distance-related as well as involving different taxes on transit merchandise and time required to cross borders. There are also potential benefits in terms of market size in each city. Incorporating these factors, the "cost" of each leg has been evaluated, and is given with the map of possible routes in Figure 9-2. The firm would like to determine the least cost route.

One possible solution approach might be a decision diagram (one with no uncertain events). The tree would be as shown in Figure 9-3. In order to put the endpoint values on the tree, one would have to calculate the costs of all possible routes, so that this method is just explicit enumeration. Another solution approach would be a network optimization model with 18 decision variables (one for each arc), 10 constraints (one for each node), 18 capacity constraints, and 14 possible routes. Dynamic programming provides another solution technique, based on the sequential nature of the problem (at each stop, the truck has to decide where to go next). Dynamic programming starts with a small piece of the problem and finds the optimal solution for this smaller problem. It then enlarges the problem and finds the optimal solution for the larger problem from the previous one.

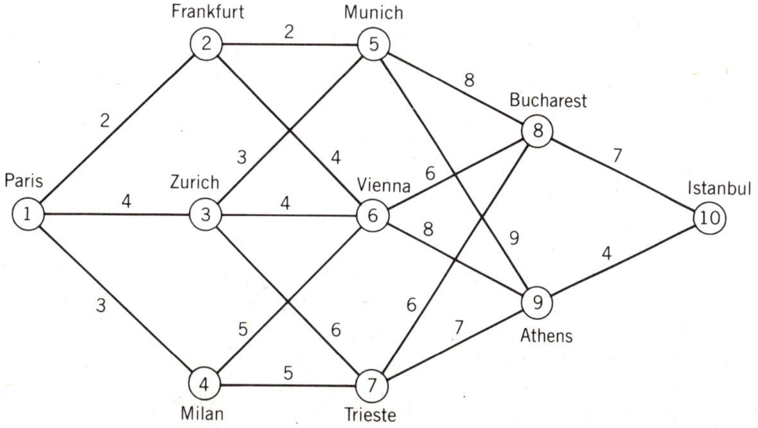

FIGURE 9-2. Truck routes.

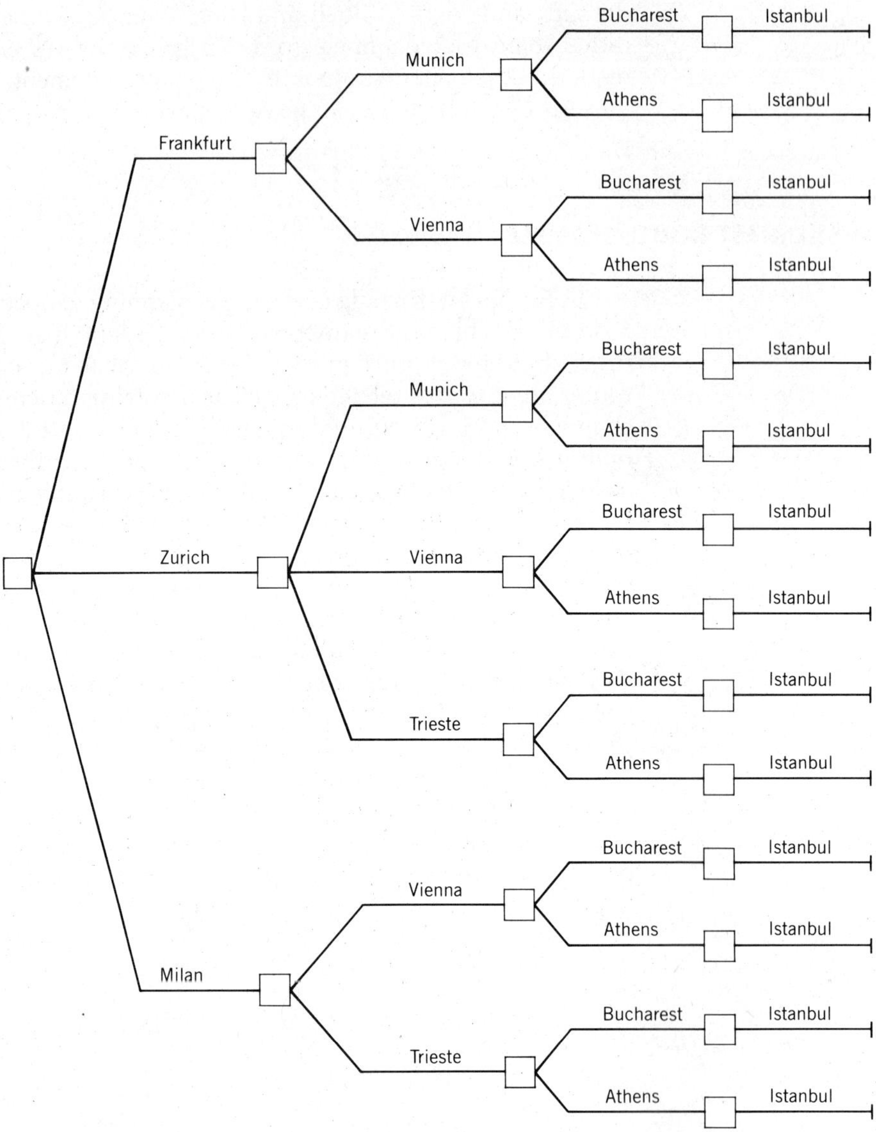

FIGURE 9-3. Complete enumeration.

Consider dividing the firm's decision problem into four time periods or *stages* (four decisions have to be made). At each stage, the truck can be in different cities; for example, at stage 2, it can be in Munich, Vienna, or Trieste (cities numbered 7, 8, or 9). Then:

- Let the decision variables x_n ($n = 1, 2, 3, 4$) be the intermediate destination at stage n, and let us call those cities the *state* in which the truck is.[1] Thus, the route selected would be

$$1 \rightarrow x_1 \rightarrow x_2 \rightarrow x_3 \rightarrow x_4 \text{ with } x_4 = 10$$

- Given that the truck is in state s at stage n and has selected x_n as the next destination, let $f_n(s,x_n)$ denote *the total cost of the best overall policy for the remaining stages.*

- Given stage n and state s, let x_n^* be the value of x_n that minimizes $f_n(s,x_n)$. In other words, given the truck's current location, x_n^* is the optimal next city—that is, the one that minimizes the total cost over the remaining part of the trip. Let $f_n^*(s)$ be the corresponding minimum value that depends only on stage n and state s; that is,

$$f_n^*(s) = f_n(s, x_n^*) = \min_{x_n} f_n(s, x_n)$$

Our objective is to find $f_1^*(1)$—the minimum cost—and the corresponding optimal route. In this example, we will use dynamic programming to find it by working backward through $f_4^*(s)$, $f_3^*(s)$, $f_2^*(s)$, and $f_1^*(s)$.[2] At stage 4, there is no real decision involved. Regardless of whether the truck is in state 8 or state 9, the optimal next stop, x_4^*, is 10. See Table 9-1.

At stage 3, in each state (city) s, the truck driver has the choice between two destinations, 8 or 9. We have to evaluate $f_3(s, x_3)$, the minimum *additional* cost of each decision. This can be separated into two elements: (1) the cost of going from state s to state 8 or 9 plus the minimum cost attached to state 8 or state 9 for the remaining stages (in this case only one). Table 9-2 has been set up to provide the necessary computations. If the truck is in state (city) 5 (Munich) and goes to state 8 (Bucharest), the minimum additional cost of this decision is \$8 for the cost of the decision itself plus \$7 for the minimum cost of traveling from Bucharest onward which we read from the previous table as $f_4^*(8)$. Similarly, if the truck driver chooses to go to state 9 from state 5, the cost is \$9 (from 5 to 9) plus $f_4^*(9)$ (the minimum cost of 9 onward). Thus, from 5, the truck driver's best action is to go to 9 and incur a cost from 5 onward [$f_3^*(5)$] of

TABLE 9-1

(State) s	$f_4^*(s)$	x_4^*
(Bucharest) 8	\$7	10
(Athens) 9	\$4	10

[1]In this example, the *decision* and the *state* are the same. This will often not be true.
[2]Dynamic programming often proceeds forward through the stages of the problem at hand. But there is little conceptual difference in that approach, so we will not consider it in this chapter.

\$13. In a similar fashion, we can then determine the minimum cost attached to each state and the corresponding decision.

Note that x_3^* is always 9; that is, the optimal decision at stage 3 is to go to state 9 (Athens), regardless of where the truck is starting from.

At stage 2, the procedure is exactly the same using the results $[f_3^*(s)]$ of stage 3. The computations are performed in Table 9-3.

Note here that x_2^* depends on the truck's starting point. If at state 2 (Frankfurt), the truck should go to stage 5 (Munich); if at state 3 (Zurich) the truck should go to either Munich or state 6 (Vienna); if at state 4 (Milan), to state 7 (Trieste).

At stage 1, we obtain the optimal solution for the overall problem by the same process (Table 9-4). Thus, the overall cost of the trip is \$17. The optimal route can be constructed by tracing back through the tables. The first step (at stage 1) is to go to state 2, since $x_1^* = 2$. At stage 2 ($s = 2$), the best move from state 2 is to state 5 ($x_2^*(2) = 5$), and so on. The optimal route is $1 \rightarrow 2 \rightarrow 5 \rightarrow 9 \rightarrow 10$.

In the next section we shall examine the general principles of dynamic programming formulation and solution that underlie this example.

TABLE 9-2

	$f_3(s, x_3) = [\text{cost from } s \text{ to } x_3] + f_4^*(x_3)$			
	x_3			
s	8 (Bucharest)	9 (Athens)	$f_3^*(s)$	x_3^*
5 (Munich)	\$15 = 8 + 7	\$13 = 9 + 4	\$13	9
6 (Vienna)	\$13 = 6 + 7	\$12 = 8 + 4	\$12	9
7 (Trieste)	\$13 = 6 + 7	\$11 = 7 + 4	\$11	9

TABLE 9-3

	$f_2(s, x_2) = [\text{cost from } s \text{ to } x_2] + f_3^*(x_2)$				
	x_2				
s	5	6	7	$f_2^*(s)$	x_2^*
2	\$15	\$16		\$15	5
3	\$16	\$16	\$17	\$16	5 or 6
4		\$17	\$16	\$16	7

TABLE 9-4

	$f_1(s, x_1) = [\text{cost from } s \text{ to } x_1] + f_2^*(x_1)$				
	x_1				
s	2	3	4	$f_1^*(s)$	x_1^*
1	\$17	\$20	\$19	\$17	2

DYNAMIC PROGRAMMING FORMULATION STEPS

There are five basic elements in the formulation of a dynamic program:

1. *Stages* (usually denoted $n = 1, 2, 3, \ldots$). Decompose the problem into a series of stages with policy decisions required at each stage.
2. *Decision Variables* (usually denoted x_n). Determine the set of decisions required at each stage.
3. *States* (usually denote s_n). Each stage has a number of associated states. The states are the various possible conditions of the system at that stage of the problem. The number of states may be finite (as in the preceding example) or infinite (state described by a continuous variable such as number of items in an inventory). The key characteristic to remember in state definition is that the state *completely* specifies the condition of the system. This "memorylessness" implies that given the current state, an optimal policy for the remaining stages is independent of the policy adopted in previous stages. In other words, *knowledge of the current state carries all the information necessary for determining the subsequent optimal policy*. In the previous example, once in Munich it does not matter how the truck got to Munich to determine if it is now preferable to go through Athens or Bucharest.
4. *State Transformation Equation*. The decision made at the current stage transforms the system in some way. Since the state carries all relevant information about the condition of the system, we can describe this transformation with an equation that relates the state of the system at the start of the next stage to the decision just made and/or the current state. In the preceding network example,

   $$s_{n+1} = x_n$$

   That is, the state at stage $n + 1$ is simply the destination chosen at stage n. Rarely will the state transformation equation be so simple.
5. *Recursive Relationship* (usually written $f_n^*(s) = \ldots$). The recursive relationship embodies the objective function, and solution identifies the optimal policy for each state at stage n, given the optimal policy for each state at stage $(n + 1)$. In the previous example, this recursive relationship could be written algebraically as

   $$f_n^*(s_n) = \min_{x_n} \{c(s_n, x_n) + f_{n+1}^*(s_{n+1})\}$$

   where $f_{n+1}^*(s_{n+1})$ is the cost of the optimal policy from stage $(n + 1)$ onward given the system is in state s_{n+1}, which is the state the system enters at stage $(n + 1)$—defined by the state transformation equation; that is, $s_{n+1} = x_n$. $c(s_n, x_n)$ is the cost associated with the decision x_n given that the current state is s_n. The cost, in the preceding example, was the cost of traversing the arc from s to x_n. $f_n^*(s)$ is the cost of the optimal policy from stage n onward given the system is in state s_n.

The concept embodied in this recursive relationship (called the *optimality principle*) is the cornerstone of dynamic programming. It says that the cost (or value) of the optimal policy is the cost of traveling to the next state plus the cost of the *optimal continuation from there on.* Also, the optimal policy is found by determining that intermediate state that optimizes the sum of those costs.

The solution procedure, often called backward induction, steps backward from f_N^* (where N denotes the last stage) through $f_{N-1}^*, f_{N-2}^*, \ldots, f_2^*$ to f_1^*. f_1^* represents the optimal objective function value and the optimal levels of the decision variables may be found at each of the steps: x_N^* is determined in obtaining f_N^*; x_{N-1}^* is determined in solving for f_{N-1}^*, and so forth.

In order to start the procedure, we need an end point or terminal value. Often, we can initiate the process by writing down x_N^* and f_N^* for which the solution is obvious. In our previous example, $x_4^* = 10$ (Athens) since that was the only possible final destination, so f_4^* could be read directly from the arcs leading to Athens; that is,

$$f_4^*(s) = \begin{cases} 7 & \text{for } s = 8 \\ 4 & \text{for } s = 9 \end{cases}$$

Then we solved

$$f_3^*(s) = \min_{x_3} \{c(s,x_3) + f_4^*(x_3)\}$$

via Table 9-2, and so forth.

A second example will help clarify the five formulation steps as well as the solution procedure.

A NEW PRODUCT PROMOTION CAMPAIGN WITH NONLINEAR OBJECTIVE FUNCTION

A company will soon be introducing a new product into a very competitive market and so is currently planning its marketing strategy. The decision has been made to introduce the product in three phases. Phase 1 will feature making a special introductory offer of the product to the public at a greatly reduced price to attract first-time buyers. Phase 2 will involve an intensive advertising campaign to persuade these first-time buyers to continue purchasing the product at a regular price. It is known that another company will be introducing a new competitive product at about the time phase 2 will end. Therefore, phase 3 will involve a follow-up advertising and promotion campaign to try to keep the regular purchasers from switching to the competitive product. A total of $5 million has been budgeted for this marketing campaign. The problem now is to determine how to most effectively allocate this money to the three phases.

The tables below show the estimated effect of expenditures in each phase.

Phase 1

Expenditure ($M)	Resultant Market Share
0	0%
1	10%
2	15%
3	22%
4	27%
5	30%

Phase 2

Expenditure ($M)	Fraction of Market Share that Is Retained
0	.30
1	.55
2	.70
3	.80
4	.85
5	.90

Phase 3

Expenditure ($M)	Fraction of Market Share that Is Retained
0	.50
1	.70
2	.85
3	.90
4	.93
5	.95

If we denote by m_1 the market share obtained after Phase 1 of the campaign and by m_2 and m_3 the fraction of that share that is retained in Phases 2 and 3, then we can write the firm's problem as

"Determine the expenditures to be made in each of the three phases so as to

$$\text{Max} \quad m_1 m_2 m_3$$

subject to total expenditure $\leq$ \$5 million."

Note that not only is the objective function nonlinear in the m_i's, but each m_i is nonlinear (showing marginally decreasing returns) in the expenditure levels.

A dynamic programming formulation would be:

Stages: $n = 1,2,3$ to correspond to each of the three phases.

Decision variables: x_n, the expenditure to be made at stage n, in millions of dollars. As usual, x_n^* will denote the optimal x_n.

States: s_n, the state when entering stage n, to represent the money (in millions) *remaining* to be spent. For example, $s_1 = 5$.

State transformation: $s_{n+1} = s_n - x_n$. That is, the state at the next stage is simply the current state minus the money spent during this stage.

Recursive relationship:

$$f_n^*(s_n) = \max_{x_n} [m_n(x_n) \times f_{n+1}^*(s_{n+1})]$$

Or, using the state transformation equation,

$$f_n^*(s_n) = \max_{x_n} [m_n(x_n) \times f_{n+1}^*(s_n - x_n)]$$

Note that this recursive relationship is a *product* in contrast to the *sum* encountered in the preceding example. Fortunately, the dynamic programming (induction) procedure is indifferent to such nonlinearities.

Solution

At stage 3, the solution to our decision problem is trivial: spend whatever money remains and reap the benefits; that is,

s_3	x_3^*	$f_3^*(s_3)$
0	0	.50
1	1	.70
2	2	.85
3	3	.90
4	4	.93
5	5	.95

At stage 2, the recursive relation is

$$f_2^*(s_2) = \max_{x_2 \leq s_2} [m_2(x_2) \times f_3^*(s_2 - x_2)]$$

Note that the maximization is to be performed only over those values of x_2 such that $x_2 \leq s_2$; that is, the optimization of this stage 2 subproblem involves a *constraint*. This is usually the case.[3] Table 9-5 contains the computations necessary to optimize at stage 2.

Note that the solution at stage 2 is of the form of a policy statement rather than a single answer. By "policy statement" we mean a sequence of decisions such that the specific one to be implemented depends on the state of the system at the time. That is, to the manager in charge of implementing phase 2, the solution says "If \$3 million or more remains in the budget, spend \$2 million; if only \$2 million remains, spend \$1 million; if less than \$2 million remain, spend it."

We now move backward to stage 1. Since the starting budget is \$5 million($s_1 = 5$), only $f_1^*(5)$ need be computed, as done in Table 9-6.

From Table 9-6, we see that the optimal promotion policy should enable us to capture 8.47% of the market, and that optimal policy is

Phase 1: Spend \$3 million (x_1^* from Table 9-6).

Phase 2: Spend \$1 million ($x_1^* = 3$ implies $s_2 = 2$ [\$2 million left], since $s_{n+1} = s_n - x_n^*$, and x_2^* [at $s_2 = 2$] is 1, from Table 9-5).

Phase 3: Spend \$1 million (the remainder).

TABLE 9-5

	x_2							
s_2	0	1	2	3	4	5	x_2^*	$f_2^*(s_2)$
0	.1500						0	.1500
1	.2100	.2750		*Infeasible*			1	.2750
2	.2550	.3850	.3500				1	.3850
3	.2700	.4675	.4900	.4000			2	.4900
4	.2790	.4950	.5950	.5600	.4250		2	.5950
5	.2850	.5115	.7650	.6800	.5950	.4500	2	.7650

TABLE 9-6

	x_1							
s_1	0	1	2	3	4	5	x_1^*	$f_1^*(s_1)$
5	0	.0595	.0735	.0847	.07425	.0450	3	.0847

[3]For illustration, the examples in the chapter are simple enough that subproblems may be optimized by complete enumeration using tables. In practice, this may not be possible. For large-scale real problems, the optimization of each stage-by-stage constrained subproblem may be a major task, involving some of the methods we have encountered earlier in this text.

RELATIONSHIP TO NETWORK OPTIMIZATION

A deterministic dynamic programming problem can be interpreted as a network optimization model:

- Each node corresponds to a state.
- Nodes can be grouped pictorially in columns that correspond to a stage so that flow from a node can go only to a node in the next column to its right.
- The value assigned to each arc can be interpreted as the contribution to the objective function made by going from one state to the other.

Observe that Figure 9-2 has this form naturally. A state described by a continuous variable can be incorporated by treating it as a discrete variable with a large number of possible values. Such is the case for our second example. Its corresponding network is shown as Figure 9-4. Unfortunately, due to the form of its objective function, this is a multiplicative longest route problem, we could not solve it by means of any of the *linear* network optimization techniques we have considered. In this type of situation especially, dynamic programming offers a particularly great computational advantage over the network model.

ANOTHER EXAMPLE: INVENTORY AND PRODUCTION SCHEDULING

A firm wishes to establish a production schedule for an item for the next N periods. Due to a long-standing contract, the demand is known with certainty and is level at 3 units per period. This demand must be met exactly. The manufacturing time to produce a batch is negligible. Thus the production of one period may be used to satisfy the demand of that same period. The production costs consist of a \$13 set-up cost and a \$2 variable cost/unit. There is an inventory holding cost equal to \$1 per item in ending inventory. The firm does not want to have any inventory at the end of the last period. To simplify the calculations, let us further assume that the firm cannot produce more than 5 units in a period and cannot hold more than 4 units in inventory.

This problem can be solved as a network. The nodes are the beginning inventories and the arcs are the production decision. The cost associated with each arc is the cost of production plus the inventory holding cost.

The network for three periods is shown schematically in Figure 9-5. The pair of numbers on each arc represents the number of items produced and the resultant cost; the node numbers represent the inventory level.

Apart from the obvious computational savings, dynamic programming offers another advantage in this case: since the solution works backward, we are not obliged to choose in advance the number of periods to be analyzed. To add one period we need not go through all the calculations again (as for

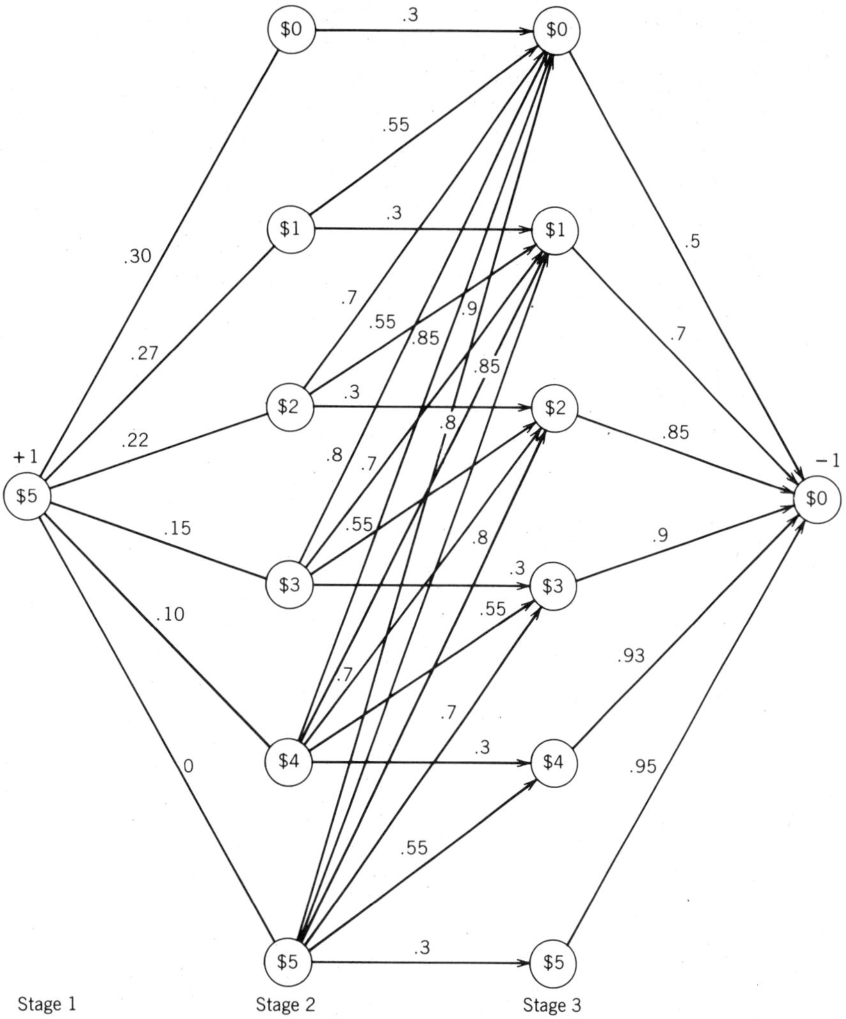

The dollar figures inside each node represent the remaining budget.

FIGURE 9-4. The product promotion campaign network.

networks), just one more stage. This will be made clear in the discussion of sensitivity analysis.

Again, for a dynamic programming solution, we must define the particular characteristics of this dynamic problem. First, what are the stages, decision variables, and states of this problem? The stages are the periods—at each stage a decision must be made. The decision is x_n—the number of items to be produced at stage n. The state of the system can be defined by the inventory at the beginning of each period, which we will denote i_n. Given this state, the impact of a production decision is independent of the production or inventory deci-

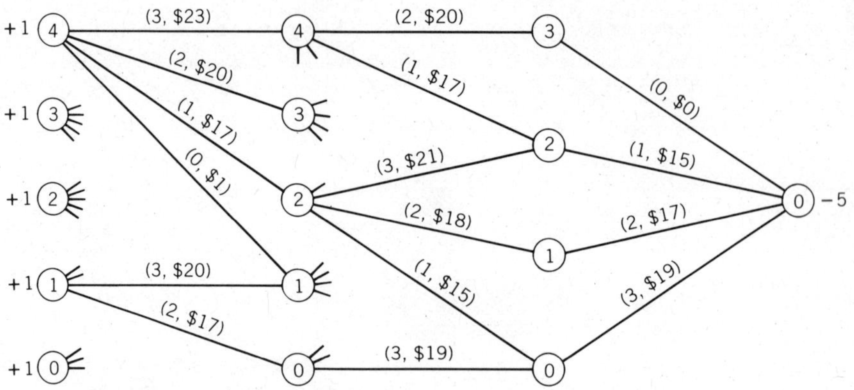

FIGURE 9-5. The production planning network.

sions made before, and the current production decision together with the current state determine completely the beginning inventory (the state) for the next stage.[4]

The relation between the beginning inventory of period $n + 1$ (which also is the ending inventory of period n) and i_n defines the state transformation:

$$i_{n+1} = i_n + x_n - 3$$

As before, we define $f^*_{n+1}(i_{n+1})$ to be the value of the optimal policy for the remaining stages given the state i_{n+1} at stage $n + 1$. We want to find $f^*_n(x_n, i_n)$. This is equal to the cost of the decision x_n *itself* plus the optimal value associated with the state where this decision leads; that is, $f^*_{n+1}(i_n + x_n - 3)$.

The cost of the decision itself is the production cost plus the inventory holding cost:

$$C(x_n,i_n) = \text{production cost} + \text{inventory holding cost}$$

$$= \left[\begin{Bmatrix} 0 & \text{if } x_n = 0 \\ 13 + 2x_n & \text{if } x_n \neq 0 \end{Bmatrix} + \$1 \cdot i_n + x_n - 3\right]$$

So we now have our recursive relationship:

$$f^*_n(i_n) = \min_{x_n} \{C(x_n,i_n) + f^*_{n+1}(i_n + x_n - 3)\}$$

Given this relationship, we have now to compute the optimal policy beginning at stage N.

[4]Determining the appropriate state descriptor for a given problem is often quite difficult and takes practice.

Stage N

Production decisions are imposed at this stage by the zero ending inventory requirements.

Beginning Inventory

i	x_N^*	$f_N^*(i)$
0	3	19
1	2	17
2	1	15
3	0	0

Stage N-1

$$f_{N-1}(x, i) = C(x, i) + f_N^*(i + x - 3)$$

	x							
i	0	1	2	3	4	5	$f_{N-1}^*(i)$	x_{N-1}^*
0	Demand not			19+0+19	21+1+17	23+2+15	38	3
1	satisfied		17+0+19	19+1+17	21+2+15	23+3+0	26	5
2		15+0+19	17+1+17	19+2+15	21+3+0		24	4
3	0+0+19	15+1+17	17+2+15	19+3+0	Too much		19	0
4	0+1+17	15+2+15	17+3+0		ending inventory		18	0

Observe, as before, that the table above incorporates two implicit constraints:

$$i_n \geqslant 0 \quad \text{(to satisfy demand)}$$

$$i_n \leqslant 4 \quad \text{(storage constraint)}$$

Also, note again that the last column of the table is a policy statement: "With two periods to go to the planning horizon, produce 3 items if there are none in stock; produce 5 items if there is one in stock; produce 4 if there are two in stock; otherwise produce none."

Stage N-2

$$f_{N-2}(x, i) = C(x, i) + f_{N-1}^*(i + x - 3)$$

	x							
i	0	1	2	3	4	5	$f_{N-2}^*(i)$	x_{N-2}^*
0	Demand not			57	48	49	48	4
1		satisfied	55	46	47	45	45	5
2		53	44	45	43	45	43	4
3	38	42	43	41	43		38	0
4	27	41	39	41	Inventory > 4		27	0

The process can be so continued from stage to stage. The table below summarizes the results for six periods.

Beginning Inventory

Period	N		N-1		N-2		N-3		N-4		N-5	
i	x_N^*	$f_N^*(i)$	x_{N-1}^*	$f_{N-1}^*(i)$	x_{N-2}^*	$f_{N-2}^*(i)$	x_{N-3}^*	$f_{N-3}^*(i)$	x_{N-4}^*	$f_{N-4}^*(i)$	x_{N-5}^*	$f_{N-5}^*(i)$
0	3	19	3	38	4	48	4	67	5	79	4	96
1	2	17	5	26	5	45	5	64	5	74	5	93
2	1	15	4	24	4	43	5	54	4	72	4	91
3	0	0	0	19	0	38	0	48	0	67	0	79
4			0	18	0	27	0	46	0	65	0	75

This table shows the optimal policy to follow and its cost, given the planning horizon and the beginning inventory. For example, planning for 5 periods with a beginning inventory of 0 requires us to produce 5 units in the first period. The cost of the overall optimal policy is \$79 (column N-4, first row). The ending inventory of the 1st period will be $5 - 3 = 2$, thus the production for the 2nd period should be 5 again (column N-3, third row). The complete optimal policy is

$$i = 0 \xrightarrow{x=5} i = 2 \xrightarrow{x=5} i = 4 \xrightarrow{x=0} i = 1 \xrightarrow{x=5} i = 3 \xrightarrow{x=0} i = 0$$

SENSITIVITY TO THE HORIZON

Increasing the planning horizon by one period only requires that we calculate $f_{N-6}^*(i)$, which, in turn, depends only on the value in the last two columns of the above table. This ease in calculating the optimal policy's sensitivity to the horizon chosen makes dynamic programming particularly useful.[5] For example, with no beginning inventory, we can derive from the above table the optimal production amounts in each period given the planning horizon, as shown in the next table:

Planning Horizon	Period 1	2	3	4	5	6	Cost	Cost/Unit
1	3						19	6.33
2	3	3					38	6.33
3	4	5	0				48	5.33
4[a]	3	4	5	0			67	5.58
4[a]	4	5	0	3			67	5.58
5	5	5	0	5	0		79	5.27
6	4	5	0	4	5	0	96	5.33

[a]Given a 4-period horizon, there are two equally optimal production patterns.

[5]In fact, most of the theoretical literature concerns infinite horizon dynamic programs, and gives insight into solution techniques when the horizon is far away.

PROBABILISTIC DYNAMIC PROGRAMMING

So far we have not considered any uncertainty in the problems studied. Usually in practice, however, the state of the system at the next stage is not completely determined by the state and policy decision at the current stage. Instead, we might think of assessing a probability distribution for what the next state will be. This probability distribution is completely determined by the state and policy decision at the current state. The basic structure is described by the diagram below. The solution methodology for this type of problem is essentially the same as that used for deterministic dynamic programming. In this case, however, the objective function is an expected value.[6]

Again, probabilistic dynamic programming allows more efficient solution than a decision tree approach whenever there is a repeating pattern of decisions and events. The analysis of a probabilistic decision problem by means of the decision tree solution procedure requires the determination of the end point values, quantifying every outcome of each strategy. One then folds back the tree by calculating the expected value of each full strategy and comparing one to the other. Dynamic programming takes into account the fact that in certain decision problems, *one does not need to know what the previous decisions were to make the next one. Knowing the current state is enough.*

Example

A firm has received an order to supply one unit of a complex electronic component that is particularly difficult to manufacture. There is only a ⅔ probability that an item produced will be acceptable. The manufacturer will have time for at most three production runs. The set-up cost is $500 with a variable cost of $200/unit. Management has assessed a penalty cost of $3,000 for not completing the sale. The objective is to determine the batch size on each of the production runs that will minimize the total expected cost. The stages are the production runs. The decision variables are the production lot size x_n on production run $n(n = 1, 2, 3)$. The state of the system can be described by the

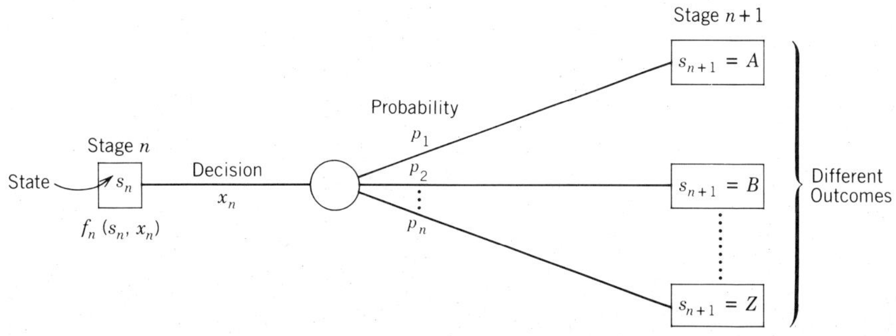

[6]For risky decisions, nonlinear utility may be used instead of expected value, but we will ignore such complications.

number of acceptable items still needed. Thus $s_1 = 1$ at stage 1 and is equal to 0 or 1 subsequently. We have to compute $f_n(s, x_n)$, the total expected cost for stage n onward, given s and x_n, and, from that, $f_n^*(s)$.

First, if at any time, the state is 0, then an acceptable item has been obtained, the state will remain 0 from there onward, and there will be no further cost incurred. Thus $f_n^*(0) = 0$ for any stage n. On the other hand, $f_n(1, x_n)$ is equal to the cost of the production run itself plus the expected value associated with the next state:

$$\begin{aligned} f_n(1, x_n) &= \text{cost of producing } x_n \\ &\quad + f_{n+1}^*(1) \cdot \text{Prob}\{\text{no acceptable items given } x_n \text{ went into production}\} \\ &\quad + f_{n+1}^*(0) \cdot \text{Prob}\{\text{at least one acceptable item was produced given } x_n \text{ went into production}\} \\ &= \begin{Bmatrix} 0 & \text{if } x_n = 0 \\ 500 + 200x_n & \text{if } x_n \neq 0 \end{Bmatrix} + \left(\frac{1}{3}\right)^{x_n} \cdot f_{n+1}^*(1) \end{aligned}$$

since $f_{n+1}^*(0) = 0$. Hence, the recursive relation is

$$f_n^*(1) = \min_{x_n} \left[\begin{Bmatrix} 0 & \text{if } x_n = 0 \\ 500 + 200x_n & \text{if } x_n \neq 0 \end{Bmatrix} + \left(\frac{1}{3}\right)^{x_n} f_{n+1}^*(1) \right]$$

The state transformation relationship is embodied in Figure 9-6.

Recognizing that $f_4^*(1) = \$3000$ (if before a fourth production run, we still have no acceptable output, we incur the penalty cost), we can now go through the calculations. Assume for simplicity that at most three items may be put into production.

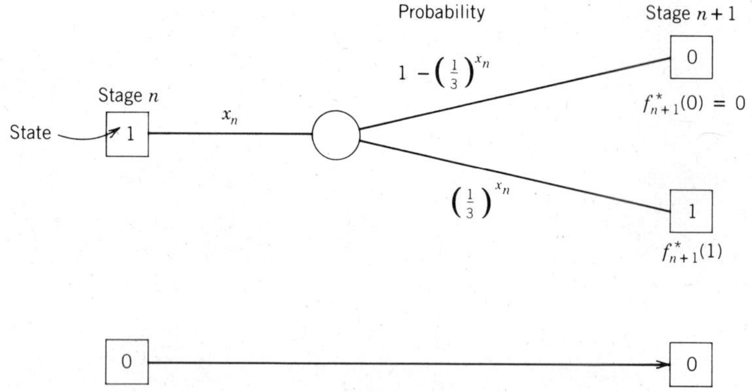

FIGURE 9-6. Probabilistic state transformation.

Stage 3

$$f_3(1, x_3) = \text{production cost} + \$3000\left(\frac{1}{3}\right)^{x_3}$$

	x_3					
s	0	1	2	3	$f_3^*(1)$	x_3^*
1	3000	1700	1233	1211	1211	3

Stage 2

$$f_2(1, x_2) = \text{production cost} + \$1211\left(\frac{1}{3}\right)^{x_2}$$

	x_2					
s	0	1	2	3	$f_2^*(1)$	x_3^*
1	1211	1103	1035	1145	1035	2

Stage 1

$$f_1(1, x_1) = \text{production cost} + \$1035\left(\frac{1}{3}\right)^{x_1}$$

	x_1					
s	0	1	2	3	$f_1^*(1)$	x_1^*
1	1035	1045	1015	1138	1015	2

Thus the optimal policy is to produce 2 items on the first run, then 2 again if none of the first batch are acceptable and then 3 if one is still needed. The total expected cost of this policy is $1,015.

CONCLUSION

As the examples in this chapter show, dynamic programming can be of tremendous computational value when making a sequence of interrelated decisions, especially in contrast to the enumeration of all possibilities, and especially when uncertainty is introduced. For example, if a problem has 20 stages, 10 possible decisions and 5 possible outcomes for each decision, a network must consider 50^{20} possible paths, whereas dynamic programming need make no more than 10 × 50 × 20 calculations (50 for each of 10 states at each stage). For a computer solution, this poses no tremendous difficulty. Where before it might have been difficult to integrate a great number of individual actions in a coherent fashion, a dynamic programming model allows us to do so.

The key to success with dynamic programming depends on skill in formulating the state variables. Notice that the examples of this chapter have *one-dimensional* state variables. Since the state variables have to completely specify the condition of the system, this is not always possible. In fact, two-dimensional state variables (e.g., in inventory planning problems with two products characterized by shared costs) are not uncommon. Unfortunately, solving a dynamic program requires the consideration of *all possible combinations* of values of the state variables, so that as the dimensionality of the state variables increases above two or three, the combinations that have to be considered (possibly as many as the product of the number of values of each of the state variables) skyrocket. This so-called *curse of dimensionality* limits the computational effectiveness of dynamic programming.

EXERCISES

1. The text describes a decision problem facing a firm that produces a particularly hard-to-produce electronic component. The production manager of that firm has discovered that there is time for *four* production runs rather than only the three runs considered previously. What is the optimal production strategy (i.e., how many items are to be produced on each run)? If the production manager could schedule as many production runs as desired, how many would he or she schedule? Try to be efficient with your computations.

2. In Chapter 5, the equipment replacement problem was formulated as a network. The example was:

Consider a machine shop that uses an expensive extrusion press which must be maintained and replaced periodically as it wears out. The purchase of a new model has been authorized, and the production foreman has been asked to determine the optimal replacement plan for the next 6 years after which the press will no longer be needed. The necessary data are given below:

k	Purchase Price at the Beginning of Year k	Salvage Value at End of k Years of Use	Operating Cost of Equip. During kth Year of Use
1	100	50	30
2	110	25	40
3	105	10	45
4	90	5	50
5	90	2	70
6	90	0	100

For example, if the press were purchased at the beginning of year 1, used for 3 years, and another purchase made at the beginning of year 4, the total cost would be the discounted sum of the costs shown below:

Beginning of Year:	1	2	3	4	5	6	7
Purchase:	100			90			
Salvage:				−10			−10
Operation:	30	40	45	30	40	45	
Total Annual Cost:	130	40	45	110	40	45	−10

This is a sequential decision problem: At the beginning of each year, the production foreman must decide to either replace the press or not. Thus, the problem can be solved effectively by dynamic programming. Do so.

3. In Chapter 6, we introduced the knapsack problem with the following example:

Suppose a planner has five possible expenditures under consideration as cost-saving projects, and has calculated the Net Present Value of each at the company's hurdle rate. A budget of $100,000 has been allocated for the cost reduction program. Given below are the necessary figures.

Projects	NPV	Cost	NPV/$ invested
1	$ 3,000	$10,000	.30
2	$18,000	$50,000	.36
3	$10,000	$40,000	.25
4	$23,000	$60,000	.38
5	$16,000	$50,000	.32

Dynamic programming is one of the most computationally efficient ways of solving a problem with this special structure (0−1 variables, one constraint). Use dynamic programming to solve the above example.

4. A firm is planning the weekly production at one of its chemical plants. The series of processes is as shown below:

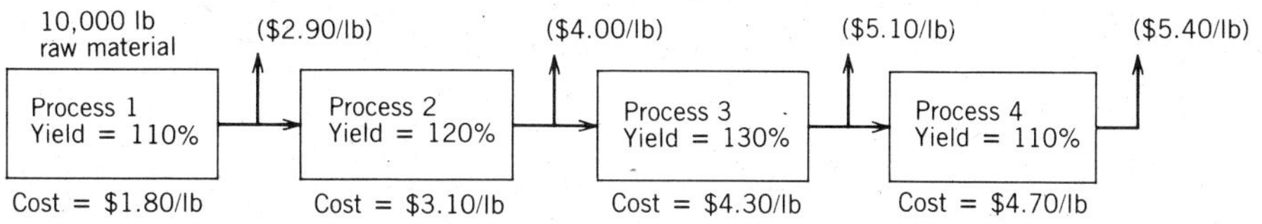

As is illustrated, each of the intermediates can be sold to the outside market (at the price shown in the diagram) as well as the final product. The variable cost (per *output* pound) of operating each process is also given. Ten thousand pounds of raw material are input (during the week) to process 1, and since that process has a 110% yield, 11,000 pounds of the first intermediate are produced. The yields of the other processes are given in the above diagram.

Formulate and solve a dynamic programming model to determine how these processes should be operated to maximize weekly contribution. Speculate on the solution to the general problem with N processes.

5. A firm has received an order for ten special-design diesel engines. In order to meet the promised delivery date, these ten must be produced within the next three weeks. In view of the firm's prior commitments as well as storage, labor, and equipment costs and limitations, the total production cost for these engines is $300p_1^2 + 200p_2^2 + 100p_3^2$, where p_i represents the production in week i, $i = 1, 2, 3$.

a. Assuming that the p_i can only take on integer values, find the production schedule which minimizes costs.

b. (Requires calculus) A noninteger p_i *can be thought of as a production rate.* For example, $p_1 = 4\frac{1}{2}$ means production is set at a rate that would produce 9 units in two weeks, but only run for one week. At the end of that week, one engine is half completed. In view of this interpretation, find the optimal production schedule.

6. A rapidly growing firm is concerned with workforce planning for a new plant now in construction and soon to open. The workforce would grow over the first 3 years of the plant's life and then stabilize at a lower level in year 4 when another plant will be opened in a neighboring state. In particular, the firm's planners estimate the workforce requirements to be:

Year 1	200
Year 2	600
Year 3	1000
Year 4	400

The labor force in the surrounding area is sufficiently large that there is no difficulty hiring workers. Company policy guarantees employment to its "permanent" workers. When necessary, the firm will hire "temporaries" who are automatically let go at the end of the year. The problem is that training costs are quite high and thus hiring permanent workers is more cost-efficient. Unfortunately, these workers cannot be let go.

To be specific, each worker's annual salary is \$15,000. Training costs, including lost productivity are \$5000 per worker. Assuming that workers are hired only at the beginning of each year, use dynamic programming to determine how many permanent and temporary workers to hire in each of the first 4 years.

7. Imagine that you have \$1000 to invest and that you will have the opportunity to invest that amount in either of two investments (A or B) at the beginning of each of the next 4 years. Both investments have uncertain returns. For investment A you will either lose your money entirely or (with higher probability) get \$2000 back (a profit of \$1000) at the end of the year. For investment B you will either get just your \$1000 back or (with low probability) \$2000 back at the end of the year. The probabilities for these events are:

Investment	Amount Returned (\$)	Probability
A	0	0.4
	2,000	0.6
B	1,000	0.9
	2,000	0.1

You are only allowed to make (at most) *one* investment each year and can only invest \$1000 each time. Of course, if you lose all your money, you cannot invest in either A or B. Any additional money accumulated is left idle during the period. As long as you have at least \$1000, you can keep investing. Use dynamic programming to find the investment policy that maximizes the expected amount of money you will have at the end of 4 years.

8. Resolve the production planning problem solved as an example in the text taking demand to be uncertain—3 units with probability ½; 4 units with probability ½—rather than 3 units with certainty.

9. A retailer experiences demand for 1000 units of an item per month. Two weeks before the start of each month, he orders the item from either or both of two manufacturers. Unfortunately, these manufacturers have often been unreliable in the past, and if the order is not filled, the retailer loses the \$20 contribution per item, which he would normally experience. Past indications are, however, that both manufacturers become more reliable as the total volume of past orders from the retailer increases. For example, if before month 6, the retailer has ordered all 5000 of his past volume from manufacturer A, then an order placed with that manufacturer in month 6 will have a far greater probability of being filled than if the retailer had only ordered 2500 items during the past five months. The rationale is that these manufacturers "try harder" or give special preference to proven good customers. In particular, for manufacturer A, the probability that an order is filled is

$$1 - \frac{1}{2}e^{-v_A/1000}$$

where v_A is the cumulative past volume ordered from A. For manufacturer B, the probability that an order is filled is

$$1 - \frac{1}{3}e^{-v_B/2000}$$

where v_B is the cumulative past volume order from B. Initially, both v_A and v_B are zero; that is, B is more reliable initially (the probability that B will fill an order for month 1 is ⅔ to A's ½), but A becomes reliable (with volume) more quickly. (See the following illustration.) Neither manufacturer will ship a partial order. Either an order is filled or not. To hedge his bets, the retailer can order from both manufacturers, even order a full month's demand from each.

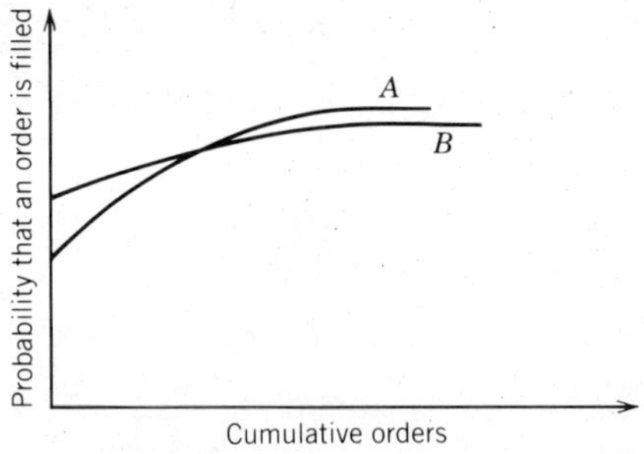

Determine the optimal ordering policy for the retailer for the next six months, assuming that the retailer loses \$10 per unit for any items left over after six months.

10. Any linear program can be written as

$$\text{Max} \quad \sum c_j x_j$$

$$\text{Subject to} \quad \sum_j a_{ij} x_j \leq b_i \quad \text{for } i = 1, 2, \ldots, m$$

$$x_j \geq 0 \quad \text{for } j = 1, 2, \ldots, n$$

Although solution of this problem by dynamic programming would be computationally inefficient, it can be done. Formulate the dynamic program to do so.

The Yankee Telephone Company

The Yankee Telephone Company is an independent telephone company operating in northern New England along the Massachusetts-New Hampshire border. Like all telephone companies, Yankee Telephone has to anticipate far in advance patterns of population growth which will affect the need for "outside plant," that is, the network of cables through which telephone service is provided. This planning problem is particularly acute for Yankee Telephone, which had been serving largely rural areas, and now anticipates a substantial migration of professional people working and around Boston (and for whom Boston is the cultural center) to southern New Hampshire over the next thirty years. A variety of circumstances conspire to bring about this spreading of the Boston suburbs to New Hampshire: the quality of roads leading north has improved markedly, Massachusetts state and local taxes have become increasingly heavy for the upper income brackets, the popularity of winter sports continues to grow, and the development of industry along the "northern" section of Route 128 is expected to accelerate.

As a result, the Yankee Telephone Company forecasts an exceptional rate of growth of population in the 25-square mile area northwest of Methuen, Mass. and east of Nashua, N.H. (see Exhibit 1). Although at present essentially undeveloped, this area is expected to contain by the end of the year 2000 some 8000 dwelling units, all requiring telephone service. Although some of these dwelling units will require multiple telephone numbers, these are expected to be roughly equal in number to the dwelling units requiring two-party lines, so that it is approximately correct to infer that demand for telephone service for this area will require by the end of the year 2000 approximately 8000 pairs.[1] The area shown in Exhibit 1 is at present being serviced by rural telephone service from Nashua, N.H. The new pairs for the suburban developments, however, should be brought to the area from the Methuen central office since calling patterns are expected to be predominantly southward. As a result, four feeder cables are planned, each of 2000 pairs, to serve four equi-population sections as shown in Exhibit 1.

EXHIBIT 1
Map Showing Feeder Cables to the Four Rapidly Growing Sections

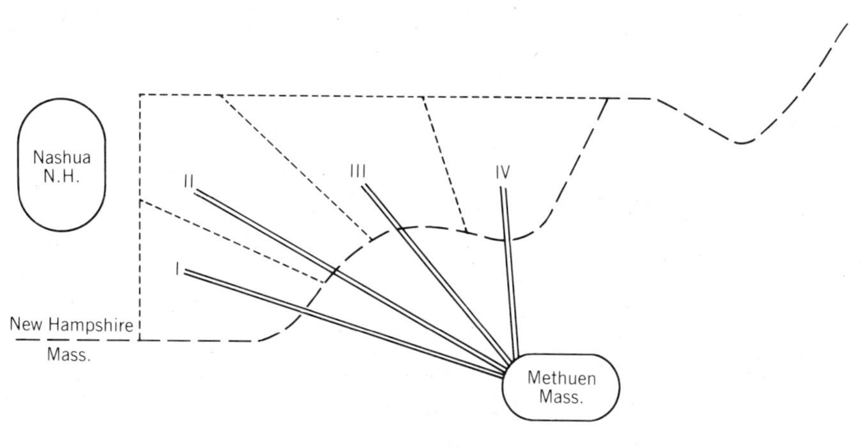

[1] A "pair" denotes in telephone parlance a pair of copper wires running from the central office (the location of a telephone exchange) to the customer. A "pair" can service a single one-party customer or two two-party customers.

The demand for telephone service will, of course, grow gradually as the region is developed. A forecast of demand for a typical section (Section 1) is shown in Exhibit 2. Demand is expected to grow starting in January 1971, by 40 pairs per year for the first 5 years. The growth rate will then increase to 80 pairs per year during a 20-year development period lasting through the year 1995. In the 5 years from 1996 through 2000, the growth rate is expected to drop again to 40 pairs per year as the region's development slows down, and by the beginning of the year 2001, the region is expected to be mature and exhibit no further growth. A numerical tabulation of these demand forecasts is given in Exhibit 3.

Since the area contains considerable natural resources, conservationist groups have lobbied hard and successfully to prevent above-ground cables or

EXHIBIT 2
Forecast of Demand for Section I

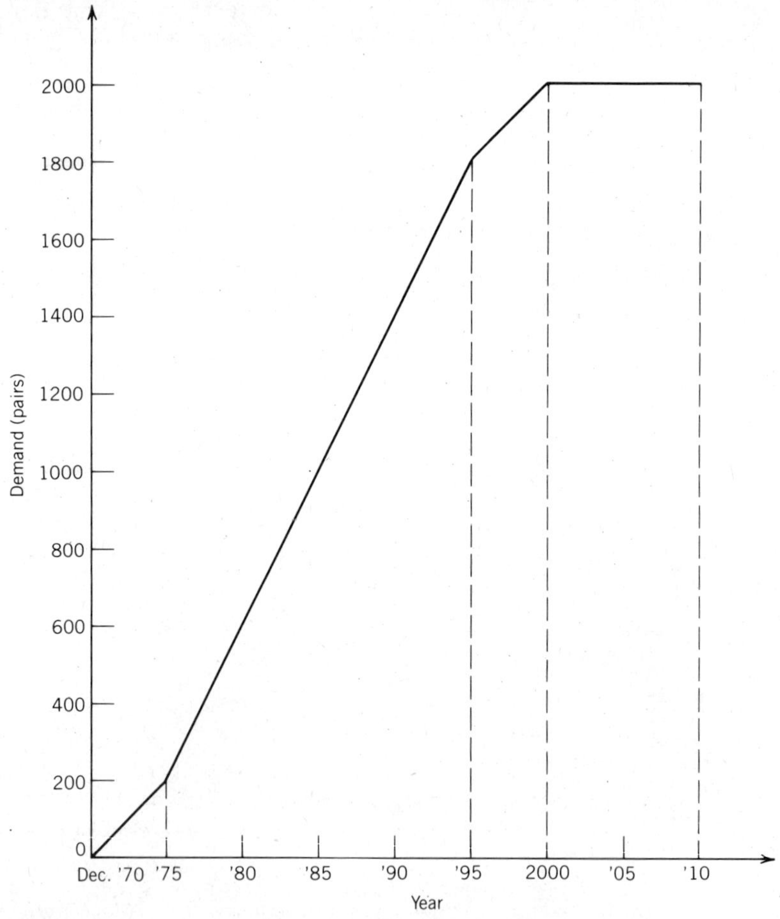

EXHIBIT 3
Section I Demand and Feeder Cable Costs as a Function of Time

Time	Demand (pairs)	Cost of Laying Cable During Preceding Year: Fixed Cost (per installment)	Cost of Laying Cable During Preceding Year: Variable Cost (per 100 pairs)
Dec. 31, 1970	0	$ 36,000	$ 40,000
Dec. 31, 1971	40	37,080	41,200
Dec. 31, 1972	80	38,192	42,436
Dec. 31, 1973	120	39,338	43,709
Dec. 31, 1974	160	40,518	45,020
Dec. 31, 1975	200	41,734	46,371
Dec. 31, 1976	280	42,926	47,762
Dec. 31, 1977	360	44,275	49,195
Dec. 31, 1978	440	45,604	50,671
Dec. 31, 1979	520	46,972	52,191
Dec. 31, 1980	600	48,381	53,757
Dec. 31, 1981	680	49,832	55,369
Dec. 31, 1982	760	51,327	57,030
Dec. 31, 1983	840	52,867	58,741
Dec. 31, 1984	920	54,453	60,504
Dec. 31, 1985	1000	56,087	62,319
Dec. 31, 1986	1080	57,769	64,188
Dec. 31, 1987	1160	59,503	66,114
Dec. 31, 1988	1240	61,288	68,097
Dec. 31, 1989	1320	63,126	70,140
Dec. 31, 1990	1400	65,020	72,244
Dec. 31, 1991	1480	66,971	74,412
Dec. 31, 1992	1560	68,980	76,644
Dec. 31, 1993	1640	71,049	78,943
Dec. 31, 1994	1720	73,181	81,312
Dec. 31, 1995	1800	75,376	83,751
Dec. 31, 1996	1840	77,637	86,264
Dec. 31, 1997	1880	79,966	88,852
Dec. 31, 1998	1920	82,365	91,517
Dec. 31, 1999	1960	84,836	94,263
Dec. 31, 2000	2000	87,381	97,090

overhead lines from being constructed and marring the landscape. As a result, Yankee Telephone has been forced to lay feeder cables in underground ducts. The cost of laying such ducts is almost independent of the duct capacity, so that ducts adequate for 2000 pairs should be provided from the outset. There is, however, considerable question whether all 2000 pairs should be installed at once. The ducts have to be easily accessible for possible repair work, so that the phased introduction of feeder cable is perfectly feasible. Feeder cable is produced in multiples of 100 pairs; in other words, cables containing integral multiples of 100 pairs are available ranging from 100 to 2000 pairs. There are no price discounts for the purchase of larger cables, since Yankee Telephone already purchases cable in maximum purchase quantities.

The following are some data relevant to the decision of how large a cable should initially be laid to Section 1. All costs are based on the work being done in 1970. After 1970, an inflation rate of 3 percent per year is expected both for materials and labor, and this rate of inflation is expected to continue unabated until the year 2000. To lay a cable through existing ducts from the central office to Section 1 would cost \$36,000, excluding the cost of the cable and other variable costs. This cost is irrespective of the number of pairs being laid and applies to the first slug of feeder cable, as well as to later slugs if the cable is laid in successive installments. The cost of the feeder cable itself and the cost of splicing and testing the feeder cable (i.e., the variable costs) amounts to \$400

EXHIBIT 4

Cost of Materials and Labor Required to Lay an Installment of Cable to Section I

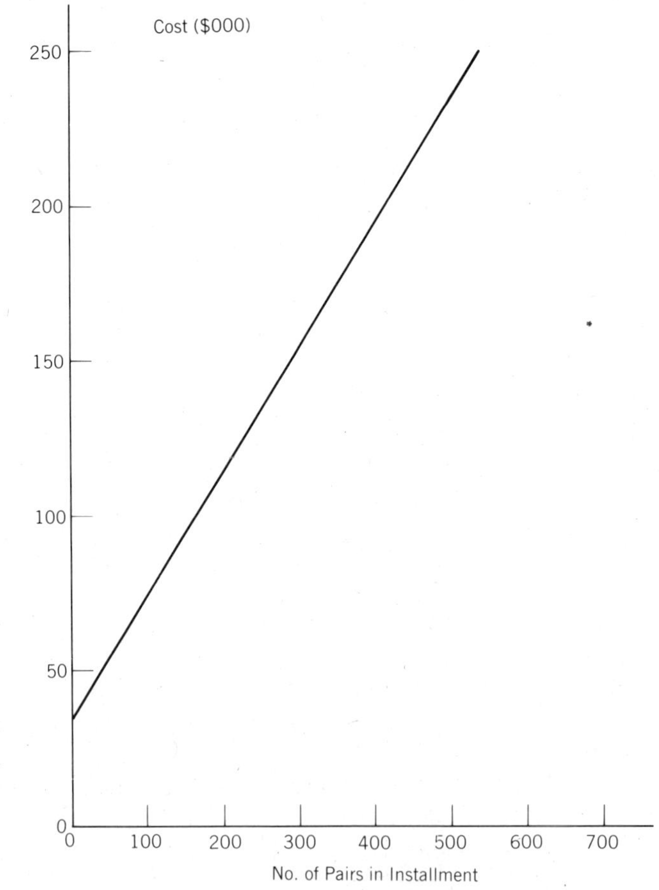

per pair. These two kinds of cost are the only costs involved in adding an installment of cable, so that the cost of adding an installment as a function of the number of pairs added is as shown in Exhibit 4. The inflationary increase in these costs is given in Exhibit 3.

The useful life of this type of feeder cable is almost indefinite. Virtually all the feeder cable laid in the United States during the past hundred years is still in use. For tax purposes, however, feeder cable is depreciated over 30 years using a double-declining balance method of depreciation. The Yankee Telephone Company is allowed by the FCC a rate structure which provides for fair earnings on the invested capital. In practice, the FCC has generally approved rates that allowed a 9 percent return on the book value of investments after income taxes but before interest payments. The company maintained a debt/equity ratio of 3 to 2 and intended to maintain such a ratio for the foreseeable future. Straight debt financing costs were forecast for the decades ahead to average around 7 percent, and the company used a hurdle rate of 18 percent (before taxes) to analyze the desirability of investment projects. Historically, this rate had cleared roughly the right number of projects to consume available capital without requiring an undue rate of new financing. The company expected an 18 percent hurdle rate before taxes to continue to be appropriate in the coming years.

Although it was theoretically possible to provide in a given year fewer pairs than were demanded, the resulting poor quality of customer service and customer complaints, some of which would surely go to the FCC, would make it very difficult for Yankee Telephone to obtain necessary rate increases and general cooperation from the FCC. Because of this, and because of the relatively long lead times required in ordering and laying cable, Yankee Telephone had a policy that all the cable required for a particular year would have to be fully installed by the beginning of that year.

APPENDIX **Consultant's Report**

In the hope that a mathematical approach would help clarify their problem, Yankee Telephone engaged early in 1970 an MIT Operations Research graduate student as a consultant. The consultant spent a day discussing the problem with the Plant Planning Section and went back to his MIT computer console. A few weeks later Yankee Telephone received the following cryptic report, which they found unintelligible.

PHASED FEEDER CABLE INSTALLATION

The proper solution of the phased feeder cable installation problem may be achieved trivially through the use of dynamic programming techniques. The approach is as follows.

Let $V(N,t)$ denote the present discounted value of all future costs up to the horizon time (Dec. 2000) *if* at time t we have installed N 100-pair subcables and *if* from time t on we follow the optimal policy. We know that $V(N, 2000)$ is infinite unless $N = 20$, in which case $V(20,2000) = 0$.

Suppose we know $V(N, t)$ for a particular t and for $N = 0, 1, \ldots, 20$; then we can obtain $V(N, t\text{-}1)$ by noting that:

1. $V(N,t\text{-}1) = \infty$ if $100N$ pairs is insufficient to meet demand at time (t-1).
2. If $100N$ pairs is sufficient to meet demand at time (t-1), then $V(N, t\text{-}1)$ is the smallest of:

$$\begin{cases} D\cdot V(N,t) & \text{(add nothing)} \\ D\cdot V(N+1,t) + F(t-1) + C(t-1) & \text{(add one 100-pair subcable)} \\ D\cdot V(N+2,t) + F(t-1) + 2C(t-1) & \text{(add two subcables)} \\ \quad\vdots & \quad\vdots \\ D\cdot V(20,t) + F(t-1) + (20-N)C(t-1) & \text{(add } (20-N) \text{ subcables)} \end{cases}$$

where:

$$D = \text{discount factor} = 1/1.18$$

$$F(t-1) = \text{fixed cost of installation in year } (t-1) \text{ (i.e., \$36,000 in 1970)}$$

$$C(t-1) = \text{variable cost of installation per subcable (i.e., \$40,000 in 1970)}$$

The result of this calculation for the data appropriate to Section I is attached. (See Exhibit 5.) Note that an infinite value of $V(N, t)$ is denoted by 10^{10}, and always shows a "dummy" increment of 9900, which has no further meaning.

EXHIBIT 5

PAIRS	P.V. COSTS	INCREMENT	P.V. COSTS	INCREMENT	P.V. COSTS	INCREMENT
YEAR:	1970		1971		1972	
0	307948.	200	10000000000.	9900	10000000000.	9900
100	249354.	0	294828.	0	347897.	100
200	191948.	0	226499.	0	267268.	0
300	161630.	0	190723.	0	225054.	0
400	137415.	0	162150.	0	191337.	0
500	116359.	0	137304.	0	162018.	0
600	92629.	0	109302.	0	128977.	0
700	77266.	0	91174.	0	107585.	0
800	64996.	0	76695.	0	90500.	0
900	54326.	0	64104.	0	75643.	0
1000	42301.	0	49915.	0	58900.	0
1100	34516.	0	40729.	0	48060.	0
1200	28293.	0	33392.	0	39403.	0
1300	22892.	0	27012.	0	31875.	0
1400	16799.	0	19822.	0	23390.	0
1500	12854.	0	15167.	0	17897.	0
1600	9703.	0	11450.	0	13510.	0
1700	6841.	0	8072.	0	9525.	0
1800	3876.	0	4573.	0	5396.	0
1900	1935.	0	2283.	0	2694.	0
2000	0	0	0	0	0	
YEAR:	1973		1974		1975	
0	10000000000.	9900	10000000000.	9900	10000000000.	9900
100	10000000000.	9900	10000000000.	9900	10000000000.	9900
200	315377.	0	372144.	0	439130.	400
300	265563.	0	313365.	0	369770.	0
400	225778.	0	266417.	0	314373.	0
500	191182.	0	225594.	0	266201.	0
600	152192.	0	179587.	0	211913.	0
700	126950.	0	149802.	0	176766.	0
800	106790.	0	126012.	0	148694.	0
900	89259.	0	105326.	0	124284.	0
1000	69502.	0	82012.	0	96775.	0
1100	56711.	0	66919.	0	78965.	0
1200	46495.	0	54864.	0	64740.	0
1300	37612.	0	44382.	0	52371.	0
1400	27600.	0	32569.	0	38431.	0
1500	21119.	0	24920.	0	29406.	0
1600	15942.	0	18812.	0	22198.	0
1700	11239.	0	13262.	0	15650.	0
1800	6368.	0	7514.	0	8866.	0
1900	3179.	0	3751.	0	4426.	0
2000	0	0	0	0	0	0

(continued)

EXHIBIT 5 (Continued)

PAIRS	P.V. COSTS	INCREMENT	P.V. COSTS	INCREMENT	P.V. COSTS	INCREMENT
YEAR:	1976		1977		1978	
0	10000000000.	9900	10000000000.	9900	10000000000.	9900
100	10000000000.	9900	10000000000.	9900	10000000000.	9900
200	10000000000.	9900	10000000000.	9900	10000000000.	9900
300	436329.	300	10000000000.	9900	10000000000.	9900
400	370960.	0	437732.	200	10000000000.	9900
500	314118.	0	370659.	0	437377.	200
600	250057.	0	295067.	0	348179.	0
700	208584.	0	246129.	0	290432.	0
800	175459.	0	207042.	0	244309.	0
900	146656.	0	173054.	0	204203.	0
1000	114194.	0	134749.	0	159004.	0
1100	93178.	0	109951.	0	129742.	0
1200	76393.	0	93144.	0	106370.	0
1300	61798.	0	72921.	0	86047.	0
1400	45348.	0	53511.	0	63143.	0
1500	34699.	0	40945.	0	48315.	0
1600	26194.	0	30909.	0	36472.	0
1700	18467.	0	21791.	0	25713.	0
1800	10462.	0	12346.	0	14568.	0
1900	5223.	0	6163.	0	7272.	0
2000	0	0	0	0	0	0
YEAR:	1979		1980		1881	
0	10000000000.	9900	10000000000.	9900	10000000000.	9900
100	10000000000.	9900	10000000000.	9900	10000000000.	9900
200	10000000000.	9900	10000000000.	9900	10000000000.	9900
300	10000000000.	9900	10000000000.	9900	10000000000.	9900
400	10000000000.	9900	10000000000.	9900	10000000000.	9900
500	10000000000.	9900	10000000000.	9900	10000000000.	9900
600	410851.	0	484805.	400	10000000000.	9900
700	342710.	0	404397.	0	477189.	300
800	288285.	0	340176.	0	401408.	0
900	240960.	0	284333.	0	335512.	0
1000	187625.	0	221397.	0	261248.	0
1100	153095.	0	180652.	0	213170.	0
1200	125517.	0	148110.	0	174769.	0
1300	101536.	0	119812.	0	141378.	0
1400	74509.	0	87921.	0	103746.	0
1500	57012.	0	67274.	0	79383.	0
1600	43037.	0	50784.	0	59925.	0
1700	30341.	0	35803.	0	42247.	0
1800	17190.	0	20284.	0	23935.	0
1900	8581.	0	10126.	0	11948.	0
2000	0	0	0	0	0	0

	PAIRS	P.V. COSTS	INCREMENT	P.V. COSTS	INCREMENT	P.V. COSTS	INCREMENT
YEAR:		1982		1983		1984	
	0	10000000000.	9900	10000000000.	9900	10000000000.	9900
	100	10000000000.	9900	10000000000.	9900	10000000000.	9900
	200	10000000000.	9900	10000000000.	9900	10000000000.	9900
	300	10000000000.	9900	10000000000.	9900	10000000000.	9900
	400	10000000000.	9900	10000000000.	9900	10000000000.	9900
	500	10000000000.	9900	10000000000.	9900	10000000000.	9900
	600	10000000000.	9900	10000000000.	9900	10000000000.	9900
	700	10000000000.	9900	10000000000.	9900	10000000000.	9900
	800	473661.	200	10000000000.	9900	10000000000.	9900
	900	395905.	0	467167.	200	10000000000.	9900
	1000	308273.	0	363762.	0	429240.	0
	1100	251540.	0	296818.	0	350245.	0
	1200	206228.	0	243349.	0	287152.	0
	1300	166826.	0	196855.	0	232289.	0
	1400	122421.	0	144456.	0	170459.	0
	1500	93672.	0	110533.	0	130430.	0
	1600	70711.	0	83439.	0	98458.	0
	1700	49852.	0	58825.	0	69413.	0
	1800	28244.	0	33328.	0	39327.	0
	1900	14099.	0	16637.	0	19631.	0
	2000	0	0	0	0	0	0
YEAR:		1985		1986		1987	
	0	10000000000.	9900	10000000000.	9900	10000000000.	9900
	100	10000000000.	9900	10000000000.	9900	10000000000.	9900
	200	10000000000.	9900	10000000000.	9900	10000000000.	9900
	300	10000000000.	9900	10000000000.	9900	10000000000.	9900
	400	10000000000.	9900	10000000000.	9900	10000000000.	9900
	500	10000000000.	9900	10000000000.	9900	10000000000.	9900
	600	10000000000.	9900	10000000000.	9900	10000000000.	9900
	700	10000000000.	9900	10000000000.	9900	10000000000.	9900
	800	10000000000.	9900	10000000000.	9900	10000000000.	9900
	900	10000000000.	9900	10000000000.	9900	10000000000.	9900
	1000	506503.	400	10000000000.	9900	10000000000.	9900
	1100	413289.	0	487681.	300	10000000000.	9900
	1200	338839.	0	399830.	0	471799.	200
	1300	274101.	0	323439.	0	381658.	0
	1400	201141.	0	237347.	0	284069.	0
	1500	153907.	0	181610.	0	214300.	0
	1600	116181.	0	137093.	0	161770.	0
	1700	81908.	0	96651.	0	114048.	0
	1800	46405.	0	54758.	0	64615.	0
	1900	23165.	0	27335.	0	32255.	0
	2000	0	0	0	0	0	0

(continued)

EXHIBIT 5 (Continued)

	PAIRS	P.V. COSTS	INCREMENT	P.V. COSTS	INCREMENT	P.V. COSTS	INCREMENT
YEAR:		1988		1989		1990	
	0	10000000000.	9900	10000000000.	9900	10000000000.	9900
	100	10000000000.	9900	10000000000.	9900	10000000000.	9900
	200	10000000000.	9900	10000000000.	9900	10000000000.	9900
	300	10000000000.	9900	10000000000.	9900	10000000000.	9=00
	400	10000000000.	9900	10000000000.	9900	10000000000.	9900
	500	10000000000.	9900	10000000000.	9900	10000000000.	9900
	600	10000000000.	9900	10000000000.	9900	10000000000.	9900
	700	10000000000.	9900	10000000000.	9900	10000000000.	9900
	800	10000000000.	9900	10000000000.	9900	10000000000.	9900
	900	10000000000.	9900	10000000000.	9900	10000000000.	9900
	1000	10000000000.	9900	10000000000.	9900	10000000000.	9900
	1100	10000000000.	9900	10000000000.	9900	10000000000.	9900
	1200	10000000000.	9900	10000000000.	9900	10000000000.	9900
	1300	450356.	200	10000000000.	9900	10000000000.	9900
	1400	330481.	0	389968.	0	460162.	400
	1500	252874.	0	298391.	0	352102.	0
	1600	190889.	0	225249.	0	265794.	0
	1700	134577.	0	158801.	0	187385.	0
	1800	76245.	0	89970.	0	106164.	0
	1900	38061.	0	44912.	0	52996.	0
	2000	0	0	0	0	0	0
YEAR:		1991		1992		1993	
	0	10000000000.	9900	10000000000.	9900	10000000000.	9900
	100	10000000000.	9900	10000000000.	9900	10000000000.	9900
	200	10000000000.	9900	10000000000.	9900	10000000000.	9900
	300	10000000000.	9900	10000000000.	9900	10000000000.	9900
	400	10000000000.	9900	10000000000.	9900	10000000000.	9900
	500	10000000000.	9900	10000000000.	9900	10000000000.	9900
	600	10000000000.	9900	10000000000.	9900	10000000000.	9900
	700	10000000000.	9900	10000000000.	9900	10000000000.	9900
	800	10000000000.	9900	10000000000.	9900	10000000000.	9900
	900	10000000000.	9900	10000000000.	9900	10000000000.	9900
	1000	10000000000.	9900	10000000000.	9900	10000000000.	9900
	1100	10000000000.	9900	10000000000.	9900	10000000000.	9900
	1200	10000000000.	9900	10000000000.	9900	10000000000.	9900
	1300	10000000000.	9900	10000000000.	9900	10000000000.	9900
	1400	10000000000.	9900	10000000000.	9900	10000000000.	9900
	1500	415480.	300	10000000000.	9900	10000000000.	9900
	1600	313637.	0	370091.	200	10000000000.	9900
	1700	221114.	0	260915.	0	307879.	300
	1800	125274.	0	147823.	0	174431.	0
	1900	62535.	0	73792.	0	87074.	0
	2000	0	0	0	0	0	0

PAIRS	P.V. COSTS	INCREMENT	P.V. COSTS	INCREMENT	P.V. COSTS	INCREMENT
YEAR:	1994		1995		1996	
0	10000000000.	9900	10000000000.	9900	10000000000.	9900
100	10000000000.	9900	10000000000.	9900	10000000000.	9900
200	10000000000.	9900	10000000000.	9900	10000000000.	9900
300	10000000000.	9900	10000000000.	9900	10000000000.	9900
400	10000000000.	9900	10000000000.	9900	10000000000.	9900
500	10000000000.	9900	10000000000.	9900	10000000000.	9900
600	10000000000.	9900	10000000000.	9900	10000000000.	9900
700	10000000000.	9900	10000000000.	9900	10000000000.	9900
800	10000000000.	9900	10000000000.	9900	10000000000.	9900
900	10000000000.	9900	10000000000.	9900	10000000000.	9900
1000	10000000000.	9900	10000000000.	9900	10000000000.	9900
1100	10000000000.	9900	10000000000.	9900	10000000000.	9900
1200	10000000000.	9900	10000000000.	9900	10000000000.	9900
1300	10000000000.	9900	10000000000.	9900	10000000000.	9900
1400	10000000000.	9900	10000000000.	9900	10000000000.	9900
1500	10000000000.	9900	10000000000.	9900	10000000000.	9900
1600	10000000000.	9900	10000000000.	9900	10000000000.	9900
1700	10000000000.	9900	10000000000.	9900	10000000000.	9900
1800	205829.	0	242878.	200	10000000000.	9900
1900	102748.	0	121242.		143466.	0
2000	0	0	0	0	0	0
YEAR:	1997		1998		1999	
0	10000000000.	9900	10000000000.	9900	10000000000.	9900
100	10000000000.	9900	10000000000.	9900	10000000000.	9900
200	10000000000.	9900	10000000000.	9900	10000000000.	9900
300	10000000000.	9900	10000000000.	9900	10000000000.	9900
400	10000000000.	9900	10000000000.	9900	10000000000.	9900
500	10000000000.	9900	10000000000.	9900	10000000000.	9900
600	10000000000.	9900	10000000000.	9900	10000000000.	9900
700	10000000000.	9900	10000000000.	9900	10000000000.	9900
800	10000000000.	9900	10000000000.	9900	10000000000.	9900
900	10000000000.	9900	10000000000.	9900	10000000000.	9900
1000	10000000000.	9900	10000000000.	9900	10000000000.	9900
1100	10000000000.	9900	10000000000.	9900	10000000000.	9900
1200	10000000000.	9900	10000000000.	9900	10000000000.	9900
1300	10000000000.	9900	10000000000.	9900	10000000000.	9900
1400	10000000000.	9900	10000000000.	9900	10000000000.	9900
1500	10000000000.	9900	10000000000.	9900	10000000000.	9900
1600	10000000000.	9900	10000000000.	9900	10000000000.	9900
1700	10000000000.	9900	10000000000.	9900	10000000000.	9900
1800	10000000000.	9900	10000000000.	9900	10000000000.	9900
1900	168818.	100	10000000000.	9900	10000000000.	9900
2000	0	0	0	0	0	0

WCVB-TV

Bill Mockbee, Director of Research and Sales Promotion for Station WCVB-TV of Boston, was studying the results of the latest Arbitron and Nielsen audience rating surveys. Based on his interpretation of these and other inputs, he, in conjunction with other members of the sales management team, would prepare the new rate card for WCVB-TV commercial time spots.

INDUSTRY BACKGROUND

The television industry is dominated by three major networks: the American Broadcasting Company (ABC), CBS, Inc., and National Broadcasting Company (NBC). 1978 network broadcasting revenues were \$1.5, \$1.3, and \$1.2 billion, respectively, in a \$10 billion dollar industry. The networks produce or purchase shows for broadcast for their own and affiliated stations.[1] Revenues come from selling commercial time spots on their nationally televised shows and from local shows carried by their wholly owned stations. The control over and sale of these spots has evolved considerably over the years.

During the early years of commercial television, network attention was directed towards developing affiliated stations and promoting the use of television by the populace. Less attention was paid toward programming. Many of the early shows were produced directly by the sponsors, as the radio shows had been. Sponsorship of a show generally ran an entire season, or, in the case of shows produced by the sponsor, for the "life" of the show. Shows such as "Lux Video Theatre" and "The Hallmark Hall of Fame" became household favorites. However, the result of sole sponsorship was limited access for advertisers to television. Only a relatively few companies could afford to sponsor an entire season. Also, the "all or nothing" sponsorship made it difficult to get broad exposure to a product. If a viewer regularly missed a particular show, he probably would not see the product advertised elsewhere on television.

Television spot sales underwent a major change in the mid-1950s. Seeking to gain more control over television programming and to increase revenues, networks turned their attention from developing stations to producing shows. As the networks became the dominant force in production, they were able to market shows to a variety of sponsors, providing greater flexibility for national

[1]Regulations of the Federal Communications Commission (FCC), the regulatory agency overseeing the television industry, restrict the number of stations a single company may own to a total of five. Affiliated stations are owned by companies other than the major networks and negotiate regular agreements with the networks to relay certain shows broadcast by the latter.

advertisers. To achieve broader coverage, companies could spread their advertising over a wider range of time slots. Specific audiences could be targeted more easily and for specific periods of time. Firms could now concentrate their advertising efforts to coincide with major promotional campaigns, product changes, and so on instead of spreading their efforts throughout an entire season. Additionally, the change opened up television advertising to smaller advertisers who could afford only a relatively few spots.

During the 1960s, the networks were faced with production cost increases averaging 9.7% annually. Revenue increases averaged 8.5% annually. To counter this trend, the networks reduced the proportion of income from network advertising paid to affiliated stations. This fostered a shift by affiliated stations toward more local programming and nationally syndicated shows[2] so they could obtain a greater share of the advertising income. These program sources offered a greater number of commercial spots to the non-network advertisers. The result was a growth in the amount of television advertising by local companies.

Commercial air time is normally sold in 30 second blocks. The number of blocks available per hour is currently restricted by agreement within the National Association of Broadcasters (NAB).[3] The number varies depending on whether a station is a network affiliate or an independent, whether the programming is considered "prime" time, "children's," or "other," and other factors. Pricing for the blocks vary based on audience demographics and psychographics. Demographics refers to certain measurable quantities such as the audience size, age, and sex. Psychographics refers to quantities more difficult to measure such as an audience's "life style" and its propensity to spend on certain groups of products. Two audiences of the same demographics could have different psychographics if one group was composed of lower middle class males while the other was upper middle class males. As Mockbee commented, "What we're really selling is people (*sic* audience demographics and psychographics), not air time."

Commercial television air time is sold in three markets—Network, National Spot, and Local Spot. Network time is sold by the parent networks to national advertisers for broadcast on shows produced and marketed by the networks. These shows generally include national news shows, "prime" time between 8:00 and 11:00 P.M., and nationally televised talk shows, soap operas, and so on. Typical messages of network advertising are those promoting general image ("Fly the friendly skies of United Airlines") and those promoting national sales ("United is back in service and offering 50% discount coupons"). Network affiliated stations receive a percentage of network advertising income and receive a specified number of 30-second spots to sell privately, generally

[2]Nationally syndicated shows include first-run shows produced and marketed by independent producers, such as Norman Lear's "Mary Hartman, Mary Hartman," and re-runs of former network or independent shows, such as "The Mary Tyler Moore Show."
[3]This agreement is currently being challenged in an antitrust suit filed against the NAB by the U.S. Department of Justice.

one or two per hour of network time. Income from the network for their shows amounts to 5%–10% of total station income.

Unlike network advertising, where a company broadcasts its message only through network affiliated stations, National Spot advertising allows an advertiser to select the best station in each market to reach its audience, regardless of network affiliation.[4] National Spot can be aired on all local programming and in the selected spots made available in network shows to the individual stations. Messages are generally the same as network advertising or tend to focus on regions. ("New England Air Lines offers 30% discounts on all flights.") This time is sold by individual stations, either directly to the advertiser or through an agency representing numerous stations. Agency commissions for this service are approximately 15%. Station income from National Spot amounts to 50%–65% of total income.

Local Spot advertising is sold by individual stations to local advertisers. Broadcast occurs in the same time frame as National Spot. Since National Spot and Local Spot are both competing for the same air time, information on available inventory must be updated constantly by all sales personnel. Spots are sold on a first-come, first-served basis. Typical messages include those of local business or of a national company with a local sale. ("XYZ Airline offers half fare on all flights from Boston to Miami June 1–30.") This time is sold directly through the stations and accounts for 30%–40% of total station revenues.

Within all three markets, commercial spots are sold in basically two different ways:

1. Individuals—spots are sold for specific shows, generally for multiple broadcasts.

2. Rotations, orbits, and packages—allows the stations to guarantee access to the most popular shows to more advertisers and helps the stations fill up time slots on less popular shows.

Spots for one-time broadcasts on specific shows can be purchased, but most stations prefer not to make such sales, especially for the more popular shows.

Spots are generally sold on a "fixed/preemptable" basis. Advertisers purchase spots according to a "ladder" of prices, sometimes called sections. For its fall 1979 season, WCVB-TV offered three possible sections to an advertiser. The lowest, Section 3, carried the lowest price for a given spot and offered no priority over other customers bidding for the same spot. Section 2 sold at a premium over a "3" and would "bump" a "3" on a given show if all other slots in that time frame had been filled. (That "3" customer would then have the option of paying more for a Section 1 to regain the old spot, switching to a spot

[4]This could include independents, as well, who are not affiliated with any networks. Their programming sources include non-network, nationally syndicated and locally produced shows. Approximately 20% of the U.S. television stations are independents.

EXHIBIT 1
Rate Card—Fall 1979

WCVB FALL RATE CARD eff 9-10-79

PROGRAM INFORMATION	DAY	TIME	I	II	III
EYEOPENER NEWS	MON/FRI	600- 700A	175	150	125
GOOD MORNING, AMERICA	MON/FRI	700- 900A	350	300	250
GOOD DAY, NEW ENGLAND	MON/FRI	900-1030A	275	225	175
AM ROTATION	MON/FRI	1030-1200N	400	350	300
NEWSCENTER 5 - MIDDAY	MON/FRI	1200-1230P	450	400	350
AFTERNOON ROTATION	MON/FRI	1230- 400P	600	550	500
RHODA/THE FAMILY FEUD	MON/FRI	400- 500P	450	400	350
CAROL BURNETT/ALL IN THE FAMILY	MON/FRI	500- 600P	1000	900	800
ALL IN THE FAMILY	MON/FRI	530- 600P	1400	1200	1000
NEWSCENTER 5 AT 6PM - 7 DAY ROTATION	MON/SUN	600- 730P	1100	1000	900
PARK STREET UNDER	MON	730- 800P	1100	900	700
THE MUPPETS	TUE	730- 800P	1700	1500	1300
THE BIG MONEY GAME	WED	730- 800P	1100	900	700
HOUSECALL/THIRD WORLD	THR	730- 800P	500	400	300
CALENDAR	FRI	730- 800P	900	800	700
LAWRENCE WELK	SAT	700- 800P	2200	2000	1800
THE BAXTERS	SUN	630- 700P	700	600	500
THE LATE NEWS - MONDAY FOLLOWS NFL FOOTBALL	MON	1200-1230A	400	350	300
NEWSCENTER 5 AT 11PM - 6 DAY ROTATION	TUE/SUN	1100-1130P	2800	2500	2200
ABC LATE NIGHT PROGRAMMING	TUE/FRI	1130- 105A	450	400	350
ABC LATE NIGHT PROGRAMMING	TUE/FRI	1130- 135A	400	350	300
FIVE ALL NIGHT - AS SCHEDULED	MON/FRI	135-CONCL	65	50	35
CHARLIE'S ANGELS	FRI	1130-1235A	700	600	500
GREAT ENTERTAINMENT	SAT/SUN	1130- 105A	500	450	400
GREAT ENTERTAINMENT	SAT/SUN	1130- 135A	450	400	350
FIVE ALL NIGHT - AS SCHEDULED	TUE/SAT	135-CONCL	65	50	35
NEWS FINAL (+OPEN HSE-R, SUNDAYS 135A)	MON/FRI	AS SCHEDUD	65	50	35
240 ROBERT	MON	800- 900P	2200	2000	1800
NFL MONDAY NIGHT FOOTBALL	MON	900-CONCL	3100	2800	2500
(PATRIOT'S FOOTBALL)	MON	900-CONCL	7000	6000	
HAPPY DAYS/NOBODY'S PERFECT/3'S CO./TAXI	TUE	800-1000P	5500	5000	4500
LAZARUS SYNDROME	TUE	1000-1100P	3500	3200	2900
EIGHT IS ENOUGH	WED	800- 900P	5500	5000	4500
CHARLIE'S ANGELS/VEGA$	WED	900-1100P	3900	3600	3300
LAVERNE & SHIRLEY/BENSON/BARNEY MILLER/SOAP	THR	800-1000P	5500	5000	4500
20/20 WITH HUGH DOWNS	THR	1000-1100P	2800	2500	2200
FANTASY ISLAND	FRI	800- 900P	4100	3800	3500
ABC FRI/SUN MOVIE (SPEC. MOVIES=SPEC. RATES)	FRI/SUN	900-CONCL	4000	3500	3000
THE ROPERS/ANGIE	SAT	800- 900P	3000	2700	2400
LOVE BOAT/HART TO HART	SAT	900-1100P	5000	4500	4000
OUT OF THE BLUE/NEW KIND OF FAMILY	SUN	700- 800P	2200	2000	1800
MORK AND MINDY	SUN	800- 830P	5500	5000	4500
JABBERWOCKY	SAT	630- 700A	250	200	150
KIDS ROTATION-SATURDAY	SAT	700-1200N	900	800	700
CANDLEPIN BOWLING/CANDLE. SUPER BOWL	SAT	1200- 130P	650	550	450
NCAA FOOTBALL	SAT	1230-CONCL	1900	1600	1300
WIDE WORLD OF SPORTS	SAT	500- 630P	1500	1300	1100
CAPTAIN BOB	SUN	630- 700A	200	150	100
KIDS ROTATION- SUNDAY	SUN	730- 930A	700	600	500
SUNDAY HOUSE/NEW HEAV./AQUI	SUN	930- 100P	100	75	50
ISSUES AND ANSWERS	SUN	1200-1230P	350	250	150
THIS WEEK	SUN	100- 130P	150	125	100
NEXT STEP BEYOND	SUN	130- 200P	200	175	150
GREAT ENTERTAINMENT MOVIE	SUN	200- 400P	300	250	200
JACQUES COUSTEAU/MUSICAL SPECIALS	SUN	400- 600P	500	400	300

COMBOS/PACKAGES - OVER

(continued)

NOTES, ETC..

NEWS COMBO - 30 SECONDS

EARLY NEWS (NEWSCENTER 5 AT 6 PM, 7 DAY ROTATION, MON/SUN 600-700PM)
1X $800
LATE NEWS (NEWSCENTER 5 AT 11PM, 6 DAY ROTATION, TUE/SUN 1100-1130PM)
1X 2200 TOTAL $3000

NEWS INSERTS

TUESDAY, WEDNESDAY, THURSDAY, SATURDAY...APPROXIMATELY 1030 PM
40 SECOND NEWS PROGRAM W/BILLBOARD AND EIGHT SECOND ID WITHIN:

1X	$2600
13X	2300@
26X	2000@
39X	1800@

NCAA/NFL FOOTBALL - 30 SECONDS

SATURDAY 1230PM-CONCLUSION, NCAA FOOTBALL (AS SCHEDULED)
1X
NFL MONDAY NIGHT FOOTBALL, 900PM-CONCLUSION (AS SCHEDULED)
1X TOTAL $3500

WIDE WORLD + MONDAY NIGHT FOOTBALL - 30 SECONDS

SATURDAY 500-630PM, WIDE WORLD OF SPORTS (OR AS SCHEDULED)
1X
NFL MONDAY NIGHT FOOTBALL, 900PM-CONCLUSION (AS SCHEDULED)
1X TOTAL $3500

NOTES: RATES SUBJECT TO CHANGE. PROGRAMMING SUBJECT TO CHANGE.
THIS RATE CARD IS ISSUED FOR THE USE OF AGENCIES, ADVERTISERS AND CLIENTS. IT IS NOT INTENDED TO BE AN OFFER OF FACILITIES. WCVB-TV, CHANNEL 5, OWNED AND OPERATED BY BOSTON BROADCASTERS, INC., 5 TV PLACE, NEEDHAM BR., BOSTON, MA. 02192. TELEPHONE 617-449-0400. TWX # 710-325-0773.

WCVB TV IS REPRESENTED NATIONALLY BY HARRINGTON, RIGHTER AND PARSONS, INC., NEW YORK AND OTHER PRINCIPAL CITIES. WCVB TV WILL TELECAST THE 1980 WINTER OLYMPICS FROM LAKE PLACID, NEW YORK. SPECIAL RATES APPLY.

WCVB TV IS AFFILIATED WITH THE ABC TELEVISION NETWORK. THE STATION OPERATES TWENTY FOUR HOURS A DAY (EXCEPT FOR ROUTINE MAINTENANCE FOR 4 HOURS WEEKLY).

America's Number One Television Station is right here in Boston.

with similar demographics/psychographics, or cancelling the order all together.) A Section 2 customer had to be given a two-telecast warning before being bumped by a "1," whereas a "3" could be preempted by a "1" or "2" at any time prior to broadcast. The only way an advertiser could be absolutely certain of broadcasting a message on a certain spot was to purchase a Section 1 spot. Exhibit 1 gives an example of WCVB-TV's rate card for the 1979 fall season. A minority of stations do not use a preemptable system, instead offering a single price for specific spots and guaranteeing broadcast. In regard to this policy, Mockbee commented "Any station that has a fixed only rate card may as well put up a sign saying 'going out of business;' they aren't controlling inventory, they're letting it control them."

Individual stations differ in their willingness to change prices as the air time of a given spot approaches. Prices are often set by the stations for a given period and do not change, *regardless of the number of unsold spots*. Changes to adapt to market demand are generally made only at the start of each new sales period. However, some stations recognize that an unsold spot is worthless after its broadcast time and attempt to adapt prices to reflect market demand. As broadcast time approaches, prices are lowered to foster sales. The incentive to an advertiser for early purchase at the highest price is the guarantee of broadcast in a specific time period.

WCVB-TV

WCVB-TV, an ABC affiliate station, operates in the greater Boston area. Owned and operated by Boston Broadcasters, Inc., it is based in Needham, Massachusetts, and serves the sixth largest television market in the United States. (See Exhibit 2.)

The genesis of this broadcasting enterprise was in 1963, when Boston Broadcasters, Inc. accepted an FCC invitation to apply for Boston's Channel 5 license. After a historic legal battle that reached the U.S. Supreme Court, the franchise was awarded to BBI, and WCVB-TV went on the air March 19, 1972.

Among the principles on which WCVB-TV was founded are quality locally produced programming, development of syndicated programs, round-the-clock operation, live telecasting, free political time, and, outside the realm of programming, local ownership and the integration of stockholders/management in the day-to-day operation of the station. It operates 24 hours daily except for two hours each on Sunday and Monday nights. The area of dominant influence, an ARB measure of market size, is depicted in Exhibit 2.

WCVB-TV is proud of its reputation as a company responsible to the public. A recent survey by the FCC listed WCVB-TV as the station broadcasting the most hours of local news, public affairs programming, and local programming. As part of its licensing agreement with the FCC, it broadcasts 50 public

EXHIBIT 2
Market Size

ARBITRON TELEVISION

Audience Estimates in the Arbitron Market of

Boston

▲ City of License of Satellite Station

The "Total Survey Area" of this market is shown in white on the accompanying map. Where appropriate, the "Area of Dominant Influence" is indicated by coarse cross-hatching and the Arbitron "Metro (or Home County) Rating Area" by fine cross-hatching.

Survey Period: May 2 - May 29, 1979

Survey Months:

OCT NOV JAN FEB MAR MAY JUL

This report is furnished for the exclusive use of network, advertiser, advertising agency, and film company clients, plus these subscribing stations—

WBZ WCVB WNAC WXNE

WSBK WLVI

Schedule of Survey Dates 1978-79

October	Sept. 27 - Oct. 24, 1978
November	Nov. 1 - Nov. 28, 1978
January	Jan. 3 - Jan. 30, 1979
February	Jan. 31 - Feb. 27, 1979
March	Feb. 28 - March 27, 1979
May	May 2 - May 29, 1979
July	July 11 - Aug. 7, 1979

Estimates of Households in Market

	TSA	Pct TV HH	ADI	Pct TV HH	Metro Rating Area	Pct TV HH
TOTAL HOUSEHOLDS	3,871,400		1,838,900		1,163,300	
TV HOUSEHOLDS	3,761,300	100	1,788,300	100	1,132,700	100
COLOR TV HH	3,115,400	83	1,452,700	81	906,600	80
MULTI-SET TVHH	1,981,400	53	942,900	53	644,600	57
CATV SUBSCRIBERS	580,400	15	185,700	10	60,900	5
UHF TV HH	3,600,400	96	1,722,000	96	1,099,900	97

service announcements daily.[5] The station has received over 200 international, national, regional, and local awards, including 30 regional Emmys (see Exhibit 3), two national Emmy nominations and numerous others.

[5]Because stations broadcast over "public" airwaves, Congress mandated that all stations must serve the public interest. Stations are required to renew their licenses every 3 years with the FCC. At that time, they must demonstrate how they have fulfilled this requirement. One method of doing so is to allocate a certain portion of their commercial time spots to public service announcements.

America's Number One Television Station
is right here in Boston.
NewsCenter 5
More hours of news than any other station in America.
Sunday Open House, Boston's only live Sunday morning public affairs program.
JERRY LEWIS TELETHON
New England's largest telethon effort, raising more than one million dollars each year for the Muscular Dystrophy Association.
The prestigious Peabody Award won for total station excellence.
America's pioneer medical program, *House Call* with Dr. Timothy Johnson.
WCVB-TV won 14 Emmys in the first annual N.E. competition.
GOOD DAY!
Good Day!
New England's most popular local morning show since 1974.
Heritage of Power, the first documentary on Edward M. Kennedy.
CPR
Aquí
The Baxters, Boston's own TV comedy about family and personal problems.
An ongoing effort to train 100,000 people in cardiopulmonary resuscitation.
Marc Holmes conducts a meaningful dialogue with New England's *Third World* communities.
Information, entertainment about and for Boston's Hispanic community.
The Federal Communications Commission has recently published its findings showing WCVB-TV Channel 5 is the number one station in America. In its recent survey of this country's more than six hundred network affiliated stations, Channel 5 ranked number one in total hours of news, number one in public affairs programs, and number one in local programming. More local programming to serve you better . . . promising, from our first broadcast in 1972 to do more, and living up to that promise every year since. Innovative programming . . . pioneering new forms to better fulfill the varied needs of the community and continually refining them. Ambitious programming . . . regularly producing specials rarely attempted by other local stations and presenting them in prime time. Quality programming . . . receiving more than two hundred awards to date and repeatedly winning some of the industry's most prestigious honors. Our commitment to bring you more quality programming is stronger than ever, because putting you first has made us number one in Boston . . . and number one in America.
WCVB-TV 5
setting a new standard

WCVB-TV originates a great deal of regular live telecasting in Boston. In addition to live remote inserts in newscasts and other programs, WCVB-TV schedules several daily and weekly series live and has produced local shows, such as "The Baxters," which are now syndicated and sold nationally. WCVB-TV's successes with local programming has had an impact on advertisers' perceptions of and willingness to sponsor "public affairs" shows. As Mockbee commented, "Because of this station's commitment to public affairs programs, we've brought other stations along and advertisers are willing to listen. The day of the advertiser turning turkey when you mention public affairs is gone."

SALES AND PRICING

Spot sales at WCVB-TV are sold on a short-term basis as opposed to the long-term commitments that had characterized early television. For example, by July 17, 1979, no Section 1 spots and few Section 2 spots had been sold for August. Most sales had been Section 3s and 4s (four sections were offered for the summer season). Exhibit 4 lists the number of unsold spots for several representative time slots at various times prior to broadcast date. Spot sales varied by individual show and were cyclical, with peaks in May and November, and typically slow sales during the summer season.

A typical media buyer will negotiate with all six Boston stations before making a purchase. The buyer communicates the specific needs of his client as well as the proposed budget to each of the stations, who then respond with competitive bids on packages to fit those needs. A period of negotiation follows, after which the final purchase is made. A buyer for a large client will typically split the budget among several stations rather than allowing a single station to carry all that client's advertising.

Except for those who have to ensure broadcast on certain dates, such as a retailer prior to a weekend sale or a toy manufacturer during the peak pre-Christmas season, most customers monitor the demand level for spot sales. They generally buy the lowest priority/priced sections which they feel will have a reasonable chance for broadcast. To this end, the station salespersons work with customers in recommending specific sections and monitoring demand up to broadcast time. If demand accelerates and preemptions appear probable, they may contact customers to see if they wish to move to a higher priority/priced section.

WCVB-TV has traditionally held firm on prices for its time slots, filling unused space with public service announcements and program promotions. Explaining this policy, Mockbee commented:

> *Selling spots requires a high level of trust and professional ethics between both parties. There is a lot of give and take. If an advertiser gets bumped by a higher priority, we try very hard to accommodate them, switching them to other similar shows. If there's an error in the commercial that's our fault, we'll make good by broadcasting the next spot free. You don't find that in all advertising media. We*

EXHIBIT 4
Selected Spot Sales, Summer of 1979[a]

	No. of 30-Second Spots Available	For Week/ Day of	Spots Remaining as of	
Weekdays 4:00–4:30 P.M.			June 25	July 13
CONCENTRATION	75	July 16	35	0
		July 23	34	1
		July 30	31	19
		Aug. 6	NA	19
Weekdays 5:30–6:00 P.M.			June 25	July 13
RHODA	40	July 16	28	5
		July 23	34	27
		July 30	37	29
		Aug. 6	NA	29
Tuesday, 8:00–10:00 P.M.			July 2	July 16
HAPPY DAYS/LAVERNE & SHIRLEY/THREE'S COMPANY/ TAXI	8	July 17	3	1
		July 24	7	5
		July 31	7	5
		Aug. 7	4	3
Weekly 11:00–11:30 P.M.			June 25	July 13
LATE NEWS	105	July 16	86	61
		July 23	84	72
		July 30	94	80
		Aug. 6	NA	83

[a]Since the summer season is typically slow, the inventory of spots available for sale was somewhat larger than would be expected in the fall or spring seasons.

don't want to cut prices and have one advertiser get a better deal than another. Everyone knows our prices and expects to get what they've paid for. They know they won't pay $1,000 for a spot and find someone else got the same spot for $200.

Also, a small amount of our advertising already comes as last minute customer decisions. A store may burn down, a strike may end suddenly and a company wants to tell its customers they are open for business again, or a store may need to sell off its inventory right away to generate cash or make room for a new shipment. Cutting prices based on demand would mean a discount to that percentage of our business that is "last minute" anyway. If we lowered prices to try and sell all of our spots, we wouldn't have any left over for last minute "emergency" use. If a good customer needs to get something on the air fast and we can't accommodate him because we're booked up, he'll go to our competition. Not only will we be forcing him to deal with someone else, but he'll have a psychological bond with our competitor that says, "They helped me when I needed it most." We may lose an account because we tried to sell every spot.

In a separate conversation, Val Conte, Director of Traffic and Operations, added, "I used to work for a station in Los Angeles which sometimes cut prices just before broadcast. Every Friday at 4:00 P.M., all the car dealers would start calling to buy time on the weekend movies. It was like a fire sale—dealers buying spots for half price and less."

Mockbee feels there is an additional reason: "Once you've gained a reputation for cutting prices, you can never get the premium price for a spot."

APPENDIX A

THE SIMPLEX METHOD

The Simplex algorithm is the cornerstone of linear programming, and also of a wide variety of solution approaches for other types of mathematical programming problems. These include many algorithms for network problems, branch and bound and cutting plane methods for solving integer programming problems, the dominant solution technique for quadratic programs, and a host of others. For this reason, we present here a brief overview of the Simplex method. Many variants of the original algorithm devised by George Dantzig in 1947 have been successfully implemented and are available in commercial software packages. Especially of interest in the solution of large-scale problems are techniques that make it easier to deal with large matrices of data.[1] For purposes of illustration, however, we will ignore these complexities and concentrate on the basic method.

The Simplex method, a technique for solving linear programs, proceeds by evaluating, one after another, the corner points of the feasible region, much like the "vertex-hopping" algorithm illustrated in Chapter 3. This evaluation does not proceed at random; instead, a technique is used which guarantees that for a maximization (minimization) problem a corner point will never be

[1]For example, a medium size problem with 600 variables and 400 constraints has a tableau of nearly ¼ million entries. Certainly most are zero. Techniques that store and can efficiently recall only nonzero entries are clearly necessary here.

Source: Tom Walker provided invaluable assistance in the writing of Appendix A evaluated if it produces as objective function value lower (higher) than the previously evaluated point. This strategy of successive improvements allows the efficient solution of problems which would be too large to solve if *every* possible corner point had to be evaluated. Below we will present a step-by-step example illustrating the operation of the Simplex method.

THE ALGEBRA OF THE SIMPLEX METHOD

Consider again the printer's calendar problem solved graphically in Chapter 3:

$$\text{Max} \quad 3x_1 + 5x_2$$

Subject to

$$\begin{aligned} 8x_1 + 4x_2 &\leq 4800 \\ 10x_1 + 70x_2 &\leq 38500 \\ x_1 \qquad &\leq 500 \end{aligned} \tag{1}$$

We have replaced the variables S and D with x_1 and x_2, for ease of notation. To solve this problem using the Simplex method, the equations must be restated in *canonical form*. In canonical form, every constraint must be written as an equality, and in each equality constraint there must be one variable with a +1 coefficient. This special variable may appear in no other constraint.

A technique for transforming the inequalities of equation (1) into equalities is to add a new *slack variable* to each of the equations. For example, consider the following constraint:

$$8x_1 + 4x_2 \leq 4800$$

The value of the left-hand side of this inequality will never be greater than 4800, but could be less. Rewriting this constraint with a slack variable x_3, we obtain an equivalent expression:

$$8x_1 + 4x_2 + x_3 = 4800$$

When $(8x_1 + 4x_2)$ is less than 4800, x_3 will equal the difference. When $(8x_1 + 4x_2)$ exactly equals 4800, x_3 is zero. Note that as long as the original constraint "$8x_1 + 4x_2 \leq 4800$" holds, x_3 cannot be negative. Turning this statement around: As long as x_3 is nonnegative, the constraint will hold. As will be seen as we proceed, the Simplex method never allows a variable to take on a negative value. In particular, the value of the slack variables may never be negative. Adding slack variables to *each* constraint transforms the original problem (1) to its canonical form.[2] By convention, the objective function is set equal to a variable z:

$$\begin{aligned}
8x_1 + 4x_2 + x_3 \qquad\qquad &= 4{,}800 \\
10x_1 + 70x_2 \qquad + x_4 \qquad &= 38{,}500 \\
x_1 \qquad\qquad\qquad + x_5 &= 500 \\
3x_1 + 5x_2 \qquad\qquad\qquad &= z \\
x_1, x_2, x_3, x_4, x_5 \geqslant 0 &
\end{aligned} \tag{2}$$

Note that the values of both the original *decision variables* (x_1 and x_2) and of the slack variables must all be nonnegative. The slack variables (x_3, x_4, x_5) each have a coefficient of +1 and each appear in only one of the constraints. Any variable that meets these conditions is called a *basic variable*, and together the set of basic variables is called the *basis*. For example, the basis for the set of equations (2) is (x_3, x_4, x_5). The number of variables in the basis will never exceed the number of constraints on the problem. Otherwise, there would be more unknowns than equations and the system could not be uniquely solved. The remaining variables are called *nonbasic*.

The importance of the canonical form is that a *unique feasible solution may be read directly from the system*. Rearranging the terms in each of the equations so as to isolate the basic variables on the left-hand side of each equation, we get:

$$\begin{aligned}
x_3 &= 4{,}800 - 8x_1 - 4x_2 \\
x_4 &= 38{,}500 - 10x_1 - 70x_2 \\
x_5 &= 500 - x_1 \\
z &= 3x_1 + 5x_2
\end{aligned}$$

[2]If the constraint was of the "greater-than-or-equal-to" type, we would subtract a *surplus variable*. For example:

$$8x_1 + 4x_2 \geqslant 4800$$

would become

$$8x_1 + 4x_2 - x_3 = 4800$$

Note that this will not automatically produce a canonical form since no variable has a coefficient of +1 in the equality constraint. This case will be discussed in more detail below.

A feasible solution can be easily generated by setting the variables x_1 and x_2, the nonbasic variables, equal to zero. When this is the case in our example problem, the solution ($x_1 = 0$, $x_2 = 0$, $x_3 = 4{,}800$, $x_4 = 38{,}500$, $x_5 = 500$) can be read directly from the system. This is our initial feasible solution, and it serves as a starting point for the Simplex method.

A feasible solution obtained this way (from a set of equations in canonical form) is called a *basic feasible solution*. In any basic feasible solution, the nonbasic variables equal zero; typically (although not always) the basic variables are strictly positive. As we will show below, *each basic feasible solution corresponds uniquely to a corner point*. Thus, since from Chapter 3 we know that an optimal solution will occur at a corner point, we can be assured that some basic feasible solution is optimal.

Graphically, we can track only the original decision variables x_1 and x_2 directly, as we did in Chapter 2. The initial feasible solution corresponds to the origin point "0" of the feasible region in Figure A-1. The Simplex method operates by moving from one corner point to another, the initial feasible solution providing us with a starting point. Moving from one corner point to another is accomplished by transforming the system of equations (2) into an equivalent system from which the basic feasible solution that corresponds to that second corner point can be read directly. The elegance and efficiency of the Simplex method derives from the fact that the Simplex method will never lead to the evaluation of a corner point that produces an objective function value worse than the one for the previous corner point evaluated. This method of successive improvements almost always leads to an optimum solution far more rapidly than a strategy of evaluating every possible corner point.

To understand the operation of the Simplex method it is important to know how a system of equations in canonical form may be transformed to represent the solution at another corner point. The algebra of this involves transforming the system so that a new variable becomes basic and one of the variables currently in the basis becomes nonbasic. To do this, we must first determine which variable is to *enter* the basis (i.e., which variable is to become basic). Referring to the last equation of (2), we see that, for each unit of x_1 produced, the objective function value will increase by 3 dollars, and for each unit of x_2 it will increase by 5 dollars. Thus, if we can increase the value of either x_1 or x_2 over its current (nonbasic) value of zero, the objective function value will increase. Although several techniques are available to determine which of these variables should be the *entering variable*, the most straightforward method selects the nonbasic variable with the largest objective function coefficient, in this case x_2.

Since the solution can never have more basic variables than constraints, placing x_2 in the basis implies removing either x_3, x_4, or x_5 from it. The technique for selecting the *leaving variable* involves determining how much we can increase x_2 (keeping $x_1 = 0$) while maintaining feasibility ($x_3, x_4, x_5 \geq 0$). In the first equation of (2):

$$x_3 = 4800 - 8x_1 - 4x_2$$

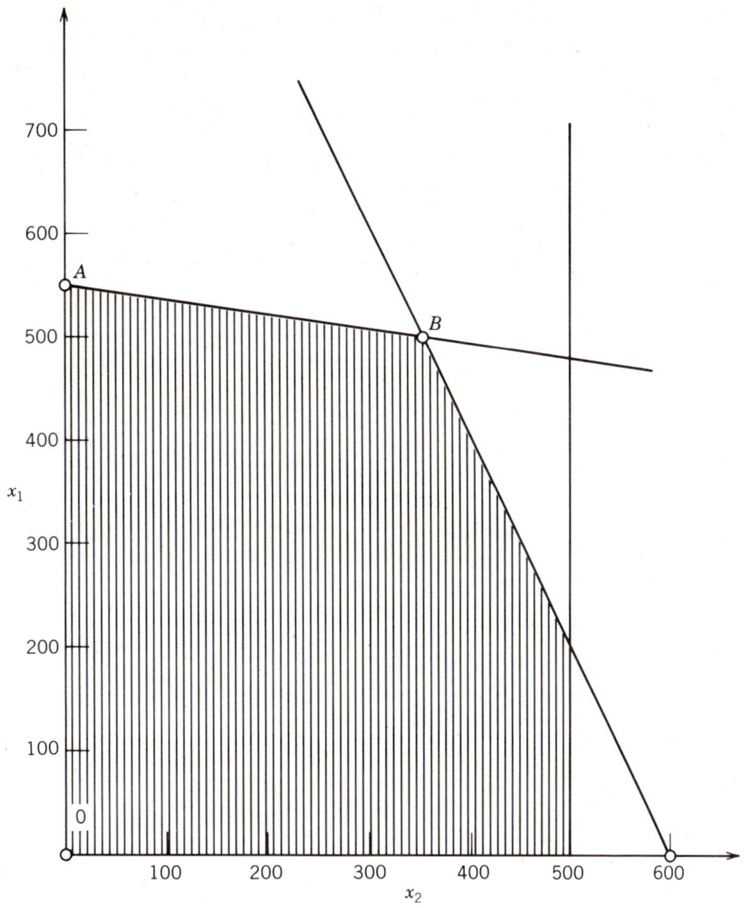

FIGURE A-1. The feasible region.

we could increase x_2 up to 1200 before x_3 becomes a negative number (holding $x_1 = 0$). Similarly, in the second equation,

$$x_4 = 38{,}500 - 10x_1 - 70x_2$$

we can increase x_2 up to 550 before x_4 goes negative. There is no limit on the increase in x_2 required by the third equation since x_2 does not appear there. Thus, the increase in x_2 is most limited by the second equation. When x_2 is increased up to its limit (550), x_4 drops to zero and leaves the basis. x_4 is called the *leaving variable* in this case.

This limiting equation may be easily determined mechanically using the *ratio test*. For each equation where the coefficient of the entering variable is negative, determine the ratio of the current value of that equation's basic variable to the coefficient of the entering variable. The limiting equation is then the one which yields to the smallest ratio.

In our example, this procedure leads to the ratios given in Table A-1. Thus, it is the ratio 38500/70 (550) that is the first to impose a limit on the growth of the entering variable x_2.

To transform the equations in (2) so that in the new solution, $x_2 = 550$ and $x_4 = 0$, we rearrange the second equation so x_2 is on the left-hand side and x_4 is on the right-hand side:

$$70x_2 = 38{,}500 - 10x_1 - x_4$$

Then we divide both sides of this row by 70 so that x_2 has a coefficient of +1, as is required to keep the set of equations in canonical form:

$$x_2 = 550 - (1/7)x_1 - (1/70)x_4$$

Finally, substituting the expression "$550 - (1/7)x_1 - (1/70)x_4$" for x_2 in the right-hand sides of all three other equations (including the objective function) produces a new canonical form from which we can read off a new basic feasible solution:

$$\begin{aligned}
x_3 &= 4800 - 8x_1 - 4[550 - (1/7)x_1 - (1/70)x_4] \\
x_2 &= 550 - (1/7)x_1 - (1/70)x_4 \\
x_5 &= 500 - x_1 \\
z &= 3x_1 + 5[550 - (1/7)x_1 - (1/70)x_4]
\end{aligned}$$

Simplifying the system, we get:

$$\begin{aligned}
x_3 &= 2600 - (52/7)x_1 + (4/70)x_4 \\
x_2 &= 550 - (1/7)x_1 - (1/70)x_4 \\
x_5 &= 500 - x_1 \\
z &= 2750 + (16/7)x_1 - (5/70)x_4 \qquad (3)
\end{aligned}$$

The new basis is (x_2, x_3, x_5), and the solution to the system becomes ($x_1 = 0$, $x_2 = 550$, $x_3 = 2600$, $x_4 = 4$, $x_5 = 500$). The value of the objective function has increased from 0 to 2750. Having completed the first *interation* of the Simplex algorithm, we now have an equivalent system, still in canonical form, from which we can read the solution which corresponds to another corner point. Referring to Figure A-1, this solution corresponds to corner point A.

TABLE A-1

Equation No.	Basic Variable for That Equation	Current Value of Basic Variable	Coefficient of Entering Variable (x_2) (absolute value)	Ratio
(1)	x_3	4,800	4	1200
(2)	x_4	38,500	70	550
(3)	x_5	500	0	NA

In order to begin the second interation we must determine whether entering a new variable into the basis would lead to any further increase in the objective function. Looking at the coefficients of the nonbasic variables in the objective function of (3), we see that increasing x_1 by one unit would add 16/7 dollars to the objective value whereas increasing x_4 by a unit would decrease it by 5/70 dollars. Therefore, introducing x_1 into the basis will yield an objective function value higher than the current 2750 dollars: The entering variable for the next interation therefore is x_1.

Next we must determine which constraint is most binding on increases in x_1. Using the ratio test (see Table A-2), we see that in equation (1) we can increase x_1 up to 350 (calculated as the current value of the basic variable x_3 divided by the absolute value of the coefficient of the entering variable x_1, i.e., 2600/(52/7)) before x_3 becomes negative. In equation (2), x_1 can be increased to 3850, and in equation (3), x_1 can be increased to 500 before any other variables are forced negative. Thus equation (1) is the most limiting on increases in x_1, with the result that x_3 is the new leaving variable. Rearranging equation (1), dividing both sides of it by 52/7 to isolate x_1 on the left-hand side, and substituting for x_1 in the other equations, we get:

$$x_1 = 350 - (7/52)x_3 - (1/130)x_4$$

$$x_2 = 550 - (1/7)\,[350 - (7/52)x_3 - (1/130)x_4] - (1/70)x_4$$

$$x_5 = 500 - [350 - (7/52)x_3 - (1/130)x_4]$$

$$z = 2750 + (16/7)\,[350 - (7/52)x_3 - (1/130)x_4] - (5/70)x_4$$

TABLE A-2

Equation No.	Basic Variable for That Equation	Current Value of Basic Variable	Coefficient of Entering Variable (x_1) (absolute value)	Ratio
(1)	x_3	2600	52/7	350
(2)	x_2	550	1/7	3850
(3)	x_5	500	1	500

Simplifying,

$$x_1 = 350 - (7/52)x_3 - (1/130)x_4$$

$$x_2 = 550 + (1/52)x_3 - (12/910)x_4$$

$$x_5 = 150 + (7/52)x_3 + (1/130)x_4$$

$$z = 3550 - (16/52)x_3 - (81/910)x_4 \tag{4}$$

The value of the objective function has now increased to 3550 and the basic feasible solution to the system is ($x_1 = 350$, $x_2 = 550$, $x_3 = 0$, $x_4 = 0$, $x_5 = 150$). This solution is represented by point B on the graph in Figure A-1.

After the second iteration, the coefficients of the nonbasic variables in the objective function (x_3 and x_4) are both negative. Introducing either of these variables into the basis, thereby increasing its value above its current nonbasic zero, would only cause a reduction in the value of the objective. This observation gives rise to the required test for optimality.

Test for Optimality (for a maximization problem): The basic feasible solution to a system of equations in canonical form is optimal if and only if all objective function coefficients are nonpositive.

For a minimization problem, the solution can be improved if there exists a *negative* coefficient in the objective function row. In that case, the objective function value can be further *reduced* by increasing the level of that (currently) nonbasic variable above zero. This notion gives rise to an equivalent optimality test for a *minimization* problem:

The basic feasible solution to a system of equations in canonical form is optimal if and only if all objective function coefficients are nonnegative.

The canonical set of equations (3) passes the test for optimality. Thus, its basic feasible solution ($x_1 = 350$, $x_2 = 550$, $x_3 = 0$, $x_4 = 0$, $x_5 = 150$), corresponding to corner point B, is optimal.

BASIC FEASIBLE SOLUTIONS AND CORNER POINTS

It was stated above that each corner point corresponds uniquely to basic feasible solution read from a set of equations in canonical form. Let us pursue this further, referring to the feasible region graphed in Figure A-2*a*. Although this illustration allows us to track only x_1 and x_2 directly, we can deduce the values of the slack variables x_3, x_4, and x_5 from the picture. For example, any point *on* constraint number 1 satisfies

$$8x_1 + 4x_2 = 4800$$

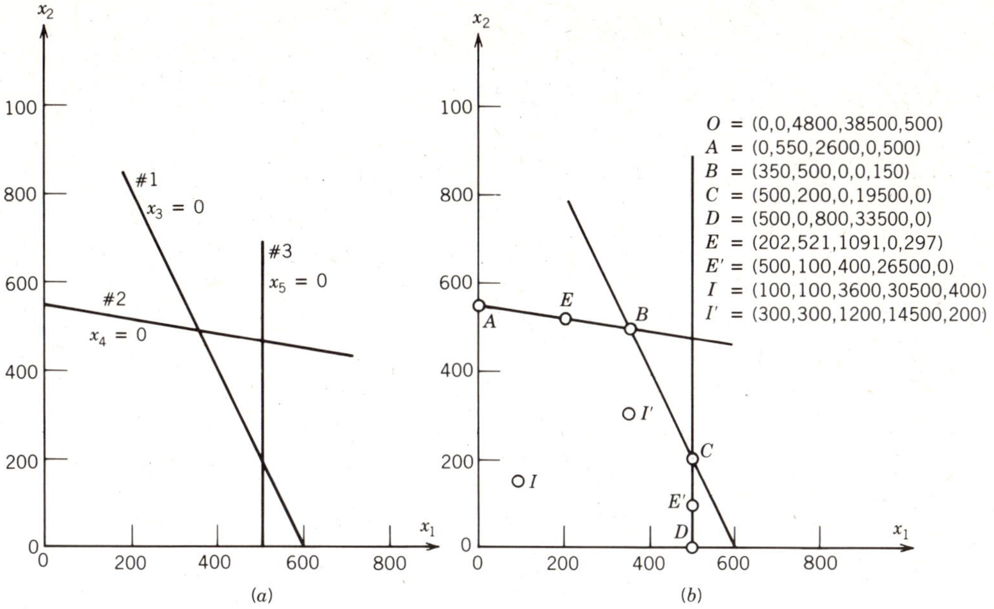

FIGURE A-2. Basic feasible solutions occur at corner points.

Thus the slack variable x_3 equals zero. For any *feasible* point *off* constraint number 1,

$$8x_1 + 4x_2 < 4800$$

and x_3 is positive. Similarly, x_4 is zero for all points on constraint line number 2 and positive for all other feasible points. For all points on constraint line number 3, x_5 is zero. For all other feasible points it is strictly positive.

Now, recall that each basis we examined had exactly three elements, the same as the number of constraints. For each basis, we were able to determine a basic feasible solution, generated from its canonical form by setting nonbasic variables to zero; basic variables took on positive values. In other words, each basic feasible solution was characterized by three positive values and two zero values. Refer to Figure A-2*b*. At I and I', interior points, all five variables are positive; at E and E', edge points, four of the five variables are positive—the other is zero. Only at the corner points O, A, B, C and D, are *three and only three* variables positive. Basic feasible solutions correspond only to corner points. Also, note from Figure A-2 that at each corner point, a *different* set of three variables are positive than at all other corner points; that is, each corner point corresponds to a unique basis. This correspondence is summarized in Table A-3.

Note also, that in moving from one corner point to an *adjacent* one, the basis remains the same except that *exactly one* variable is replaced by another.

TABLE A-3

Corner Point	Basis	Basic Feasible Solution $(x_1, x_2, x_3, x_4, x_5)$
O	$\{x_3, x_4, x_5\}$	(0, 0, 4800, 38500, 500)
A	$\{x_2, x_3, x_5\}$	(0, 550, 2600, 0, 500)
B	$\{x_1, x_2, x_5\}$	(350, 500, 0, 0, 150)
C	$\{x_1, x_2, x_4\}$	(500, 200, 0, 19500, 0)
D	$\{x_1, x_3, x_4\}$	(500, 0, 800, 33500, 0)

This is precisely what the Simplex algorithm accomplishes: In each iteration, one variable leaves the basis; another enters; a new basic feasible solution is determined. That new solution corresponds to a corner point *adjacent* to the one before it. Indeed, the Simplex method is an algebraic copy of the "vertex-hopping" algorithm we illustrated in Chapter 3. For the printer's problem, the corner points visited and their corresponding basic feasible solutions as read from equations sets (2), (3), and (4) are as shown by Table A-4.

OTHER ISSUES

Although they did not arise in the printer's calendar problem, two further issues deserve mention here. First, in problems with equality or "greater-than-or-equal-to" constraints, generating the initial canonical form will be more difficult than for our example problem. In these types of problems, the origin may not be feasible. Simply adding slack variables and using the origin as a starting point (i.e., assuming all the original decision variables are equal to zero) may not produce an initial feasible solution. Graphically, suppose the printer had already agreed to deliver 250 deluxe calendars to one of his customers. This would introduce the new constraint

$$x_2 \geqslant 250$$

into the problem. Note in Figure A-3 that this constraint removes the origin from the feasible region. In cases like this, a variant of the Simplex method called Phase 1 is used to generate an initial canonical form for the problem,

TABLE A-4

	Corner Point	Canonical Form	Basic Feasible Solution $(x_1, x_2, x_3, x_4, x_5)$
"Initial"	O	(2)	(0, 0, 4800, 38500, 500)
	A	(3)	(0, 550, 2600, 0, 500)
"Optimal"	B	(4)	(350, 500, 0, 0, 150)

simultaneously determining an initial feasible solution. The Phase 1 method will also determine if a problem is infeasible.[3]

Second, in this problem, each corner point corresponded to a unique basis and basic feasible solution. Note also that each corner point was defined by the intersection of *exactly two* constraint lines (including the axes—the graphical representations of the nonnegativity constraints). Consider instead the situation in which the printer increases the limit on standard calendars from 500 to 600. The third constraint becomes

$$x_1 \leqslant 600$$

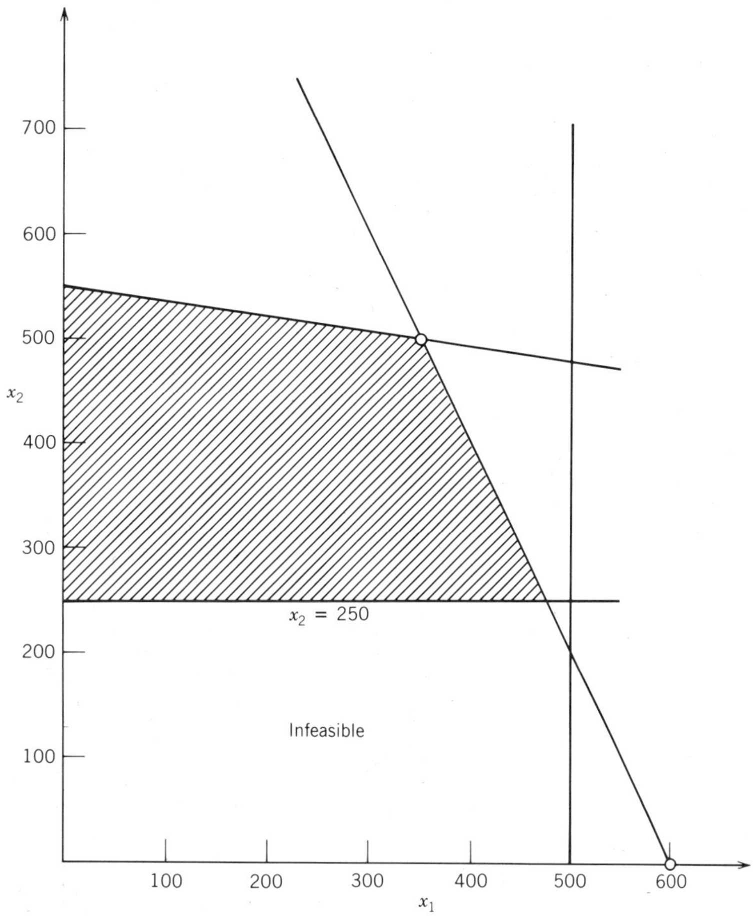

FIGURE A-3. A feasible region that excludes the origin.

[3]For an in-depth description of Phase 1, see Bradley, Hax, and Magnanti, *Applied Linear Programming*, Addison-Wesley, Reading, MA, 1978.

and the feasible region is graphed as shown in Figure A-4. Observe that corner point D is now the intersection of *three* lines. More relevant for the Simplex algorithm, the value of the five variables at D are:

$$x_1 = 600$$

$$x_2 = 0$$

$$x_3 = 0$$

$$x_4 = 32500$$

$$x_5 = 0$$

Only Two Variables are Positive. Above, we saw that every basic feasible solution was characterized by three positive variables. In this special case, one of the basic variables is itself zero. In the language of linear programming, this situation is called primal *degeneracy*. Theoretically, degeneracy can present a serious problem for the Simplex method. Let us see why.

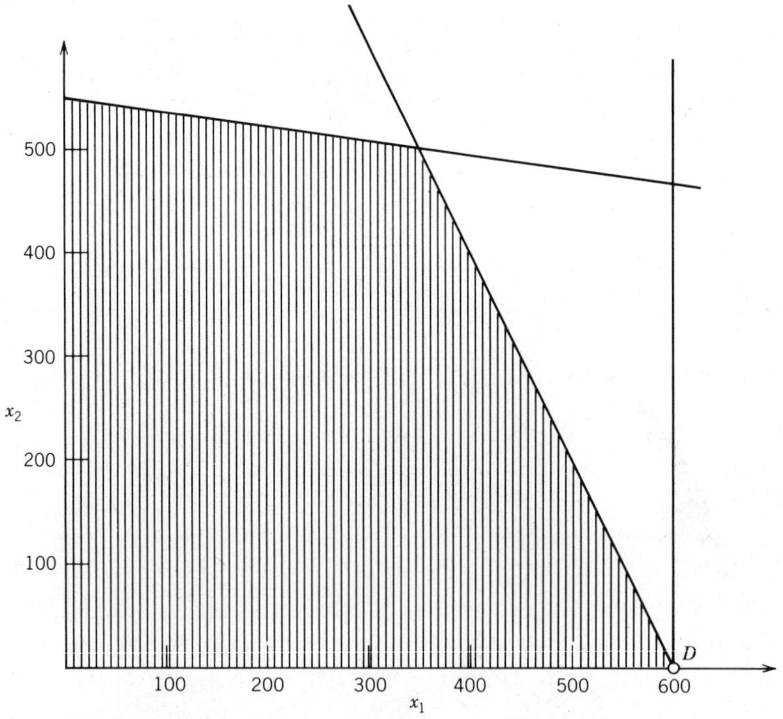

FIGURE A-4. Primal degeneracy.

The fact that we must set a basic variable to zero, means that corner point D, in fact, corresponds to three different basic feasible solutions:

1. Basis = $\{x_1, x_2, x_4\}$
2. Basis = $\{x_1, x_3, x_4\}$
3. Basis = $\{x_1, x_4, x_5\}$

As far as the Simplex algorithm is concerned, corner point D is three *different* vertices, each defined by a pair of intersecting constraint lines. The problem with this "outlook" is that, in its travel from one corner point to "an adjacent one," the Simplex method might move from point D, defined in terms of one basis, to D, defined in terms of another basis—in effect, not moving at all. In this special situation a Simplex iteration will not result in a better objective function value.

If this happens, the Simplex algorithm could cycle; for example,

$$\begin{array}{c} D \text{ with basis } \{x_1, x_2, x_4\} \leftarrow \\ \downarrow \\ D \text{ with basis } \{x_1, x_3, x_4\} \\ \downarrow \\ D \text{ with basis } \{x_1, x_4, x_5\} \end{array}$$

(the last basis returns to the first)

In recognition of this theoretical possibility, adjustments to the Simplex method have been developed to prevent cycling in case of degeneracy. In fact, three decades of empirical evidence indicate that, in practice, cycling has never been found to be a problem.

THE SIMPLEX IN TABLEAU FORMAT

By convention, and for ease in the required arithmetic calculations, LP problems are generally presented and solved in tableau format. In this section we will illustrate the Simplex solution to the printer's calendar problem using the tableau. As noted previously, the canonical form of this problem is:

$$\begin{aligned} 8x_1 + 4x_2 + x_3 &= 4{,}800 \\ 10x_1 + 70x_2 + x_4 &= 38{,}500 \\ x_1 + x_5 &= 500 \\ -z + 3x_1 + 5x_2 &= 0 \end{aligned}$$

The initial tableau is derived by writing down only the coefficients for the problem. Each column in the tableau represents the coefficients for a single variable and each row represents an individual equation. By convention, the final row displays the objective function coefficients. In the four-step procedure outlined below, we will iterate through the Simplex method, in tableau format, to the optimal solution. The tableau format provides a short-cut way of performing the algebraic substitutions illustrated earlier.

TABLEAU 1

Basic Variable	x_1	x_2	x_3	x_4	x_5	
x_3	8	4	1	0	0	4800
x_4	10	70	0	1	0	3850
x_5	1	0	0	0	1	500
z	3	5	0	0	0	0

Step 1: Testing for Optimality

To execute the Simplex method in tableau format, we must check each tableau, representing the problem in canonical form, to determine whether an optimum solution has been found. Examining the objective function, the last row of the tableau, we see that two variables have positive coefficients. The test for optimality insures that as long as at least one variable has a positive objective function coefficient, we have not yet reached the optimal solution. Therefore, we should enter one of these variables into the basis.

Step 2: Choosing the Pivot Column

Next we must select the entering variable. As before, we will enter into the basis the variable with the largest positive objective function coefficient. In the tableau, the column in which this variable appears is known as the *pivot column*.

TABLEAU 1

Basic Variable	x_1	x_2	x_3	x_4	x_5	
x_3	8	4	1	0	0	4,800
x_4	10	70	0	1	0	38,500
x_5	1	0	0	0	1	500
z	3	5	0	0	0	0
		Pivot Column				

Step 3: Choosing the Pivot Row

Next we must determine which constraint is most limiting on increases in the entering variable. Using the ratio test, we identify the row with the smallest ratio of the value in the right-most column to the value in the pivot column.

This test needs to be done only for rows where the value in the pivot column is greater than zero.[4] If the pivot point column value is negative or zero, no limit will be imposed on the entering variable by that constraint. If *all* coefficients are nonpositive, then that variable can be set to *any* positive level without causing infeasibility. Thus the objective function can be increased indefinitely—the problem is unbounded.[5] Otherwise, the row with the smallest ratio becomes the *pivot row*. In our example, the ratio is 1200 for the first row, 550 for the second row, and not applicable for the third row (since the pivot column entry is zero). Thus, row 2 is the pivot row.

TABLEAU 1

Basic Variable	x_1	x_2	x_3	x_4	x_5		
x_3	8	4	1	0	0	4,800	
x_4	10	70	0	1	0	38,500	Pivot Row
x_5	1	0	0	0	1	500	
z	3	5	0	0	0	0	

The intersection of the pivot row and the pivot column is called the pivot number, or simply the *pivot*.

Step 4: Performing the First Pivot

This step is the equivalent of the algebraic substitution used in the first iteration on page 586. However, the Simplex algorithm in tableau form uses a more abstract technique involving *elementary row operations*. To begin the elementary row operations, we first divide every value in the pivot row by the pivot number. In our example, the pivot number is 70. Dividing all values in the pivot row by the pivot number produces an equivalent equation with which we replace the old pivot row:

$$1/7 \quad 1 \quad 0 \quad 1/70 \quad 0 \quad 550$$

[4]Since the equations in the tableau have all variables on the left-hand side, the signs of the coefficients for the nonbasic variables are opposite from what was presented in the previous algebraic discussion of the Simplex method. Thus, we need to perform the ratio test only on those variables with *positive* coefficients.

[5]This realization gives rise to a formal test:

> *Test for Unboundedness (for a maximization problem): If any nonbasic variable has a positive objective function coefficient and, simultaneously, all constraint coefficients in the tableau are nonpositive, then the objective function value may be increased without bound by increasing the value of that variable.*

For a minimization problem, unboundedness corresponds to a *negative* objective function coefficient occurring in the same tableau column with only nonpositive constraint entries.

Now, to get to a new canonical form in which all entries in the x_2 column (including the objective row) are zero except for the +1 entry in the pivot row, we subtract a multiple of the new pivot row from each other row; the multiple is chosen to be that number which will cause zeros to appear in each of the other pivot column entries. To produce a zero coefficient in the pivot column of row 1, we multiply each of the entries in the new pivot row by 4 and subtract these from row 1.

	x_1	x_2	x_3	x_4	x_5	
Original Row 1	8	4	1	0	0	4800
−4 × (New Pivot Row)	−4/7	−4	−0	−4/70	0	−2200
New Row 1	52/7	0	1	−4/70	0	2600

Row 3 already has a zero in the x_2 column, so we need not alter it. Next, we multiply the new pivot row by 5 and subtract it from the final row to transform the objective function row. This yields the completed second tableau:

TABLEAU 2

Basic Variable	x_1	x_2	x_3	x_4	x_5	
x_3	52/7	0	1	−4/70	0	2600
x_2	1/7	1	0	1/70	0	550
x_5	1	0	0	0	0	500
z	16/7	0	0	−5/70	0	−2750

Note that x_2 has entered the basis, replacing x_4, as anticipated. Rearranging the terms, we see that this is exactly the same result we got after the first iteration using algebraic substitution (see Equations 3 on page 586):

$$x_3 = 2600 - (52/7)x_1 + (4/70)x_4$$

$$x_2 = 550 - (1/7)x_1 - (1/70)x_4$$

$$x_5 = 500 - x_1$$

$$z = 2750 + (16/7)x_1 - (5/70)x_4$$

We have now completed the first iteration of the Simplex method, and we return to step 1 to check whether we have an optimal solution. Referring to the last row of the Tableau 2, we see that the first column still has a positive value. Thus, this variable (x_1) should be entered into the basis. Column 1 becomes the new pivot column. Then, performing the ratio test on each of the constraint rows 1, 2, and 3, we find ratios 350, 3850, and 500, respectively. Therefore, row 1 becomes the pivot row.

TABLEAU 2

Basic Variable	x_1	x_2	x_3	x_4	x_5	
x_3	52/7	0	1	−4/70	0	2600
x_2	1/7	1	0	1/70	0	550
x_5	1	0	0	0	0	500
z	16/7	0	0	−5/70	0	−2750

To transform this tableau into an equivalent form with 1 as the pivot number, we divide each entry in row 1 by 52/7. This operation generates the new pivot row. Then, to get a zero entry in the pivot column of row 2, we multiply each entry of the new pivot row by 1/7 and subtract this from the second row. To get a zero entry for the pivot column in row 3, we simply subtract the new pivot row from it. Finally, we multiply the new pivot row by 16/7 and subtract row 3 from the final row to complete the third tableau.

TABLEAU 3

Basic Variable	x_1	x_2	x_3	x_4	x_5	
x_1	1	0	7/52	−1/130	0	350
x_2	0	1	−1/52	12/910	0	500
x_5	0	0	−7/52	−1/130	1	150
z	0	0	−16/52	−81/910	0	−3550

Having completed the third iteration, we return to step 1 and test for optimality. Since all coefficients in the final row are negative, introducing any of the currently nonbasic variables into the basis will only reduce the value of the objective function. Hence, we have again arrived at the optimal solution to the printer's calendar problem. Because x_3 and x_4, the only variables with coefficients other than 0 or 1, are not in the basis in the optimal solution and hence are 0, the solution values for the basic variables can be read directly from the right-most column—$x_1 = 350$, $x_2 = 500$, and $x_5 = 150$. Because the objective function was expressed as $(-z + 3x_1 + 5x_2 = 0)$ in setting up the tableau, the −3550 value in row 4 implies +3550 as the final value for z, the objective function.

Figure A-5 summarizes the steps of the Simplex algorithm. This procedure is, in practice, extremely efficient. Each iteration produces a new feasible solution whose objective function value is higher than before.[6] Since each feasible solution produced this way corresponds uniquely to a corner point, and since there are a finite number of corner points, the method is guaranteed to find (and due to the test for optimality, *recognize*) an optimal solution. Although the algebra of the Simplex method is tedious when calculations are done by hand, it is a well-defined algorithm easily programmed into the

[6]Barring degeneracy.

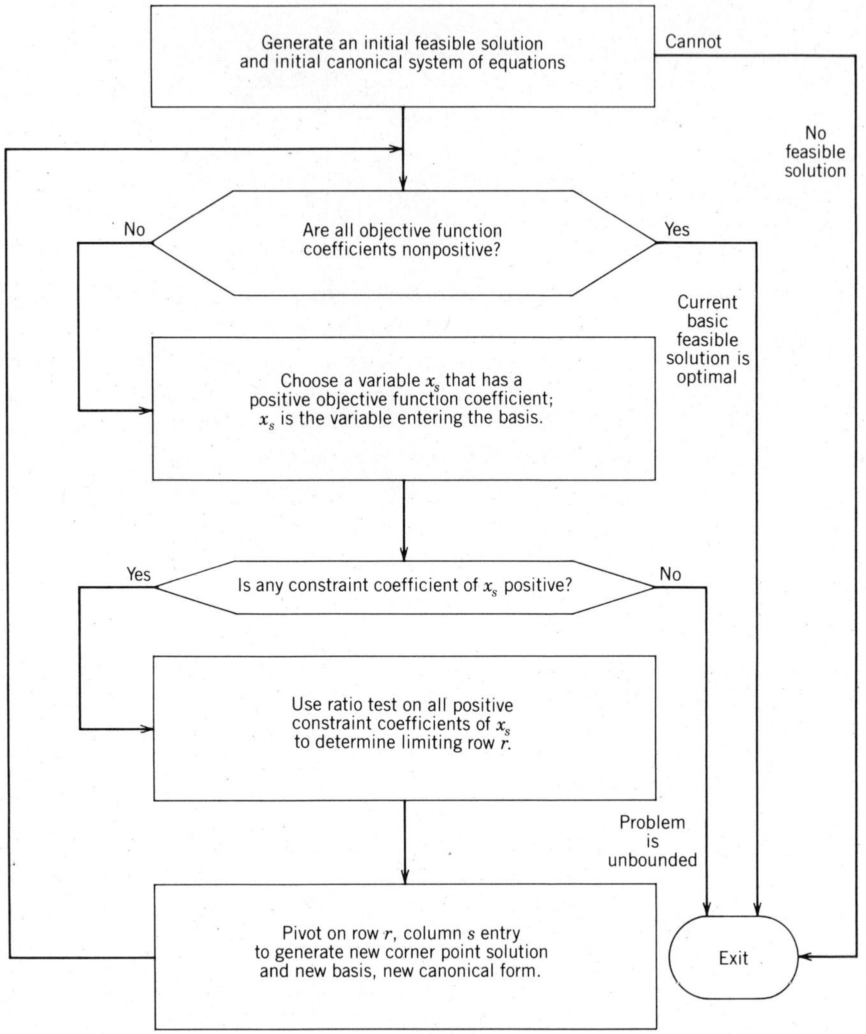

FIGURE A-5. The Simplex algorithm.

computer. Computer evaluation of hundreds of corner points can be accomplished very quickly.

POSTOPTIMALITY ANALYSIS

The lasting value to a manager of the Simplex method lies not only in the efficient solution technique but also in the fact that the final system of equations holds a wealth of additional information. The shadow prices and reduced costs for each of the problem's constraints can be read directly from the final tableau. Also, based on a quick calculation, the user can determine from the final tableau how much an objective function coefficient or right-hand side value can be increased or decreased without changing the optimal solution. Using the printer's calendar problem, we will illustrate how this information can be derived from the final tableau.

Shadow Prices and Reduced Costs

Recall from Chapter 3 that the *shadow price* of a constraint represents the increase in the objective function associated with a one unit increase in the right-hand side of that constraint, holding constant all other data in the problem. In the case of the paper availability constraint in the example problem, it is the most the printer would pay for another square foot of paper. If a constraint is not binding, its shadow price will be zero because increasing the value of the right-hand side will not result in a higher value for the objective function.

As stated above, one advantage of the Simplex method is that these shadow prices can be read directly from the final tableau for any problem. Consider again the final tableau for the printer's calendar problem:

TABLEAU 3

Basic Variable	x_1	x_2	x_3	x_4	x_5	
x_1	1	0	7/52	−1/130	0	350
x_2	0	1	−1/52	12/910	0	500
x_5	0	0	−7/52	−1/130	1	150
z	0	0	−16/52	−81/910	0	−3550

The shadow price of the paper availability constraint may be read off as the negative of the coefficient of x_3 in the objective function row, namely 16/52. To see why this is true, suppose we knew we wanted this piece of information before we began solving the problem. In that case, we might have written the paper availability constraint as

$$8x_1 + 4x_2 + x_3 = 4800 + \Delta$$

where Δ is equivalent to some increase in the right-hand side above 4800. Rearranging this equation, we get

$$8x_1 + 4x_2 + (x_3 - \Delta) = 4800$$

Carrying $(x_3 - \Delta)$ through the two iterations of the Simplex method we end up with the following objective function:

$$-z = -3350 + (16/52)(x_3 - \Delta) + (81/910)x_4$$

or

$$z = 3350 - (16/52)x_3 - (81/910)x_4 + (16/52)\Delta$$

Since x_3 and x_4 are nonbasic and are therefore set to zero, this becomes

$$z = 3350 + (16/52)\Delta$$

Recall that z denotes the objective function value. Hence, if paper availability were to increase by one unit, corresponding to $\Delta = 1$, z would increase by 16/52. Thus, the shadow price for a constraint is seen to be the *negative of the coefficient in the objective function row for that constraint's slack* variable. In this case the printer would be willing to pay up to an additional 16/52 of a dollar (or approximately 31¢) for an additional square foot of paper.[7] When the slack variable for a particular constraint is in the basis, its objective function coefficient is, by the canonical form rules, zero. This is logical—if the slack variable is basic, the constraint is not binding, and therefore further increases in the right-hand side will have no effect on the optimal solution. For example, since the slack variable (x_5) for the constraint on demand for the regular calendars is basic in the optimal solution, its coefficient in the objective row of the final tableau is 0. Even if demand increases above 500 calendars, the printer would not want to produce any more regular calendars. Hence, he is indifferent to any increase in demand.

Reduced costs are the shadow prices of the nonnegativity constraints in the problem. If an activity is at a zero level in the optimal solution, its reduced cost represents the cost per unit of deviating from the optimal solution by running the activity at a nonzero level. Formally, the reduced cost is the change which would occur if the value of that activity were increased from 0 to 1, corresponding to changing the nonnegativity constraint from $x \geqslant 0$ to $x \geqslant 1$. Because none of the original activities, x_1 or x_2, shows up as zero in the optimal solution to the printer's calendar problem, this is not a particularly good example for illustrating the concept. In general, however, we can say that the reduced cost of an original decision variable that is nonbasic in the optimal solution is equal to *its coefficient in the objective function row of the final tableau.* Requiring that one unit of this activity enter the basis will reduce the objective function value by the amount of the coefficient.

Variations in the Objective Function Coefficients

After solving the original calendar problem, the printer may want to find out how much he could raise his prices before the current optimal solution would change. Recall from the graphical solution to this problem that changing

[7]This 31¢ is a premium, over and above the usual paper cost that has already been taken into account in the calculation of the objective function coefficients.

one of the objective function coefficients changes the slope of the objective function. Small changes in the slope may not change the optimal solution, but larger changes can result in a new optimal corner point. Another advantage of the Simplex method is that one can calculate from the final tableau exactly how much one of the objective function coefficients can change before a new solution becomes optimal.

Consider again the printer's calendar problem, but instead of contributing 3 dollars, the regular calendars now contribute $(3 + \Delta c_1)$, where Δc_1 is some arbitrary, yet-to-be-defined change in the calendar price. Working through the Simplex tableaus, substituting $(3 + \Delta c_1)$ gives:

Initial Feasible Solution

8	4	1	0	0	4800
10	70	0	1	0	38500
1	0	0	0	1	500
$3 + \Delta c_1$	5	0	0	0	0

First Iteration

52/7	0	1	−4/70	0	2600
1/7	1	0	1/70	0	550
1	0	0	0	1	500
$16/7 + \Delta c_1$	0	0	−5/70	0	−2250

Second Iteration

1	0	7/52	−1/130	0	350
0	1	−1/52	12/910	0	500
0	0	−7/52	−1/130	1	150
Δc_1	0	−16/52	−81/910	0	−3550

Because the Δc_1 appears as the coefficient for x_1 in the final row, the problem is not in canonical form after the second iteration. To put the final tableau into canonical form, we multiple row 1 by Δc_1 and subtract this from row 4, replacing row 4 with the difference:

1	0	7/52	−1/130	0	350
0	1	−1/52	12/910	0	500
0	0	−7/52	−1/130	1	150
0	$0-(16/52) - (7/52)\Delta c_1$	$-(81/910) - (1/130)\Delta c_1$	0		$-3550 - 350\Delta c_1$

For the current optimal solution to remain the same, the objective function row coefficients for the nonbasic variables, x_3 and x_4, must remain nonpositive. Only then will the test for optimality be satisfied. For it to be the case that the objective function row coefficients for both x_3 and x_4 be nonpositive,

$$-16/52 - 7/52\Delta c_1 \leqslant 0 \quad (\text{or } \Delta c_1 \geqslant -16/7)$$

and

$$-81/910 + 1/130\Delta c_1 \leqslant 0 \quad (\text{or } \Delta c_1 \leqslant 81/7)$$

Combining these restrictions on Δc_1, we get:

$$-16/7 \leqslant \Delta c_1 \leqslant 81/7$$

Since the original objective function coefficient for x_1 was 3, this means that the initial optimal solution would remain valid as long as

$$3 - 16/7 \leqslant C_{\text{new}} \leqslant 3 + 81/7$$

or

$$5/7 \leqslant C_{\text{new}} \leqslant 14\ 4/7$$

Remember that this range of values within which the optimal solution will not change is valid only if the other objective function coefficients are held constant.

This tedious set of new iterations in order to get

$$-16/7 \leqslant \Delta c_1 \leqslant 81/7$$

was, in fact, unnecessary. These values could have been calculated directly from the final tableau, repeated below:

Basic Variable	x_1	x_2	x_3	x_4	x_5	
x_1	1	0	7/52	−1/130	0	350
x_2	0	1	−1/52	12/910	0	500
x_5	0	0	−7/52	−1/130	1	150
z	0	0	−16/52	−81/910	0	3550

Note from the above discussion that the net effect of increasing x_1's objective function coefficient by Δc_1 was to change the coefficients of the nonbasic variables x_3 and x_4 from −16/52 to −16/52 −7/52 Δc_1 and from −81/910 to

$-81/910 - 1/130\ \Delta c_1$, respectively. That is, they were changed by *subtracting* row 1's coefficients multiplied by Δc_1 from the objective function row's coefficients in the final tableau. Row 1 was the special row because x_1 is the basic variable for that row.

We can generalize this procedure to determine the limits on Δc for any basic variable x:

Step 1: Find the row for which x *is the basic variable. Denote it row R.*

Step 2: For each of the nonbasic variables in the objective function row of the final tableau,

(2A) pick out two entries from the final tableau:
entry 1: the entry in that nonbasic variable's column and row R,
entry 2: the entry in that nonbasic variable's column and the objective function row.

(2B) If Δc had been included in the initial tableau, then the entry in the objective function row for that nonbasic variable in the final tableau would be

$$(\text{entry } 1)\Delta c + (\text{entry } 2)$$

This expression must be nonpositive in order for this new final tableau to pass the test for optimality. Thus, Δc is limited so that

$$-(\text{entry } 1)\Delta c + (\text{entry } 2) \leq 0$$

or

$$\Delta c \geq \frac{\text{entry } 2}{\text{entry } 1} \text{ if entry } 1 \geq 0$$

$$\Delta c \leq \frac{\text{entry } 2}{\text{entry } 1} \text{ if entry } 1 \leq 0$$

Step 3: Find the most limiting of these inequalities on Δc. These define the range on Δc within which the optimal decision variables will remain unchanged.

Repeating this process for Δc_1:

Step 1: Row R *is row 1.*

Step 2: Nonbasic variable x_3:

$$\text{entry } 1 = 7/52$$

$$\text{entry } 2 = -16/52$$

Hence $\Delta c \geq -16/7$
Nonbasic variable x_4:

$$\text{entry 1} = -1/130$$

$$\text{entry 2} = -81/910$$

Hence $\Delta c \leq 81/7$.

Step 3: $-16/7 \leq \Delta c \leq 81/7$, *as discovered above.*

Variations in Right-hand-Side Values

As noted in Chapter 3, the right-hand-side range is the interval over which a right-hand side can be varied, holding all the other problem data constant, such that the variables that constitute the basis will remain the same. From the manager's point of view, this is important because as long as the basis does not change, the shadow prices and reduced costs for the solution remain valid. In the context of the calendar problem, the printer could use the shadow price of the paper availability constraint to evaluate changes in paper availability only for changes within the specified right-hand-side range for this constraint.

If prior to performing the Simplex method on the printer's problem, we had known that we wanted to find out how much we could vary the right-hand side of the paper availability constraint without changing the optimal solution, we could have written this constraint as:

$$8x_1 + 4x_2 \leq 4800 + \Delta b_1$$

where Δb_1 is some, as yet unknown, change in paper availability. Thus our original tableau would be:

8	4	1	0	0	4800 + Δb_1
10	70	0	1	0	38500
1	0	0	0	1	500
3	5	0	0	0	0

After the first iteration, we get:

52/7	0	1	−4/70	0	2600 + Δb_1
1/7	1	0	1/70	0	550
1	0	0	0	1	500
16/7	0	0	−5/70	0	−2250

Continuing:

1	0	7/52	−1/130	0	350 + (7/52)Δb_1
0	1	−1/52	12/910	0	500 − (1/52)Δb_1
0	0	−7/52	−1/130	1	150 − (7/52)Δb_1
0	0	−16/52	−81/910	0	−3550 − (16/52)Δb_1

This tableau is in canonical form *only* if the basic variables are nonnegative. Since the basic variables are to be set equal to the right-most column values, they must, in turn, remain nonnegative. Thus, in order to maintain the nonnegativity of the basic variables (x_1, x_2, and x_5):

$$x_1 = 350 + 7/52\Delta b_1 \geq 0 \quad \text{(that is, } \Delta b_1 \geq -1560\text{)}$$

$$x_2 = 500 - 1/52\Delta b_1 \geq 0 \quad \text{(that is, } \Delta b_1 \leq 26{,}000\text{)}$$

$$x_5 = 150 - 7/52\Delta b_1 \geq 0 \quad \text{(that is, } \Delta b_1 \leq 1114\ \ 2/7\text{)}$$

Taking the most restrictive bounds on Δb_1, we find that:

$$-1560 \leq \Delta b_1 \leq 1114\ \ 2/7$$

Adding the original right-hand side (4800) to both sides of the above relation we find the range of right-hand sides for this constraint within which the shadow prices remain valid:

$$3240 \leq B_{\text{new}} \leq 5914\ \ 2/7$$

Where B_{new} denotes the new right-hand-side value.

Note that we could have derived the final tableau for the problem with Δb_1 added to the right-hand side of the paper availability constraint by simply adding the coefficient for this constraint's slack variable (x_3 in this case) in each row to the current right-hand side for that row.

As was done for objective function coefficients, we can generalize this process, and derive an algorithm that allows the calculation of the range on any right-hand side change Δb *directly* from the entries in the final tableau:

Step 1: Find the column corresponding to the slack variable of the row whose right-hand-side value is to be changed by Δb*. Denote it column C.*

Step 2: For each *row (not including the objective function row),*
(2A) pick out two entries from the final tableau:
entry 1: the entry in that row and column C
entry 2: the entry in that row and the right-most column.

(2B) *If Δb had been included in the initial tableau, then the entry in the right-most column for that row in the final tableau would be*

$$(\text{entry } 1)\Delta b + (\text{entry } 2)$$

This expression must be nonnegative in order for the new final tableau to give nonnegative values to the basic variables. Thus, Δb is limited so that

$$(\text{entry } 1)\Delta b + (\text{entry } 2) \geqslant 0$$

or

$$\Delta b \geqslant -\frac{(\text{entry } 2)}{\text{entry } 1} \text{ if entry } 1 \geqslant 0$$

$$\Delta b \leqslant -\frac{(\text{entry } 2)}{\text{entry } 1} \text{ if entry } 1 \leqslant 0$$

Step 3: Find the most limiting of these inequalities on Δb. These define the range within which Δb may vary without altering the optimal basis and thus without altering the optimal shadow prices.

Repeating this process for Δb_1 for row 1:

Step 1: Since x_3 was the slack variable defined for row 1, column C is the third column.

Step 2: row 1:

$$\text{entry } 1 = 7/52$$

$$\text{entry } 2 = 350$$

Hence

$$\Delta b_1 \geqslant -\frac{350}{7/52} = -1560$$

row 2:

$$\text{entry } 1 = -1/52$$

$$\text{entry } 2 = 500$$

Hence

$$\Delta b_1 \leqslant 26000$$

row 3:

$$\text{entry } 1 = -7/52$$

$$\text{entry } 2 = 150$$

Hence

$$\Delta b_1 \leqslant \frac{150}{7/52} = 1114 \ \ 2/7.$$

Step 3: $-1560 \leqslant \Delta b_1 \leqslant 1114\ 2/7$.

These calculations, as well as those described earlier, can be performed mechanically and easily by computer. As a result, all commercially available codes (see Appendix D for examples) provide this useful information.

Appendix B

THE TRANSPORTATION METHOD

Let us return to the transportation example of Chapter 5:

> *A certain rent-a-car company has to fulfill the excess demand for cars in four different cities on a specific future day. Two cars are requested in city* A, *three in city* B, *five in city* C, *and seven cars in city* D. *The company has offices in three cities* E, F, *and* G, *each of which has excess supply. Six cars are available in city* E *on this specific day, one in city* F, *and* 10 *cars in city* G. *The distances between each of the company's offices and any destination city are given in Table B-1 (in hundreds of miles).*

What is in the optimal distribution policy, that minimizes total traveling distance and meets the demand at each city? We will find it convenient to use the format of Table B-2 to capture distance or costs (in the upper right-hand corner of each cell), supplies (in the right-most column) and demands (in the bottom row). Note that each row corresponds to an origin, each column to a destination. Our objective is to fill in the cells of Table B-2 with the least-cost (in this case, minimum distance) flows.

The first step (much like that of the Simplex method) is to generate one feasible solution. Many heuristics have been devised to do this; perhaps the fastest and easiest is the northwest-corner method, so called because we start assigning flow in the northwest, or upper left corner of our table. The basic procedure is to fill in as much as possible, dictated by demand and supply limits, and proceed south or east, allocating remaining demand or supply.

TABLE B-1

	Destinations			
Origins	*A*	*B*	*C*	*D*
E	7	11	3	2
F	7	6	10	1
G	9	15	8	5

TABLE B-2

Origins	*A*	*B*	*C*	*D*	Total Supply
E	7	11	3	2	6
F	7	6	0	1	1
G	9	15	8	5	10
Total Demand	2	3	5	7	17

Hence, we assign two cars from *E* to *A* (as much as possible considering *A*'s demand). We still have not allocated all of *E*'s supply, so we move eastward, assigning three cars from *E* to *B* (all of *B*'s demand). That leaves *E* with one car still available, so we again move eastward, assigning that car to *C*, thus exhausting *E*'s supply. However, we have not satisfied *C*'s requirements for five cars; *C* is still four short. Hence, we move southward, and assign all of *F*'s supply (one car) to *C*. *C* is still three short, so we again move southward, assigning three cars from *G* to *C*. Since supply and demand were balanced (if we had had to, we would have created a dummy source or sink to guarantee this balance), we are left with seven cars to be assigned from *G* to *D*, simultaneously exhausting *G*'s supply and fulfilling *D*'s needs. This process has led to the initial feasible solution shown in Table B-3.

We could find the cost (distance) of this solution by multiplying the cell entries (representing flow) by the distances in the upper right-hand corner of each cell:

Total car-miles (hundreds)

$$= 2 \times 7 + 3 \times 11 + 1 \times 3 + 1 \times 0 + 3 \times 8 + 7 \times 5 = 109$$

Although the northwest-corner method is a quick and easy way to find an initial solution, note that it totally ignores the costs. Thus the initial solution

TABLE B-3

Origins	*A*	*B*	*C*	*D*	Total Supply
E	7 2	11 3	3 1	2	6
F	7	6	0 1	1	1
G	9	15	8 3	5 7	10
Total Demand	2	3	5	7	17

may be a high-cost solution.[1] Since, as we will see, the full solution procedure itself is so quick, this is of little concern, especially with today's high-speed computers.

Now that we have an initial solution, we can proceed, trying with each iteration of the solution method to improve the current solution. Since each of the cells which currently have flow are constrained by either supply or demand restrictions and hence can not be assigned any more flow without other changes, an easy way to change the flow pattern is to look at each of the cells with no flow, asking "Can we improve the solution by assigning a unit of flow to this (now empty) cell, taking that unit of flow away from another cell (in the same row or column) with a greater cost?"

For example, consider the $F \rightarrow A$ cell, currently empty. Its cost is 7. Is there another cell with a higher cost in *F*'s row or *A*'s column that currently has flow? *No*. So we proceed to another empty cell, say, $F \rightarrow B$. It's cost is 6. Is there another (full) cell with a higher cost in *F*'s row or *B*'s column? Yes, $E \rightarrow B$. Let us try to reallocate flow so as to move one or more units from $E \rightarrow B$ to $F \rightarrow B$. In such a reallocation, we have to be careful to insure that each row (supply) and each column (demand) suffers no net change. That is, for *every* flow change in a row, there has to be an equal and opposite change in that same row; for every change in a column there has to be an equal and opposite change in that same column.

For our desired reallocation, let us see what the implication of the above is, moving a single unit for illustration. If, as we wish, we move one unit from $E \rightarrow B$ to $F \rightarrow B$, *B*'s column is all right, but we have to deal with *E*'s and *F*'s rows. Decreasing $E \rightarrow B$ by one means increasing the flow in some other cell in that row by one, say $E \rightarrow D$. Increasing $F \rightarrow B$ by one means we have to decrease the flow in another cell in *F*'s row by one. Decreasing $F \rightarrow C$'s flow does the trick, and also balances *C*'s column. Table B-4 illustrates this loop. Any reallocation will follow such a loop.

[1] As an exercise, devise a *northeast corner* procedure and determine an initial feasible solution using it. How does it compare with the solution above?

TABLE B-4

	Destinations				
Origins	*A*	*B*	*C*	*D*	Total Supply
E	7 2	11 3 (−1)	3 1 (+1)	2	6
F	7	6 (+1)	0 (−1) 1	1	1
G	9	15	8 3	5 7	10
Total Demand	2	3	5	7	17

For problems larger than that treated above, tracing the pattern of changes required by a reallocation is sometimes difficult. One method for doing this is to trace through the increases and decreases necessary to keep the solution feasible, as we did above. This method is often called the *stepping stone method*.

The allocation squares can be thought of as areas in a pond. Those squares that *have* allocations, representing flows along a particular route, can be thought of as stepping stones. Imagine that a frog is sitting on a lily pad in the middle of the square being evaluated ($F \rightarrow B$ in this case). The frog wishes to exercise by hopping around the pond but realizes that takeoff and landing are confined to the stepping stones. The frog is constrained to move only in right-angle directions. As soon as the frog lands on a stone, it will leave that stone by jumping directly to the left or to the right, never straight ahead or diagonally.

Given these simple rules, there is *one and only one path* that the frog can trace that will bring him back to his current location. In the case of the $F \rightarrow B$ route, it is the one shown in Table B-4. In fact, for any currently unoccupied cell that needs to be evaluated, the path of the jumping frog (that is, the stepping stone path) can be used to determine the cost tradeoffs caused by a reallocation. After tracing the path, the routes need to be identified as representing positive or negative changes. We can easily do this by placing a plus one in the cell being evaluated (a reallocation will involve an increase along this route) and starting around the path, placing a minus one in the first stepping stone square and then alternating plus ones and minus ones as the path is traced (to reflect the changes needed to keep the solution feasible). It makes no difference in which direction the patch is traced, because the two "stones" adjacent to the square being evaluated will both contain minus ones.

Route	*Change*	*Impact*
$F \to B$	$+1$	$+6$
$E \to B$	-1	-11
$E \to C$	$+1$	$+3$
$F \to C$	-1	-2
		-2

Since the net cost impact is negative we can lower overall costs by making this change. Why stop there? Since all relationships in this problem are linear, if a one-unit reallocation around the loop is beneficial, why not a two-unit or three-unit change, or more? How far can we go? The answer is based on recalling the implicit nonnegativity constraints $x_{ij} \geq 0$ for the problem. We cannot decrease the flow on any route below 0. Hence, look at the cells in our loop of Table B-4 for which the proposed reallocation decreases flow. How far can we go before one of these reaches zero? Unfortunately, the flow on $F \to C$ cannot be decreased more than 1. That defines the size of the change. Table B-5 then gives the new solution after reallocation.

Before we go on, note that any change that looks good *initially* is not necessarily beneficial. For example, returning to Table B-3, it seems as if we can lower cost by four units by transfering a unit of flow from route $G \to D$ (cost = 5) to route $F \to D$ (cost = 1). If we trace out the entire loop as shown in Table B-6, we can calculate the net impact:

Route	*Change*	*Impact*
$F \to D$	$+1$	$+1$
$G \to D$	-1	-5
$G \to C$	$+1$	$+8$
$F \to C$	-1	-0
		$+4$

TABLE B-5

Origins	A	B	C	D	Total Supply
E	7 / 2	11 / 2	3 / 2	2	6
F	7	6 / 1	0	1	1
G	9	15	8 / 3	5 / 7	10
Total Demand	2	3	5	7	17

TABLE B-6

	Destinations				
Origins	*A*	*B*	*C*	*D*	Total Supply
E	7 2	11 3	3 1	2	6
F	7	6	0 1 (−1)	1 (+1)	1
G	9	15	8 3 (+1)	5 7 (−1)	10
Total Demand	2	3	5	7	17

Hence, reallocation within this loop will *increase* costs by 4 units for every unit reallocated.

Now, back to our current solution, indicated by Table B-5. Its cost (car-miles) is

$$2 \times 7 + 2 \times 11 + 2 \times 3 + 1 \times 6 + 3 \times 8 + 7 \times 5 = 107$$

a decrease of 2 units as we had predicted. Can we improve further? Tables B-7(a)–(f) indicate the possible reallocations. (Remember we start by assigning a unit of flow to any empty cell and then follow the stepping stone method.) Note that B-7(d) represents the reverse of the reallocation we just made, so of course it would increase costs. Note also, from B-7(f), that "loops" generated by the stepping stone method are not always simple. Reallocations B-7(b) and B-7(c) will offer improvements. Let us try B-7(b) since it will provide the largest per unit cost reduction (not necessarily the largest total cost reduction, since we do not yet know how many units can be reallocated around each loop). How many units can we shift? Flow decreases are to be made on routes $D \rightarrow A$ and $G \rightarrow C$, so we are limited to 2 units, so that flow on $D \rightarrow A$ not be brought below 0.

Table B-8(a) gives the new solutions after this second reallocation, with the new cost, 101 car-miles. Table B-8(b) again shows a fruitful reallocation: Only one unit can be shifted to insure that the flow on route $G \rightarrow C$ not go below zero, leading to the solution of Table B-9 with objective function value 100 car-miles. At this point, checking the stepping stone loops associated with each of the empty cells shows that there exists no further reallocation that can further decrease the cost.

TABLE B-7(a)

Destinations

Origins	A	B	C	D	Total Supply
E	7 · 2 (−1)	11 · 2 (+1)	3 · 2	2	6
F	7 · (+1)	6 · 1 (−1)	0	1	1
G	9	15	8 · 3	5 · 7	10
Total Demand	2	3	5	7	17

Loop change: +7 −7 +11 −6 = +5 (NO)

As you can see, the stepping stone method is rather tedious and time consuming. In practice, the solution procedure often employed for the transportation problem operates by using dual variables to *identify* the empty cell to be filled rather than going through the stepping stone enumeration embodied in Tables B-7(a)–(f). The stepping stone method was presented because it is

TABLE B-7(b)

Destinations

Origins	A	B	C	D	Total Supply
E	7 · 2 (−1)	11 · 2	3 · 2 (+1)	2	6
F	7	6 · 1	0	1	1
G	9 · (+1)	15	8 · 3 (−1)	5 · 7	10
Total Demand	2	3	5	7	17

Loop change: +9 −7 +3 −8 = −3 (YES)

TABLE B-7(c)

Destinations

Origins	*A*	*B*	*C*	*D*	Total Supply
E	cost 7; 2	cost 11; 2 (−1)	cost 3; 2 (+1)	cost 2	6
F	cost 7	cost 6; 1	cost 0	cost 1	1
G	cost 9	cost 15; (+1)	cost 8; 3 (−1)	cost 5; 7	10
Total Demand	2	3	5	7	17

Loop change: +15 −11 +3 −8 = −1 (YES)

TABLE B-7(d)

Destinations

Origins	*A*	*B*	*C*	*D*	Total Supply
E	cost 7; 2	cost 11; 2 (+1)	cost 3; 2 (−1)	cost 2	6
F	cost 7	cost 6; 1 (−1)	cost 0; (+1)	cost 1	1
G	cost 9	cost 15	cost 8; 3	cost 5; 7	10
Total Demand	2	3	5	7	17

Loop change: +0 −6 +11 −3 = −2 (NO)

TABLE B-7(e)

Destinations

Origins	A	B	C	D	Total Supply
E	2 [7]	2 [11]	2 [3] (−1)	[2] (+1)	6
F	[7]	1 [6]	[0]	[1]	1
G	[9]	[15]	3 [8] (+1)	7 [5] (−1)	10
Total Demand	2	3	5	7	17

Loop change: +2 −5 +8 −3 = +2 (NO)

TABLE B-7(f)

Destinations

Origins	A	B	C	D	Total Suppy
E	2 [7]	2 [11] (+1)	2 [3] (−1)	[2]	6
F	[7]	1 [6] (−1)	[0]	[1] (+1)	1
G	[9]	[15]	3 [8] (+1)	7 [5] (−1)	10
Total Demand	2	3	5	7	17

Loop change: +1 −5 +8 −3 +11 −6 = +6 (NO)

TABLE B-8(a)

Origins	A	B	C	D	Total Supply
E	[7]	[11] 2	[3] 4	[2]	6
F	[7]	[6] 1	[0]	[1]	1
G	[9] 2	[15]	[8] 1	[5] 7	10
Total Demand	2	3	5	7	17

TABLE B-8(b)

Destinations

Origins	A	B	C	D	Total Supply
E	[7]	2 [11] (+1)	4 [3] (−1)	[2]	6
F	[7]	1 [6]	[0]	[1]	1
G	2 [9]	[15] (+1)	1 [8] (−1)	7 [5]	10
Total Demand	2	3	5	7	17

Loop change: $+15 - 11 + 3 - 8 = -1$ (YES)

easier to understand and because a stepping stone path-tracing technique *is*, in practice, used to determine the shipment pattern changes whenever a reallocation is required to be made.

The standard transportation problem is a special case of a more general network flow model called the *transshipment model*. In a transshipment problem the flow from a given origin to any destination can be routed through one or more *transshipment nodes*. Furthermore, any source (origin) or sink (destination) point can be also an intermediate point in the sense that flow is allowed (1) between sources and (2) between sinks, either directly or indirectly.

TABLE B-9

	Destinations				
Origins	*A*	*B*	*C*	*D*	Total Supply
E	[7]	1 [11]	5 [3]	[2]	6
F	[7]	1 [6]	[0]	[1]	1
G	2 [9]	1 [15]	[8]	7 [5]	10
Total Demand	2	3	5	7	17

Appendix C

DUALITY IN LINEAR PROGRAMMING

One of the most important early developments in linear programming was the discovery that to each linear program corresponds another, called the dual, whose characteristics offer insights to the manager in a variety of ways. Perhaps the most valuable aspect of the relationship between a linear program and its dual is the fact that solution of the dual generates shadow prices for the original problem. In that shadow prices are the primary tool in a manager's analysis and use of a linear programming model, this appendix briefly explains the origin of shadow prices and interprets some of the ramifications of the theory behind them. Before going into detail, however, let us motivate the discussion of that theory with two examples which underscore the underlying economics.

THE DIET PROBLEM

One of the most well-known linear programming problems is the diet problem, first solved by linear programming in the year 1947. Stated briefly, a shopper desires to purchase a least-cost selection of food items such that minimum nutritional requirements are satisfied. As a simple example, our shopper desires to choose among five food items (Table C-1) so as to guarantee a minimum weekly intake of calories, Vitamin A, and iron.

Our shopper could easily design a least-cost menu by solving the follow-

TABLE C-1
Nutrient Content of Five Foods

	Kilo-calories	Iron (mg)	Vitamin A (thousand I.U.)	Cost
Eggs (dozen)	.96	13.2	4.1	75¢
Milk (quart)	.64	0.4	1.4	46¢
Meat (lb)	1.75	12.5	.2	128¢
Rice (lb)	.90	4.0	0	65¢
Carrots (lb)	.18	3.6	50	43¢
Required Amount/Week	19.6	126	35	

ing linear program: The decision variables ($x_1, x_2, \ldots, x_5$) are the amounts to be purchased, weekly, of the five foods, respectively (x_1 = number of dozen eggs to be purchased, x_2 = number of quarts of milk, etc.). For each nutrient, there must be a constraint which says that the purchased foods provide at least the weekly requirement of that nutrient.

$$\text{Min} \quad 75x_1 + 46x_2 + 128x_3 + 65x_4 + 43x_5$$

$$\text{Subject to} \quad .96x_1 + .64x_2 + 1.75x_3 + .90x_4 + .18x_5 \geq 19.6$$

$$13.2x_1 + .4x_2 + 12.5x_3 + 4x_4 + 3.6x_5 \geq 126$$

$$4.1x_1 + 1.4x_2 + .2x_3 \qquad 50x_5 \geq 35$$

The solution (Exhibit C-1) of this model indicates that at a cost of $14.47, the shopper can satisfy all calorie, iron and vitamin A requirements by purchasing, for the week, 3.67 quarts of milk, 9.8 pounds of meat, and .56 pounds of carrots.

Consider, now, a nutrition supplement manufacturer who wishes to produce dietary supplements to be sold at drug and health food stores. He would certainly be interested in ascertaining the *value* which shoppers place on each unit of nutrient (kilocalorie, milligram of iron, and thousand international units of vitamin A) since he plans to place on the market a pure calorie tablet, an iron tablet, and a vitamin A tablet which will compete with food items. To set the prices of these nutrients, he might formulate a linear program which is the "dual" of our shopper's. He wishes to determine prices, p_1, p_2, and p_3, so as to maximize his total return (equal to the total revenue received if the shopper buys his tablets to achieve the weekly requirements of each nutrient) consistent with the restriction that if the shopper puts together nutrient supplements to exactly equal the nutritional value of any of the foods, their cost should not be greater than the cost of that food itself. (If it was, the shopper could simply purchase the food item rather than the supplement.)

EXHIBIT C-1

```
LINEAR: SHOPPER'S PROBLEM

OBJECTIVES     EGGS        MILK        MEAT        RICE     CARROTS

  COST          75          46          128         65         43

CONSTRAINTS    EGGS        MILK        MEAT        RICE     CARROTS    RELAT   CONSTRAINTS REQUIRED

  CALORIES      .96         .64        1.75         .9        .18       GE       CALORIES    19.6
  IRON         13.2          .4        12.5         4         3.6       GE       IRON         126
  VIT.-A        4.1         1.4          .2         0          50       GE       VIT.-A        35
MAXIMIZE OR MINIMIZE? MIN

OPTIMAL SOLUTION FOUND.
     COST           1447.28

OUTPUT OPTION? E


ALL ITEMS NOT LISTED IN SECTIONS 1 - 4 HAVE THE VALUE ZERO.

*1* DECISION VARIABLES
  2. MILK        3.66582
  3. MEAT        9.80195
  5. CARROTS     .558149

*2* SLACK(+) AND SURPLUS(-) IN CONSTRAINTS

*3* SHADOW PRICES FOR CONSTRAINTS
  1. CALORIES  70.3708
  2. IRON       .378814
  3. VIT.-A     .579390

*4* REDUCED COSTS FOR DECISION VARIABLES
  1. EGGS       .681583E-01
  4. RICE       .151001

*5* RANGES ON COEFFICIENTS OF OBJECTIVE COST
     VARIABLE   LOWER BOUND   CURRENT VALUE   UPPER BOUND
  1. EGGS          74.932        75.000        UNBOUNDED
  2. MILK          45.953        46.000          46.264
  3. MEAT          123.58        128.00          128.06
  4. RICE          64.849        65.000        UNBOUNDED
  5. CARROTS       34.251        43.000          43.575

*6* RANGES ON VALUES OF RIGHT-HAND-SIDE REQUIRED
     CONSTRNT   LOWER BOUND   CURRENT VALUE   UPPER BOUND
  1.  CALORIES    17.426         19.600          31.462
  2.  IRON        37.008         126.00          141.52
  3.  VIT.-A      6.6932         35.000          1660.9
```

$$
\begin{array}{llllll}
\text{Max} & 19.6p_1 + & 126p_2 + & 35p_3 & & \\
\text{Subject to} & .96p_1 + & 13.2p_2 + & 4.1p_3 & \leq & 75 \\
 & .64p_1 + & .4p_2 + & 1.4p_3 & \leq & 46 \\
 & 1.75p_1 + & 12.5p_2 + & .2p_3 & \leq & 128 \\
 & .9p_1 + & 4p_2 & & \leq & 65 \\
 & .18p_1 + & 3.6p_2 + & 50p_3 & \leq & 43
\end{array}
$$

The solution (Exhibit C-2) to the two problems show that if both the shopper and the manufacturer act optimally, the manufacturer's return exactly equals the shopper's cost.

If we compare the solution results for the shopper's problem with that of the production problem, we notice a striking symmetry. As we might expect, the optimal values of the price variables are the shadow prices of the shopper's problem. Also true is the fact that the optimal amounts of each food for the shopper are the shadow prices for the supplement manufacturer. Notice too, the symmetry between slack/surplus for each problem and the decision variables at a zero level for the other. For example, as Section *2* of the manufacturer's output indicates, the constraint which requires that the cost of the nutrients to be found in eggs not be greater than the cost of eggs is not binding (it has slack). The consequence of this, in the shopper's eyes, is that eggs are not cost-effective. (They cost more than the value of the nutrients they supply.) This is true also of rice, and we observe that neither eggs nor rice are part of the shopper's optimal menu. Notice, thirdly, the correspondence between Sections *5* and *6* for the two problems: The ranges of objective function coefficients within which the decision variables of one problem remain constant are the same as the ranges of right-hand-side values within which the shadow prices of the corresponding problem stay constant.

THE INGENIOUS ENTREPRENEUR

Consider a government plant, recently furbished to assemble three types of complex electronic components needed for an important secret government project. In order to assemble these components, it is necessary to hire several specialized government workers. The civil servants assigned to operate the plant have collected data (Table C-2) on the rate at which the various workers assemble the different components, the number of components needed per month, and the labor rates. These components are strongly interrelated, so that using a given amount of raw materials, certain components will be byproducts of the assembly of others. Thus, an engineer can produce 2.2 Z-1s per hour and, *simultaneously* 2.7 Z-2s and 3.2 Z-3s.

TABLE C-2
Components Produced per Hour for Three Labor Grades

	Components			
	Z-1	Z-2	Z-3	Labor Rate
Engineer	2.2	2.7	3.2	$6.30
Technician	0	4.2	4.1	5.80
Apprentice	2.1	5.8	0	5.20
Monthly Demand	12,000	15,000	8,000	

Somewhat confused as to what to do with all these data, the civil servants turn to a wealthy entrepreneur to help them decide who to hire. He happens to own several plants of a similar nature, and in this request he sees an opportunity to keep busy some of his idle workers so as to avoid having to lay them off. He suggests that for a monthly fee (in addition, of course, to his usual consultant's fee), he would be willing to run the plant if they provide all necessary raw materials. They agree that if he convinces them that the government's cost will be no more this way than if the plant were to be government run, they will accept the proposal.

The entrepreneur goes back to his office and formulates for the government's sake, a linear program which minimizes labor costs and yet meets the monthly demand for components.

$$\text{Min} \quad 6.30x_1 + 5.80x_2 + 5.20x_3$$

$$\begin{aligned} \text{Subject to} \quad 2.2x_1 + \; 2.1x_3 &\geqslant 12{,}000 \\ 2.7x_1 + 4.2x_2 + 5.8x_3 &\geqslant 15{,}000 \\ 3.2x_1 + 4.1x_2 &\geqslant 8{,}000 \end{aligned}$$

where x_1 is the number of hours of labor required from engineers, x_2 the number required from technicians, and x_3 the number required from apprentices.

After running this model (Exhibit C-3), he finds that 2500 hours of labor are needed from engineers, none from technicians, and 3095 hours from apprentices at a total labor cost of $31,845.20. He is delighted since he has available at least that amount of idle-worker time. He thinks, at first, of showing the formulation and results to the bureaucrats and asking for $31,845 monthly to operate the plant but realizes that if he does this, they will simply pay him his consultant's fee, procure the necessary workers from civil service pools, and then operate the plant themselves. Instead, he invents the ingenious scheme of *establishing prices* for the various components and rather than a fixed monthly fee, charging these amounts for the products he will promise to deliver. To do

EXHIBIT C-2

```
LINEAR: SUPPLEMENT MANUFACTURER'S PROBLEM

OBJECTIVES   PRICE-CA   PRICE-IR   PRICE-A

  REVENUE      19.6        126        35

CONSTRAINTS PRICE-CA   PRICE-IR   PRICE-A    RELATION    COST

  EGGS           .96       13.2       4.1        LE         75
  MILK           .64         .4       1.4        LE         46
  MEAT          1.75       12.5        .2        LE        128
  RICE            .9          4         0        LE         65
  CARROTS        .18        3.6        50        LE         43

MAXIMIZE OR MINIMIZE? MAX

OPTIMAL SOLUTION FOUND.
     REVENUE          1447.28

OUTPUT OPTION? E


ALL ITEMS NOT LISTED IN SECTIONS 1 - 4 HAVE THE VALUE ZERO

*1* DECISION VARIABLES
  1. PRICE-CA   70.3708
  2. PRICE-IR   .378814
  3. PRICE-A    .579390

*2* SLACK(+) AND SURPLUS(-) IN CONSTRAINTS
  1. +EGGS        .681595E-01
  4. +RICE        .151001

*3* SHADOW PRICES FOR CONSTRAINTS
  2. MILK         3.66582
  3. MEAT         9.80195
  5. CARROTS      .558149

*4* REDUCED COSTS FOR DECISION VARIABLES

*5* RANGES ON COEFFICIENTS OF OBJECTIVE REVENUE
     VARIABLE   LOWER BOUND   CURRENT VALUE   UPPER BOUND
  1. PRICE-CA     17.426         19.600         31.462
  2. PRICE-IR     37.008         126.00         141.52
  3. PRICE-A      6.6932         35.000         1660.9

*6* RANGES ON VALUES OF RIGHT-HAND-SIDE COST
     CONSTRNT   LOWER BOUND   CURRENT VALUE   UPPER BOUND
  1.  EGGS        74.932         75.000        UNBOUNDED
  2.  MILK        45.953         46.000         46.264
  3.  MEAT        123.58         128.00         128.06
  4.  RICE        64.849         65.000        UNBOUNDED
  5.  CARROTS     34.251         43.000         43.575
```

EXHIBIT C-3

```
LINEAR: GOVERNMENT'S PROBLEM


OBJECTIVES     ENG         TECH       APPRENT

  COST         6.3         5.8         5.2


CONSTRAINTS    ENG         TECH       APPRENT    RELATION    DEMAND

  Z1           2.2           0         2.1          GE        12000
  Z2           2.7         4.2         5.8          GE        15000
  Z3           3.2         4.1           0          GE         8000

MAXIMIZE OR MINIMIZE? MIN

OPTIMAL SOLUTION FOUND.
     COST           31845.2

OUTPUT OPTION? E


ALL ITEMS NOT LISTED IN SECTIONS 1 - 4 HAVE THE VALUE ZERO.

*1* DECISION VARIABLES
  1. ENG          2500.00
  3. APPRENT      3095.24

*2* SLACK(+) AND SURPLUS(-) IN CONSTRAINTS
  2. -Z2          9702.38

*3* SHADOW PRICES FOR CONSTRAINTS
  1. Z1           2.47619
  3. Z3           .266369

*4* REDUCED COSTS FOR DECISION VARIABLES
  2. TECH         4.70789

*5* RANGES ON COEFFICIENTS OF OBJECTIVE COST
     VARIABLE    LOWER BOUND   CURRENT VALUE   UPPER BOUND
  1. ENG          5.4476         6.3000          9.9744
  2. TECH         1.0921         5.8000         UNBOUNDED
  3. APPRENT      1.6926         5.2000          6.0136

*6* RANGES ON VALUES OF RIGHT-HAND-SIDE DEMAND
     CONSTRNT    LOWER BOUND   CURRENT VALUE   UPPER BOUND
  1.  Z1          8487.1         12000.          .33554E+12
  2.  Z2         UNBOUNDED       15000.          24702.
  3.  Z3          .12207E-03     8000.0          17196.
```

this, he formulates a linear program which is the "dual" of the first, to maximize total revenue subject to the restriction that the revenue received from the components produced by any one worker not exceed that worker's pay. (If it did, the government could replace that worker by one of its own and thus reduce its cost.) For example, consider a certain engineer. This worker, in an hour, will produce 2.2 Z-1s, 2.7 Z-2s, and 3.2 Z-3s. Thus, in an hour's work revenue produced by the worker is 2.2 p_1 + 2.7 p_2 + 3.2 p_3 where p_1 is the price to be charged for component Z-i. This amount cannot exceed $6.30, the worker's pay for that hour. There need be a similar constraint for each worker classification. Then the entrepreneur desires to

$$\begin{aligned}
\text{Max} \quad & 12{,}000p_1 + 15{,}000p_2 + 8{,}000p_3 \\
\text{Subject to} \quad & 2.2p_1 + 2.7p_2 + 3.2p_3 \leqslant 6.30 \\
& 4.2p_2 + 4.1p_3 \leqslant 5.80 \\
& 2.1p_1 + 5.8p_2 \leqslant 5.20
\end{aligned}$$

The wealthy entrepreneur solves this problem and is happy to find that at the optimal prices, his monthly revenue will be the same amount he would have requested as his fee for operating the plant. (See Exhibit C-4.)

At the next meeting with the government negotiators he presents his offer in price terms. They check, of course, to see that no worker brings in more revenue than his or her pay and, since they can find no way to operate the plant at a lesser cost (he can be sure of this from the first linear program), they agree to sign the contract. Of course, they try to get him to reveal his plan for operation of the plant, but he pretends not to be concerned with such details.

If the government negotiators had understood the details of linear programming, they might have known that not only are the optimal prices the shadow prices of the first formulation, but also that the optimal mix of workers may be obtained from the shadow prices of the second formulation. Also notice, as with our first example, that slack/surplus in a constraint in one model implies a zero-level of the corresponding variable in the dual formulation. For example, the optimal prices are set so that the revenue produced by a technician in an hour are far below ($4.71 below) that worker's hourly wage. Thus, it makes sense that in the original formulation, no technicians are to be hired. In order for a worker to be hired cost-effectively, that worker must produce revenue equal to his or her wage rate.

DUALITY IN GENERAL

We might well ask at this point, "What is it about these two situations that give rise to the formulation of two complementary linear programs?" Both

EXHIBIT C-4

```
LINEAR: ENTREPRENEUR'S PROBLEM


OBJECTIVES      P1          P2          P3

  REVENUE     12000       15000       8000

CONSTRAINTS     P1          P2          P3       RELATION   LAB-RATE

  ENG          2.2         2.7         3.2         LE         6.3
  TECH           0         4.2         4.1         LE         5.8
  APPRENT      2.1         5.8           0         LE         5.2


MAXIMIZE OR MINIMIZE? MAX

OPTIMAL SOLUTION FOUND.
     REVENUE        31845.2

OUTPUT OPTION? E



ALL ITEMS NOT LISTED IN SECTIONS 1 - 4 HAVE THE VALUE ZERO.

*1* DECISION VARIABLES
  1. P1          2.47619
  3. P3          .266369


*2* SLACK(+) AND SURPLUS(-) IN CONSTRAINTS
  2. +TECH       4.70789


*3* SHADOW PRICES FOR CONSTRAINTS
  1. ENG         2500.00
  3. APPRENT     3095.24


*4* REDUCED COSTS FOR DECISION VARIABLES
  2. P2         -9702.38


*5* RANGES ON COEFFICIENTS OF OBJECTIVE REVENUE
     VARIABLE   LOWER BOUND   CURRENT VALUE   UPPER BOUND
  1. P1          8487.1         12000.        UNBOUNDED
  2. P2         UNBOUNDED       15000.         24702.
  3. P3          .00000         8000.0         17196.


*6* RANGES ON VALUES OF RIGHT-HAND-SIDE LAB-RATE
     CONSTRNT   LOWER BOUND   CURRENT VALUE   UPPER BOUND
  1.  ENG        5.4476         6.3000         9.9744
  2.  TECH       1.0921         5.8000        UNBOUNDED
  3.  APPRENT    1.6926         5.2000         6.0136
```

cases unfold in a competitive environment involving a cost minimizer and a revenue maximizer. In addition, both are cases of "perfect" competition in a self/contained system with no external influences. Profits gained by one party are gained at the expense of the remainder of the system. In the usual production setting, this corresponds to a manufacturer maximizing profits subject to restrictions on the use of scarce resources, and the rest of the system setting prices so as to minimize the value of resources consumed subject to the restriction that, at those prices, the manufacturer makes no excess profits in any of the production processes. In problems where this competitive environment exists, the complementary linear program provides valuable insight into the original problem. Often, however, such a natural economic interpretation is not justified. Nevertheless, every linear program has associated with it another called its *dual* (the original is called the *primal*), and its formulation and solution may provide computational advantage as well as fresh insight even when the dual is interpretable only as a mathematical construct rather than as a natural economic entity.

To construct the dual in the general case, we can represent any primal problem with n variables and m constraints as follows:

Primal: Find $x_1, x_2, \ldots, x_n$ to

$$\text{Min} \quad \sum_{j=1}^{n} c_j x_j$$

Such that

$$\sum_{j=1}^{n} a_{ij} x_j \geq b_i \quad \text{for each } i = 1, 2, \ldots, K$$

$$\sum_{j=1}^{n} a_{ij} x_j = b_i \quad \text{for each } i = K+1, K+2, \ldots, m$$

$$x_j \geq 0 \qquad \text{for each } j = 1, 2, \ldots, L \quad (L \leq n)$$

In other words, the problem (to be minimized) has m constraints, of which K are inequalities and the rest are equalities, and n decision variables, of which L (which may or may not be all n) are restricted to be nonnegative and the rest unrestricted as to sign. Note that any formulation can be put into this form:

- A maximization can be transformed to an equivalent minimization by reversing the sign of all the coefficients of its objective function (Max $3x_1 + 2x_2 - 5x_3$ is the same as Min $-3x_1 - 2x_2 + 5x_3$).
- An inequality formulated to be "$\leq$" can be transformed to one of the "$\geq$" – type by reversing the signs of all the coefficients of the constraint *and* by reversing the sign of the appropriate right-hand-side value. ($2x_1 - x_2 + 4x_3 \leq 3$ is equivalent to $-2x_1 + x_2 - 4x_3 \geq -3$.)

The dual program may then be formulated as follows:

Dual: Find $y_1, y_2, \ldots, y_m$ to

$$\text{Max} \quad \sum_{i=1}^{m} b_i y_i$$

Such that

$$\sum_{i=1}^{m} a_{ij} y_i \leq c_j \quad \text{for each } j = 1, 2, \ldots, L$$

$$\sum_{i=1}^{m} a_{ij} y_i = c_j \quad \text{for each } j = L, L+1, \ldots, n$$

$$y_i \geq 0 \qquad \text{for each } i = 1, 2, \ldots, K$$

Note the symmetry between the primal and dual programs:

- Corresponding to each variable of the primal is a constraint of the dual, and to each constraint of the primal corresponds a dual variable.
- The coefficient of the *j*th primal variable appearing in the *i*th primal constraint becomes the coefficient of the *i*th dual variable in the *j*th dual constraint. (In matrix terminology, the dual tableau is called the *transpose* of the primal tableau.)
- The objective function coefficients of one program become the right-hand-side values of the other; the right-hand-side values of one becomes the objective function coefficients of the other.
- Each primal inequality corresponds to a dual variable restricted to be nonnegative; each primal equality corresponds to a dual variable unrestricted in sign. The same symmetry holds between primal variables and dual constraints.
- The dual to the dual problem is the primal.

INTERPRETATION OF THE DUAL SOLUTION RESULTS

The solution of the dual is an automatic extra which accompanies the primal solution; the Simplex method generates the solution to one as it solves the other. The primary usefulness of the dual problem lies in the fact that its solution allows us to impute prices to the resources/products of the primal; the primal's shadow prices are the optimal values of the dual variables. Thus, since a nonbinding constraint carries a zero shadow price we find a correspondence between nonbinding constraints and dual variables at a zero level; binding constraints and dual variables at a nonzero level. This principle, remarked in the examples above, is known as *complementary slackness*.

Because of the correspondence between variables of one program and constraints of the other, sensitivity questions may be posed in either framework. Since, for any program, the shadow prices remain unchanged as long as the right-hand-side values stay within specified limits (as given in Output Section *6*), the decision variables of its dual will remain unchanged as long as the dual's objective function coefficients remain within those same ranges (as given in Output Section *5*). Thus, Sections *5* and *6* will correspond exactly to the dual's *6* and *5*.

The essential feature of duality (called the *Strong Duality Theorem*) is that the values of the two objective functions will be equal.[1] The economic interpretation in the perfect competition setting discussed earlier is that the lowest possible feasible cost to the consumer is the maximum feasible profit for the seller. In this setting, there is a state of pure conflict between the supplier of scarce resources and the user of those resources, and whatever one gains the other must give up.[2]

In situations where this perfect competition outlook is not appropriate, the fact that the dual exists and has the aforementioned properties may, nevertheless, be of great use, both computationally and for gaining insight into the solution results. The next section briefly discusses two of these applications.

USES OF DUALITY

The answers to most sensitivity questions require calculations with dual variables. The most commonly asked is, "What happens to the objective function if the ith right-hand-side value is increased by one unit?" Since this unit increase in the primal's right-hand side corresponds to a unit increase in the ith dual objective function coefficient, this change causes the dual's objective function value to increase by y_i,[3] the ith dual decision variable. Since, however, the Strong Duality Theorem guarantees that the dual objective function is equal to the primal objective function, it must be that a unit change in the ith right-hand-side value causes a change of y_i in the objective function. In terms of the primal, y_i is the shadow price, and the above argument derives the definition of the shadow price which we have been using. If the constraint in

[1]This is true as long as both programs have some feasible solution.

[2]In game-theoretic terms, this corresponds to the Zero-Sum Game environment, and the Strong Duality Theorem corresponds to the existence of a minimax equilibrium point. In fact, every linear program can be formulated as a zero-sum game, and every zero-sum game can be formulated as a linear program (whose decision variables, restricted to be between zero and one, are the frequencies with which the various strategies should be chosen).

[3]This happens as long as the increase does not take the objective function coefficient outside of the specified ranges.

question is a resource limitation, then the shadow price has a natural "value of one unit of that resource" interpretation, consistent with the interpretation of the dual decision variables. Even if the constraint does not give rise to such a natural economic interpretation, duality theory provides us with a way of investigating the sensitivity of the objective to changes in the data.

Most major computer manufacturers offer to supply their customers with sophisticated linear programming packages which make extensive use of the theory associated with the dual. It so happens that computation time is far more sensitive to the number of constraints in a linear program than it is to the number of variables. Roughly speaking, computation time increases as the cube of the number of rows in the tableau, but less than linearly with the number of columns. For example, doubling the number of constraints would increase the computation time roughly by a factor of 8; doubling the number of variables would increase computation time by a factor smaller than 2. The simplex method, however, can solve either the dual or primal problem and, simultaneously, identify the solution to the other as well. Thus, if a problem has far more constraints than variables, it will be more efficient to solve its dual.

Another important computational use of duality comes into play when large-scale linear programs need to be solved. There are certain particular structures which permit quick solution by special means. (Some of these, for example, networks, are dealt with in the text.) If it is noted that the dual of a problem to be solved is of some special type, the dual can be solved more easily than the primal. The Critical Path Method, for example, may be formulated as a linear program. As is shown in Chapter 5, its dual has that special structure that is associated with a network and, thus, special computational advantage may be gained by exploiting this structure.

EXERCISES

1. In Appendix A, we use a two-variable example so we could demonstrate graphically what the Simplex method does. Unfortunately, if a problem has more than two variables, this approach is usually impossible. Consider the 5-variable problem:

$$\begin{array}{ll} \text{Max} & -4x_2 + 3x_3 + 2x_4 - 8x_5 \\ \text{Subject to} & 3x_1 + x_2 + 2x_3 + x_4 \leq 3 \\ & x_1 - x_2 + x_4 - x_5 \geq 2 \\ & x_1, \ldots, x_5 \geq 0 \end{array}$$

Be creative and use a graphical approach to solve this problem.

2. Formulate the dual of the following LP:

$$\text{Max} \quad x_1 + 2x_2 + 2x_3$$

$$\begin{aligned} \text{Subject to} \quad x_1 + x_2 + 2x_3 &\leq 12 \\ 2x_1 + x_2 + 5x_3 &= 20 \\ x_1 + x_2 - x_3 &\geq 8 \end{aligned}$$

$$x_1, x_2, \geq 0; \quad x_3 \text{ unrestricted in sign}$$

3. A manufacturer produces three models of a popular product line. Each model goes through fabrication, subassembly and assembly, and production is limited by the capacity in each of the departments. After assembly, the product is stored in a small (4000 square foot) in-plant warehouse area. The daily labor availabilities and the requirements of each model are shown below:

	Model			Available
	1	2	3	Daily
Fabrication Labor (hr)	2.1	3.2	1.9	120
Subassembly Labor (hr)	4.2	2.8	1.3	160
Assembly Labor (hr)	4.2	2.1	4.4	100
Storage (sq. ft)	82	94	79	4000
Contribution ($) unit	8.00	7.00	9.50	

An LP has been formulated and run to help the production scheduler plan operations. The model and output are shown below.

a. Formulate the dual to this problem.
b. What do the dual variables mean?
c. What would Sections *1*–*6* of the CMP output for the dual look like?

```
OBJECTIVES   MODEL-1     MODEL-2     MODEL-3

  CONTRIB          8           7         9.5

CONSTRAINTS MODEL-1     MODEL-2     MODEL-3     RELATION     AVAIL

  FAB            2.1         3.2         1.9         LE          120
  SUBASS'Y       4.2         2.8         1.3         LE          160
  ASS'Y          2.7         2.1         4.4         LE          100
  STORAGE         82          94          79         LE         4000
```

```
MAXIMIZE OR MINIMIZE? MAX

OPTIMAL SOLUTION FOUND.
     CONTRIB       317.258

OUTPUT OPTION? E


ALL ITEMS NOT LISTED IN SECTIONS 1 - 4 HAVE THE VALUE ZERO.

*1* DECISION VARIABLES
  1. MODEL-1    16.0756
  2. MODEL-2    26.9504

*2* SLACK(+) AND SURPLUS(-) IN CONSTRAINTS
  2. +SUBASS'Y 17.0213
  4. +STORAGE  148.463

*3* SHADOW PRICES FOR CONSTRAINTS
  1. FAB        .496454
  3. ASS'Y      2.57683

*4* REDUCED COSTS FOR DECISION VARIABLES
  3. MODEL-3   -2.78132

*5* RANGES ON COEFFICIENTS OF OBJECTIVE CONTRIB
     VARIABLE  LOWER BOUND  CURRENT VALUE  UPPER BOUND
  1. MODEL-1     6.8340        8.0000         9.0000
  2. MODEL-2     6.2222        7.0000         9.8625
  3. MODEL-3   UNBOUNDED       9.5000         12.281

*6* RANGES ON VALUES OF RIGHT-HAND-SIDE AVAIL
     CONSTRNT  LOWER BOUND  CURRENT VALUE  UPPER BOUND
  1.  FAB        77.778         120.00        127.70
  2.  SUBASS'Y   142.98         160.00       UNBOUNDED
  3.  ASS'Y      78.750         100.00        109.52
  4.  STORAGE    3851.5         4000.0       UNBOUNDED
```

Appendix D

PRACTICAL ADVICE FOR THE MATHEMATICAL PROGRAMMER

This appendix introduces some practical information about the use and operation of mathematical programming computer software. The discussion is designed to provide the mathematical programmer with a basic familiarity with the more common options available on commercially available software packages. These options often involve complex computations or manipulation of the data, and for that reason will not be discussed in great detail. Often these options, however, provide substantial savings in time and computer costs and a user of a commercial package should, at the very least, understand how and when they may prove helpful.

Before embarking on our description of available mathematical programming software options, we will provide a general discussion of the relationship between problem size and solution time. Because it is very easy to spend large sums of money solving mathematical programming problems, this is an important consideration in model building for the problem solver on a limited budget.

PROBLEM SIZE AND SOLUTION TIME

With the Simplex algorithm there is no theoretical guarantee that a problem will be solved faster than it would be if every corner point if the feasible

Source: This Appendix was contributed by Tom Walker.

region were enumerated and evaluated. Klee and Minty have constructed a hypothetical problem where the Simplex method is forced to evaluate every corner point before discovering the optimal one.[1] In this problem, the number of Simplex iterations is 2^n-1, where n is the number of variables. A computer performing 100 iterations per second would take approximately 11 years to solve a problem with 35 variables and 300,000 years to solve a 50 variable problem. While possible in theory, these kinds of problems have never been found to appear in practice and are commonly referred to as "pathological."

Although no theoretical basis exists for stating that the Simplex algorithm will solve problems in fewer iterations than complete enumeration, Monte Carlo simulation studies suggest that, on average, the number of iterations is a linear function of the number of constraints on the problem. Dantzig reported that for problems with less than 50 constraints and less than 200 variables, the number of iterations was usually less than $3m/2$, where m is the number of constraints.[2] Holding the number of constraints constant and varying the number of variables had very little effect on the number of iterations needed to determine the optimal solution. In fact, as a general rule, to make an LP problem solve more efficiently and hence reduce computing costs, it is more important to reduce the number of constraints than it is to reduce the number of variables.

For integer (IP) and mixed integer (MIP) programming problems, it is more difficult to relate solution time to problem size. Sometimes adding additional constraints will *reduce* the number of branch and bound iterations needed to find the optimal solution. This might occur if the addition of new constraints reduces the size of the feasible region (even though the number of corner points is increased). This in turn can reduce the number of branches the algorithm must consider. In general, however, a small problem will solve more rapidly than a large one.

As a general rule, MIPs and IPs will be far more difficult to solve than LPs with the same number of rows and columns. First, the continuous version of the problem must be solved. Only then is the branch and bound algorithm used to get an integer solution. Furthermore, removing constraints from a MIP or IP will not necessarily reduce the solution time for the problem. Although no simple rules are available for identifying the kinds of constraints that might be added or deleted to reduce the number of branch and bound iterations, this type of knowledge can be gained from experience with models that have a structure similar to the one the user is trying to solve.

MAJOR OPTIONS COMMONLY AVAILABLE

As mentioned previously, commercial mathematical programming packages often have available a number of advanced computational and data

[1]Klee, V. and G. Minty, "How Good is the Simplex Algorithm?", in *Inequalities-III*, ed. O. Shisha, Academic Press, New York, 1972.

[2]Dantzig, G. B., *Linear Programming and Extensions*, Princeton University Press, Princeton, N.J., 1963.

management techniques that can save the user significant time and money. However, most textbooks do not provide any guidance on what these options are or why they might be useful. Even worse, the user's guides that accompany these packages often assume the user is already familiar with the standard available options and does not need detailed guidance in determining when to use them. To fill this information gap, we will describe below some of the standard options found in commercially available packages. Because most involve advanced techniques beyond the scope of this appendix, little will be said about underlying mechanics. Instead, the emphasis will be on describing what the more important options do and when they may prove useful.

Scaling

Since computers can only carry a limited number of significant digits in executing the mathematical computations required by the pivoting process, the Simplex algorithm inevitably leads to rounding errors. As the iteration process proceeds, rounding errors begin to accumulate. Potentially, these errors might be of a magnitude great enough to generate infeasibilities or to produce an optimal solution bearing little resemblance to the actual optimal solution to the original problem.

Rounding errors are especially a problem when the coefficients (including the right-hand sides) of a mathematical program differ from each other by more than two or three orders of magnitude. When this occurs, division involving numbers that differ by several orders of magnitude can generate large rounding errors quite rapidly.

One way of avoiding this problem is to ensure that a problem is well *scaled*. A well-scaled problem is one in which the coefficients do not differ by more than approximately one order of magnitude. Most commercial packages have an option that automatically generates a well-scaled version of the problem formulation. Solving the scaled version will reduce the magnitude of the rounding errors and will often cause the Simplex to solve the problem more rapidly.

Removal of Redundant Constraints

Often in complicated mathematical programming formulations, some constraints or variables will be redundant. Since the solution time is generally a function of the size of the problem, eliminating redundancies will often yield more easily solved formulations.

The occurance of redundant constraints is most easily understood in the context of a two-dimensional graphical presentation. Consider the following problem:

$$\text{Max} \quad 4x + 2y$$

$$\text{Subject to:} \quad 30x + 20y \leq 1200$$

$$50x + 40y \leqslant 10{,}000$$

$$x, y \geqslant 0$$

The constraints for this problem are graphed in Figure D-1. Note that the feasible area is completely defined by the first constraint. The second constraint is superfluous. This is obvious in the graphical representation of this problem, but for larger multidimensional mathematical programs these kinds of redundancies can be very difficult to recognize.

The existence of redundant variables is more difficult to visualize. These occur, for example, when the objective function coefficients are such that, at optimality, the value of a variable will be pushed to zero, and no constraint requires the variable to be greater than zero. In a case like this, the variable could simply be dropped from all equations in the model.

Most of the major mathematical programming packages contain an option that finds and removes from the formulation redundant variables and constraints. For large, complex problems where redundancies are not immediately identifiable, users may find it advantageous to use the reduction option. For models that will be run many times, the savings in computing costs could be substantial.

Simple and Generalized Upper Bounding

Another technique for reducing the size of a mathematical programming problem is to substitute simple or generalized bounds on a variable's value for

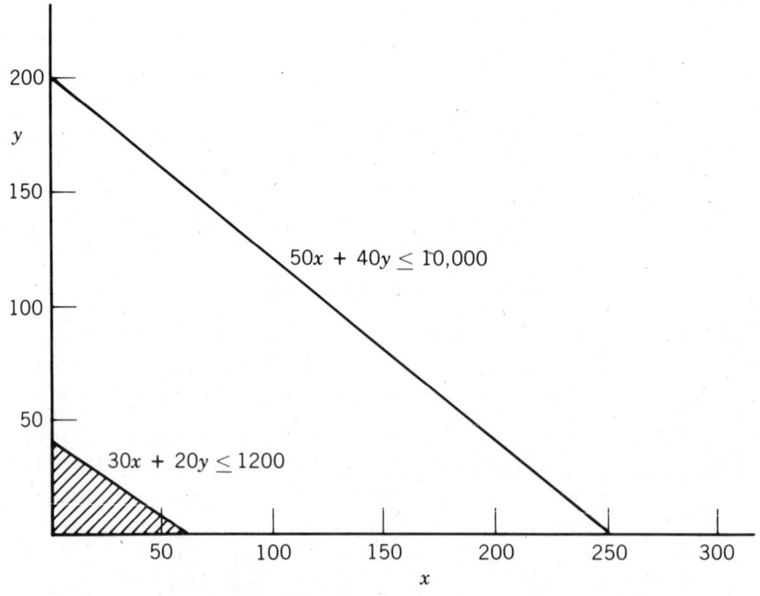

FIGURE D-1. A redundant constraint.

explicit constraints on its value. Again, by reducing the number of constraints in the problem, we may achieve greater solution efficiency.

To see how simple bounds can be used to eliminate constraints, consider the following case where the production of x is restricted to be less than 1000 units. Initially, we might write this as a constraint:

$$x \leq 1000$$

However, by using the simple bound option, we can specify, *separate* from the formal constraint set, that x never be allowed to take on a value greater than 1000 and eliminate the constraint from the problem. Often, the simple bounding option will eliminate a number of contracts from a problem.

Generalized upper bounds (GUB) are similar to simple bounds except that they set an upper bound on the sum of a set of variables. For example, consider the constraint

$$x + y + z \leq 5000$$

This set of variables (x, y, z) are said to have a generalized upper bound of 5000 as long as none of the individual variables are members of any other GUB set. We may then set a generalized upper bound on them, again separate from the formal constraint set, using the GUB option appearing in many commercial mathematical programming packages and eliminate the constraint from the model.

The generalized upper bounding modification of the Simplex algorithm will usually only prove useful where a number of GUB sets can be isolated. Mathematical programmers can recognize problems where the GUB option might be useful by looking at the original problem's tableau. For example, when one sees structures like the following,

$$\begin{array}{lllr}
x_1 + x_2 + x_3 & & & \leq 100 \\
& x_4 + x_5 + x_6 & & \leq 200 \\
& & x_7 + x_8 + x_9 & \leq 150
\end{array}$$

generalized upper bounding is useful. In this case, each row constitutes a GUB set because none of the variables in one GUB set appears in any other GUB set.

Starting from an Advanced Basis

Savings in computing time can sometimes be realized by requiring the Simplex algorithm to begin at an *advanced basis*. An advanced basis is one the user believes is nearer to the optimal solution than the initial feasible solution automatically generated by the computer. Hence, fewer iterations will be required to solve the problem than if the Simplex method began with its internally generated initial feasible solution.

Where does a user find an advanced basis? Consider the situation where after solving an LP, the user decides to simultaneously relax some of the constraints on the problem. If this were a maximization problem, relaxing these constraints would mean that the optimal basis for the original problem would still be feasible for the revised problem but might not be optimal. Since we know that the old optimal solution is feasible, we could begin iterations of the Simplex algorithm starting from the old optimal basis. In many cases, the new optimal solution will be determined in many fewer iterations than it would be if the Simplex method had commenced at its internally generated initial feasible solution.

Mathematical programming software that lets the user start from an advanced basis generally has an accompanying option which saves the basis during the solution procedure each time a given number of iterations is completed. This "save option" is useful when running a large program that might not reach an optimal solution in the time allocated to solve it. In such a case, if the user had specified that the basis be saved each time a certain number of iternations is completed, then the problem could be restarted from the most recently saved basis, avoiding the need to restart the Simplex algorithm from the initial feasible solution.

Other Available Solution Algorithms

Commercial mathematical programming packages will generally offer the user a choice of solution algorithms for linear programs. These may include the standard implementation of the Simplex method (often called the PRIMAL algorithm), the DUAL algorithm, and sometimes the DEVEX algorithm. For integer or mixed integer programs the branch and bound algorithm is usually the only available method, although some packages let the user specify which algorithm should be used to get the optimal solution to the continuous problem, that is needed to initiate the branch and bound process.

Generally, for a totally unfamiliar problem, on should use the PRIMAL algorithm since it is most likely to produce an efficient solution to the problem. However, if the problem is to be run more than a couple of times, it is probably worthwhile for the user to experiment with the DEVEX algorithm. The DEVEX almost always solves a problem in fewer iterations than the PRIMAL, but each iteration takes longer. In some cases, the savings in iterations will be more than enough to offset the additional time required to complete each iteration.

The DUAL algorithm should be used when the user knows that he or she has a dual feasible initial basis. This would occur, for example, if after initially solving a maximization problem, the user decides to tighten some of the constraints. As a result, the optimal solution to the original problem might no longer be feasible. However, the original optimal solution would be a feasible solution to the revised problem's dual. Thus, starting the DUAL algorithm from the optimal basis for the original problem might yield an optimal solution more quickly than solving the revised since the primal's right-hand sides become the dual's objective function coefficients problem using the PRIMAL method.

Finally, some packages offer *quadratic programming* algorithms. These can be used to solve problems where the objective function contains squared terms for some of the decision variables.[3] This is quite common in problems where downward sloping demand functions indicate that the quantity demanded is a function of the price. Consider the following example:

$$P = 500 - .05x_1$$

where P is the price of product x_1. If the objective of an LP is to maximize total revenues from sales of product x_1:

$$\text{Max} \quad Px_1,$$

then

$$Px_1 = (500 - .05x_1)\, x_1$$

The equivalent objective function is:

$$\text{Max} \quad 500x_1 - .05x_1^2$$

The objective function for this problem is nonlinear because of the x_1 squared term. In this case, quadratic programming could be used to solve the problem. This solution technique is, however, less efficient than the Simplex algorithm. Optimal solutions to quadratic programming problems will often be significantly more expensive to compute.

Matrix Generators and Report Writers

In building large mathematical programming models it is possible to spend enormous amounts of time inputting the constraints in the required format. Not only is this process very time consuming and tedious, but it is also prone to error. Models with 5,000 to 10,000 constraints and 50,000 to 100,000 variables can be solved by existing commercially available packages. In these cases, entering the data manually would be prohibitively expensive.

Fortunately, most very large mathematical programs exhibit what is referred to as *parallel structure*. This occurs when certain constraint types are repeated over and over again in the model. This frequently happens in multiple

[3]See Chapter 8 for more detail.

time period models where the constraints on activities in one time period have the identical structure as constraints on activities in another period.[4]

When this type of parallel structure is present in the formulation of a problem, it is often possible to make use of the *matrix generators* found on the most commercial mathematical programming packages. Matrix generators allow the user to write a program that will automatically produce all or some of the equations for the model in the required format. This reduces the chance of input errors in the model's coefficients and speeds the process of entering all the constraints into the computer.

Also included in most packages are *report writers*. Each package will have a standard format in which the output from the model is presented. Often, these results are not in a form that the nontechnical user can easily understand and interpret. Report writers allow the user to customize the format of the final results. This is especially useful where the results must be presented to persons unfamiliar with the terminology of mathematical programming.

COMMERCIAL PACKAGES

Of the several commercially available large-scale, general purpose mathematical programming packages, Table D-1 lists the perhaps four most popular in the United States. Of these, IBM's MPSX and CDC's APEX IV are the most commonly used.

Each of the packages includes solution algorithms for linear, integer, and mixed-integer programs. As can be seen from the table, they all have options for identifying redundant constraints and variables, for placing simple and generalized bounds on a problem, for scaling, and for specifying that the solution algorithm start from an advanced basis.

Indeed the striking feature of Table D-1 is the similarity between these packages. The only real difference is the structure of the matrix generators and report writers available with each of these packages. Analysts who are familiar with each will argue that their particular favorite is more user friendly or easy to use. For the uninitiated user, any one is likely as good as any other.

[4]For an example of this commonly occurring structure, refer to Example 5 of Chapters 2 and 3.

TABLE D-1

	Linear Programming	Mixed Integer	Primal	Dual	Devex	Branch and Bond	Quadratic	GUB	Advanced Basis	Report Writer	Matrix Generator	Scaling	Reduction
APEX IV Control Data Corp.	X	X	X	X	X	X		X	X	X	X	X	X
MPSX IBM	X	X	X	X	X	X		X	X	X	X	X	X
MPS III Ketron Inc.	X	X	X	X	X	X	X	X	X	X	X	X	X
FMPS Sperry Univac	X	X	X	X	X	X	X	X	X	X	X	X	X

INDEX